AF900239

OXIDATIVE STRESS – ENVIRONMENTAL INDUCTION AND DIETARY ANTIOXIDANTS

Edited by **Volodymyr I. Lushchak**

Oxidative Stress – Environmental Induction and Dietary Antioxidants
Edited by Volodymyr I. Lushchak

Published by InTech

Janeza Trdine 9, 51000 Rijeka, Croatia

Copyright © 2012 InTech

All chapters are distributed under the Creative Commons Attribution 3.0 license, which allows users to download, copy and build upon published articles even for commercial purposes, as long as the author and publisher are properly credited, which ensures maximum dissemination and a wider impact of our publications. After this work has been published by InTech, authors have the right to republish it, in whole or part, in any publication of which they are the author, and to make other personal use of the work. Any republication, referencing or personal use of the work must explicitly identify the original source.

As for readers, this license allows users to download, copy and build upon published chapters even for commercial purposes, as long as the author and publisher are properly credited, which ensures maximum dissemination and a wider impact of our publications.

Notice

Statements and opinions expressed in the chapters are these of the individual contributors and not necessarily those of the editors or publisher. No responsibility is accepted for the accuracy of information contained in the published chapters. The publisher assumes no responsibility for any damage or injury to persons or property arising out of the use of any materials, instructions, methods or ideas contained in the book.

Publishing Process Manager Sasa Leporic

Technical Editor Teodora Smiljanic

Cover Designer InTech Design Team

Additional hard copies can be obtained from orders@intechopen.com

Oxidative Stress – Environmental Induction and Dietary Antioxidants,
Edited by Volodymyr I. Lushchak
p. cm.
ISBN 978-953-51-0553-4

We are IntechOpen, the world's leading publisher of Open Access books

Built by scientists, for scientists

4,100+
Open access books available

116,000+
International authors and editors

120M+
Downloads

151
Countries delivered to

Our authors are among the
Top 1%
most cited scientists

12.2%
Contributors from top 500 universities

CLARIVATE ANALYTICS BOOK CITATION INDEX INDEXED

WEB OF SCIENCE™

Selection of our books indexed in the Book Citation Index in Web of Science™ Core Collection (BKCI)

Interested in publishing with us?
Contact book.department@intechopen.com

Numbers displayed above are based on latest data collected.
For more information visit www.intechopen.com

For EU product safety concerns:
IN TECH d.o.o., Prolaz Marije Krucifikse Kozulić 3, 51000 Rijeka, Croatia,
info@intechopen.com or visit our website at intechopen.com.

Meet the editor

Dr. Volodymyr I. Lushchak graduated from the Lomonosov Moscow State University (Soviet Union) in 1982. In 1986 he defended his Ph.D. Theisis and in 2002 his D.Sc. Dissertation. He obtained his full professor title in 2004. From 1987 to 1998 he worked in Southern Seas Biology Institute of Academy of Sciences of Ukraine. In 1998 Dr. Lushchak joined Vassyl Stefanyk Precarpathian National University in Ivano-Frankivsk, Ukraine) where he currently works as the head of the Department of Biochemistry and Biotechnology. He also worked at Universities of Canada, Brazil, United Kingdom, Finland, Poland, Sweden and Germany. Dr. Lushchak's scientific interests are related to effects of environmental conditions and molecular aspects of adaptive responses in bacteria, fungi, plants and animals. Although interested in general aspects of metabolism regulation, his interests are mainly focused on the homeostasis of reactive species of oxygen, nitrogen and carbon.

Contents

Preface

Free radicals discovered in biological systems in 1950es were immediately suggested to be involved in diseases and aging (Harman, 1956; 1985). The term "free radicals" was later extended to denote a wider group of activated oxygen forms whose activity is higher than molecular oxygen, and were collectively named reactive oxygen species (ROS), which include singlet oxygen, superoxide anion radical, hydrogen peroxide, hydroxyl radical, and many of their derivatives. In 1969, J. McCord and I. Fridovich described the catalytic function for erythrocuprein (hemocuprein) as superoxide dismutase responsible for elimination of the superoxide anion. The information on free radical processes in biological systems allowed Helmut Sies (1985) to systematize "Oxidative stress" and came to denote a disturbance in the prooxidant-antioxidant balance in favor of the former. Recently, we modified this definition as "Oxidative stress is a situation when steady-state ROS concentration is transiently or chronically enhanced, disturbing cellular metabolism and its regulation, and damaging cellular constituents" (Lushchak, 2011b). The last definition included accumulated the up-to-date knowledge on the effects of ROS on core and regulatory processes, and underlined the idea on their steady-state level in biological systems. Our understanding of the ROS roles in biological systems has gone through three phases: their appreciation as damaging ones, protection against infections and, finally, signaling and regulatory molecules in diverse biological processes. We can now state that all listed components operate in organisms in concert and are absolutely necessary for realization of biological functions.

Intensive research was invested into discovering whether the environmental factors can affect intracellular ROS steady-state levels. That resulted in understanding that this level may be modified by many external physical, chemical and biological factors. Since it is difficult to register ROS levels *in situ*, these data were mainly gained through indirect methods with the evaluation of levels of ROS-modified molecules of both external and internal origin. Therefore, this book mainly contains the information on oxidative stress induced by physical and chemical factors and a portion of the book includes the information on antioxidants capable to modify ROS levels.

On January 2, 2012, a Google Scholar search for "oxidative stress environment" yielded about 589,000 publication hits, whereas in Scopus and Pubmed databases it yielded 4,428 and 6,302 hits, respectively. We have presented 17 chapters in this book, covering several important aspects of environmentally induced oxidative stress and its prevention by antioxidants. Since oxidative stress seems to be an inevitable component of virtually all stresses that are strong enough, the book provides the interested readers with information needed to recognize this.

The Introduction section (V. I. Lushchak) covers general aspects of oxidative stress theory and briefly analyses potential ways of oxidative stress induction by environmental factors – stimulation of ROS production and depletion of antioxidants. The role of antioxidants is also highlighted.

The book is divided into four parts. The first section, entitled "Physical Factors" demonstrates the induction of oxidative stress by exercise, light and temperature fluctuations. The chapter written by V. Lj. Jakovljevic and colleagues extensively introduces the biology of reactive oxygen and nitrogen species, measurement of redox status, levels of superoxide anion radical, hydrogen peroxide, glutathione, lipid peroxides, activities of superoxide dismutase and catalase, and then demonstrates that exercise may increase the production of ROS and modify redox status. Interestingly, it has been demonstrated that perturbations of free radical processes depend on the intensity and type of exercise, as well as specialization of athletes and their physical state. Different light types possessing high energy can also induce damage to cellular components, even in specialized organs. K. Engelmann et al. described the operation of human retina, ROS-related processes, protective role of specific parts of the light spectrum and retina protection by tinted intraocular lenses in detail. The next two experimental chapters deal with oxidative stress induced by temperature changes – in fungi and plants. Using two Antarctic fungi, *Penicillium sp.* and *Aspergillus glaucus*, N. Kostadinova et al. demonstrated a relationship between cold shock and oxidative stress evidenced by an increased level of oxidized proteins and activation of antioxidant enzymes. Since abscisic acid may increase freezing tolerance of plants, M. E. Mora-Herrera et al. were able to demonstrate that ther decrease in temperature affected the level of hydrogen peroxide and catalase isoforms in potato microplants, which was related to tolerance to low temperatures.

The induction of oxidative stress by chemical factors is presented in the second section of the book. Ions of metals may induce oxidative stress in at least two ways – entering Fenton reaction and replacing other metal ions in their binding centers (Valko et al., 2007). The detailed description of toxicokinetics of lead and cadmium, induction and role of oxidative stress in neurochemical changes in the hypothalamus and pituitary of F1 generation PND 56 male and female rats are presented by P. Pillai et al. Herbicides are well known inducers of oxidative stress and many mechanisms were described in this case. 2,4-Dichlorophenoxyacetic herbicide is one of the broadly used ones, and W. Tayeb et al. describe the general phenomenology and potential mechanisms of induction of oxidative stress in different organisms. The chapter by O. B. Stoliar and V. I. Lushchak is devoted to analysis of oxidative stress induced in fish by different environmental pollutants.

The next section is devoted to induction of oxidative stress by biological factors. Diverse pathogens invading the host organism are attacked by the immune system equipped by machinery to produce reactive species. R. C. Ebel and N. Kumar investigated the involvement of reactive oxygen species in combating *Xanthomonas citri pv citri* (Xcc), causing citrus canker in *Citrus sp.* and found that pathogen-induced oxidative stress was differently expressed in different representatives of the genera studied. K. Okabayashi et al. were able to demonstrate that ethacrynic acid, a thiol-modulating reagent, inhibited amylase release induced by β-adrenergic agonist in rat parotid acinar cells and the effect was independent of depletion of glutathione in the cells. The authors concluded that the inhibitory effect of ethacrynic acid on amylase release induced by β-adrenergic agonist was caused by the thiol-modulation of β-adrenergic receptors.

It is very attractive to use antioxidants to prevent ROS-induced modification of organisms' functions. Intuitively developed at the beginning of ROS investigation in living organisms, it looked promising to use them for prophylactics and treatment of ROS-modulated damages. However, the promises were not realized and it became clear that there are no absolutely direct links between ROS-induced changes and pathologies. The last section of the book presents a broad discussion of positive effects of diverse antioxidants. The Estonian team led by T. Kullisaar provides an interesting topic – after short surveys on probiotics and oxidative stress they share extensive information on the potential use of different probiotics in functional foods and capsules that may be helpful to combat oxidative stress related to many pathologies, like cardiovascular diseases, metabolic syndrome, allergy, atopic dermatitis, radiation-induced problems in the intestinal tract. Diabetes is a very common human disease, which, in addition to health problems caused, is accompanied by many complications related with oxidative stress and the system character of the pathology therefore clearly needs specific approaches. It is very attractive to use a food stuff instead drugs and B. Alipoor et al. describe the potential of one of the most common drinks, tea, with health benefits particularly for diabetes and related complications. Sulphur mustard as a bifunctional alkylating agent readily reacts with a variety of macromolecules including nucleic acids, proteins and lipids, as well as small molecular mass metabolites such as glutathione, which is in the focus of chapter written by R. Vijayaraghavan and A. Gautam. Since sulphur mustard also induces oxidative stress, antioxidants can be useful and the authors analyze available data on the use of flavonoids, particularly from *Hippophae rhamnoides.* Bee products accompanied people since ancient times and only now do we start to understand the molecular mechanisms of many processes modulated by these products. Therefore, P. Tatli Seven provide an extensive analysis of beneficial properties of propolis with the focus on its antioxidant, antimicrobial, anti-inflammatory and antitumor effects. The antioxidant potential of 152 samples of Thai fruits, vegetables and herbs, and 33 brands of tea was measured by W. Sangkitikomol and this study shows that the products are a good source of compounds with health benefits. Since the toxicity of cyanide is associated with the induction of oxidative stress, F. G. Elsaid suggests and proves that it can be reduced by the application of aqueous extracts of *Allium kurrat* and *Ricinus communis* which possess antioxidant properties. Due to high sugar and fat diets and sedentary lifestyles, modern people are frequently subjected to atherosclerosis and obesity, which are important risk factors for metabolic syndrome and greatly predispose individuals to liver diseases, cardiovascular disease, type 2 diabetes, dyslipidemia, hypertension and numerous cancers, and is associated with markedly diminished life expectancy. The French team (S. Gaillet, D. Lacan, J.-M. Rouanet) presents results of titanic systematic work to identify the beneficial diets and find a broad set of diary foods and beverages possessing antioxidant properties and helping to combat the mentioned pathologies. These products are fresh and possessed fruits grapes, and berries, preparations from them as well as selenium-enriched microalgae, algal and fungal polysaccharides. Recently, while screening more than 250 cyanine dyes for their neurotrophin-like activity, the compound called NK-4 and some related compounds were found to be potent neurotrophic agents for the promotion of growth and differentiation of neuronal rat adrenal pheochromocytoma cell line PC12. NK-4 is a divalent cationic pentamethine trinuclear cyanine dye that contains three quinolinium rings, N-alkyl side chains, and two iodine anions. In the last chapter of the book, the Japanese team (H. Ohta, K. Akita & T. Ohta) summarized the data on the biological effects in different models and found that NK-4 possesses free radical-scavenging activity, neuroprotective against various cytotoxic stresses, neuroprotective effects against β-amyloid (Aβ) toxicity, and intracellular signaling. Therefore, the authors suggest that this dye can be used to protect animal organisms against neurodegeneration.

This book is expected to be interesting to experts in the field of basic investigations of reactive oxygen species and oxidative stress, as well as to practical users in the diverse fields like environmental sciences, medicine, and toxicology.

Prof. Dr. Volodymyr I. Lushchak
PhD, DSc, Department of Biochemistry and Biotechnology,
Vassyl Stefanyk Precarpathian National University,
Ivano-Frankivsk,
Ukraine

Section 1

Introduction

1

Introductory Chapter

Volodymyr I. Lushchak
Vassyl Stefanyk Precarpathian National University, Ukraine

1. Introduction

Oxidative stress, which will be defined and described in details below, is inevitable attribute of most strong stresses. In this book, the induction of oxidative stress by environmental challenges like physical, chemical as well as biological factors is described. These factors can induce oxidative stress in direct and non-direct ways, which will be covered by several chapters. Substantial bulk of chapters will describe the defensive mechanisms against deleterious effects of reactive species in different organisms. The book gives a broad description of the processes related to production of reactive species and their elimination. Particular attention will be given to natural and chemically synthesised antioxidants.

2. Introduction in oxidative stress theory

Free radicals are relatively unstable particles with one or more unpaired electrons on outer atomic or molecular orbitals. Many of them have as short life time and they can exist for only microseconds or even less. That is why most scientists for long time believed that free radicals were too unstable to exist in biological systems. The presence of free radicals in biological systems was discovered about 60 years ago and was virtually immediately implicated by Rebecca Gerschman and colleagues (1954) in human diseases. Two years later Denham Harman (1956) suggested that free radicals could be involved in pathologies as well as animal and human aging, and he first proposed free radical hypothesis of aging. Since 1950th critically important discoveries on roles of free radicals in living organisms promoted deep understanding that they are involved in many pathologies of animal and human organisms. D. Harman also specified later mitochondria as a place in the cell principally determining lifespan and proposed that mitochondria could be the "biological clock" and in this manner govern longevity, and further the hypothesis proposed was developed in mitochondrial theory of aging with key role of free radicals (Harman, 1972). Investigations on ROS roles in living organisms, particularly, in organisms' aging culminated by the formulation of free radical theory of aging (Harman, 1983), which in different formulations has been applied to all organisms - bacteria, fungi, plants and animals (Lushchak, 2011a). In 1995, D. Harman was nominated for the Nobel Prize in medicine for his works on the role of free radicals in diseases and aging. It seems that among all theories of aging, the Harman's one has the most consistent experimental support to date. The development of the theory extended it to age-related pathologies and also disturbances not directly related to aging.

It should be noted that now the term "reactive oxygen species" (ROS), which include oxygen free radicals along with some other activated oxygen forms like peroxides (e.g. H_2O_2), is more commonly used than "oxygen free radicals" to underline the existence of activated oxygen forms with non-radical nature. The investigation with many organisms resulted in disclosing of molecular mechanisms leading to increased ROS production, corruption of defense systems and different combinations of these routs. The interest to free radical processes was stimulated by the discovery of enzymatic mechanism of ROS elimination by the enzyme superoxide dismutase in 1969 by Irvin Fridovich and Joe McCord (1969). Several years later, nitric oxide as one more reactive form was found to play important regulatory roles in muscle relaxation and many other processes (Gruetter et al., 1979). This led to discovery of nitric oxide synthase (NOS). Reactive species were also found to be involved in defense mechanisms of immune system for attack of invaders (Klebanoff, 1967). Identification of enzymatic finely controlled systems of ROS production like NADPH-oxidases producing $O_2^{\bullet -}$ and H_2O_2, and NOS producing $^{\bullet}NO$, filled up the gape to view free radical processes as controlled ones. Helmut Sies (1985) was the first who defined "oxidative stress" as "Oxidative stress" came to denote a disturbance in the prooxidant-antioxidant balance in favor of the former". Extensive investigations in the field of free radical processes and their role in living organisms as well as ROS dynamics, regulation and consequences of imbalance between production and elimination let me propose the next definition of oxidative stress: "Oxidative stress is a situation when steady-state ROS concentration is transiently or chronically enhanced, disturbing cellular metabolism and its regulation and damaging cellular constituents" (Lushchak, 2011b). In this definition, the dynamic character of ROS-involving processes and their effects on core and regulatory processes in living organisms are underlined.

To date, development of oxidative stress was described in all phyla of organisms - bacteria, fungi, plants and animals. Although ROS are mainly supposed to play negative roles in living organisms, more and more data accumulated demonstrate their involvement in regulation of many physiologically important processes such as development, metamorphosis, morphogenesis, aging, etc. Reactive species do that either directly affecting certain systems or influencing specific regulatory pathways. The question on the specificity of ROS-involving processes is very important and to now it is responded in complicated way as the concerting type, spatio-temporal production, available direct targets and sensors. In many cases, these issues have been described in details, although the chemical instability of reactive species dictates specific rules in the "game" with them.

3. Induction of redox disbalance

3.1 Stimulation of ROS production

High production of ROS is usually implicated as the main mechanisms for oxidative stress induction. Therefore, here I suppose to characterize briefly the main known to date sources of reactive species. They are electron transport chains (ETC) of mitochondria, endoplasmic reticulum (ER), plasmatic and nuclear membranes, photosynthetic apparatus in plants; certain oxidative enzymatic reactions catalysed by specific oxidases; and autooxidation of endogenous and exogenous (xenobiotics) compounds.

Reactive species may be generated due to "leakage" of electrons from electron transport chains. In mitochondria electrons can escape the electron transport chain in several places, but mainly at the level of coenzyme Q and complex III. In this case, electrons interact with molecular oxygen resulting in formation of superoxide anion radical, which further spontaneously or enzymatically at operation of superoxide dismutase can be converted to hydrogen peroxide. Similarly to mitochondria, in photosynthetic apparatus, leakage of electrons also leads to production of superoxide anion radical and hydrogen peroxide. However, here the light energy absorbed may result in formation of other ROS, for instance singlet oxygen (Hideg et al., 2011). In electron transport chain of endoplasmic reticulum, the electrons transported may also escape to oxygen with the production of corresponding ROS. Here, this process is catalyzed by the enzymes of cytochrome P450 family. It should be noted that ER may be a place of ROS production not only as the result of direct operation of cytochromes. Compounds transformed here not being initially ROS generators may become them after transformation followed by entrance in reversible autooxidation. The nuclear membrane, particularly nuclear pore complex, can also be ROS producer (Hahn et al., 2011). Xantine oxidase and glucose oxidase are the best known oxidases generating ROS during catalytic acts. Xantine oxidase can produce superoxide anion radical via NADH-oxidase activity and nitric oxide via nitrate and nitrite reductase activities (Berry and Hare, 2004), whereas glucose oxidase catalyses the oxidation of glucose to D-glucono-δ-lactone with co-production of hydrogen peroxide (Raba and Mottola, 1995). Reactive species may also be produced by certain oxidases of amino acids and polyamines.

NADPH oxidase of plasmatic membranes is a specific enzymatic system known to produce reactive species (Sirker et al., 2011). Using NADPH the enzyme adds electrons to molecular oxygen that was first found in phagocytic cells and implicated to be responsible for killing of microorganisms either intra- or extracellularly. The enzymes of this class were found in most animals and plants. Now it is known that they are not only responsible for attack of invaders, but also generate ROS for signaling purposes (Sirker et al., 2011). The system is under strict control, because ROS overproduction is harmful for the cell. The second group of enzymes, NOS produce •NO in very well controlled manner similarly to NADPH oxidase. Nitric oxide is used not only for signaling purposes, but also to kill microorganisms (Vazquez-Torres et al., 2008). Moreover, in phagocytic cells two abovementioned enzymes cooperate to enhance the antimicrobial effects. The products of these enzymes namely, superoxide anion radical and nitric oxide, interact with the formation of very powerful oxidant peroxinitrite. Although the latter is not a free radical, it was found to be capable to enter nitrosylation reactions modifying in this manner proteins and nucleic acids. Moreover, it can spontaneously decompose with the formation of one of the most active oxidants – hydroxyl radical. These two enzymatic systems, in cooperation with myeloperoxidase, producing very strong oxidizing agent hypochlorite ion (ClO^-), also known as chlorate (I) anion, are responsible for antimicrobial activity of phagocytic cells (Arnhold and Flemmig, 2010).

Finally, different small molecules may enter autooxidation reactions and being capable of revesible oxidation can donate electrons to molecular oxygen and other compounds. Catecholamines, polyamines, polyphenols and some other endogenous compounds are known to enter autooxidation. However, most attention in this direction is paid to exogenous compounds (xenobiotics) capable to generate ROS in the organisms via autooxidation process. Xenobiotics affecting living organisms via generation of reactive species include number of pesticides, ions of metals with changeable valence, some industrial chemicals, pollutants, drugs, etc. (Lushchak, 2011b). It is important to note, that many xenobiotics may initially not be capable to enter autooxidation, but after certain reactions carried out by enzymatic systems may become ROS generators. For example, some chlorinated phenolic compounds, which are not ROS generators, after hydroxylation in ER by cytochrome P450 become potential ROS sources (Dreiem et al., 2009).

As we could see, there are number routs of ROS generation in living organisms. So, there are also many potential possibilities to increase ROS production. In electron transport chains, it may be reached by the inhibition of electron flow through the transport chains in different manners. For instance, mitochondrial ETC operation may be inhibited by the limitation of oxygen supply, or presence of cyanides and other respiratory toxins, which inhibit cytochrome oxidase. In the case of plastid ETC in plants, high intensity illumination can significantly increase production of singlet oxygen, $O_2^{\bullet-}$, and H_2O_2. The stimulation of general oxygen consumption due to increased energy needs at the change of physiological state of organisms may also enhance electron flux through the ETC resulting in extra ROS production. The increment of ROS production in ER may be related to the presence of substrates for oxidases like at ethanol oxidation in liver of animals (Yang et al., 2010), or methanol oxidation in certain yeasts (Ozimek et al., 2005), and after oxidation the formed products may enter autooxidation.

Some microorganisms, components of their bodies or excreted products can stimulate ROS production by animal immune system (Langermans et al., 1994). The process is tightly controlled by the immune system cells via reversible phosphorylation of NAPH oxidase and NOS, or by second messengers like calcium ions. Concerning the most chapters in this book, it is worthy to note that environmental factors can be very powerful inducers of ROS production in all living organisms. They may do this via different mechanisms. But according to materials of this subsection, we have to mention mainly the introduction of xenobiotics, which may enhance ROS generation. Of course, organisms possesses powerful and efficient antioxidant systems defending them against ROS.

3.2 Depletion of antioxidants

The second principal way to increase the steady-state ROS level is connected with depletion of antioxidant system, which consists of both enzymatic and non-enzymatic components. The first includes so-called antioxidant enzymes directly dealing with ROS and are represented by superoxide dismutases, catalases, peroxidases including glutathione-dependent ones, thioredioxine reductases, etc., and associated ones supplying reductive equivalents, building blocks for antioxidant synthesis, and energy sources (Hermes-Lima, 2004a,b).

The activity of antioxidant enzymes can be decreased in different ways. First of all, they can be inactivated in direct and non-direct ways. For example, certain pesticides may extract from enzyme molecules metal ions needed for catalytic activity. For example, copper ions may be removed from Cu,Zn-SOD by diethyldithiocarbamate (Lushchak et al., 2005). The activity of catalases can be decreased due to interaction of aminotriasole pesticides with iron ions in active centre of the enzymes (Bayliak et al., 2008). The second way leading to decreased activities of antioxidant enzymes is connected with direct chemical modification, for example, by oxidation (Wedgwood et al., 2011) or interaction with diverse compounds like carbohydrates (Shin et al., 2006). Finally, the activity of antioxidant enzymes can be decreased due to suppressed expression of corresponding genes or stimulated degradation.

Depletion of reserves of low molecular mass antioxidants also can result in the development of oxidative stress. This group of antioxidants consists of tocopherols, carotenoids, antocyanes, ascorbic and uric acids, etc. Glutathione, a cysteine-containing tripeptide (γ-glutamyl-cysteinyl-glycine) is important endogenous antioxidant, level of which is tightly controlled by the organisms at stages of biosynthesis, transport and consumption (Lushchak, 2011c). In any case, depletion of reserves of low molecular mass antioxidants may decrease the efficiency of elimination of reactive species that can result in increased steady-state ROS levels and lead to development of oxidative stress. Once oxidized by reactive species, cellular components usually became not effective components of living organisms. Therefore, there are two principal routs to deal with them: reparation or elimination.

Cells actively fix ROS-caused damages to DNA (Lu et al., 2001) and some oxidized amino acid residues in proteins can be also repaired (Lushchak, 2007). That needs operation of very efficient specific reparation mechanisms. After oxidation carbohydrates, lipids, proteins, RNA and free nucleotides are further mainly degraded with very few exceptions described for proteins. The necessity to degrade nonfunctional constituents is not only dictated by their useless, but also potential hazard due to disruption of cellular structures like membranes and cytoskeletons. In addition, in many cases the products of ROS-induced modification of lipids, carbohydrates, proteins and nucleic acids can themselves generate reactive species. It is absolutely clear, that oxidatively modified cellular components should be degraded, and this work is mainly carried out by diverse hydrolases like lipases, proteases, nucleases, etc.

4. Induction of oxidative stress

The factors, which induce oxidative stress, can be grouped in external (physical and chemical) and internal. The physical factors include variation of temperature, light and irradiation. The chemical factors consist of diverse compounds of various natures, which entering organisms cause increase in levels of reactive species. Finally, internal factors may not be directly related to metabolism of reactive species, but induce oxidative stress in non-direct way like energy depletion.

The potential mechanisms of oxidative stress induction by physical factors include both activation of ROS production and corruption of ROS-eliminating routs. Increased temperature may disturb membrane structure enhancing electron leakage from electron-transport chains and their interaction with molecular oxygen. Illumination by visible light may transform some photosensibilizators entered organisms like quercetin via excitation to activated electron donors. Another mechanism of ROS generation by extensive illumination can be connected with light absorbtion by specific cellular compounds like chlorophylls of thylacoids or eye retina. Radiation dependently on the type and intensity may either corrupt defense mechanisms or at extensive irradiation promote homolytic fission of covalent bonds followed by ROS formation.

Due to many reasons, most attention in environmentally induced oxidative stress field is paid to chemicals. The compounds can enter organisms via different routs - with food and beverages, through lungs, skin, and gills. There are several groups of mechanisms of oxidative stress induction by exogenous compounds (xenobiotics): (*i*) compounds once entered the organism may be directly involved in redox processes yielding ROS; (*ii*) in organism some chemicals may be converted to redox active compounds due to metabolism; and (*iii*) the compounds entering organisms may non-directly stimulate ROS production or corrupt defense systems. Certain compounds may realize their effects via several mechanisms simultaneously.

This book provides the information on induction of oxidative stress in diverse living organisms by physical and chemical factors. Substantial part of the book is devoted to antioxidants, i.e. compounds protecting an organism against deleterious ROS effects.

5. Acknowledgments

The editor would like to thank all authors who participated in this project for their contributions and hard work to prepare interesting chapters on the induction of oxidative stress by physical and chemical factors as well as protection of organisms against deleterious effects of reactive species by antioxidants. I also thank to colleagues from Precarpathian National University, who helped to develop the ideology of this book during many years of collaboration, helpful, creative, and sometimes "hot" discussions, which stimulated to perfect my knowledge on the role of reactive species in diverse living processes. I am also grateful to the "In-Tech" Publisher personnel, especially to Ms. Sasa Leporic who excellently assisted me in the arrangement of the book and scheduling the activities.

6. References

Arnhold, J. & Flemmig, J. (2010). Human myeloperoxidase in innate and acquired immunity. *Archives of Biochemistry and Biophysics*, Vol.500, No.1, pp. 92-106.

Bayliak, M.; Gospodaryov, D.; Semchyshyn, H. & Lushchak, V. (2008). Inhibition of catalase by aminotriazole *in vivo* results in reduction of glucose-6-phosphate dehydrogenase activity in *Saccharomyces cerevisiae* cells. *Biochemistry,* Moscow,. Vol.73, No.4, pp. 420-426.

Berry, C. & Hare, J. (2004). Xanthine oxidoreductase and cardiovascular disease: molecular mechanisms and pathophysiological implications. *Journal of Physiology,* Vol.555, No.3, pp. 589-606.

Daff, S. (2010). NO synthase: structures and mechanisms. *Nitric Oxide,* Vol.23, No.1, pp. 1-11.

Dreiem, A.; Rykken, S.; Lehmler H.; Robertson, L. & Fonnum F. (2009). Hydroxylated polychlorinated biphenyls increase reactive oxygen species formation and induce cell death in cultured cerebellar granule cells. *Toxicology and Applied Pharmacology,* Vol.240, No.2, pp. 306-313.

Gerschman, R.; Gilbert, D.; Nye, S.; Dwyer, P. & Fenn, W. (1954). Oxygen poisoning and x-irradiation - A mechanism in common. *Science,* Vol. 119, pp. 623–626.

Gruetter, C.; Barry, B.; McNamara, D.; Gruetter, D.; Kadowitz, P. & Ignarro, L. (1979). Relaxation of bovine coronary artery and activation of coronary arterial guanylate cyclase by nitric oxide, nitroprusside and a carcinogenic nitrosoamine. *Journal of cyclic nucleotide research*, Vol.5, No.3, pp. 211-224.

Hahn, N.; Meischl, C.; Wijnker, P.; Musters, R.; Fornerod, M.; Janssen, H.; Paulus, W.; van Rossum, A.;Niessen, H. & Krijnen, P. (2011). NOX2, p22phox and p47phox are targeted to the nuclear pore complex in ischemic cardiomyocytes colocalizing with local reactive oxygen species. *Cellular Physiology and Biochemistry*, Vol.27, No.5, pp. 471-478.

Harman, D. (1956). Aging: a theory based on free radical and radiation chemistry. *Journal of Gerontology*, Vol.11, No.3, pp. 298-300.

Harman, D. (1972). The biologic clock: The mitochondria? *Journal of the American Geriatrics Society*, Vol.20, pp. 145-147.

Harman, D. (1983). Free radical theory of aging: Consequences of mitochondrial aging. *Age*, Vol.6, pp. 86-94.

Hermes-Lima, M. (2004a). Oxidative stress and medical sciences, In: *Functional Metabolism: Regulation and Adaptation*, Storey, K., (Ed.), pp. 369-382, Wiley-Liss, NY.

Hermes-Lima, M. (2004b). Oxygen in biology and biochemistry: role of free radicals, In: *Functional Metabolism: Regulation and Adaptatio,*, Storey, K., (Ed.), pp. 319-368, Wiley-Liss, NY.

Hideg, E. ; Kálai, T. & Hideg, K. (2011). Direct detection of free radicals and reactive oxygen species in thylakoids. *Methods in Molecular Biology*, Vol.684, pp. 187-200.

Julio Raba, J. & Mottola, H. (1995). Glucose Oxidase as an Analytical Reagent. *Critical Reviews in Analytical Chemistry*, Vol.25, No.1, pp. 1–42.

Klebanoff, S. (1967). Iodination of bacteria: a bactericidal mechanism. *Journal of Experimental Medicine*, Vol.126, No.6, pp. 1063-1078.

Langermans, J.; Hazenbos, W. & van Furth R. (1994). Antimicrobial functions of mononuclear phagocytes. *Journal of Immunological Methods*, Vol.174, No.1-2, pp. 185-194.

Lu, A. , Li, X. ; Gu, Y. ; Wright. P. & Chang, D. (2001). Repair of oxidative DNA damage: mechanisms and functions. *Cell Biochemistry and Biophysics*, Vol.35, No.2, pp. 141-170.

Lushchak, V.; Semchyshyn, H. ; Lushchak, O. & Mandryk, S. (2005). Diethyldithiocarbamate inhibits *in vivo* Cu,Zn-superoxide dismutase and perturbs free radical processes in the yeast *Saccharomyces cerevisiae* cells. *Biochemical and Biophysical Research Communications*, Vol.338, No.8, pp. 1739-1744.

Lushchak, V. (2007). Free radical oxidation of proteins and its relationship with functional state of organisms. *Biochemistry* Moscow, Vol.72, No.8, pp.:809-827.

Lushchak, V. (2011a). Adaptive response to oxidative stress: Bacteria, fungi, plants and animals. *Comparative Biochemistry and Physiology - Part C Toxicology & Pharmacology*, Vol.153, pp. 175-190.

Lushchak, V. (2011b). Environmentally induced oxidative stress in aquatic animals. *Aquatic Toxicology*,Vol.101, pp. 13-30.

Lushchak. V. (2011c). Glutathione homeostasis and functions: potential targets for medical interventions. *Journal of Amino Acids*, in press.

McCord, J. & Fridovich, I. (1969). Superoxide dismutase. An enzymic function for erythrocuprein (hemocuprein). *Journal of Biological Chemistry*, Vol.244, pp. 6049–6055.

Ozimek, P.; Veenhuis, M. & van der Klei, I. (2005). Alcohol oxidase: a complex peroxisomal, oligomeric flavoprotein. *FEMS Yeast Research*, Vol.5, No.11, pp. 975-983.

Sies, H. (1985). Oxidative stress: Introductory remarks. In: *Oxidative stress*, Sies H, (Ed.), pp. 1-8, Academic Press, London.

Shin, A.; Oh, C. , & Park, J. (2006). Glycation-induced inactivation of antioxidant enzymes and modulation of cellular redox status in lens cells. *Archives of Pharmacal Research*, Vol.29, No.7, pp. 577-581.

Sirker, A.; Zhang, M. & Shah. A. (2011). NADPH oxidases in cardiovascular disease: insights from *in vivo* models and clinical studies. *Basic Research in Cardiology*, Vol.106, No.5, pp. 735-747.

Vazquez-Torres, A.; Stevanin, T.; Jones-Carson, J.; Castor, M.; Read, R. & Fang, F. (2008). Analysis of nitric oxide-dependent antimicrobial actions in macrophages and mice. *Methods in Enzymology*, Vol.437, pp. 521-538.

Wedgwood, S.; Lakshminrusimha, S.; Fukai, T.; Russell, J.; Schumacker, P. & Steinhorn, R. (2011). Hydrogen peroxide regulates extracellular superoxide dismutase activity and expression in neonatal pulmonary hypertension. *Antioxidants and Redox Signaling*, Vol.15, No.6, pp. 1497-1506.

Yang, L.; Latchoumycandane, C.; McMullen, M.; Pratt, B.; Zhang, R.; Papouchado. B.; Nagy, L.; Feldstein, A. & McIntyre, T. (2010). Chronic alcohol exposure increases circulating bioactive oxidized phospholipids. *Journal of Biological Chemistry*, Vol.285, No.29, pp. 22211-22220.

Section 2

Physical Factors

2

Oxidative Stress Induced Damage of the Human Retina: Overview of Mechanisms and Preventional Strategies

Katrin Engelmann[1,2,3],
Klio Ai Becker[1] and Richard Funk[2,3]
[1]*Augenklinik, Klinikum Chemnitz gGmbH,*
[2]*Institut für Anatomie, Medizinische Fakultät der TU Dresden,*
[3]*CRTD, Zentrum für Regenerative Therapien Dresden,*
DFG-Forschungszentrum und Exzellenzcluster, TU Dresden,
Germany

1. Introduction

A number of studies have shown that excessive visible light or a special wavelength (blue light) can induce damage to photoreceptor and retinal pigment epithelial cells of the retina, inducing apoptosis. Most of these studies were performed in experimental animal models. However, the mechanisms which lead to damage and subsequently to degenerative diseases like age related macular degeneration (ARMD) remain still unclear. Whether direct interaction of light with retinal cells or a secondary mechanism of transport or circulation of the retinal pigment epithelium or the choroid causes this retinal damage is currently under debate. Cellular mechanisms involved are lipid peroxidation, production of reactive oxygen species (ROS), apoptosis, DNA-damage and others. Clinical or epidemiological studies on this topic are rare and publications about light damage of retinal cells *in vivo* are difficult to achieve. Nevertheless, the clinical practise to implant yellow artificial lenses during cataract surgery is a common practise. These implants are expected to prevent blue light damage to the aging retina. We will address the fact that numerous basic scientific publications point to a rationale for this practice, although it is often difficult to derive clear-cut evidence from clinical epidemiological studies for the preventive use of yellow tinted artificial lenses. We refer to studies showing that the shortwave part of the visible spectrum of light can be harmful to the retina, especially to the macula and optic nerve. For this, we have screened the literature for the major sources of radical production and for the targets of oxidative stress after impingement of "blue light" on the retina. Furthermore, we can show that many studies in cell and molecular biology, animal experiments and first clinical trials point to a preferential use of yellow tinted lenses especially in the elderly and ARMD patients. As in several other fields, so too in this field does "cell biological knowledge" exceed clinical knowledge. Thus, prevention strategies and therapies are still missing. It is important that clinicians should become more aware of this topic so that more informed treatments decisions can be made.

2. Anatomical features of the macula and photoreceptors: Possible mechanism of blue light damage

During passage through the eye the spectrum of electromagnetic radiation (ultraviolet light (UV) UVA 280 - 315 nm and UVB 315 – 400 nm), visible light (400 – 780 nm) and infrared light (>780 nm)) undergoes different modifications: the cornea absorbs mostly short and longer wavelengths (UVA 280 - 315 nm and UVB 315 – 400 nm). Parts of UV light around 320 nm reach as far as into the lens, where they are finally absorbed. The visible part of the spectrum and only very few of the 320 nm fraction are transmitted through the vitreous body and reach the receptors of the retina. Only 1 % of the full spectrum of sunlight, or a comparable continuous spectrum, actually reaches the retina. The spectrum above 1400 nm is absorbed mainly by the water molecules (Barker & Brainard, 1991; Boettner & Wolter, 1962). With increasing age, the lens blocks more and more of the blue (short wave) fraction of light (Bron et al., 2000).

This is why some authors suggest adjusting the spectral transparency of artificial intraocular lenses to that of natural lenses of elderly persons. Indeed, some of the artificial lenses allow more passage of short wavelength light than it is found in lenses of newborn babies (van Norren & van de Kraats, 2007). The retina of elderly persons, however, is not comparable to that of a newborn.

The macula as the site of maximal retinal vision degenerates fastest because (a) it is located directly in the focus of an envisioned light source and (b) there are no other layers situated more centrally in the path of ray of lights. In the peripheral retina layers of nerve and glial cells normally filter out the short wavelengths with their cytochromes and other pigments (Algvere & Seregard 2002).

The discovery of antioxidative molecules within the macula gave a first hint that this direct impact of light onto the macula might cause oxidative damage. The antioxidative molecules are lutein (luteus, latin, means yellowish and gave the name lutea to the macula) and zeaxanthin respectively. These molecules filter out blue light due to their yellow colour. The fact that these radical scavengers are concentrated thousand-fold at this location compared with elsewhere in the retina is a real clue that too much blue light and also oxidative damages may be prevented. Indeed, many animal and cell culture experiments have shown that short wavelength light can enhance the fraction of free radicals and reactive oxygen species (Wu et al., 2006).

This is especially true for the photoreceptors. The photopigment rhodopsin is located in their outer segments, which can be induced by blue light to react in photochemical processes. This leads to intermediates, which produce radicals. That the visual cycle of the photopigments is involved in these reactions can be probed by depletion of the protein RPE65 (a protein involved in the regeneration of rhodopsin): after depletion of this protein blue light has no impact on the retina (Grimm et al., 2000a). Additionally the narcotic gas halothane can block the regeneration of rhodopsin and makes the retina insensitive to blue light impact (Keller et al., 2001). In contrast to green light, which can regenerate the bleached rhodopsin completely, blue light is only able to regenerate 30 % of it. That means that a large fraction of rhodopsin remains unbleached and is absorbing further photons and creating radical producing intermediates ("photoreversal" of rhodopsin) (Grimm et al., 2000b; Grimm et al., 2001, Organisciak et al., 1990; Wu et al., 1999a).

All-trans-retinal is a candidate of these intermediates because it is the most photosensitizing molecule (Delmelle, 1978). A triplet state can be created there by blue light, which releases free radicals (Rozanowskaet al., 1998). Thus, an excited electron can fall back into its ground state and the extra energy transfer into e.g. "reactive oxygen species" (ROS), superoxide radicals, hydrogenperoxide, hydroxyl radicals and other metabolites (Foote, 1968; Spikes & Macknight, 1972; Witting, 1965).

The radicals which originate in the rhodopsin cycle transform all-trans-retinal into di-retinoid-pyridinium-ethanolamine (A2E, see below). This metabolite then accumulates as most dangerous component of lipofuscin in the retinal pigment epithelium (RPE) (Katz et al., 1994; Katz & Gao, 1995; Katz et al., 1996; Wassel & Boulton, 1997).

Moreover, the highest concentration of polyunsaturated fatty acids within the human body is found within the outer segments of the photoreceptors. These lipids are oxidized along with the outer segments of the photoreceptors also by impinging blue light.

Furthermore, carboxyethylpyrrol-modified proteins (CEP, derivates of the non enzymatic oxidation of docosahexanoid acid) are in discussion as very harmful components (see below).

The regeneration of the outer segments by renewal and shedding of discs prevents the accumulation of too many products of oxidation in the outer segments. About 10 of the 100 disci in the outer segments shed per day. Then they are phagocytosed by the RPE - this means 3 billion times in the eyes of a 70 year old person over his or her lifetime (Birch et al., 1984; Marshall, 1987;Young, 1971).

3. Blood retinal barriers, retinal capillaries and choriocapillaries

It is a further peculiarity that the photoreceptors, as specialized nerve cells, reach out with the outer segment into a micro-milieu which is totally different from that of neural (inner, ellipsoid, perikaryon and neurite with the synapses) part of the cell: the neural part is supplied by a microcirculatory unit (the retinal capillaries) which is typical for the central nervous system. Here, capillaries with a small lumen and tight endothelium are characteristic of glial cells (Müller cells) in the immediate vicinity.

In contrast to this, the outer segments are embedded within the interphotoreceptor matrix. This contains special proteins and hyaluronic acid (Acharya et al., 2000; Hollyfield, 1999; Hollyfield et al., 2001) and the outer segments "bathe" in a sea of plasma, which is supplied by the sea of blood within the choriocapillaries (fenestrated capillaries) and choroid (Funk, 1997). The membrana limitans externa serves as watershed zone between both regions.

The choroid is regulated only minimally via the concentration of oxygen, thus, very high concentrations of oxygen can occur in the outer segments which are independent of the oxygen consumption, a fact which makes this system prone to oxidative stress (Wu et al., 2006). The mitochondria deliver the vast amount of energy which is needed for the steady synthesis of the outer segment disci. The photoreceptors consume via mitochondria 3-4 times more energy than all other retinal neurons or cells in the central nervous system. They are probably the cells with the highest oxygen consumption of all (Alder et al., 1990; Linsenmeier et al., 1998). Moreover, the mitochondria are the organelles which are preferentially susceptible to oxidative stress (Field et al., 2011): they harbour the enzymes of the respiratory chain which handle electrons. Under normal circumstances, this works with only a small leakage of free radicals. However, if the mitochondria are under stress or if they are pre-damaged by multiple small genetic failures then radicals can spread out into the cell (Jang & Remmen, 2009). Therefore damage to mitochondrial DNA can occur with increasing frequency as age advances.

The effect of short wavelength light on the metabolism of mitochondria has been an important topic of experimental *in vitro and in vivo* studies. Indeed, the studies of King et al. (2004) could show that blue light impact leads to an enhanced production of radicals in mitochondria. Molecules of the respiratory chain like flavins and cytochromoxidases can absorb at wavelength of 440 - 450 nm and they can cause the production of ROS and oxidative stress (Lascaratos et al., 2007).

What does this mean for the retina as a whole? The photoreceptors are stuffed with mitochondria in their inner segment, especially in the ellipsoid. The disci of the outer segments probably get loaded with radicals by these mitochondria. In addition they are sources of radical production and indeed, vast amounts of radicals are produced if photoreceptors are loaded with blue light (Yang et al., 2003).

Not only the photoreceptors but also the retinal ganglion cells, which contain numerous mitochondria, are prone to blue light damage. Studies of Osborne et al. (2008) showed that blue light was ineffective regarding radical damage in cells which are depleted of mitochondria.

An important new aspect linking blue light damage and genesis of glaucoma should be noted (Osborne et al., 2006): The axons of the retinal ganglion cells possess no myelin sheath because lipid sheets would not allow the light to pass through the retina. So a myelin sheath is not built before the passage through the sclera via the lamina cribrosa. This is unique amongst the body's neurons because normally the neurons are only non-myelinated at the end of their processes. This causes a so-called "impendance mismatch", which leads to an enormous additional energy consumption. This additional energy is delivered by small clusters of mitochondria located in bulges along the axons from the ganglion cells till the optic papilla.

4. Experimental studies regarding light-induced damage of the retina

Regarding the retina as a whole, experimental studies have revealed the layers which are damaged by intense light (Noell, 1965; Noell et al., 1966).

Wenzel et al. (2005) showed damage and apoptotic processes especially in the photoreceptors. This fact is believed to be the main cause of the light induced cell stress.

Several animal studies demonstrated that light exposure leads to lipid oxidation. So Wiegand et al. (1983) assumed that the peroxidation of polyunsaturated fatty acids due to light is a cause for light-induced retinal degeneration. Here, antioxidative substances could prevent this effect (Tanito, et al., 2006).

Both short intense exposure to light and longer continous low-light exposure (e.g. light bulb emissions for several weeks or months) have been shown to lead to retinal damages in rat retina (Kuwabara & Gom, 1968; O`Steen et al., 1972). Interestingly, the first damages took place in the outer segments of the photoreceptors thereafter the mitochondria in the inner segment began to swell. Also in monkeys, similar photoreceptor damage occurs after irradiation with light emission bulbs (Sykes et al., 1981). Here again the central part of the retina, the macula, is affected.

It is noteworthy that after a very high but short (1000 sec) dose the retinal pigment epithelium (RPE) but not the photoreceptors is damaged following irradiation (Ham et al., 1978). The RPE has to digest daily about 1/10 of the photoreceptor mass - together with this all the oxidation products and damaged molecules (see above; "disc shedding") (Bok, 1993).

As previously noted, blue light may induce damage by induction of intermediate reactive species, which act in the outer segments of the photoreceptors. These intermediates produce oxidated photopigments, proteins and probably also products of the lipid oxidation, substances which than are phagocytosed by the RPE. For this purpose RPE-cells posses besides of lysosomes also microperoxisomes, delivering peroxides for intracellular digestion and have a function for detoxification and antioxidation (Bok, 1993). They also regenerate the visual pigments (Bok, 1990).

All these enumerated metabolic products together build up the age-related pigment lipofuscin. Lipofuscin accumulates during life time in the RPE especially in ARMD. It leads to many damaging effects, including generation of ROS (Boulton et al., 1993) and phototoxicity (Davies et al., 2001). One specific lipofuscin fraction is A2E. This orange-reflecting pyridinium bisretinoid is a metabolite of the retinoid cycle. Data implicates that lipofuscin is an agent that makes RPE cells more sensitive to photooxidative stress. The action curve of blue light damages, the so-called blue light hazard, has a peak around 440 nm. Here, it seems very probable that the impact at this wavelength light is dominated by the chromophor A2E (Sparrow & Cai, 2001). If A2E has absorbed a photon, especially of the wavelength 430 - 440, then free radicals are generated, mostly ROS as mentioned above (Boulton et al., 1993; Gaillard et al., 1995). So Wielgus et al. (2010) were able to show that if albino rats were exposed to blue light (450 nm, 3,1 mW cm $^{-2}$), especially the oxidized form of A2E increased. This seems to be especially responsible for the damaging process of retinal cells. Recently it has been shown that A2E generates toxic oxidative products after adsorption of blue light (for review, see (Holz et al., 2004)). This results in a damaging cascade of cell function and the expression of inflammatory and angiogenic substances (Wihlmark et al., 1997; Rezai et al., 2008; Schutt et al., 2000; Sparrow et al., 2000). So A2E inhibits important functions of the cell and is able to increase the apoptosis of the RPE.

It has been shown that a significantly higher rate of cell death occurs in lipofuscin or chromophor A2E loaded retinal pigmented epithelial cells *in vitro*, when these cells were exposed to blue light (430 ± 30 nm) than when they are exposed to white light (390 till 750 nm) (Sparrow et al., 2004). But Tanito et al. (2005) found that an intensified exposure to white light induced also protein modifications. This reaction is mediated by 4-HNE and 4-hydroxyhexanal. Both are reactive aldehydes, which are produced during enzymatic oxidation of n-6 und n-3 nonsaturated fatty acids. The protein modifications did not occur if radical scavengers like phenyl-N-*TERT*-butylnitrone (PBN) were used in this *in vitro* system. Additionally, apoptosis of photoreceptors did not occur (Tanito et al., 2005; Ranchon et al., 2003). Thus, it was speculated, that the 4-HNE-based protein modifications may function as an indicator for oxidative stress which could be detected also in hereditary diseases like Retinitis Pigmentosa (Shen et al., 2005). Another possible marker for oxidative stress is the carboxyethylpyrrol (CEP)-modified protein, a derivate of the non–enzymatic oxidation of the docosahexanoid acid. This protein modification could also be demonstrated in an ARMD eye (Crabb et al., 2002; Gu et al., 2003). A CEP – modification could also be identified after irradiation with blue light of shorter and longer wavelengths (Dunaief et al., 2002).

So we can summarize that chemical reaction of lipids and proteins induced by radical actions can be induced by oxidation as well as "blue light". This may lead to products (adducts) like "advanced lipid end products" (ALEs). This reaction is analogue to the reaction which induces a cross linking of proteins and carbohydrates (advanced glycation end products, AGEs, processed in the so-called "Maillard" reaction). These AGEs accumulate together with lipid oxidation products in extra cellular space (e.g. the Bruch Membrane) as well as within cells e.g. within the RPE (Glenn et al., 2009; Howes et al., 2004). Protein-sugar products or the protein – lipid oxidation products (e.g. CEP) can accumulate also in the intra cellular space and build up an important component of lipofuscin (see also (Schmidt et al., 2008).

The experimental data regarding "blue light damage" to photoreceptors shows that the recycling of the visual pigments in the retinoid cycle can be stressed by bright blue light. In doing so reactive intermediate are formed, which can generate radicals by themselves (Grimm et al., 2000a). Furthermore, the high concentration of polyunsaturated fatty acids favours the oxidation of lipids. In addition, advanced glycation end products enhance the formation of radicals.

Pigmented epithelial cells suffer from the overload of oxidized discs e.g. A2E in the outer segments because RPE cells have to phagocytise these products of oxidation (Wu et al., 1999a; Wu et al., 1999b).

Both radical sources the photoreceptor outer segment with their lipid membranes and the mitochondria can potentiate mutually: e.g. A2E can block the transfer of cytochrome C to complex IV in the respiratory chain; by this a deviation of electrons and cytochrome C takes place. The latter can induce apoptosis via typical signalling cascades (Shaban & Richter, 2002).

5. Light intensity and animal studies

Young primates were used to investigate the mechanisms of damage by specific parts of the spectrum (violet and blue-green) (Ham et al., 1976; Ham et al., 1979). It was found that light with damaging wavelengths does not correlate with light adsorption lines of the photo pigments like rhodopsin. That is why this group assumes other mechanisms of electron excitation and followed radical damage (Ham et al., 1976). On the other hand, other authors demonstrate that also low dosages of light can induce significant amounts of radicals (Lawwill et al., 1977). A cumulative damage occurs in the retina during this kind of irradiation. Here, fractionated doses of light are acting with higher intensity then comparable - although continuous – actions. This effect does not occur if the retina is allowed to regenerate in a longer dark period (Noell et al., 1966; Ham et al., 1979; Lawwill et al., 1977; Tsò et al., 1972).

When considering translation of these observations into a better understanding of human eyes, the following factors are important: the light dose, the duration and the time points of actions (also during day – night cycle).

6. Quality of light and adaptation

Sunlight possesses a continuous spectrum also in the long wavelength range (with a few dips due to water absorption, see below). Neon-strip lamps and energy saving bulbs have discontinuous spectra (only several peaks in the short- and middle wavelength part). This artificial light often is not very bright; however, the eye perceives this in a relative way. The eye adjusts its sensitivity over the whole spectral range as an integral over many wavelengths. If there are too few peaks e.g. due to the absence of some wavelengths, then the sensitivity of the eye increases. The retina produces more photopigment and a mydriasis occurs. So damaging wavelengths can be more harmful than under bright sunlight. Many experimental studies proof the capability of the photoreceptors to adapt by the mechanism mentioned above. Rats which were reared in darkness have an enhanced amount of rhodopsin (Noell, 1979). This can lead to an increased loss of photoreceptors after light exposure compared to animals reared under a normal day-night cycle (Battelle & LaVail, 1978; Organisciak & Noell, 1977; Organisciak et al., 1985; Penn & Anderson, 1987; Penn et al., 1987). Furthermore, the retinal cells can adapt in the antioxidative capacity, too.

6.1 Time of exposure

Nowadays we spend most of our time under relatively bright artificial light, especially at night times e.g. in shift working. In prior centuries people were working under dim candle or incandescent lamp at night time.

The human body is much more vulnerable to environmental stress in times of activation of the parasympathetic tone and in times of regeneration. More melatonin is released in the night than in times under the sympathetic tone due to activity.

Finally, an important factor cannot be mimicked correctly in cell- and animal experiments: the absolute duration of light impact and of other additional stressors, which can last for years and decades in a human lifetime.

7. Protective role of defined parts of the light spectrum

Opposite to the action of blue light, red or infrared light can have positive effects - a fact which is described in more and more recent studies (Eells et al., 2004; Wong-Riley et al., 2005; Albarracin et al., 2011)

These parts of the light spectrum are present in all continuous spectra of natural light sources like sun or fire but also in incandescent or halogen lamps. Only in recent years have studies shown the positive effects of red or infrared light for regeneration processes in the retina. Here also, the mitochondrium seems to play a major role (Liang et al., 2008).

8. Pathogenesis of ARMD – The role of short wave light

The age related macular degeneration (ARMD) has become a leading cause for blindness in elderly persons (> 60 years) in the industrial world (Klein et al., 1992). ARMD is a degenerative disease caused by multiple factors. It seems that the kind of light to which a person`s eyes has been exposed may play a role. Over the last decades industrialisation makes a night a day. So the intensity and life-long duration of high light dosages increased

(Mainster et al., 1983; Margrain et al., 2004). This interferes with the sensitivity of the macula to light damage as explained above by the anatomical and cell-biological considerations. Another important factor for degenerative diseases is the increasing lifespan of people (Schrader, 2006).

The late form of ARMD - wet or exsudative ARMD - is mainly caused by angiogenesis. Fortunately anti-angiogenetic therapies became available for such patients during the last years (Holz et al., 2004). But therapeutic strategies for the early stages of ARMD are missing up to now. One reason is the poor understanding of key mechanism which results in degeneration of the different cell types of the macula. Also, specific pathologies of ARMD like detachment of the pigment epithelium or geographic atrophy are still poorly understood, although models based on cellular mechanisms are beginning to be discussed. It has been shown that in the case of geographic atrophy degeneration started in all cell types, the RPE, the photoreceptors and in the choroidea. Previously it was assumed that the degeneration started in the RPE (Grebe et al., 2009).

During the last years a genetic predisposition for ARMD came into focus. Two gene loci were identified which are related to ARMD and which can be both used to explain the above mentioned pathogenetic concept. These loci are the complement factor H (CFH) and C3 which normally down-regulate inflammatory processes. Other candidates are the high temperature requirement factor A1 (HTRA1) and LOC387715/ARMS2 (Age-related maculopathy susceptibility 2) and additionally a locus that is responsible for the synthesis of the mitochondrial membranes. Furthermore two mutations of the locus ABCA4 were found. ABCA4 regulates the ATP - binding cassette reporter in the discs of the photoreceptor outer segments. This reporter replaces worn out molecules of the visual pigment and impedes an accumulation of toxic metabolites (Scholl et al., 2007; Swaroop et al., 2007).

It is interesting that Gu et al. (2009) found out that modifications (CEP adducts) and antibodies against CEP-proteins were found in higher concentration in the blood plasma of AMD patients. Patients with the ARMS2 and HTRA1 allele, which leads to a higher AMD risk, showed especially elevated CEP-markers.

There are some hints and observations that in the living human eye radicals may be produced also in mitochondria. Mitochondrial DNA deletions and deficiencies of cytochrome c oxidase (complex IV of the respiratory chain) were detected preferentially in the cones of the fovea centralis of aging retina (Barron et al., 2001).

9. Experimental studies on blue light action and on the use of tinted intraocular lenses

If the hypothesis is true that an increase in the overall amount of irradiation dose and especially a higher percentage of blue light may trigger the ARMD process towards higher stages after removal of the natural lens, it seems logical to examine light effects on the known pathomechanisms for early and late ARMD. Only few valid data from epidemiological studies can currently be generated. In contrast multiple cell-based and animal studies were performed to investigate the effect of yellow tinted intraocular lenses:

In cell cultures of retinal pigmented epithelial cells toxicity tests were performed (Rezai et al., 2008). It could be shown for fetal RPE cells that exposure to blue light (430 - 450 nm) up to 10 days was accompanied by an increasing rate of apoptosis (up to 85 % cell death). If the cell culture dishes were covered with yellow tinted artificial lenses (Acryl-Soft-Natural-Filter) the apoptosis rate could be reduced to 37% (Rezai et al., 2008).

Nilsson et al. (1989) investigated the reaction of Xenon light exposure over 3.5 hours to rabbit eyes. Untinted or yellow tinted lenses were used to protect the eyes. In the eyes treated with clear lenses a reduction of the b- and c waves in the electroretinogram (ERG) became visible in contrast to the tinted lenses. This experiment was one of the first that gave hints to a possible light damage of retinal tissue.

Tanito et al. (2006) demonstrated the damaging effect of blue light (both short and longer waves) using rats. The animals were exposed for 7 days to blue light with and without yellow light filter. Especially in case of short wavelengths of the blue light a reduction in the cell count of the outer nuclear cell layer (ONL) was found. In addition the a- and b-waves in the ERG were reduced in these rats. Postmortally the retinal tissue of the irradiated eyes was examined with respect to the protein modification 4-HNE and CEP. Western blot and enzymometric analysis showed a stronger reaction in the eyes which were not protected with yellow lenses. The relatively short exposure time to blue light was a disadvantage of the here described animal experiments. Another fact is that the experiments were performed on "healthy" retinas. Therefore it can be suggested that the elderly human eye would have shown much more oxidative damage due to extremely long exposure time respectively years compared to the experimental situation.

10. Evidence of light damage in epidemiological studies

Severe sclerosis of the lens nucleus seems to protect people against acquiring degenerative diseases of the macula (Sperduto et al., 1981; West et al., 1989). On the other hand few studies showed that ARMD is significantly increased in pseudophakic or aphakic eyes (Mitchell et al., 1995; Mitchell et al., 2002; van Newkirk et al., 2000; Wang et al., 2003, Wang et al., 1999). Other authors could not find a significant difference (Wang et al., 1999). In pseudophakic eyes with clear artificial lenses, blue sensitive cones are the first photoreceptors, which decrease in number - due to specific light damage (Werner et al., 1989). Moreover, in histopathological sections of ARMD eyes a higher incidence of severe stages of ARMD was observed (van der Schaft et al., 1994).

One of the first who speculated about a higher incidence of wet ARMD after cataract extraction was Pollak et al. (1996). The retrospective character of the study, the small number of patients and the short follow-up time were criticized. Up to now only non-multicenter studies were initiated and therefore only small studies can be found regarding the question: Can blue light induce wet ARMD or induce a progress of dry ARMD? Some of these studies are discussed here: Photodynamic treatment (PDT) needed for subfoveal chorioretinal neovascularisations (CNVs) after cataract surgery in comparison to a control group was investigated (Kaiserman et al., 2007). In this study data of 5913 patients after lens extraction were evaluated and compared to 29565 matched controls. Follow-up time was about four years (1/2001 to 5/2005). The average patient age was comparable in both groups at 74 years. After cataract extraction PDT was significantly higher in pseudophacic eyes of patients compared to phacic eyes during the first 6 months and 1 to 1.5 years after cataract surgery (p=0.004 and p=0.001). However, no differences were observed between both groups prior to surgery. On the other hand PDT 12 month after cataract extraction was comparable in both groups. This study showed an increased risk to develop exsudative ARMD during a "vulnerable" phase directly after cataract extraction. This might be due to a sudden drop down of the protection of the patient-own, aged, and yellow tinted natural lens. The authors also discussed that the higher treatment rate might be caused by better prerequisites for ophthalmoscopic examination after removal of an opacified lens. This argument of a better view on the retina also animated other authors to look at retrospective data of cataract patients. Baatz et al. (2008) did not find a difference between the control and treatment group. A disadvantage of this study is the relatively short follow up time and the heterogeneity between the number of patients in which a fluorescein angiography was performed (177 prior to surgery, 225 after surgery and 97 in the control group). An angiogram was only performed if the clinical examination gave a clue for ARMD.

Blue Mountain Eye Study und Beaver Dam Eye Study indicated a higher incidence of ARMD after cataract extraction (Cugati et al., 2006; Wang et al., 2003).

The Australian Prospective Study of Cataract Surgery and Age-Related Macular Degeneration Study (Cugati et al., 2007) evaluated data from 2000 patients over a follow-up time of five years and at the time of publication about 1600 patients were included. If the preoperative fundus photography was not analysable due to dense cataract the 1-month post-operative retinal photographs were set as a preoperative status. This was based on the fact that a primary documentation of the macular was missing in all prior non-comparable studies. It is assumed that this study will be a sufficient basis for further discussion. Sufficient data are not available yet.

It can be assumed that the results of the published data may lead surgeons to restrict cataract extraction in ARMD patients. However, Armbrecht et al. (2000) and Shuttleworth and Galloway (1999) demonstrated that quality of life of ARMD patients increased after cataract surgery. The data were evaluated using standardized "Quality of Life" questionnaires. Especially, the specific and differentiated visual functions improved in patients with moderate cataract and ARMD (Armbrecht et al., 2000). In a pseudophakic group, which was examined by Shuttleworth et al. (1998), 10.1% of the patients showed a progression of ARMD, in 2% a CNV developed. Nevertheless, most of these studies included too few patients and were not randomized. On the other hand no disadvantages due to the use of yellow tinted artificial lenses have yet been described. So other authors support their use for preventive purposes (Falkner-Radler et al., 2008).

All published data coming from of larger or small retro- or prospective studies as well as of epidemiological studies used different criteria for the development of ARMD. Therefore, it is not possible to draw firm conclusions from current data. Is this reason enough to choose to implant yellow tinted lenses? Efforts should be made in clinical and basic preventive research to minimize the socioeconomic costs of this widespread disease ARMD. We hypothesise that other questions should be raised independent of clinical trials: What happens during cataract extraction that could lead to a progress of ARMD? And is the use of yellow tinted lenses in cataract patients still justified?

An interesting hypothesis was raised by Wegner and Khoramnia (2011). He claimed that the age-related cataract is not a single disease, but is induced by a retinal messenger of unknown character. So beside the protection of the eye from oxidative stress through e.g. high levels of vitamin C in the anterior and posterior fluids of the eye, the yellow pigments and isomers of a hydroycarotenoid, lutein and zeaxanthin are effective in protection of the macula. Both are powerful anti-oxidants and function as a filter for short wavelength blue light, thus limiting oxidative damage and stress to the retinal cells and inhibiting apoptosis (Snodderly, 1995). The macular pigment functions as a natural filter or "protector" that commonly decreases in density throughout the years in elderly persons (Beatty et al., 2001; Hammond & Caruso-Avery, 2000). Based on these facts Wegner hypothesised that with decreasing levels of protection by the macular pigment a retinal messenger is generated. This triggers cataract-formation as a self-defence reaction. Therefore, both cataract formation and ARMD development may depend on each other (Wegner A et al., 2011). Based on this hypothesis the implantation of blue filtering artificial lenses may be justified as a substitute for the "protective" elderly natural tinted yellow lens.

11. Oxidative stress and phako emulsification

It can be assumed that only a few surgeons know about the fact that during phaco emulsification oxidative stress is induced. In the past, the focus was set to mechanical damage through ultra sonic or the rinsing process during cataract extraction (for review, see Takahashi 2005). There are many reports about the induction of free radicals by ultra sound energy. This process is described as "acoustic cavitation" (Riesz & Kondo, 1992). Water molecules are disintegrated with potential formation of hydroxyl radicals that are most effective in their biological action. This phenomenon is called sonolysis. The specification of the different species of free radicals is complicate and guides the chosen test method and handling of the probes. Free radicals were described first at the beginning of the 1950s (Heimberg et al. 1953, Beauchamp & Fridovich 1970). The influence of such free radicals as damaging agent for the corneal endothelium, the most sensitive cell layer of the cornea was evaluated first. To protect the corneal endothelium from free radical damage during phaco emulsification high viscoelastic substances supplemented with natrium hyalurate as a radical scavenger were developed. Only few are known: Has the decreasing level of vitamin C in the anterior chamber a negative role? This ascorbic acid is highly concentrated in the anterior chamber compared to the blood-levels (anterior chamber 4.3 mg/dl blood plasma 0.8 mg/dl) and it plays an important role as a radical scavenger (Miratshi et al 2005). Therefore, it is not surprising that Rubowitz et al (Nemet et al., 2007; Rubowitz 2003) demonstrated a protective effect of ascorbic acid to prevent endothelial damage. Also other molecules, which act as antioxidants are relevant. Augustin and Dick (2004) found an elevated lipid peroxide level after phakoemulsification in 130 patients. The level correlated positively with the time of ultrasonic exposure during surgery. Even if this oxidative stress can be minimized using viscoelastic substances during surgery, we do not know what happens after the removal of these substances at the end of phacoemulsification. The aqueous humour is exchanged with a salt solution which does not represent the natural liquid environment, e.g by a reduced level of natural antioxidants. In an animal model it needs more than 15 days to build up a normal ascorbic acid level in the anterior chamber after experimental surgery (De Biaggi et al., 2006) On the other hand the overall protein amount in the anterior chamber increased as a sign of stress (De Biaggi et al., 2006). It is supposed that this reconstitution is induced by ROS, which are also known to act as damaging agents (Cameron et al., 2001). Oxidative stress induced by "acoustic cavitation" should not be ignored especially if the retina or macula is impaired also. It is reported that cataract formation may be enhanced in patients with a generally reduced "antioxidative status" (Dherani et al., 2009), even if such data are difficult to evaluate with respect to different population and diseases. Nevertheless, the addition of several factors can potentiate the perioperative stress factors in real patient situations, and it is also conceivable that degeneration in the macula may be stimulated (Yagihashi et al., 2007). Even if it is a multifactorial, and therefore difficult, research field, it seems to be important to look deeper into the epidemiological field especially in times of aging populations.

12. Conclusion

Evidence-based medicine unquestionably improves the quality of the practice of medicine. However, it can often be difficult to generate sufficient evidence for the best treatment of degenerative, multi-factorial, chronic diseases like ARMD from clinical and epidemiological data alone. The implantation of artificial lenses that filter blue light is such an example, and the discussion on this topic is vigorous. There are good arguments for their use, and a few against, but strong clinical evidence is difficult to find. In the end, arguments for or against protection from blue light may be too focused: Does this discussion really matter in the treatment of a degenerative disease like ARMD? In our opinion, it is desirable that the preventive or protective aspects of treating degenerative diseases like ARMD should become an increasing focus of medical and scientific research, especially as the population ages. However, practical considerations suggest that the development of preventive and protective strategies should not be excluded in the absence of rigorous clinical studies. It is simply not possible to design and execute studies for such a complex, multifactorial disease. The state of research into the protective effects of supplements, e.g. antioxidants such as the macula pigments lutein and zeaxanthin, presents similar questions over which different clinical specialties (e.g. ophthalmologists vs. nutrition specialists) may argue. This is in contrast to the often clear results of cell-biological experiments. These reveal strong arguments for a protection against too much blue light or regarding to a deficiency of preventional factors inside the eye. We may find ourselves at the beginning of the development of preventive strategies, which must be developed from various points of view especially for such a multi-factorial disease as ARMD. However, we must also be prepared to accept a perpetual discrepancy between the rigorous scientific data obtainable from cell biology experiments and the difficulty of interpreting these data into meaningful therapeutic strategies.

There will certainly be many things to consider. For example, take the argument that blue light is important for the daily light balance for the body (sleeping-waking rhythms), while one should also be mindful of a potential blue light "overdose" due to night-time light intensity and unnaturally high blue-light rays from energy saving light bulbs, LEDs, televisions (LCD, plasma, or cathode ray), and long hours in front of the computer. These lifestyle-induced changes in people's light balance are difficult to account for and separate in current arguments. It is known that blue light (e.g. from LED diodes) reduces melatonin production and increases activity in younger people (which have relatively high levels compared to the markedly reduced levels of melatonin in elderly persons) (West et al. 1989).

Does this suggest similar activity (and also no yellow lenses) for older people? Do the elderly not already get too little sleep for proper regeneration? We already know that melatonin production is reduced in older people. Before we speculate too much on the role of the sleep-wake cycle, we should increase age-related research.

However, perhaps these considerations could help us to reflect better on our lifestyles. What external factors influence our physical and psychological conditions? If we are aware of the possible consequences of lifestyle choices, then we may pay closer attention to these influences. We may come to find potentially protective options in other fields, as tinted artificial lenses may offer in cataract surgery. In conclusion, although traditional clinical studies cannot answer such complex, multifactorial questions completely, the other experimental results discussed here may nonetheless be useful in devising new therapeutic strategies.

13. Acknowledgment

We acknowledge the funding of own studies related to this topic by the Dr. med. hc. Erwin Braun Stiftung.

14. References

Acharya, S.; Foletta, VC.; Lee, JW.; Rayborn, ME.; Rodriguez, IR.; Young WS 3rd. & Hollyfield JG. (2000). SPACRCAN, a novel human interphotoreceptor matrix hyaluronan-binding proteoglycan synthesized by photoreceptors and pinealocytes. *J Biol Chem,* Vol. 275, pp. 6945-6955

Albarracin, R.; Eells, J. & Valter, K. (2011). Photobiomodulation protects the retina from light-induced photoreceptor degeneration. *Invest Ophthalmol Vis Sci*, Vol. 52, No. 6, (June 2011), pp. 3582-92

Alder, VA.; Ben-Nun, J. & Cringle, SJ. (1990). PO2 profiles and oxygen consumption in cat retina with an occluded retinal circulation. *Invest Ophthalmol Vis Sci* , Vol. 31, pp. 1029-1034

Algvere, PV. & Seregard, S. (2002). Age-related maculopathy: pathogenetic features and new treatment modalities. *Acta Ophthalmol Scand,* Vol. 80, pp. 136-143

Armbrecht, AM.; Findlay, C.; Kaushal, S.; Aspinall, P.; Hill, AR. & Dhillon, B.(2000). Is cataract surgery justified in patients with age related macular degeneration? A visual function and quality of life assessment. *Br J Ophthalmol,* Vol. 84, pp. 1343-1348

Augustin, AJ. & Dick, HB. (2004). Oxidative tissue damage after phacoemulsification: influence of ophthalmic viscosurgical devices. *J Cataract Refract Surg,* Vol. 30, No. 2, (February 2004), pp. 424-7

Baatz, H.; Darawsha, R.; Ackermann, H.; Scharioth, GB.; de Ortueta, D.; Pavlidis, M. & Hattenbach, LO. (2008). Phacoemulsification does not induce neovascular age-related macular degeneration. *Invest Ophthalmol Vis Sci,* Vol. 49, pp. 1079-1083

Barker, F.; Brainard, G. (1991). The direct spectral transmittance of excised human lens as a function of age. US Food and Drug Administration Report 1991

Barron, MJ.; Johnson, MA.; Andrews, RM.; Clarke, MP.; Griffiths, PG.; Bristow, E.; He, LP.; Durham, S. & Turnbull DM. (2001). Mitochondrial abnormalities in ageing macular photoreceptors. *Invest Ophthalmol Vis Sci,* Vol. 42, pp. 3016-3022

Battelle, BA. & LaVail, MM. (1978). Rhodopsin content and rod outer segment length in albino rat eyes: modification by dark adaptation. *Exp Eye Res*, Vol. 26, pp.487-497

Beatty, S.; Murray, IJ.; Henson, DB.; Carden, D.; Koh, H. & Boulton, ME. (2001). Macular pigment and risk for age-related macular degeneration in subjects from a Northern European population. *Invest Ophthalmol Vis Sci*, Vol. 42, No. 2, (February 2001), pp.439-46

Beauchamp, C. & Fridovich I. (1970). A mechanism for the production of ethylene from methional. The generation of the hydroxyl radical by xanthine oxidase. *Journal of Biological Chemisry. Vol.* 245, pp. 4641-4646

Birch, DG.; Berson, EL. & Sandberg, MA. (1984). Diurnal rhythm in the human rod ERG. *Invest Ophthalmol Vis Sci*, Vol. 25, pp.236-238

Boettner, EA. & Wolter, JR. (1962). Transmission of the ocular media. *Invest Ophthalmol,* Vol. 1, pp. 776-783

Bok, D. (1990). Processing and transport of retinoids by the retinal pigment epithelium. *Eye*, Vol. 4, No. 2, pp. 326-332

Bok, D. (1993). The retinal pigment epithelium: a versatile partner in vision. *J Cell Sci*, Suppl, 17, pp. 189-195

Boulton, M.; Dontsov, A.; Jarvis-Evans, .J; Ostrovsky, M. & Svistunenko, D. (1993). Lipofuscin is a photoinducible free radical generator. *J Photochem Photobiol B*, Vol. 19, pp. 201-204

Bron, AJ.; Vrensen, GF.; Koretz, J.; Maraini, G. & Harding JJ. (2000). The ageing lens. *Ophthalmologica*, Vol. 214, pp. 86-104

Cameron, MD.; Poyer, JF. & Aust, SD. (2001). Identification of free radicals produced during phacoemulsification. *J Cataract Refract Surg*, Vol. 27, No. 3, (March 2001), pp. 463-70

Crabb, JW.; Miyagi, M,.;Gu, X.; Shadrach, K.; West, KA.; Sakaguchi, H.; Kamei, M.; Hasan, A.; Yan, L.; Rayborn, ME.; Salomon, RG. & Hollyfield, JG. (2002). Drusen proteome analysis: an approach to the etiology of age-related macular degeneration. *Proc Natl Acad Sci U S A*, Vol. 99, pp. 14682-14687

Cugati, S.; de Loryn, T.; Pham, T.; Arnold, J.; Mitchell, P. & Wang, JJ. (2007). Australian prospective study of cataract surgery and age-related macular degeneration: rationale and methodology. *Ophthalmic Epidemiol*, Vol. 14, pp. 408-414

Cugati, S.; Mitchell, P.; Rochtchina, E.; Tan, AG.; Smith, W. & Wang, JJ. (2006). Cataract surgery and the 10-year incidence of age-related maculopathy: the Blue Mountains Eye Study. *Ophthalmology*, Vol. 113, pp. 2020-2025

Davies, S.; Elliott, MH.; Floor, E.; Truscott, TG.; Zareba, M.; Sarna, T.; Shamsi, FA. & Boulton, ME. (2001). Photocytotoxicity of lipofuscin in human retinal pigment epithelial cells. *Free Radic Biol Med*, Vol. 31, pp. 256-265

De Biaggi, CP.; Barros, PS.; Silva, VV.; Brooks, DE. & Barros, SB. (2006). Ascorbic acid levels of aqueous humour of dogs after experimental phacoemulsification. *Vet Ophthalmol,* Vol. 9, No. 5, (September-October 2006), pp. 299-302

Delmelle, M. (1978). Retinal sensitized photodynamic damage to liposomes. *Photochem Photobiol*, Vol. 28, pp. 357-360

Dherani, M.; Murthy, GV.; Gupta, SK.; Young, IS.; Maraini, G.; Camparini, M.; Price, GM.; John, N.; Chakravarthy, U. & Fletcher, AE. (2008). Blood levels of vitamin C, carotenoids and retinol are inversely associated with cataract in a North Indian population. *Invest Ophthalmol Vis Sci*, Aug, Vol. 49, No. 8, (August 2008), pp. 3328- 35

Dunaief, JL.; Dentchev, T.; Ying, GS. & Milam, AH. (2002). The role of apoptosis in age-related macular degeneration. *Arch Ophthalmol*, Vol. 120, pp.1435-1442

Eells, JT.; Wong-Riley, MT.; VerHoeve, J.; Henry, M.; Buchman, EV.; Kane, MP.; Gould, LJ.; Das, R.; Jett, M.; Hodgson, BD.; Margolis, D. & Whelan, HT. (2004). Mitochondrial signal transduction in accelerated wound and retinal healing by near-infrared light therapy. *Mitochondrion*, Vol. 4, pp. 559-567

Falkner-Radler, CI.; Benesch, T. & Binder, S. (2008). Blue light-filter intraocular lenses in vitrectomy combined with cataract surgery: results of a randomized controlled clinical trial. *Am J Ophthalmol*, Vol. 145, pp. 499-503

Field, MG.; Yang, D.; Bian, ZM.; Petty, HR. & Elner, VM. (2011). Retinal flavoprotein fluorescence correlates with mitochondrial stress, apoptosis, and chemokine expression. *Exp Eye Res*, Jul 13. [Epub ahead of print]

Foote, CS. (1968). Mechanisms of photosensitized oxidation. There are several different types of photosensitized oxidation which may be important in biological systems. *Science*, Vol. 162, pp. 963-970

Funk, RH. (1997). Blood supply of the retina. Ophthalmic Res, Vol. 29, pp. 320-325

Gaillard, ER.; Atherton, SJ.; Eldred, G. 6 Dillon, J. (1995). Photophysical studies on human retinal lipofuscin. *Photochem Photobiol*, Vol. 61, pp. 448-453

Glenn, JV.; Mahaffy, H.; Wu, K.; Smith, G.; Nagai, R.; Simpson, DA.; Boulton, ME. & Stitt, AW. (2009). Advanced glycation end product (AGE) accumulation on Bruch's membrane: links to age-related RPE dysfunction. *Invest Ophthalmol Vis Sci*, Vol. 50, pp. 441-451

Grebe, R.; Bhutto, I.; Merges, C.; Baba, T. & Lutty, GA. (2009). Relationship between RPE and choriocapillaris in age-related macular degeneration. *Invest Ophthalmol Vis Sci*, Oct, Vol. 50, No. 10, (October 2009), pp. 4982-91

Grimm, C.; Reme, C.; Rol, PO. & Williams, TP. (2000). Blue light's effects on rhodopsin: photoreversal of bleaching in living rat Eyes. *Invest Ophthalmol Vis Sci*, Vol. 41, pp. 3984-3990

Grimm, C.; Wenzel, A.; Hafezi, F.; Yu, S.; Redmond, TM. & Remé, CE.. (2000a). Protection of Rpe65-deficient mice identifies rhodopsin as a mediator of light-induced retinal degeneration. *Nat Genet*, Vol. 25, pp. 63-66

Grimm, C.; Wenzel, A.; Williams, T.; Rol, P.; Hafezi, F. & Remé, C. (2001). Rhodopsin-mediated blue-light damage to the rat retina: effect of photoreversal of bleaching. *Invest Ophthalmol Vis Sci*, Vol. 42, pp. 497-505

Gu, J.; Paeur, GJ.; Yue, X.; Narendra, U.; Sturgill, GM.; Bena, J.; Gu, X.; Peachey, NS.; Salomon, RG.; Hagstrom, SA. & Crabb, JW.; Clinical Genomic and Proteomic AMD Study Group. (2009). Assessing susceptibility to age-related macular degeneration with proteomic and genomic biomarkers. *Mol Cell Proteomics*, Vol. 8, No. 6, (June 2009), pp. 1338-49.

Gu, X.; Meer, SG.; Miyagi, M.; Rayborn, ME.; Hollyfield, JG.; Crabb, JW. & Salomon, RG. (2003). Carboxyethylpyrrole protein adducts and autoantibodies, biomarkers for age-related macular degeneration. *J Biol Chem*, Vol. 278, pp. 42027-42035

Ham, WT Jr.; Mueller, HA.; Ruffolo, JJ Jr. & Clarke, AM. (1979). Sensitivity of the retina to radiation damage as a function of wavelength. *Photochem Photobiol*, Vol. 29, pp. 735-743

Ham, WT Jr.; Mueller, HA. & Sliney, DH. (1976). Retinal sensitivity to damage from short wavelength light. *Nature*, Vol. 260, pp. 153-155

Ham, WT Jr.; Ruffolo, JJ Jr.; Mueller, HA,.; Clarke, AM. & Moon, ME. (1978). Histologic analysis of photochemical lesions produced in rhesus retina by short-wave-length light. *Invest Ophthalmol Vis Sci*, Vol. 17, pp. 1029-1035

Hammond, BR Jr. & Caruso-Avery, M. (2000). Macular pigment optical density in a Southwestern sample. *Invest Ophthalmol Vis Sci*, Vol. 41, No. 6, (May 2000), pp. 1492-7

Heimberg, M.; Fridovich, I. & Handler, P. (1953). The enzymatic oxidation of sulfite. *Journal of Biological Chemisry.*, Vol. 211, pp 913-926

Hollyfield, JG.; Rayborn, ME.; Nishiyama, K.; Shadrach, KG.; Miyagi, M.; Crabb, JW. & Rodriguez, IR. (2001), Interphotoreceptor matrix in the fovea and peripheral retina of the primate Macaca mulatta: distribution and glycoforms of SPACR and SPACRCAN. *Exp Eye Res*, Vol. 72, pp. 49-61

Hollyfield JG.(1999). Hyaluronan and the functional organization of the interphotoreceptor matrix. *Invest Ophthalmol Vis Sci*, Vol. 40, pp. 2767-2769

Holz, FG.; Pauleikhoff, D.; Klein, R. & Bird, AC. (2004). Pathogenesis of lesions in late age-related macular disease. *Am J Ophthalmol*, Vol. 137, pp. 504-510

Howes, KA.; Liu, Y.; Dunaief, JL.; Milam, A.; Frederick, JM.; Marks, A. & Baehr, W. (2004). Receptor for advanced glycation end products and age-related macular degeneration. *Invest Ophthalmol Vis Sci*, Vol. 45, pp. 3713-3720

Jang, YC. & Remmen, HV. (2009). The mitochondrial theory of aging: Insight from t ransgenic and knockout mouse models. *Exp Gerontol*, Vol. 44, No. 4, (April 2009), pp. 256-60

Kaiserman, I.; Kaiserman, N.; Elhayany, A. 6 Vinker, S. (2007). Cataract surgery is associated with a higher rate of photodynamic therapy for age-related macular degeneration. *Ophthalmology*, Vol. 114, pp. 278-282

Katz, ML.; Christianson, JS.; Gao, CL. & Handelman, GJ.(1994). Iron-induced fluorescence in the retina: dependence on vitamin A. *Invest Ophthalmol Vis Sci*, Vol. 35, pp. 3613- 3624

Katz, ML.; Gao, CL. & Rice, LM. (1996). Formation of lipofuscin-like fluorophores by reaction of retinal with photoreceptor outer segments and liposomes. *Mech Ageing Dev*, Vol. 92, pp. 159-174

Katz, ML. & Gao, CL. (1995). Vitamin A incorporation into lipofuscin-like inclusions in the retinal pigment epithelium. *Mech Ageing Dev*, Vol. 84, pp. 29-38

Keller, C.; Grimm, C.; Wenzel, A.; Hafezi, F. & Remé, C. (2001). Protective effect of halothane anesthesia on retinal light damage: inhibition of metabolic rhodopsin regeneration. *Invest Ophthalmol Vis Sci*, Vol. 42, No. 2, (February 2001), pp. 476-80

King, A.; Gottlieb, E.; Brooks, D.G.; Murphy, M.P. & Dunaief, J.L. (2004) Mitochondria-derived reactive oxygen species mediate blue light-induced death of retinal pigment epithelial cells. *Photochem. Photobiol., Vol. 79, pp.* 470–475

Klein, R.; Klein, BE. & Linton, KL. (1992). Prevalence of age-related maculopathy. The Beaver Dam Eye Study. *Ophthalmology*, Vol. 99, pp. 933-943

Kuwabara, T. & Gom, RA. (1968). Retina damage by visible light. An electron microscopic study. *Arch Ophthalmol*, Vol. 79, pp. 69-78

Lascaratos, G.; Ji, D.; Wood, JP. & Osborne, NN. (2007). Visible light affects mitochondrial f unction and induces neuronal death in retinal cell cultures. *Vision Res*, Vol. 47, pp. 1191-1201

Lawwill, T.; Crocket, S. & Currier, G. (1977). Retinal damage secondary to chronic light exposure, thresholds and mechanisms. *Doc Ophthalmol*, Vol. 44, pp. 379-402

Liang, HL.; Whelan, HT.; Eells, JT. & Wong-Riley, MT. (2008). Near-infrared light via light-emitting diode treatment is therapeutic against rotenone- and 1-methyl-4-phenylpyridinium ion-induced neurotoxicity. *NeuroScience*, Vol. 153, pp. 963-974

Linsenmeier, RA.; Braun, RD.; McRipley, MA.; Padnick, LB.; Ahmed, J.; Hatchell, DL.; McLeod, DS. & Lutty, GA. (1998). Retinal hypoxia in long-term diabetic cats. *Invest Ophthalmol Vis Sci*, Vol. 39, pp. 1647-1657

Mainster, MA.; Ham, WT Jr. & Delori, FC. (1983). Potential retinal hazards. Instrument and environmental light sources. *Ophthalmology*, Vol. 90, pp. 927-932

Margrain, TH.; Boulton, M.; Marshall, J. & Sliney, DH. (2004). Do blue light filters confer protection against age-related macular degeneration? *Prog Retin Eye Res*, Vol. 23, 523-531

Marshall, J. (1987). The ageing retina: physiology or pathology. *Eye*, Vol. 1, No 2, pp. 282-295

Mitchell, P.; Smith, W.; Attebo, K. & Wang, JJ. (1995). Prevalence of age-related maculopathy in Australia. The Blue Mountains Eye Study. *Ophthalmology*, Vol. 102, pp. 1450-1460

Mitchell, P.; Wang, JJ.; Foran, S. & Smith, W. (2002). Five-year incidence of age-related maculopathy lesions: the Blue Mountains Eye Study. *Ophthalmology*, Vol. 109, pp. 1092-1097

Nemet, AY.; Assia, EI.; Meyerstein, D.; Meyerstein, N.; Gedanken, A. & Topaz, M. (2007). Protective effect of free-radical scavengers on corneal endothelial damage in phacoemulsification. *J Cataract Refract Surg*, Vol. 33, No. 2, (February 2007), pp. 310- 5

Nilsson, SE.; Textorius, O.; Andersson, BE. & Swenson, B. (1989). Clear PMMA versus yellow intraocular lens material. An electrophysiologic study on pigmented rabbits regarding "the blue light hazard". Prog Clin Biol Res, Vol. 314, p. 539-553

Noell, WK.; Walker, VS.; Kang, BS. & Berman, S. (1966). Retinal damage by light in rats. *Invest Ophthalmol*, Vol. 5, pp. 450-473

Noell, WK. (1965). Aspects of experimental and hereditary degeneration. In: *Biochemistry of the retina*, C Graymore, (Ed.), 51-72, Academic Press London, ISBN xxxxxxxxx, London , England

Noell WK. (1979). Effects of environmental lighting and dietary vitamin A on the vulnerability of the retina to light damage. *Photochem Photobiol*, Vol. 29.; 717-723

O'Steen, WK.; Shear, CR.; &Anderson, KV. (1972). Retinal damage after prolonged exposure to visible light. A light and electron microscopic study. *Am J Anat*, Vol. 134, pp. 5-21

Organisciak, DT.; Jiang, YL.; Wang, HM. 6 Bicknell, I. (1990). The protective effect of ascorbic acid in retinal light damage of rats exposed to intermittent light. *Invest Ophthalmol Vis Sci*, Vol. 31, pp. 1195-1202

Organisciak, DT. & Noell, WK.(1977). The rod outer segment phospholipid/opsin ratio of rats maintained in darkness or cyclic light. *Invest Ophthalmol Vis Sci*, Vol. 16, pp. 188-190

Organisciak, DT.; Wang, HM.; Li, ZY. & Tso, MO. (1985). The protective effect of ascorbate in retinal light damage of rats. *Invest Ophthalmol Vis Sci*, Vol. 26, pp. 1580-1588

Osborne, NN.; Lascaratos, G.; Bron, AJ.; Chidlow, G. & Wood, JP. (2006), A hypothesis to suggest that light is a risk factor in glaucoma and the mitochondrial optic neuropathies. *Br J Ophthalmol*, Vol. 90, pp. 237-241

Osborne, NN.; Li, GY.; Ji, D.; Mortiboys, HJ. & Jackson, S. (2008). Light affects mitochondria to cause apoptosis to cultured cells: possible relevance to ganglion cell death in certain optic neuropathies. *J Neurochem*, Vol. 105, pp. 2013-2028

Penn, JS. & Anderson, RE. (1987). Effect of light history on rod outer-segment membrane composition in the rat. *Exp Eye Res*, Vol. 44, pp. 767-778

Penn, JS.; Naash, MI. & Anderson, RE. (1987). Effect of light history on retinal antioxidants and light damage susceptibility in the rat. *Exp Eye Res*, Vol. 44, pp. 779-788

Pollack, A.; Marcovich, A.; Bukelman, A. & Oliver, M. (1996). Age-related macular degeneration after extracapsular cataract extraction with intraocular lens implantation. *Ophthalmology*, Vol. 103, pp. 1546-1554

Ranchon, I.; LaVail, MM.; Kotake, Y. & Anderson, RE. (2003). Free radical trap phenyl-N-tert-butylnitrone protects against light damage but does not rescue P23H and S334ter rhodopsin transgenic rats from inherited retinal degeneration. *J Neurosci*, Vol. 23, pp. 6050-6057

Rezai, KA.; Gasyna, E.; Seagle, BL.; Norris, JR Jr. & Rezaei, KA. (2008). AcrySof Natural filter decreases blue light-induced apoptosis in human retinal pigment epithelium. *Graefes Arch Clin Exp Ophthalmol*, Vol. 246, pp. 671-676

Riesz, P. & Kondo, T. (1992). Free radical formation induced by ultrasound and its biological implications. *Free Radic Biol Med*, Vol. 13, No. 3, (September 1992), pp. 247-70

Rozanowska, M.; Wessels, J.; Boulton, M.; Burke, JM.; Rodgers, MA.; Truscott, TG. & Sarna, T. (1998). Blue light-induced singlet oxygen generation by retinal lipofuscin in non-polar media. *Free Radic Biol Med*, Vol. 24, pp. 1107-1112

Rubowitz, A.; Assia, EI.; Rosner, M. 6 Topaz, M. (2003). Antioxidant protection against corneal damage by free radicals during phacoemulsification. *Invest Ophthalmol Vis Sci*, May, Vol. 44, No. 5, (May 2003), 1866-70

Schmidt, K-G.; Bergert, H. & Funk, RHW. (2008). Neurodegenerative Diseases of the Retina and Potential for Protection and Recovery. *Current Neuropharmacology*, Vol. 6, 164- 178

Scholl, HP.; Fleckenstein, M.; Charbel Issa, P.; Keilhauer, C.; Holz, FG. & Weber, BH. (2007). An update on the genetics of age-related macular degeneration. *Mol Vis*, Vol. 13, pp. 196-205

Schrader, WF. (2006). Age-related macular degeneration: a socioeconomic time bomb in our aging society. *Ophthalmologe*, Vol. 103, pp. 742-748

Schutt, F.; Davies, S.; Kopitz, J.; Holz, FG. & Boulton, ME. (2000). Photodamage to human RPE cells by A2-E, a retinoid component of lipofuscin. *Invest Ophthalmol Vis Sci*, Vol. 41, pp. 2303-2308

Shen, J.; Yang, X.; Dong, A.; Petters, RM.; Peng, YW. & Wong, F.; Campochiaro, PA. (2005). Oxidative damage is a potential cause of cone cell death in retinitis pigmentosa. *J Cell Physiol*, Vol. 203, pp. 457-464

Shimmura, S.; Tsubota, K.; Oguchi, Y.; Fukumura, D.; Suematsu, M. & Tsuchiya, M. (1992). Oxiradical-dependent photoemission induced by a phacoemulsification probe. *Invest Ophthalmol Vis Sci*, Vol. 33, No. 10, (September 1992), pp. 2904-7

Shuttleworth, GN. & Galloway, PH. (2002). Analysis of the United Kingdom solar eclipse public health campaign 1999. *Clin Experiment Ophthalmol*, Vol. 30, pp- 308-310

Shuttleworth, GN.; Luhishi, EA. & Harrad, RA. (1998). Do patients with age related maculopathy and cataract benefit from cataract surgery? *Br J Ophthalmol*, Vol. 82, pp. 611-616

Snodderly, DM. (1995). Evidence for protection against age-related macular degeneration by carotenoids and antioxidant vitamins. *Am J Clin Nutr*, Vol. 62, No. 6 Suppl, December 1995), pp. 1448S-1461S

Sparrow, JR.; Miller, AS. & Zhou, J. (2004). Blue light-absorbing intraocular lens and retinal pigment epithelium protection in vitro. *J Cataract Refract Surg*, Vol. 30, pp. 873-878

Sparrow, JR.; Nakanishi, K. & Parish, CA. (2000). The lipofuscin fluorophore A2E mediates blue light-induced damage to retinal pigmented epithelial cells. *Invest Ophthalmol Vis Sci*, Vol. 41, pp. 1981-1989

Sperduto, RD.; Hiller, R. 6 Seigel, D. (1981). Lens opacities and senile maculopathy. *Arch Ophthalmol*, Vol. 99, pp. 1004-1008

Spikes, JD. & Macknight, ML. (1972). Photodynamic effects on molecules of biological importance: amino acids, peptides and proteins. *Res Prog Org Biol Med Chem*, Vol. 3, No. 1, pp. 124-136

Swaroop, A.; Branham, KE.; Chen, W. & Abecasis, G. (2007). Genetic susceptibility to age-related macular degeneration: a paradigm for dissecting complex disease traits. *Hum Mol Genet*, Vol. 16, Spec No. 2, pp. R174-182

Sykes, SM.; Robison, WG Jr.; Waxler, M. & Kuwabara, T. (1981). Damage to the monkey retina by broad-spectrum fluorescent light. *Invest Ophthalmol Vis Sci*, Vol. 20, pp. 425-434

Takahashi, H. (2005). Free radical development in phacoemulsification cataract surgery. *J Nihon Med Sch*, Vol. 72, No. 1, (February 2005), pp- 4-12

Tanito, M.; Elliott, MH.; Kotake, Y. 6 Anderson, RE. (2005). Protein modifications by 4-hydroxynonenal and 4-hydroxyhexenal in light-exposed rat retina. *Invest Ophthalmol Vis Sci*, Vol. 46, pp. 3859-3868

Tanito, M.; Kaidzu, S. & Anderson, RE. (2006). Protective effects of soft acrylic yellow filter against blue light-induced retinal damage in rats. *Exp Eye Res*, Vol. 83, pp. 1493- 1504

Tanito, M.; Yoshida, Y.; Kaidzu, S.; Ohira, A. & Niki, E. (2006). Detection of lipid peroxidation in light-exposed mouse retina assessed by oxidative stress markers, total hydroxyoctadecadienoic acid and 8-iso-prostaglandin F2alpha. *Neurosci Lett*, Vol. 398, pp. 63-68

Ts'o, MO.; Fine, BS. & Zimmerman, LE. (1972). Photic maculopathy produced by the indirect ophthalmoscope. 1. Clinical and histopathologic study. *Am J Ophthalmol*, Vol. 73, pp. 686-699

van der Schaft, TL.; Mooy, CM.; de Bruijn, WC.; Mulder, PG.; Pameyer, JH. & de Jong, PT. (1994). Increased prevalence of disciform macular degeneration after cataract extraction with implantation of an intraocular lens. *Br J Ophthalmol*, Vol. 78, pp. 441-445

van Norren, D. & van de Kraats, J. (2007). Spectral transmission of intraocular lenses expressed as a virtual age. *Br J Ophthalmol*, Vol. 91, pp. 1374-1375

Van Newkirk, MR.; Nanjan, MB.; Wang, JJ.; Mitchell, P.; Taylor, HR. & McCarty, CA. (2000). The prevalence of age-related maculopathy: the visual impairment project. *Ophthalmology*, Vol. 107, pp. 1593-1600

Wang, JJ.; Klein, R.; Smith, W.; Klein, BE.; Tomany, S. & Mitchell, P. (2003). Cataract surgery and the 5-year incidence of late-stage age-related maculopathy: pooled findings from the Beaver Dam and Blue Mountains *Eye* studies. *Ophthalmology*, Vol. 110, pp. 1960-1967

Wang, JJ.; Mitchell, PG.; Cumming, RG. & Lim, R. (1999). Cataract and age-related maculopathy: the Blue Mountains Eye Study. Ophthalmic Epidemiol, Vol. 6, pp. 317-326

Wassell, J. & Boulton, M. (1997). A role for vitamin A in the formation of ocular lipofuscin. *Br J Ophthalmol,* Vol. 81, pp. 911-918

Wegner, A. & Khoramnia, R. (2011). Cataract is a self-defence reaction to protect the retina from oxidative damage. *Med Hypotheses,* Vol. 76, No. 5, (May 2011), pp. 741-4

Wenzel, A.; Grimm, C.; Samardzija, M. & Reme, CE. (2005). Molecular mechanisms of light-induced photoreceptor apoptosis and neuroprotection for retinal degeneration. *Prog Retin Eye Res,* Vol. 24, pp. 275-306

Werner, JS.; Steele, VG. & Pfoff, DS. (1989). Loss of human photoreceptor sensitivity associated with chronic exposure to ultraviolet radiation. *Ophthalmology,* Vol. 96, 1552-1558

West, SK.; Rosenthal, FS.; Bressler, NM.; Bressler, SB.; Munoz, B.; Fine, SL.; &Taylor, HR. (1989). Exposure to sunlight and other risk factors for age-related macular degeneration. *Arch Ophthalmol,* Vol. 107, pp. 875-879

Wiegand, RD.; Giusto, NM.; Rapp, LM. & Anderson, RE. (1983). Evidence for rod outer segment lipid peroxidation following constant illumination of the rat retina. *Invest Ophthalmol Vis Sci,* Vol. 24, pp. 1433-1435

Wielgus, AR.; Collier, RJ.; Martin, E.; Lih, FB.; Tomer, KB.; Chignell, CF. & Roberts, JE. (2010). Blue light induced A2E oxidation in rat Eyes--experimental animal model of dry AMD. *Photochem Photobiol* Sci, Vol. 9, No. 11, (November 2010), pp. 1505-12

Wihlmark, U.; Wrigstad, A.; Roberg, K.; Nilsson, SE. & Brunk. UT. (1997). Lipofuscin accumulation in cultured retinal pigment epithelial cells causes enhanced sensitivity to blue light irradiation. *Free Radic Biol Med,* Vol. 22, pp.1229-1234

Witting, LA. (1965). Lipid peroxidation in vivo. *J Am Oil Chem Soc,* Vol. 42, pp. 908-913

Wong-Riley, MT.; Liang, HL.; Eells, JT..; Chance, B..; Henry, MM.; Buchmann, E..; Kane, M. & Whelan, HT. (2005). Photobiomodulation directly benefits primary neurons functionally inactivated by toxins: role of cytochrome c oxidase. *J Bio. Chem,* Vol. 280, pp. 4761-4771

Wu, J.; Chen, E. & Soderberg, PG. (1999a). Failure of ascorbate to protect against broadband blue light-induced retinal damage in rat. *Graefes Arch Clin Exp Ophthalmol,* Vol. 237, pp. 855-860

Wu, J.; Seregard, S. & Algvere, PV. (2006). Photochemical damage of the retina. *Surv Ophthalmol,* Vol. 51, pp. 461-481

Wu, J.; Seregard, S.; Spangberg, B.; Oskarsson, M. & Chen, E. (1999b). Blue light induced apoptosis in rat retina. *Eye,* Vol. 13, No. 4, pp. 577-583

Yagihashi, T.; Wakabayashi, Y.; Kezuka, J.; Usui, M. & Iwasaki, T. (2007). Changes in vitreous amino acid concentrations in a rabbit model of cataract surgery. *Acta Ophthalmol Scand,* Vol. 85, No. 3, (May 2007), pp. 303-8

Yang, JH.; Basinger, SF.; Gross, RL. & Wu, SM. (2003). Blue light-induced generation of reactive oxygen species in photoreceptor ellipsoids requires mitochondrial electron transport. *Invest Ophthalmol Vis Sci,* Vol. 44, pp. 1312-1319

Young, RW. (1971). Shedding of discs from rod outer segments in the rhesus monkey. *J Ultrastruct Res,* Vol. 34, pp. 190-203

3

Exercise and Oxidative Stress

Vladimir Lj. Jakovljevic[1], Dejan Cubrilo[1], Vladimir Zivkovic[1], Dusica Djordjevic[2] and Dragan Djuric[3]
[1]Department of Physiology, Faculty of Medicine, University of Kragujevac,
[2]Department of Anthropometry, Faculty of Medicine, University of Kragujevac,
[3]Institute of Physiology „Richard Burian", School of Medicine, University of Belgrade,
Serbia

1. Introduction

Reactive oxygen species (ROS), as well as reactive nitrogen species (RNS), are products of normal cellular metabolism. ROS and RNS are well recognized for playing a dual role as both deleterious and beneficial species, since they can be either harmful or beneficial to living systems (Valko et al., 2006). Beneficial effects of ROS occur at low/moderate concentrations and involve physiological roles in cellular responses to noxia, as for example in defense against infectious agents and in the function of a number of cellular signaling systems. One further beneficial example of ROS at low/moderate concentrations is the induction of a mitogenic response.

On the other hand, high levels of free radicals may cause biological damage, which is termed oxidative stress and nitrosative stress (Ridnour et al., 2005). This occurs in biological systems when there is an overproduction of ROS/RNS or/and some kind of deficiency in antioxidant defense system (ADS). Generally, oxidative stress results from the metabolic reactions that use oxygen and represents a disturbance in the equilibrium status of prooxidant/antioxidant reactions in living organisms. The excess ROS can damage cellular lipids, proteins, or DNA and inhibit their normal function. Because of this, oxidative stress has been implicated in a number of human diseases as well as in the aging process. The delicate balance between physiological and pathophysiological effects of ROS is achieved by mechanisms called "redox regulation". The process of "redox regulation" protects living organisms from various oxidative stresses and maintains "redox homeostasis" by controlling the redox status in vivo (Droge, 2002).

Free radicals, as one form of ROS, can be defined as molecules or molecular fragments containing one or more unpaired electrons in atomic or molecular orbitals (Halliwell & Gutteridge, 1999). This unpaired electron(s) usually gives a considerable degree of reactivity to the free radical. Radicals derived from oxygen represent the most important class of radical species generated in living systems. Molecular oxygen (dioxygen) has a unique electronic configuration and itself is a radical. Superoxide anion ($O_2^{\bullet -}$), , arising either through metabolic processes or following oxygen "activation" by physical irradiation, is considered the "primary" ROS, and can further interact with other molecules to generate "secondary" ROS, either directly or indirectly via enzymatic or nonenzymatic reactions (Valko et al., 2005). The production of superoxide occurs mostly within the mitochondria of a cell. Superoxide is produced from both Complexes I and III of the electron transport chain, and once in its anionic form it readily crosses the inner mitochondrial membrane. Recently, it has been demonstrated that Complex I-produced superoxide is exclusively released into the matrix and that no detectable levels escape from intact mitochondria (Muller et al., 2004). In addition, Complex III is responsible for extramitochondrial release of superoxide, but less then 50 % of total production. Other half is directly released to the mitochondrial matrix.

The hydroxyl radical ($^{\bullet}OH$) is the neutral form of the hydroxide ion and has a high reactivity, making it a very dangerous radical with a very short *in vivo* half-life (Pastor et al., 2000). The redox state of the cell is largely linked to an iron (and copper) redox couple and is maintained within strict physiological limits. The released Fe^{2+} can participate in the Fenton reaction, generating highly reactive hydroxyl radical ($Fe^{2+} + H_2O_2 \rightarrow Fe^{3+} + {}^{\bullet}OH + OH-$). Also, the superoxide radical participates in the Haber-Weiss reaction ($O_2^{\bullet-} + H_2O_2 \rightarrow O_2 + {}^{\bullet}OH + OH-$) which combines a Fenton reaction and the reduction of Fe^{3+} by superoxide, yielding Fe^{2+} and oxygen ($Fe^{3+} + O_2^{\bullet-} \rightarrow Fe^{2+} + O_2$) (Liochev & Fridovich, 2002). The most realistic *in vivo* production of hydroxyl radical according to the Fenton reaction occurs when Mn^+ is iron, copper, chromium, or cobalt. Although Fenton chemistry is known to occur *in vitro*, its significance under physiological conditions is not clear, noting particularly the negligible availability of "free catalytic iron" due to its effective sequestration by the various metal-binding proteins (Kakhlon & Cabantchik, 2002). However, organisms overloaded by iron (as in the conditions of hemochromatosis, b-thalassemia, hemodialysis) contain higher amounts of "free available iron" and this can have deleterious effects. "Free-iron" is transported into an intermediate, labile iron pool (LIP), which represents a steady state exchangeable and readily chelatable iron compartment (Kakhlon & Cabantchik, 2002).

Additional reactive radicals derived from oxygen that can be formed in living systems are peroxyl radicals ($ROO^{\bullet}$). The simplest peroxyl radical is $HOO^{\bullet}$, which is the protonated form (conjugate acid; p*Ka* ~4.8) of superoxide ($O_2^{\bullet-}$) and is usually termed either hydroperoxyl radical or perhydroxyl radical. Hydroperoxyl radical is the most important in initiation of lipid peroxidation by two parallel pathways: fatty acid hydroperoxide (LOOH)-independent and LOOH-dependent. The LOOH-dependent pathway of $HO_2^{\bullet-}$ initiated fatty acid peroxidation may be relevant to mechanisms of lipid peroxidation initiation *in vivo*. This process is in physiological conditions mostly scavenged by action of antidoxidant enzymes, basically by superoxide dismutase (SOD), catalase (CAT), glutathione perooxidase (GPx), glutathione reductase (GR), as well as xanthine oxidoreductase (XOR) (Vorbach et al., 2003). XOR is the key enzyme in purine catabolism, by catalyzing oxidative hydroxylation of hypoxanthine to xanthine and subsequently of xanthine to uric acid. Uric acid acts as a potent antioxidant and free radical scavenger. XOR therefore has important functions as a cellular defense enzyme against oxidative stress. Thus, the synthesis of both an antioxidant (uric acid) and numerous prooxidants (ROS and RNS) makes XOR an important regulator of the cellular redox potential.

Peroxisomes are known to produce H_2O_2, but not $O_2^{\bullet-}$, under physiological conditions. Oxygen consumption in the peroxisome leads to H_2O_2 production, which is then used to oxidize a variety of molecules. H_2O_2 is physiologically produced by action of SOD, who is specifically scavenging $O_2^{\bullet -}$ and released H_2O_2 is one end product. Cytosolic CAT and GPx can prevent overproduction of H_2O_2 in physiological conditions. Peroxisomes also contain catalase, which decomposes hydrogen peroxide and presumably prevents accumulation of this toxic compound. Thus, the peroxisomes maintain a delicate balance with respect to the relative concentrations or activities of these enzymes to ensure no net production of ROS. How the organelle maintains this equilibrium is unclear. When peroxisomes are damaged and their H_2O_2 consuming enzymes downregulated, H_2O_2 releases into the cytosol which is significantly contributing to oxidative stress.

Phagocytic cells, exposed to a stimulus, have the ability to recognize the foreign particle and undergo a series of reactions called the respiratory burst (De Coursey & Ligeti, 2005). Nicotine adenine dinucleotide phosphate (NAD(P)H) oxidase, one of enzymatic components in this process, is best characterized in neutrophils, where its production of $O_2^{\bullet -}$ generates the respiratory burst necessary for bacterial destruction. The nonphagocytic NAD(P)H oxidases produce superoxide at a fraction (1–10%) of the levels produced in neutrophils and are thought to function in intracellular signaling pathways.

1.1 Reactive Nitrogen Species (RNS)

Nitric oxide (•NO) is a small molecule that contains one unpaired electron on the antibonding orbital and therefore is a radical. •NO is generated in biological tissues by specific 3 isomers of nitric oxide synthases (NOSs):

1. NOSI (neuronal NOS - nNOS) and
2. NOSIII (endothelial NOS – eNOS)

These are both constitutive and responsible for production of physiological amount of •NO, while

3. NOSII (inducible NOS – iNOS)

is inducible and responsible for cytotoxyc •NO production and acts as ROS in both physiological (i.e. immune response) or different pathophysiologocial conditions.

All NOS isoforms metabolize L-arginine to L-citrulline with the formation of •NO *via* five electron oxidative reaction (Ghafourifar & Cadenas, 2005). •NO is an abundant reactive radical that acts as an important oxidative biological signaling molecule in a large variety of diverse physiological processes, including neurotransmission, blood pressure regulation, defense mechanisms, smooth muscle relaxation and immune regulation. •NO has a half-life of only a few seconds in an aqueous environment. •NO has greater stability in an environment with a lower oxygen concentration (half-life >15 s). However, since it is soluble in both aqueous and lipid media, it readily diffuses through the cytoplasm and plasma membranes (Chiueh, 1999). •NO has effects on neuronal transmission as well as on synaptic plasticity in the central nervous system. In the extracellular milieu, •NO reacts with oxygen and water to form nitrate and nitrite anions.

Overproduction of reactive nitrogen species is called nitrosative stress (Ridnour et al., 2004). This may occur when the generation of reactive nitrogen species in a system exceeds the system's ability to neutralise and eliminate them. Nitrosative stress may lead to nitrosylation reactions that can alter the structure of proteins and so inhibit their normal function. Cells of the immune system produce both the superoxide anion and nitric oxide during the oxidative burst triggered during inflammatory processes. Under these conditions, nitric oxide and the superoxide anion may react together to produce significant amounts of a much more oxidatively active molecule, peroxynitrite anion (ONOO−), which is a potent oxidizing agent that can cause DNA fragmentation and lipid oxidation:

$$^{\bullet}NO + O_2^{\bullet -} \rightarrow ONOO^- \quad (1)$$

Nitric oxide readily binds certain transition metal ions; in fact many physiological effects of $^{\bullet}NO$ are exerted as a result of its initial binding to Fe^{2+}-Haem groups in the enzyme soluble guanylyl cyclase (sGC). The product is represented here as *{Fe^{3+}–$^{\bullet}$NO}*, however, {Fe^{3+}–NO−*}* is also commonly seen. The convention *{FeNO}*^{7}, where the superscript is the sum of the metal d electron count (here 6 or 5) and the occupancy of the relevant NO-* orbital (here 1 or 2), is 50 *M* often employed to avoid specific assignment of oxidation states.

1.2 Oxidative damage

Reactive species produced by activated blood cells can be released extracellularly and oxidize several plasma proteins. In addition, the same reactive species can attack the leukocytes generating them. The concentration of most amino acids is higher in neutrophils than in the surrounding plasma. The major proteins of neutrophils include those of the structural matrix, proteins required for its locomotion, chemotactic properties and adhesiveness, as well as the many granule proteins with bactericidal, hydrolytic and inflammatory functions (Smith et al., 2005). As it is the case with other cells, plasma membrane and membranes of the intracellular organelles of leukocytes are rich in lipids. Five percent of the neutrophils wet weight is lipid, which is distributed among various classes. Neutrophils and lymphocytes contain approximately 32 % and 28% PUFA of the total fatty acids by weight, respectively (Kew et al., 2004). A number of studies reported increases of malondialdehyde (an index of lipid peroxidation) after exercise in neutrophils (Ferrer et al., 2009) and lymphocytes (Sureda et al., 2008), whereas others reported no change (Ferrer et al., 2009; Tauler et al., 2008). Interestingly, it has been reported that exercise increased the percentage contribution of PUFA in neutrophils after exercise, supporting the idea that exercise may modulate neutrophil function through alterations in its fatty acid composition (Lagranha et al., 2008). On the other hand, it has been also suggested that changes in neutrophil fatty acid composition does not always lead to changes in neutrophil redox function, such as $O_2^{\bullet -}$ generation. The reactive species biology of platelets is not well studied and the physiological importance of reactive species produced by these blood cells (primarily $O_2^{\bullet -}$ and $^{\bullet}NO$) is uncertain (Halliwell & Gutteridge, 2007).

2. Oxidative stress and strenuous exercise

2.1 Aerobic exercise

Aerobic exercise is accompanied by increased oxygen consumption (VO_2), wand consequently increased ROS production. Aerobic exercise increases VO_2, which, in turn, may increase ROS production. Therefore, many studies suggested that such physical activity enhanced ROS production both in animals and in humans (Mastaloudis et al., 2001).

However, this phenomenon cannot occur with low exercise intensity (< 50 % of maximum oxygen consumption [VO_{2max}]). In such a case, antioxidant capacity is not overwhelmed and ROS-induced damage does not appear. Moreover, the more intense the exercise is, the more important the ROS production and the oxidative stress are. This is confirmed by some studies that show a correlation between VO_2 and oxidative stress. However, other studies show that oxidative stress does not increase after intense aerobic exercise (Chevion et al., 2003). Such contradictory results can be explained by antioxidant nutritional status. Effectively, these studies are done with trained subjects and because of this complexity, no single effects such as ROS production can be connected so simple. However, trained subjects can exhibit oxidative stress as well as sedentary subjects (Pincemail et al., 2000; Palazzetti et al 2003). Moreover, some differences can be explained by the methods used for the measurement of oxidative stress.

2.2 Anaerobic exercise

Anaerobic exercise is a type of exercise that includes a large variety of sport activities (e.g. sprints, jumps or resistance exercise). Information on the production of ROS as a result of acute anaerobic exercise is lacking compared with aerobic exercise (Groussard, 2003). However, these studies generally show an increase of the oxidative stress after supramaximal exercise such as intermittent running, sprints, jumps or sets of jumps, resistance or Wingate tests on an ergocycle (Groussard, 2003; Chen et al., 2001; Kayatekin et al., 2002; Goldfarb et al., 2005, Ramel et al., 2004).

The increase of FR production specific for the anaerobic exercise may be mediated through various pathways in addition to electron leakage which is thought to be the main source of ROS due to aerobic exercise (Groussard, 2003). In this phenomenon the dominant role belongs to xanthine oxidase and NADPH oxidase. It seems possible that ischemic reperfusion of the active muscle is greatly involved in oxidative stress during and after anaerobic exercise. Precisely, this type of exercise significantly enhances the catabolism of purins and provokes a fast deoxygenation (phenomenon of ischaemia reperfusion). These two phenomena are known to increase the activity of xanthine oxidase, which accelerates FR production. Xanthine oxidase has been demonstrated to generate FR during ischemia reperfusion, but direct evidence for xanthine oxidase as a radical generator in muscle during exercise is lacking. In ischemic tissues, it has been proposed that the xanthine dehydrogenase undergoes proteolytic conpreserve to the oxidase form, which uses O_2 as electron acceptor. It is known that xanthine oxidase in the presence of the substrates hypoxanthine or xanthine reduces molecular oxygen to $O_2^{\bullet-}$ and H_2O_2. Recently, it has been demonstrated that the enzyme can further reduce H_2O_2 to $OH^{\bullet-}$. Thus, it has been hypothesized that xanthine oxidase and its requisite substrates would be present in high concentrations in reperfused tissue and consequently would result in oxygen FR generation upon reperfusion.

Moreover, the increase of lactic acid, acidosis, catecholamine and post-exercise inflammation also represent important factors in increased ROS production. Inflammation and cellular damage often happen after traumatizing exercise such as impact sports or eccentric exercises (Childs et al., 2001).

3. Methods for exercise testing and redox status determination during exercise

Exercise testing allows measurement of basic and specific skills that are known or assumed to influence the creation of the final results in a particular sport. The testing of physical activity provides insight into the simultaneous cellular response and cardiovascular system under conditions of precisely controlled metabolic stress. The activity requires the coordinated function of the level of the heart, lungs, peripheral and pulmonary circulation with the common goal of satisfying the increased cellular respiration.

For proper testing it is necessary to provide specific conditions. Optimally equipped laboratory obtain a large number of parameters from the testing process (that must be controlled and reproducible) with continuous monitoring of blood pressure and pulse, and continuous monitoring of gas exchange, electrocardiogram and blood sampling during the test and recovery. Laboratory testing must possess certain characteristics of the microclimate with exactly defined values of temperature of 18-22 degrees Celsius and humidity of 60 %. Testing should be carried out in the morning. An athlete must have breakfast from 1 to 1.5 hour before the test, and be dressed comfortably. If blood is taken, the necessary equipment for the extraction of blood must be placed at appropriate place, in order to avoid confusion and unnecessary movement during the test. The number of people involved in testing must be limited to those who perform testing and those that are responsible for the safety of athletes.

3.1 Protocol for estimation of fitness status

The experiments performed by our investigation team usually start between 8 and 9 AM. Before breakfast the first blood sample is taken from venous blood by needle which is placed to all subjects. At the same time, the capillary blood and urine samples are taken too. After that, subjects have light and lean breakfast. The conventional medical examination is performed to make sure that all subjects are healthy without known acute and chronic diseases. During two hours of resting the participants are explained the aim and protocol of testing process. Then the written inform consent is obtained from all subjects.

Pulmonary VO_2, CO_2 production (VCO_2), and expired minute ventilation (V_E) are measured continuously using an automated metabolic cart (Quark b2, Cosmed Srl., Rome, Italy). Before each test ambient conditions are measured, and then the gas analyzer and the flow meter are calibrated with high precision gases. During submaximal and maximal exercise the VO_2 values are recorded as averages of 15 s. The participants state their subjective feeling of exhaustion by using Borg's CR10 exhaustion scale (Borg, 1982). Heart rate is monitored continuously and it is recorded as average of 15 s using a Polar Sport Tester (HRM, Finland). We hypothesize that maximal oxygen consumption VO_{2max} is reached when oxygen consumption get its plateau (the time when increasing of workload can not affect increase of oxygen consumption) (Howley et al., 1995).

Anaerobic threshold (ANT) is defined as the level of VO_2 during exercise above which aerobic energy production is supplemented by anaerobic mechanisms and is reflected by an increase in lactate and lactate/pyruvate ratio in muscle and arterial blood. During maximal exercise test, which aims to achieve maximal oxygen uptake, there are three phases and two types of anaerobic threshold. The first aerobic threshold is determined by V-slope method. The assessment of this threshold is obtained by cutting the two regression lines (S1, S2), that are the transition from aerobic metabolism, where VCO_2 increases linearly with VO_2 (curve S1), to anaerobic plus aerobic metabolism where the curve S2 receives a value greater than 1. This point represents the beginning of isocapnic buffering where values of ventilation (VE) grow proportionately to the concentration of CO_2 (VCO_2), and therefore ventilatory equivalent for CO_2 does not change at the level of aerobic threshold. The second anaerobic threshold is determined by ventilatory equivalent method by visual inspection of the breakpoints in the inflection of VE versus VCO_2, non linear increase of VE/VCO_2 versus work load and deflection point of the end tidal CO_2 pressure ($P_{ET}CO_2$) data curves (Wasserman et al., 1973; Caiozzo et al., 1982). Mean VO_2 and mean HR values are then expressed as a percentage of VO_{2max} and percentage of HR_{max} at which the anaerobic threshold occured (% ANT_{VO2}; % ANT_{HR}).

- Phase I represents a time between rest period and the first aerobic threshold. The intensity of this phase is characterized as very light physical activity. Achieved percentage of oxygen consumption in this stage is in the range of 45-55 % of VO_{2max} or 60-70 % of maximum heart rate. The concentration of lactate in this stage of the test does not exceed 2 mmol/l (BL: <2 mmol/l). The value of this level of RPE Borg replies to 6-9. In the contemporary theory of sports training this zone is referred to as E1 (Endurance Zone 1)
- Phase II represents a time between two determined thresholds. Furthermore, this phase is divided into:

a. The intensity of this interval corresponds to light physical activity. Achieved percentage of oxygen consumption is in the range of 55-75 % of VO_{2max} or 70-80 % of maximum heart rate. The concentration of lactate is between 2.0-3.0 mmol/l. The value of this level of RPE is 10-12. Applied to the training process, this zone is referred to as E2 (Endurance 2 zones).

b. The intensity of this interval is defined as a medium heavy action. Achieved percentage of oxygen consumption is in the range of 70-80 % of VO_{2max} or 80-90 % of maximum heart rate. The concentration of lactate is between 3.0-4.0 mmol/l. The value of this level of RPE Borg replies to 13-14. Applied to the training process, this zone is referred to as E3 (Endurance 3 zones).

- Phase III represents a time interval between the second anaerobic threshold and the beginning of recovery phase. The intensity of this phase is characterized as heavy physical activity. Achieved percentage of oxygen consumption of this stage is between 80 % of VO_{2max}, or 90 % of maximum heart rate. The concentration of lactate at this stage of the test exceeds 4 mmol/l (BL: > 4 mmol/l). The value of this level of RPE Borg replies to > 14. In the contemporary theory of sports training this zone is referred to as E4 (Endurance 4 zones).

The exercise protocol for rowers that were the subjects of one of our studies mentioned later in this chapter was performed on a rowing ergometer (Indoor rower Concept 2, Canada) and consisted of 15 minutes of warm-up period on the individual level intensity, following which the workload was increased during the next four stages until exhaustion. Duration of each interval was 3+3+2+1 minute. Workloads were individually graded according to 500m time lap, which was the workload for the second interval of the test. The value of the first and the third interval was 3 seconds above and below the time in the second interval. Hypothetically, the second interval time or intensity is enough for entering anaerobic metabolism.

The exercise protocol for taekwondo subjects that will also be mentioned later in this chapter was performed on treadmill (HP Cosmos, Germany) and consisted of a 15 minute of warm-up period. Intensity was measured by treadmill velocity on which participant reach the value of heart rate 150 beats/min during warm-up period. It was the starting velocity for the maximal test. Workload was increased by 2km/h every three minutes with constant elevation of 3 %, according to individual starting velocity.

The exercise protocol for cycling was performed on a cycle-ergometer (Ergo win 8008) and consisted of a 15 min warm-up period at 100W and cadence held at constant 90 rpm, following which the workload was increased by 40W every 2 min until exhaustion. The cycling position was standardized with a 160° knee angle with pedal in lowest position.

The exercise protocol for handball players that were subjects of a few investigations of our team was also performed on a cycle-ergometer using modified Astrand method for determining workload while measuring pulmonary parameters using an automated metabolic cart mentioned earlier in the text.

The exercise protocol for football players who will be mentioned later was performed on a treadmill using modified Ellestad protocol while measuring pulmonary parameters using an automated metabolic cart.

Morphological characteristics (body composition parameters) of subjects in our investigations are obtained using apparatus for bioelectrical impedance analysis In Body 720 (Biospace, Korea) whose validity was previously confirmed (Lim et al., 2009).

3.2 Biochemical methods for determination of redox status

Determination of oxidative stress parameters from blood samples begins with separation of plasma and erythrocytes from whole blood in procedure known as 'washing' of erythrocytes. In the first step, blood is centrifuged (10 min on 3000 rpm) for extraction of plasma (usual volume 1-2 ml). The rest of the plasma is aspirated in order to keep only erythrocytes. In step two, saline is added to erythrocytes (ratio 2:1) and this is mixed on the vortex machine. Then, it is centrifuged three times (10 min on 3000 rpm). After every centrifugation, supernatant is aspirated. When last centrifugation is finished, 1 ml of erythrocytes is taken and put in new test tube. Then 3 ml of cold distilled water are added (ratio 3:1). In the final step solution must be put into cold water jacket for 30 min.

3.2.1 Superoxide anion radical ($O_2^{\bullet -}$) determination

Determination of superoxide anion radical ($O_2^{\bullet -}$) plasma concentration is based on reaction of $O_2^{\bullet -}$ with Nitro Blue Tetrazolium (NBT) and forms nitro blue formazan (Auclair & Voisin, 1985). Maximum absorption for measuring is λ_{max}=550 nm. Assay mixture consists of 50 mM TRIS-HCl buffer (pH=8.6), 0.1 mM EDTA, 0.1 mg/ml of gelatin and 0.1 mM of NBT. Before using, solution needs to be gassed with nitrogen (under pressure) for 60 minutes. In the test tubes (12 x 100) 50 μl of plasma and 950 μl of assay mixture are pipetted, and the reaction starts. For blank probe (instead of plasma) adequate volume of distilled water is used. At the beginning of the reaction, extinction of mixture is measured and noted as E_1. Mixing is performed (with plastic stick) every 60 sec and extinction is noted after mixing until it is stable, which considers two consecutive, approximately the same extinctions. Last extinction is noted as E_2. The same procedure is applied for measuring the blank probe. The concentration of released $O_2^{\bullet -}$ is calculated using the following equation:

$$\Delta E_s = E_{2s} - E_{1s} \text{ (for sample)}$$

$$\Delta E_{bp} = E_{2bp} - E_{1bp} \text{ (for blank probe)}$$

$$\Delta E = \Delta E_s - \Delta E_{bp}$$

$$\text{nmol } O_2^{\bullet -}/ \text{ ml plasma} = \Delta E/0.015 \times 1/0.05$$

3.2.2 Hydrogen peroxide (H_2O_2) determination

Determination of hydrogen peroxide (H_2O_2) plasma concentration is based on oxidation of phenol red using hydrogen peroxide, in reaction catalised by enzyme peroxidase from horse radish (HRPO) (Pick & Keisari, 1980). This reaction results in forming of compound with maximum absorption of λ_{max}=610 nm. Linear dependence of H_2O_2 concentration absorbance on 610 nm is stabile for 1-60 mM of ratio concentration (1-60 nmol/ml). This method allows us opportunity to determinate forming and releasing of H_2O_2 in time interval from 5-60 min. Two hundred ml of plasma and 800 ml of fresh made phenol red solution (PRS) consisting of 140 mM NaCl, 10 mM potassium phosphate buffer (pH=7), 5.5 mM D(+) glucose and 0,28 mM phenol red are pippeted in the test tubes (12 x 100). Then 10 ml (1:20) HRPO, made *ex tempore*, is added. Samples are left on room temperature for 10 min, and pH is adjusted to >12 using 1M NaOH. For blank probe (instead of plasma) adequate volume of distilled water is used. Concentration of released H_2O_2 in venous blood is calculated using calibration diagram (standard curve) for each assay. For constructing standard curve standard (stock) H_2O_2 solution is used, after checking concentration (for 10 mM H_2O_2, A230=0.810). Five, ten and twenty ml of 1 mM H_2O_2 solution are pippeted in three test tubes (instead of plasma), together with 200 μl distilled water, 800 μl phenol red solution and 10 ml (1:20) HRPO. After 10 min on room temperature, pH>12 is adjusted using 1M NaOH (10 ml). Concentration and volume of released H_2O_2 in coronary venous effluent are calculated using factor of absorbance (F)/nmol H_2O_2:

$$F = \frac{\Delta A}{\text{nmol } H_2O_2/\text{cuv}}$$

On the basis of the sample absorbance (A_s) at λ_{max}=610 nm and comparing it with blank probe, final absorbance is calculated (ΔA) ($A = A_s - A_{bp}$). For calculating plasma concentration and volume of H_2O_2 the following formula is used:

$$\text{nmol } H_2O_2/\text{ml plasma} = \Delta A/F$$

3.2.3 Index of lipid peroxidation (TBARS) determination

Level of lipid peroxidation is determined indirectly via products of lipid peroxidation reaction with thiobarbituric acid (TBA) - (Thiobarbituric Acid Reactive Substances - TBARS). This method is based on determination of level of one of lipid peroxides (malonildialdehide - MDA) with thiobarbituric acid (Ohkawa et al., 1979). Extract is obtained by combining 400 µl of 28 % TCA (Trichloroacetic acid) and 800 µl plasma. Then it is incubated in cold water jacket (-4°C) for 10 min. After incubation the samples are centrifuged on 15000 rpm for 4 min to form supernatant. In test tubes (12 x 100) 800 µl plasma extract and 200 µl 1% TBA in 0,05 M NaOH are pippeted, and this is incubated in water jacket 100°C for 15 min. As a blank probe (instead of plasma) adequate volume of distilled water is used. Measuring is performing at λ=530 nm. The concentration of released TBARS is calculated using the following equation:

$$\text{nmol TBARS/ml plasma} = \Delta A\ (A_s - A_{bp})/1.56 \times 1.25$$

A_s = absorbance of sample
A_{bp} = absorbance of blank probe
1.56, 1.25 - correction factors for this assay

3.2.4 Nitric Oxide (•NO) determination

Nitric oxide (•NO) decomposes rapidly to form stable metabolite nitrite/nitrate products. Considering that •NO in reaction with molecular oxygen forms equimolar amount of nitrites (•NO + ½O_2 → NO_2^-), we can assert with great certainty that amount of released nitrites represent amount of released •NO. The method for detection of the plasma nitrate and nitrite levels is based on the Griess reaction. Nitrite (NO_2^-) is determined as an index of nitric oxide production with Griess reagent (forms purple diazo-complex) (Green et al., 1982). Griess reagent is prepared *ex tempore* just before the experiment by mixing equal amounts of stocks: 1 % (w/v) sulfanil-amide dissolved in 5 % HCL and 0.1% (w/v) aqueous solution of N-1-naphtyl-ethylene-diamine-dihydrochloride (N-NEDA). Extraction is obtained by combining 100 µl 3 M PCA (Perchloride acid), 400 µl 20 mM EDTA (ethylenediamonoetetraacetic acid) and 200 µl plasma, put on ice for 15 min, and then centrifuged for 15 min at 6000 rpm. After pouring off the supernatant, 220 µl K_2CO_3 is added to set pH at 7.4.

Two hundred µl plasma extract (previously described), 250 µl Griess reagent, and amonium buffer (pH=9) (containing amonium chloride - NH_4Cl and sodium tetraborat - $Na_2B_4O_7$) is then pippeted in test tubes (12 x 100) in order to stabilize diazo-complex. As a blank probe adequate amount of distilled water (instead of plasma) is used. The concentration of released nitrites is calculated using the calibration curve. Calibration curve is constructed from samples extinction (which consists of specific concentration of nitrites, after their reaction with Griess reagent and in the presence of amonium buffer). For that purpose, different amount of 1 mM $NaNO_2$ (3, 6, 12, 24 µl) in 1 ml of distilled water is pippeted. After color stabilization on room temperature (5-10 min), spectrophotometrical measuring at λ=550 nm starts. The concentration and amount of released nitrites is calculated via determination of standard factor (F):

$$\frac{\text{Extinction of blank probe}}{\text{Concentration of NaNO}_2\text{ in standard}}$$

For each standard (F1-F4), mean is calculated:

$$\text{nmol } NO_2/\text{ml extract} = \Delta E\ (E_s\text{-}E_{bp})/F$$

3.2.5 Superoxide Dismutase (SOD) activity determination

Determination of superoxide dismutase (SOD) activity is based on epinephrine method. This method belongs to 'negative' type group of methods, since it monitors decrease of autooxidation speed in alkaline medium, which is dependent of $O_2^{\bullet -}$ (Misra & Fridovich, 1972). SOD removes $O_2^{\bullet -}$ and thus inhibits autooxidation of epinephrine. Speed of epinephrine autooxidation is detectable spectrophotometrically via changing the absorbance on 480 nm. The accumulation of epinephrine induces increase of absorbance on 480 nm. The percentage of inhibition is used as unit for measuring catalitical activity of this enzyme. The speed of epinephrine autooxidation in absence of SOD, represents reference (control), while the speed of epinephrine autooxidation in presence of SOD (i.e. protein in cytosol) represents part of referent values. In 3.2 ml of reaction assay, containing 3 ml carbonate buffer (pH = 12) and 0.1 ml of epinephrine solution, add 0.01 ml of supernatant is added. Autooxidation of epinephrine is monitored for 4 min at λ=480 nm. Reaction is stable at temperature ratio 26-30°C. Control reaction is simultaneously performed. For calculating SOD activity, percentage of epinephrine autooxidation inhibition in presence of SOD from sample versus control reaction is used. The amount of SOD is expressed in units of SOD activity per gram of Hb (unit/gHb). This unit is defined as volume of proteins which induces 50% of epinephrine autooxidation speed inhibition in linear part of absorption increase. The concentration of SOD activity is calculated using following equation:

$$\text{SOD-1} = \frac{2(\Delta K - \Delta A) x R}{V \, x \, Hb \, x \, \Delta K}$$

ΔK - change of control reaction absorption per minute
ΔA - change of sample reaction absorption per minute
V - sample volume (ml)
Hb - amount of haemoglobin (g/100ml lysate)
R - dilution

3.2.6 Catalase (CAT) activity determination

Determination of catalase (CAT) activity in sonificate is based on method by Beutler (Beutler, 1982). This method considers spectophotometrical monitoring of hydrogen peroxide degradation speed in presence of catalase (CAT) at 230 nm. Concentration of hydrogen peroxide is calculated as follows: with regard to absorption of diluted buffer solution (1:10) (as zero), absorption of solution containing 0.9 ml diluted buffer and 0.1 ml diluted solution 30 % H_2O_2 (1:100), is read. Concentration of hydrogen peroxide is calculated via coefficient of extinction (for H_2O_2 - 0.071, at 230 nm), using formula:

$$C = \frac{\Delta A}{0.071}$$

This concentration is diluted until 10 mM.

Reaction assay: In quartz cuvette that contains 50 µl of buffer, 5 to 50 µl of sample (depending from CAT activity) is added. Reaction starts by adding 1 ml 10 mM hydrogen peroxide solution. The decrease of absorbance is monitored spectrophotometrically at 230 nm, for 3 min. The CAT activity is expressed in unit/mg protein, and this unit is defined as amount of reduced H_2O_2 (µM per minute). For calculation the following equation is used:

$$CAT = \frac{\Delta A \cdot R}{0.71 \cdot Low \cdot V}$$

ΔA - change of absorbance per minute
R - dilution
V - sample of volume (ml)
Low - amount of protein (mg/ml sonificate)

3.2.7 Reduced glutathione activity (GSH) determination

The level of reduced glutathione (GSH) in red blood cells (RBC) is determined spectrophotomatrically according to Beutler (Beutler, 1982) which is based on GSH oxidation via 5,5 dithio-bis-6,2-nitrobenzioc acid (DTNB). GSH extract is obtained by combining 0.1 ml 0.1 % EDTA, 400 µl and 750 µl precipitation solution (containing 1.67 g meta-phosphoric acid, 0.2 g EDTA, 30 g NaCl and filled with distilled water until 100 ml; solution is stable for 3 weeks on +4°C). After vortexing and extraction on cold ice (15 min), centrifugation on 4000 rpm takes place (10 min). For measuring, quartz cuvette (1ml) is used. Two hundred µl extract, 750 µl Na_2HPO_4 and 100 µl DTNB (1mg DTNB/ml 1 % sodium citrate) is pipppeted in test tubes. As a blank probe distilled water is used. Concentration and amount of reduced glutathione in plasma is determined on the basis of calibration diagram (standard curve) for each assay. For standard curve construction standard stock-solution of GSH (concentration 1,5 mmol/l) is used. In order to determine concentration of glutathione in standard samples (nmol/GSH/ml), in 4 test tubes (instead of plasma) 10, 20, 30 and 40 µl 1 mM GSH solutions and 300 µl distilled water are pipppeted. Measuring of absorbance (A) is performed at λ_{max}=420 nm. For obtaining final absorbance (ΔA), value of blank probe absorbance (B) is subtracted from absorbance (A). To calculate GSH concentration in plasma following formula is used:

$$\text{nmolGSH/ml RBC} = \Delta A / F$$

$$F = \frac{\Delta A}{nmol GSH / cuv}$$

4. Adaptive responses to free radical formation in exercise

4.1 The acute effects of exercise on redox status of athletes

The relationship between exercise and oxidative stress has been a topic of intensive scientific research for more than 3 decades. Since the early work of Dillard and colleagues (Dillard et al., 1978), who were the first to connect physical activity with free radical production, hundreds of original papers have been published, but the topic of exercise-induced oxidative stress still receives considerable scientific attention. Data on the acute effects of exercise on redox homeostasis in humans are equivocal because of the many types of exercise and experimental conditions used in previous studies. Although there is some inconsistency present in the literature, it is now clear that both acute aerobic and anaerobic exercises have the potential to result in increased free radical production, which may or may not result in acute oxidative stress (Fisher-Wellman & Bloomer, 2009). It is thought that the extent of oxidative stress induced by an acute bout of exercise depends on many factors, such as exercise mode, intensity and duration, the participant's state of training, gender, age, nutrition habits, etc (Bloomer & Fisher-Wellman, 2008). Having in mind the above-mentioned factors, we designed a study that had two objectives: 1) to compare the effects of acute exercise on redox state of trained and untrained subjects and 2) to compare the effects of two exercise sessions that differ by exercise mode, intensity and duration among group of athletes.

4.1.1 Comparison of blood pro/antioxidant levels before and after acute exercise in athletes and nonathletes

Many studies have compared the antioxidant status of trained and untrained subjects at rest (Ørtenblad et al., 1997; Balakrishnan & Anuradh 1998; Brites et al., 1999; Evelson et al., 2002; Cazzola et al., 2003; Gougoura et al., 2007; Yamaner, 2010), but, to our knowledge, except the study of Ørtenblad and coworkers (1997) that analyzed blood antioxidant status in untrained and jump-trained humans following six bouts of 30-s continuous jumping, no study has compared the athletes' and non-athletes' pro/antioxidant responses to acute exercise of maximal intensity. The hypothesis of the first part of that study (Djordjevic, 2011) was that athletes would have not only higher activity of antioxidants in rest, but that the extent of redox disturbance induced by a maximal progressive exercise test would be lower in athletes compared with non-athletes. A group of 58 young handball players (age 16 - 19 years) and 37 age-matched adolescents who did not perform regular physical activity took part in this research. They were subjected to a maximal progressive exercise test on a cycle ergometer and taken blood sample immediately before and after exercise test.

At rest (before the exercise test), athletes had significantly higher superoxide dismutase (SOD) and catalase (CAT) activity, higher levels of glutathione (GSH) and nitric oxide (•NO) and lower levels of lipid peroxidation (TBARS) compared with non-athletes. The results regarding redox status of athletes showed that athletes really do gain desirable changes of antioxidant defense system, which alleviates the risk of oxidative stress in rest. A maximal exercise test induced statistically significant rise of superoxide anion radical ($O_2^{\bullet-}$), hydrogen peroxide (H_2O_2) and •NO levels in non-athletes, while TBARS levels decreased. Athletes experienced the fall in •NO levels and the fall in CAT activity. After exercise, athletes had significantly lower levels of $O_2^{\bullet-}$ compared with non-athletes. Two way repeated measures ANOVA showed that exercise-induced changes of $O_2^{\bullet-}$, •NO and TBARS were dependent on combination of factors - sports engagement and exercise test.

The exercise test induced the statistically significant increase of $O_2^{\bullet-}$ levels only in non-athletes who, as already mentioned, had lower levels of SOD compared with athletes. CAT and GPx are both engaged in H_2O_2 elimination, but their affinity for H_2O_2 is different and dose dependent. Affinity of GPx for H_2O_2 is higher at low H_2O_2 levels, while CAT's affinity rises with the increase of H_2O_2 levels. Non-athletes in our study experienced a significant rise of H_2O_2 with exercise, but neither GSH levels nor CAT activity changed significantly. On the other hand, athletes did not experience the rise in H_2O_2 production, but CAT activity was decreased after exercise test. It may be that increased H_2O_2 production in non-athletes is a consequence of their less efficient ADS, while athletes' significantly higher basal GSH levels and CAT activity provided efficient elimination of excess exercise-produced H_2O_2. Subjects with a favorable blood glutathione redox status at rest maintain a more favorable redox status in response to exercise-induced oxidative stress (Laaksonen et al., 1999).

Resting TBARS levels of athletes in this study were significantly lower than resting TBARS levels of non-athletes. Since blood GSH was shown to be a determinant of plasma TBARS at rest (Laaksonen et al., 1999), we hypothesize that lower resting TBARS levels in athletes compared with non-athletes are a consequence of significantly higher GSH levels in athletes' blood. Maximal exercise test induced the fall of TBARS levels in group of non-athletes, which was quite unexpected if taken into consideration the behavior of other three prooxidative parameters. Namely, levels of $O_2^{\bullet -}$, H_2O_2 and $^{\bullet}NO$ increased after exercise test in group of non-athletes, but although the reactions between $O_2^{\bullet -}$ and other two prooxidants may lead towards formation of hydroxyl radical, a powerful inducer of lipid peroxidation, TBARS as index of lipid peroxidation was decreased.

Athletes in our study had significantly higher basal levels of $^{\bullet}NO$ compared with non-athletes. It is in accordance with numerous previous studies that showed that regular physical activity increases the bioavailability of $^{\bullet}NO$ (Kingwell et al., 1997; Jungersten et al., 1997; Maeda et al., 2001; Maiorana et al., 2003) and that physically active people have greater basal $^{\bullet}NO$ production compared with a sedentary population (Green et al., 2004; Poveda et al., 1997; Banfi et al., 2006). Studies that investigated the effects of acute exercise on $^{\bullet}NO$ production yielded various results. Some studies reported $^{\bullet}NO$ increase with exercise, some reported no change in $^{\bullet}NO$ production and some reported a decrease in $^{\bullet}NO$ production with exercise (Jungersten et al., 1997; Rassaf et al., 2007; Allen et al., 2006; Allen et al., 2009; Djordjevic et al., 2010a; Poveda et al., 1997; Djordjevic et al., 2010b; Jakovljevic et al., 2011; Cubrilo et al., 2011). Those differences are probably due to different protocols, i.e. different characteristics of subjects (age, physical activity, and health), different training and tests (type, intensity, duration of exercise tests or training), various methods of measuring RONS production, etc. The exercise test in this research induced the fall in $^{\bullet}NO$ production in athletes but the rise of $^{\bullet}NO$ levels in non-athletes. Two way repeated measures ANOVA showed that exercise-induced changes of $^{\bullet}NO$ were dependent on sports engagement of subjects. The rise of $^{\bullet}NO$ levels in non-athletes may be explained by effects of shear stress, while the response of athletes may be explained by endothelium preconditioning achieved by chronic exposure to shear stress during exercise trainings and i.e. structural and functional adaptations of endothelium (Tinken et al., 2008; Kingwell, 2000). It seems that the effects of chronic exercise on the basal NO production is more important than effects of $^{\bullet}NO$ production on tolerance of physical activity.

It should be noticed that there was statistically significant difference in total exercise test time and load between athletes and non-athletes, which may be the reason for observed biochemical responses, but the aim of our study was to assess the effects of a maximal exercise test, and the test was maximal for every participant. The differences in redox status post exercise are probably a function of the exercise-induced mechanical damage to muscle fibers and the subsequent inflammatory cascade in unaccustomed subjects.

4.1.2 Changes in athlete's redox state following sport specific and sport nonspecific bout of exercise

As previously mentioned, it is thought that the extent of redox state disturbance following acute bout of exercise depends on, among other factors, exercise mode, intensity and duration. Thus, the aim of our investigation was to compare the effects of sport non-specific exercise i.e. maximal progressive exercise test on a cycle ergometer and sport specific exercise session i.e. specific handball training (Djordjevic, 2011). Subjects (58 young handball players) were taken a blood sample immediately before and after both exercise bouts. Laboratory exercise test that lasted 10.57 ± 0.24 minutes and in which levels of athlete's heart rate corresponded to a submaximal and maximal intensity zone for 5.58 ± 1.27 minutes induced significant decrease of catalase activity, while one and a half hour handball training during which players spent 44.71 ± 10.52 minutes in submaximal and maximal zone of intensity induced significant decrease of superoxide dismutase activity and glutathione levels. Although handball training did not induce changes of prooxidative parameters, the fact that both exercise sessions induced depletion of antioxidants suggests that athletes experienced exercise-induced oxidative stress regardless of mode, intensity and duration of exercise.

4.2 Oxidative stress and nitric oxide evaluation during progressive maximal exercise test

As previously mentioned, last three decades brought hundreds of papers on relationship between acute exercise and oxidative stress, but it should be noticed that vast number of those studies were performed *in vitro*, i.e. in conditions that are not adequate to the ones in an activated muscle, so they have to be accepted with limitations. On the other hand, *in vivo* studies mostly measured free radical production after a bout of exercise, so the real extent of their production during the exercise remained unknown (Ji, 1999). The reaction between superoxide anion radical and nitric oxide represents one of the fastest reactions in the human body. This reaction not only decreases bioavailability of nitric oxide, but generates one of the most toxic reactive nitrogen species - peroxynitrite. The only limiting factor for this reaction to happen is the probability that these two species crash into one another in space. The reaction between superoxide anion radical and nitric oxide is 3 times faster than SOD-catalysed dismutation of superoxide anion radical, and the extent of their reaction increases as the levels of NO increase into the elevated nanomolar range and approach the local concentrations of SOD (Pryor et al., 1995).

Given the importance of these two species and their interaction in cardiovascular physiology, the aim of one of our previously published investigations was to assess the time-course of plasma nitric oxide ($^{\bullet}NO$) and superoxide anion radical ($O_2^{\bullet-}$) production during progressive exercise test, as well as to analyze the cause and result of changes in their production in basal conditions, during exercise and recovery. 19 elite football players were subjected to maximal progressive exercise test on treadmill. A small cathether was inserted into their antecubital vein for venous blood sampling and blood samples were taken immediately before the exercise test, in last 10 seconds of each level of the exercise testing (5 three minutes stages) without test interrupting, as well as in the 90th and 180th second of the recovery.

4.2.1 •NO and $O_2^{•-}$ concentrations during the exercise test

Statistically significant difference in •NO production (estimated through nitrites NO_2-) was found between resting state and grade I (increase in •NO production), and between stage I and stage II (decrease in •NO production). •NO stayed decreased (lower than basal) until the end of testing and recovery period. Regarding time course of $O_2^{•-}$ production, statistically significant increase in $O_2^{•-}$ production was also found between the resting state and grade I (increase in $O_2^{•-}$ production), then between the grade IV and V which was the time when athletes' anaerobic threshold was observed (increase in $O_2^{•-}$ production), and between the grade IV and the first phase of the recovery during which levels of $O_2^{•-}$ continued to rise, while in the second phase of recovery it's production started falling down.

Increased •NO production at the beginning of the exercise test can be explained by increased blood flow and shear stress (Wollin, 2000) while the subsequent drop in •NO production may be explained by effects of ROS ($O_2^{•-}$) on •NO bioavailability (Jackson et al., 2007). The increase of $O_2^{•-}$ levels with the beginning of the test and latter $O_2^{•-}$ decrease could be explained by mitochondrial properties to produce free radicals - mitochondrial free radical production increases during state 4 respiration (low VO_2; high membrane potential; low ATP production), while during state 3 respiration (high VO_2; lower membrane potential; high ATP production) it decreases (Rassaf et al., 2007). That is in accordance with results of Herrero and Barja (1997), who confirmed the decrease of radical production as VO_2 and ATP synthesis increase, and observed the lowest $O_2^{•-}$ concentration at the point of anaerobic threshold (our subjects achieved their anaerobic threshold somewhere near the end of the stage IV). After the point of respiratory compensation (anaerobic threshold), $O_2^{•-}$ concentration started to increase again, which partly can be explained by U-shaped radical formation related to the mitochondrial pO_2 and/or an alternative mechanism - mechanism of xanthine oxidase $O_2^{•-}$ production during ischaemia and reperfusion (Ji, 1999). Also, high intensity exercise is often accompanied by increased secretion of the catecholamines which can undergo autooxidation, with and without oxygen, and produce $O_2^{•-}$ (Allen et al., 2005).

4.2.2 •NO and $O_2^{•-}$ interaction during the exercise test

By following the ratio of •NO and $O_2^{•-}$ production during the exercise test, we detected the point of their most intensive reaction, which turns to be near the grade IV, at the level of anaerobic threshold. It suggests that lactate threshold could be of a crucial importance not only in anaerobic and aerobic metabolism but in mechanisms of signal transductions as well.

4.2.3 Timecourse of •NO and $O_2^{•-}$ production during the exercise test – Dependence on their basal levels

As basal •NO values of athletes varied to the great extent, we divided them into three groups (depending on their basal •NO value - B1: •NO > 20 nmol/ml, B2: •NO 10-20 nmol/ml, B3: •NO < 10 nmol/ml), in order to detect differences in time-course of •NO production between groups, i. e. to see whether basal value of •NO affects time-course of •NO production during the test. The results showed that groups produced higher or lower concentrations of •NO, but the pattern of •NO production during the test was similar. According to basal $O_2^{•-}$ production subjects were also divided into three groups (B1: $O_2^{•-} > 10$ nmol/ml, B2: $O_2^{•-} = 5\text{-}10$ nmol/ml and B3: $O_2^{•-} < 5$ nmol/ml). Results showed that $O_2^{•-}$ production was not related either to levels of effort or to belonging to one of these three groups.

4.2.4 Timecourse of •NO and O_2^- production during the exercise test – Dependence on aerobic capacity of athletes

In order to determine whether dynamics of •NO and $O_2^{\bullet}$-production during the exercise test depends on the level of aerobic capacity of athletes, we divided athletes into three groups, based on the VO_{2max} achieved on the test - G1: $VO_{2max} > 65$ ml/kg, G2: $VO_{2max} = 60\text{-}65$ ml/kg, and G3: $VO_{2max} < 60$ ml/kg. We didn't find significant variations in time-course of •NO production between groups, i.e. there wasn't significant correlation between VO_{2max} and time-course of •NO production during exercise, but that there was significant correlation between VO_{2max} and •NO basal production, since athletes who had basal •NO production > 20 nmol/ml achieved higher VO_{2max} on the test (mean value $VO_{2max} = 65.84$ ml/kg) compared to athletes who had •NO basal production < 20 nmol/ml (mean value $VO_{2max} = 60.82$ ml/kg).

Correlation between •NO bioavailability and exercise capacity was also shown in many other studies. Rassaf and colleagues showed that the capacity of the vasculature to produce •NO, and thus nitrite, predicts maximal power and duration of exercise in 55 healthy subjects (Rassaf et al, 2007). Hambrecht and colleagues suggested that improvement in •NO -mediated vascular function improves cardiac function and VO_{2max} in patients with chronic heart failure (Hambrecht et al, 1998). Allen and coworkers also showed significant relation between VO_{2max} and baseline levels of nitrite and nitrate (NO_x), but he also found correlation between VO_{2max} and the increase in NO_2^- or NOx from baseline to recovery (Allen et al., 2006; Allen et al., 2005; Allen et al., 2009), which is opposite to our results (both nitrite and nitrate increased during exercise in his studies, which is opposite to our results, too). The decrease in •NO production during the exercise test found in our study is also not in compliance with several other studies that found increased •NO production during exercise (Rassaf et al., 2007; Allen et al., 2005, 2006, 2009), or found that •NO didn't change during or after exercise bout (Poveda et al., 1997), but the differences are probably due to different protocols, i.e. different characteristics of subjects (age, physical activity, health), training and tests (type, intensity, duration of exercise tests or training), various methods of measuring RONS production.

Analysis of $O_2^{\bullet-}$ production during exercise showed that there were differences in time-course of $O_2^{\bullet-}$ production between group with the lowest VO_2max and other two groups – groups with higher aerobic capacity showed decrease in $O_2^{\bullet-}$ production after the initial increase in the beginning of the test, while in group with lower VO_{2max} concentrations of $O_2^{\bullet-}$ continued to rise until the stage III. Trend of decrease in $O_2^{\bullet-}$ production in groups with high VO_{2max} may be explained by metabolic, cardiovascular, respiratory and endocrine adaptations that developed simultaneously with aerobic capacity development (Jones & Carter, 2000). It was shown that antioxidants appear in blood within less than 5 minutes after occurrence of free radicals (Ji & Fu, 1993), i.e. during the stage II of the exercise test, so decrease in $O_2^{\bullet-}$ production may also be explained by activity of antioxidants.

4.3 Oxidative stress and nitrite dynamics under maximal load in elite athletes: Relation to sport type

Maximal workload in elite athletes may result in increased generation of reactive oxygen/nitrogen species (RONS) and oxidative stress. The primary objective of our investigation was to evaluate the effects of regular engagement in different sports on basal oxidative stress and •NO level as well as to compare their dynamics during maximal exercise testing (sport specific). The basal NO_2^- (as a marker of •NO) concentrations were significantly different between examined groups of athletes. It seems that individual training led to adaptation and establishing of a new basal levels, different for different types of sport (and type of exercise). Basal values of •NO in rowers and cyclists were similar to values observed in young handball players (Djordjevic et al., 2011), while the values of taekwondo athletes corresponded to those of senior-level soccer players (Djordjevic et al., 2010). If we assume that mean nutritional intake per kg of body weight was similar in all three groups of athletes, the observed lower basal levels of •NO in taekwondo athletes compared to levels of cyclists and rowers may be due to smaller length of training or extended training cessations (more than 6 weeks) (Djordjevic et al., 2010). Furthermore, taekwondo is an anaerobic performance, and oxygen supply comes after workload without additional systemic demands for •NO-mediated vasodilatation. In aerobic workload, demands for oxygen delivery are high and depend on blood flow and endothelium-dependent vasodilatation through an increased •NO production. In our study, rowers had the highest •NO and lowest TBARS basal levels among examined sportsmen suggesting more circulating •NO and lower oxidative pressure in these athletes. The elevation in basal •NO in rowers could be related to the main mechanism that leads to the increase in •NO bioavailability and that is the decrease in •NO inactivation by ROS (Jones & Carter, 2004). Since basal levels of superoxide are similar in examined groups it seems that circulating hydrogen peroxide and lipid peroxides are effective ROS mediators of oxidative stress in these athletes. In any group of athletes, •NO levels did not change significantly due to maximal workload. Even during recovery period, measured levels were similar to values in rest.

Baseline TBARS values in the group of rowers correspond to those obtained in our previous research on young handball players (Djordjevic et al., 2011). High-basal level of TBARS in cyclists suggests that ROS production overwhelmed antioxidative defense despite possible adaptation (Miyazaki et al., 2001). These results indicate that the values of maximal oxygen uptake can not be an important predictive factor in terms of the TBARS level of athletes engaged in different sports. Only in the case of ranking athletes in one sport, VO_2max value may indicate the specific dynamics of TBARS level (Djordjevic et al., 2011). Baseline values of taekwondo athletes compared with a group of cyclists once again call into question the significance of aerobic capacity level on the level of the TBARS, due to significantly lower values in the group of taekwondo athletes. Comparing the results of other researchers who have studied the impact of aerobic activity on markers of oxidative stress after a VO_{2max} test, we conclude that in all of them there was an increase in TBARS concentration after a maximal exercise test on a treadmill or on bicycle (Miyazaki et al., 2001; Groussard et al,. 2003). However, comparing the results of research that examines the influence anaerobic activity on markers of oxidative stress, we recognize that anaerobic activity did not lead to an increase in TBARS concentration and/or led to its decline (Groussard et al., 2003). In our study, lower VO_{2max} value in taekwondo athletes could be connected with low •NO levels supporting again notation that anaerobic and/or lower oxygen demanding performance do not request additional systemic demands for •NO-mediated vasodilatation.

Concentrations of superoxide in blood plasma were similar in rest, at the maximal intensity of exercise, and during the recovery period, in all the examined sportsman groups. Values obtained in our study were significantly higher compared to the results obtained in our previous research when analyzing the oxidative status of football and handball players (Djordjevic et al., 2011; Jakovljevic et al., 2011). It is possible that prolonged vigorous exercises cause an immediate inflammatory response and probably result in an infiltration of mononuclear cells and neutrophils in tissues. During and after exercise, this process can generate a substantial amount of ROS and, consequently, these may attack lipids and proteins (Vollaard et al., 2005). It could be that recovery period in our study was insufficient for measurable changes since examined parameters are end products of several multistep balanced processes including respiration, free radical-mediated oxidation of cellular molecules, •NO dynamics, the activity of antioxidant enzymes, and levels of circulating antioxidants and inflammation.

In summary, the data found in this study allow us to conclude that regular, long term, different training strategies (aerobic, anaerobic or aerobic-anaerobic) are able to provoke training induced upregulation of nitrite level, as well as the products of lipid peroxidation. Furthermore, maximal, progressive exercise of sport specific intensities does not influence nitrite and oxidative stress parameters level in maximal load and the first 10 minutes of recovery no matter which sport individuals perform.

4.4 The influence of pre-exercise superoxide dismutase activity on pro/antioxidant response to acute exercise

Since superoxide dismutase (SOD), the first line defense enzyme in red blood cells (RBCs), was the most commonly found to be the one that changes under the influence of both acute and chronic exercise (Miyazaki et al., 2001; Groussard et al., 2003; Ookawara et al., 2003) and the one that differentiates between well trained subjects and controls (Evelson et al., 2002; Metin et al., 2003; Brites et al, 1999), we hypothesized that the level of its pre-exercise (basal) activity may determine the extent of oxidative stress induced by acute exercise. So, one of our recent papers dealt with the differences in response to acute exercise in subjects with different basal level of SOD activity (Djordjevic et al., 2010a). 24 young handball players were subjected to maximal graded exercise test and taken blood samples immediately before and after exercise. Maximal progressive exercise test induced significant changes in five out of six investigated parameters of redox state ($O_2^{\bullet -}$ was not changed significantly while H_2O_2, •NO, TBARS, SOD, CAT were) which suggests that either this kind of exercise is a potent oxidative stress inducer, or antioxidative defense system of our subjects was not efficient enough to resist the generated prooxidants. Interestingly, when analyzing the changes of these biochemical parameters in groups of athletes with different basal SOD activity (8 athletes with the lowest, 8 athletes with average and 8 athletes with the highest levels of basal SOD activity), H_2O_2, •NO, SOD and CAT were changed after exercise only in group of athletes with the lowest basal SOD activity. In other two investigated groups of athletes only TBARS changed significantly. So, statistical significance of the changes seen in the whole group of investigated athletes had its roots in the group of athletes with the lowest basal SOD activity which points out to the role of pre-exercise SOD activity in maintaining desirable redox state, both in exercise-related and exercise-nonrelated conditions.

5. The chronic effects of exercise on redox status of athletes

5.1 Assessment of the redox state of athletes and non-athletes

During sports training athletes are continously exposed to various kind of stress. Adaptations to stress occur on numerous levels: from adaptations on subcellular, cellular and tissue level, to adaptations of organs and the whole organism of an athlete. Adaptations to stressors, i.e., structural and functional changes, enable improvement to occur in an athlete's sports performance. Since energy demands and oxygen consumption increase several-fold during exercise, it is thought that production of reactive oxygen and nitrogen species (RONS) also increases. Although much of early research viewed exercise-induced RONS production as a potential detriment to physiological function, more recent work investigates an alternative role for RONS production in regards to favorable exercise-induced adaptations (Fisher-Wellman & Bloomer, 2009). It is thought that the basic principle of exercise, stress-adaptation, also takes place in events related to exercise-induced oxidative stress. The exercise-induced increase in free radical production can be seen as no different from other responses to exercise: a certain load disturbs homeostasis, resulting in adaptations in the body to be able to cope with a similar load in the future (Vollaard et al., 2005). The oxidative challenge-related adaptive process of exercise is systemic and includes increased antioxidant/damage repair enzyme activity, lower oxidative damage, and increased resistance to oxidative stress (Radak et al., 2008a). The upregulation of antioxidant system provides adaptive protection from RONS during subsequent training sessions, as well as during non-exercise related conditions.

Previously published papers on exercise-induced changes in redox state of athletes mainly explored redox state of athletes engaged in dominantly aerobic or anaerobic training protocols, while mixed (aerobic-anaerobic) sports were not so explored. According to some studies, anaerobically trained subjects have a better antioxidant enzyme activity in blood, in tissues and especially in working muscle (Evelson et al., 2002; Marzatiko et al., 1997; (Radak et al., 2008b). It was also shown that a controlled protocol of endurance training is followed by an increase in antioxidant enzyme activity in plasma and other tissues (Tanskanen et al., 2010; Selamoglu et al., 2000; Lekhi et al., 2007). Regarding mixed (aerobic-anaerobic) training, there are only a few studies suggesting that trained football and rugby players show lower oxidative stress at rest than sedentary subjects, and that there is a correlation between redox status and fitness level (Chang et al., 2002; Metin et al., 2003; Cazzola et al., 2003). Currently, there is little information available regarding exercise-induced adaptations of the antioxidant defense system in adolescent and child athletes. The absence of studies investigating the effect of exercise on young population is surprising considering the numerous metabolic and physiologic differences between children, adolescents and adults (Cooper et al., 2004; Armstrong et al., 2008; Boisseau et al., 2000). The lack of information on young individuals' biochemical responses to exercise is mainly attributable to ethical concerns and methodological constraints that limit invasive research in children/adolescents. Thus, our team performed a set of investigations on exercise induced changes in redox state of young, adolescent handball players. Effects of acute exercise on their redox state were described earlier in this chapter. Here we'll focus on relationship between duration of training experience, aerobic capacities and biochemical parameters that represent factors of redox homeostasis. The first objective of our investigation was to explore the relationship between sports engagment and redox state so we compared redox state of 33 young handball players (age 16 - 19 years) with redox state of 14 age-matched adolescents who do not have regular physical activity (Djordjevic et al., 2011). The results showed that athletes and non-athletes did not differ significantly in levels of investigated prooxidants ($O_2^{\bullet -}$, H_2O_2, $^{\bullet}NO$, TBARS), but athletes had significantly higher levels of SOD and lower level of CAT activity. Higher SOD activity in athletes was also found in other studies, including studies involving karate athletes (Naghizadeh et al, 2009) soccer players (Briites et al., 1999; Cazzola et al., 2003), rugby players (Evelson et al., 2002;), jump-trained subjects (Ortenblad et al., 1997) and students of Physical Education and Sports Sciences (Balakrishnan & Anuradh, 1998). Regarding CAT activity, there is a number of previous studies that also found that it's acitivity is decreased in athletes compared with controls, but there are also a number of studies that reported no change in CAT activity as a consequence of exercise training (Miyazaki et al., 2001; Lekhi et al., 2007; Metin et al., 2003; Ortenblad et al., 1997; Balakrishnan & Anuradh, 1998). According to numerous results from previous papers, it can be concluded that SOD behaviour as a consequence of chronic exercise training is clear (its' activity increases), while CAT has variant tendencies. The second objective of that work was to compare redox state of athletes with different duration of training experience. The results showed that there was no significant correlation between duration of sports engagement and redox state of athletes. The final objective of this work related to the correlation between morphofunctional characteristics and redox state of subjects. Athletes were divided into 3 groups based on ther aerobic capacity expressed through maximal oxygen consumption ((1) athletes with poor cardiorespiratory fitness (VO_{2max} < 38.3 ml/kg/min), (2) athletes with average cardiorespiratory fitness (VO_{2max} = 38.4- 45.1 ml/kg/min), and (3) athletes with good cardiorespiratory fitness (VO_2max > 45.2 ml/kg/min)). Interestingly, athletes with low aerobic capacity had higher levels of H_2O_2 than athletes with average or high aerobic capacity, but lower levels of TBARS. Positive correlation was found between muscle percentage and TBARS. We hypothesize that the highest levels of TBARS in athletes with the highest aerobic power may be a consequence of the higher working capacity and consequently increased oxidative stress in working musculature of these athletes.

5.2 Systemic adaptation to oxidative stress induced by regular long term exercise

5.2.1 Morphofunctional effects of redox disturbance homeostasis induced by long term exercise

The overall positive impact of exercise on growth and development of children and youth is one of the generally accepted facts. From the aspect of physiology, adaptation to muscle activity is presented through a systematic response in order to provide the best possible performance with the lowest energy deficit. Adaptation is an universal, common characteristic of all living beings, responsible for the survival of the organism under different conditions.

Football, as a team game of simple rules without significant financial investment in equipment and space to play, is one of the most popular sports today, played by all nations, both sexes and various ages, regardless of skill level (Stolen, 2005). According to the intensity of football loads it can be classified as a highly intensive intermittent team sport (Bangsbo, 1994). For these reasons, it is evident that players are expected to have a high level of aerobic capacity and aerobic endurance. There is an evidently progressive trend of aerobic capacity in the last 20 years (Casajus, 2001), compared with results from eighties of the twentieth century (Ekblom, 1986; Faina et al., 1988). One of our studies was focused on analysis of the aerobic capacity of young soccer players of different age (14-15 versus 16-17 years old) and length of the sports experience (Cubrilo, 2009). Surveys conducted to date show that, when compared to the seniors, young players show lower values of maximum oxygen consumption, lower than 60 ml/kg/min (Stolen, 2005), which is consistent with the results of our research. In fact, comparative analysis of our research between the categories of young athletes in relation to age and length of the sports experience showed that aerobic capacity is significantly changed in terms of age, while the length of training does not affect the value of this parameter. The values of oxygen consumption of 51.05 ± 2.39 ml/kg/min in a group of older athletes with 10-11 years of sport experience approach the values achieved at the level of elite senior players rank of Serbia (Ostojić, 2000; Ponorac, 2005). These results confirm the fact that during the development of top athletes aerobic capacity reaches its maximum between 17-22 years, then linearly decreases with age (Shephard, 1999), which would practically mean that the beginning of a process of adaptation to level of stroke volume, or end-diastolic volume as a key factor of aerobic capacity (Levine, 2008), takes place only after 16 years of age. The level of aerobic capacity in the older group of athletes positively correlated with the H_2O_2 concentrations, as opposed to a group of young athletes where the level of aerobic capacity negatively correlated with the level of SOD activity. Due to the higher values of SOD in the group of young athletes, a negative correlation could be interpreted as an adaptive response to significantly lower values of aerobic capacity.

The percentages of fat mass (% FM) and fat free mass (% FFM) were not statistically different in the observed groups of athletes, while body mass index (BMI) was higher in older compared to younger group of athletes. However, correlations of measured parameters were age specific. In group of older athletes, fat mass percentage positively correlated with TBARS and H_2O_2 levels in older group while BMI positively correlated with SOD levels and negatively with CAT levels and GPx levels. Percentage of fat free mass positively correlated with H_2O_2 in younger athletes. Since there was no statistically significant differences in H_2O_2 concentrations between groups of younger and older players, observed positive correlation between the concentration of H_2O_2 in plasma and valuable asset of maximal oxygen uptake in a group of older athletes is especially interesting. Given that in the older group of athletes only TBARS plasma concentrations correlated with ECG parameters at rest, one might assume that the concentration of hydrogen peroxide in plasma indirectly mediates the induction of morphological changes of the left ventricle, with resultant effects on the regulation of myocardial inotropic properties.

Maintenance of optimal amounts of reactive species formed from molecular oxygen in the homeostatic balance is of great importance for preserving health. High levels of ROS are viewed as a toxic mediator of cell and tissue injury. Erythrocytes are the cells most exposed to possible damage caused by reactive oxygen species (Tappel, 1953). The results of correlation analysis in our work indicated that CAT and GPx significantly positively correlated in the group of young players. SOD significantly negatively correlated with CAT and GPx in older players tested. In erythrocytes, a major role in maintaining homeostatic balance of ROS belongs to enzymatic antioxidant system (AOS). Studies in healthy people have shown the expression of antioxidant enzymes is in a positive correlation with SOD activity, CAT and GR. A study of 220 healthy subjects from Danish population showed that erythrocyte SOD positively correlated with CAT, but there is no correlation between GSH-Px and other antioxidant enzymes (Andersen, 1997). The appearance of a negative correlation between SOD, CAT and GPx in the group of older players in our study resulted from the decrease in SOD activity in the older group, compared to the younger players. Reduction of SOD activity is most likely a result of inhibition of CuZn SOD by hydrogen peroxide. It is known that evidently increased antioxidant activity observed under the influence of the training process is the result of altered gene expression and influence on both the mRNA and at the level of protein synthesis. The results of these studies indicate that changes in markers of oxidative stress and antioxidant defense may occur within the normal variation in training and food intake rich in antioxidants. This means that variations in diet can affect markers of oxidative stress and antioxidant defense (for example if food is richer or poorer in antioxidants). All this indicates that under conditions of intense exercise there are numerous sources of hydrogen peroxide and nitric oxide present in the circulation and the potentiated heart and that some of these reactive species can have a detrimental effect on circulating blood elements by propagating systemic damage. Increased amounts of prooxidants in erythrocytes in the circulation may cause vascular spasm which further emphasizes the production of free radicals.

5.2.2 The cardiac conduction system and redox status

Young athletes in constant competition and training process represent a specific subgroup of healthy people with specific lifestyles, who gained the ability to overcome supramaximal physical efforts (Maron, 2003, Maron & Zipes 2005, Maron et al., 1996). "Athlete's heart" was first described by Henchen in the European literature of the 1899th. Applying percutaneus method, he found an increase in the heart of racers-runners (Henchen, 1899). The growing interest in studying the effects of intensive physical loads on the cardiovascular system, and enormous progress in the field of noninvasive cardiac diagnostics, especially echocardiography in the last 30 years, have led to significant process of observing and noticing of heart remodeling. This fact is the focus of scientists in monitoring and understanding of specific changes in the heart of an athlete, known as "athlete's heart" (Morganroth et al., 1975; Martin et al., 1986; Pelliccia et al., 1991; Pelliccia et al., 1996; Douglas et al., 1997; Sharma et al., 2002; Pelliccia et al., 2002; Fagard, 2003; Pelliccia et al., 2005; Maron & Pelliccia, 2006; Pelliccia et al., 2007). Electrophysiological changes associated with athlete's heart syndrome, reflected by changes in heart rhythm, change in heart conductivity, repolarization, and precordial voltage changes, are consequently manifested by changes on ECG. Previous studies related to the search for the causes of these changes were mainly oriented towards the study of the autonomous regulation of heart rate due to changes in terms of increased vagal tone on the one hand and the suppression of sympathetic nerve regulation on the other. It is evident that intensive training influences the autonomic control and intrinsic cardiac pacemaker function (Huston et al., 1985). For these reasons, there is growing interest in more complete understanding of systemic change in the heart of an athlete, with emphasis on differentiation of changes in terms of age, or the influence of sports training and competition. Another, no less important reason is related to the strict classification of ECG changes in terms of successful differentiation of athlete's heart from cardiovascular disease and the possible development of sudden cardiac death of athletes.

Understanding the specific adaptations of the sport and its specific mechanisms and regularity, especially in a group of young athletes, requires an interdisciplinary approach, which should answer many questions related to the optimization of training, the athletes' health and morbidity in terms of preventing injury and sudden cardiac death in the field. Regular, intense physical activity is associated with increased dimensions of the heart muscle, where these changes are interpreted as benign (Maron, 1998, Maron & Pelliccia, 2006). In addition, it was shown that cardiac dimensions and functional abilities tend to change over training cycle as an adaptation to the specific requirements of a particular sports season (Crouse et al., 1992). The degree of cardiac adaptations can be determined by various factors including age, sex, type and intensity of exercise (Pelliccia et al., 2000), as well as ethnicity and race (Crouse et al., 2009).

Analysis of electrocardiograms of athletes in our study (Cubrilo, 2009), who were included in regular, intense physical activity, showed that some ECG changes may occur in athletes under the influence of training on the one hand or the regular process of growth and development during the normal training cycle on the other. It is shown that athletes' age significantly affects the dynamics of change in P wave voltage to the lead D3, the amount of ST elevation in V1 lead, and the duration of PQ, QRS and QT intervals. On the other hand, the length of the sports experience showed a statistically significant effect on the mean QRS vector, P wave voltage to the lead D2, as well as the value of the sum of Σ (RD1 + SD3). Also, in older athletes it has been reported higher incidence in percentage of sinus arrhythmia and an incomplete right bundle branch block, regardless of the length of the sports experience. The results of our research in terms of classification of ECG findings, based on generally accepted clinical criteria (Pelliccia et al., 2000), show that athletes 57 % of both age groups can be classified as with normal ECG findings while 41 % is classified as partially normal and 2 % significantly abnormal findings with potential structural changes and clinical significance. The largest number of research on electrocardiographic changes include athletes between 18 and 35 years, while very few studies focused on the monitoring of young athletes with the aim of determining the difference in the adaptation of the heart to the training process in relation to the age. In our study ECG voltage criteria of LVH (RD1 + SD3> 25 mm) (Gubner & Ungerleider, 1943) showed statistically significant difference compared to the length of the sports experience. Consideration of the dynamics of average values of this parameter in the observed groups of athletes showed a decreasing trend with increasing length of the sports experience in both age groups. Average values of P waves in leads D2 and D3 showed statistically significant differences observed among groups of athletes, where the dynamics of P waves in the lead D2 showed a statistically significant difference compared to the length of the sports experience, as opposed to the dynamics of the P wave to the lead D3, where it was observed statistically significant differences by athletes age. By analyzing the dynamics of average values of P wave in D2 lead, voltage reduction is observed with increasing length of the sports experience. The mean QRS vector in our study shows a statistically significant upward trend in this parameter with increasing length of the sport experience in both groups.

Increased oxidative stress is involved in the pathophysiology of diverse diseases such as atherosclerosis, neurodegeneration, renal disease and cancer. Over the past 20 years, significant evidence has suggested a role for increased oxidative stress in the pathophysiology of congestive heart failure (CHF). In previous scientific work focus is mainly put on the analysis of adaptive response to redox disturbance during physical activity, where the subject matter referred to the response of skeletal muscle, liver and brain (Radak, 2008). The potential connection of ROS as second-messenger in terms of transmission of biological information through the modulation of signaling molecules, enzymes and proteins at the level of the heart muscle and heart conduction system of young athletes, represented through ECG changes, in current literature has not yet been processed. The concept of ROS as normal and necessary components of the cellular milieu is emerging as an important homeostatic mechanism that participates in the control of multiple cellular processes (Finkel, 1998; Lander, 1997). Skeletal muscles produce ROS at a rate that is activity dependent. This, plus evidence of ROS induced cellular damage, lead to idea that ROS may participate in the development of fatigue and/or activity induced injury as well as processes of cardiac contraction and relaxation (intracellular calcium cycle). More recently, various steps of the contractile process have proven to be susceptible to redox modulation. First, the opening probability of isolated sarcoplasmic reticulum (SR) Ca^{2+} release channels of the ryanodine receptors increases upon oxidation of accessible protein thiols. Moreover, SR Ca^{2+} reuptake is inhibited by high concentration of H_2O_2. Finally, oxidants alter myofibrillar Ca^{2+} sensitivity in a time and concentration dependent fashion. This so-called redox signalling function is especially true for the ROS, H_2O_2, which is more stable and diffusible than radical species such as $O_2^{\bullet -}$, but also applies to nitric oxide. An important consideration in the experimental use of oxidants such as H_2O_2 is whether the observed changes are physiological or are the results of overt oxidative stress. This issue has become topical in our research on young football players (children of 14-15 years old) where H_2O_2 positively correlated with the percentage of muscle, while in the older group of athletes demonstrated (16-17 years old) a positive correlation between H_2O_2 and the percentage of fat was found. A positive correlation between H_2O_2 and the sum of Σ(SV1 + RV5) was also observed.

The observed positive correlation between TBARS concentration and QT interval duration, and a negative correlation between the concentration of $^{\bullet}$NO and PQ interval duration can be observed in the light of altered Ca^{2+} homeostasis as an indicator of oxidative stress. Andrade and coworkers showed that continuous exposure to relatively high concentrations of H_2O_2 for extended periods of time caused resting Ca^{2+} to increase and slowed the return of Ca^{2+} to resting levels after stimulation ended (Andrade et al., 1998). Prolonged QT syndrome is a functional abnormality probably coupled with neurological severe impacts that can cause lethal arrhythmias (Schwartz et al., 1975). The acquired forms of prolonged QT interval may be due to idiosyncrasies of medicines (antiarrhythmics and psychotropic drugs), electrolyte abnormalities, hypothermia, toxic substances and injuries of the central nervous system (Bhadari & Scheinman, 1985). Correlations obtained in our study may include oxidative stress as the cause of the prolonged QT interval in young athletes. TBARS as index of lipid peroxidation could play a significant role in the modulation of cardiac systolic phase in which the mechanism of action can be linked to Ca^{2+} homeostasis and the Ca^{2+} influx through slow calcium channels. TBARS could act on the Ca^{2+} dependent slow action potential, which is its negative inotropic effect: reduced Ca^{2+} current with prolongation of the action potential slow phase.

Heart conductivity disorders are very common finding in active athletes. First-degree atrioventricular block is represented with 6-33 % (Puffer, 2001) or 35-40 % (Crisafulli et al., 2002) in athletes, which is significantly higher than in the general population (0.65 %). The observed negative correlation between the concentration of $^{\bullet}$NO and PQ interval duration in young athletes called into question a possible link with the myocardial (atrial) contractile

•NO effect. Conductivity at the AV (atrioventricular) node is mediated by increased parasympathetic tone and/or reduced sympathetic tone in rest. So far some of the most important controversies surrounding the myocardial contractile effect of •NO were if •NO exert a myocardial contractile effect under baseline conditions or only following adrenergic or cholinergic stimulation on the one hand, or can contractile effects of •NO be labeled as positively or negatively inotropic on the other. One might assume that •NO might play an important role in the contraction of the atria (presystolic phase) and the filling of the ventriculi in the physiological resting conditions of the athlete's heart.

Changes in QRS complexes voltage are very common finding in the sports population. However there is considerable variation in the percentage of literature data which goes up to 8-76 %, thus the problem of standardization of criteria for their registration exists. Intensive, long-practicing activity is associated with morphological changes in heart muscle, including increased volume, as well as increased thickness of the walls of the heart (Huston et al., 1985; Fagard, 2003). The main difference between pathologycal and physiological hypertrophy is the nature of the stimulus for the growth of cells, as well as the duration of a given stimulus. For the development of physiological hypertrophy, the most common stimulus is training process, with effect from time to time, episodic, and largely through the sympathetic neurotransmitter.

Several studies showed high incidence (80 %) sportsmen who met electrocardiographic criteria for left ventricular hypertrophy using the Sokolow and Lyon criteria (S wave in V1 + R wave in V5> 35 mm) (Sharma et al., 1999; Pelliccia et al., 2000). The percentage of the right ventricle was 18-69 %, where the sum is taken as a criterion RV1 and SV5 deflection greater than 10.5 mm. ECG voltage criteria of LVH (Gubner & Ungerleider, 1943), (RD1 + SD3> 25 mm), in our study showed a statistically significant difference compared to the length of the sports experience. Consideration of the dynamics of average values of this parameter in the observed groups of athletes observed a decreasing trend with increasing length of the sports experience in both age groups.

Electrical heart axis in sport depends on age, chest structure and position of heart in chest cavity. It is manifested by the development of left ventricular hypertrophy. In about 60-70 % of all athletes electrical cardiac axis is between 30-70 degrees, and turn 90 degrees to the right and more encounters with 10-20 % of athletes. Turning electrical axis of 30 degrees to the left and more are found in 15-25 % of athletes. Thus, it was determined that it often turns to the right of athletes involved in endurance sports type, while in athletes involved in power sports-type it often turns to the left (Sharma et al., 2002). The literature has shown that the shaft QRS complex becomes more vertical with increasing levels of well trained, often associated with right bundle branch block incomplete which can be explained by increased myocardial mass at the top of the right ventricle. The mean QRS vector in our study shows a statistically significant upward trend (verticalisation) with increasing length of sport experience of the athlete in both groups. In the group of younger athletes, negative corelation between $O_2^{\bullet -}$ concentration in plasma and P wave voltage in inferior bipolar leads was observed and/or negative correlation with the mean QRS vector. On the other hand, in the older group of athletes a positive correlation between TBARS in plasma with values voltage criteria for LVH (RD1 + SD3) was observed, and negative correlation with the mean QRS vector. The presence of these correlations confirms the assumptions about the role of ROS as signaling molecules ideal, where the level of their production during and after physical activity could be a crucial link between exercise and consequently disturbed homeostasis caused by the influence of adaptation at the level of gene transcription (Vollaard et al., 2005).

The following research could provide a clearer picture of the possible diagnostic and therapeutic importance (individual antioxidant supplementation) of these correlations in a group of young athletes of both ages. Mechanical stretch and neurohumoral factors induce changes in intracellular signaling pathways resulting in increase protein synthesis and the activation of specific genes promoting growth and potentially leading to ventricular remodeling. Our results demonstrated that oxidative stress and body (systemic) redox state influence heart functionality, but the effects are age specific.

6. Effect of different supplementation strategies on nitric oxide and oxidative stress parameters dynamics during maximal exercise testing

6.1 The influence of L-Arginin supplementation on nitric oxide and oxidative stress parameters dynamics during maximal exercise testing

Arginine is one of the 20 amino acids (AA) necessary for protein synthesis and coded by DNA. Plasma arginine concentrations are therefore maintained mostly by protein catabolism (≈ 85 %) (Morris et al., 2006) or by synthesis from other AA. Of the total synthesis, de novo synthesis accounts for < 15 %, while 60 % of arginine is derived from citrulline (Wu et al., 1998). Arginine metabolism may give rise to several other AA and molecules indispensable for life. Exogenous arginine is largely destroyed (≈ 40 %) during absorption by the gut itself. Much attention was given to arginine availability due to the discovery in the 1980's that endothelium-dependent relaxing factor' is a gas, nitric oxide (•NO), produced from arginine by endothelial •NO synthase (eNOS) (Ignarro et al., 1987) in a reaction that gives also rise to citrulline. In endothelial cells, eNOS and the two sequential enzymes argininosuccinate lyase (ASL) and argininosuccinate synthase (ASsynth), necessary for recycling citrulline to arginine, are co-localized in the caveolae (Li et al., 2005), a fraction of cell membranes that provides an efficient environment for maintaining arginine available to eNOS. The cytoplasm of endothelial cells has a concentration of arginine largely saturating eNOS, and yet the acute introduction of exogenous arginine elicits an increase in •NO production, a puzzling finding known as "the arginine paradox" (Kurz et al., 1997). Another puzzling finding is the peculiar regulation of arginine transport and its effects on •NO. The transport of arginine into cells is mediated by the cationic AA transporter 1 (CAT1). Over-expression of CAT1 enhances arginine uptake sixfold, and •NO production twofold. The presence of AA competing with arginine decreases arginine uptake, but not •NO production (Li et al., 2005). Moreover, citrulline succeeds in stimulating NO production even in a medium containing saturating levels of arginine, and extracellular citrulline does not influence intracellular arginine levels. Therefore, •NO production depends mostly on the efficient recycling of arginine-derived citrulline back to arginine, and not so much on exogenous arginine supply. Thus, the presence of a "micro-environment", where arginine metabolism and recycling are independently regulated and only partially in balance with plasma arginine concentrations, is the most consistent explanation for the endothelial arginine paradox (Flam et al., 2007).

6.1.1 Endothelial vasomotion and the exogenous arginine paradox

To understand the clinical paradox of arginine, i.e. the fact that the much needed arginine is ineffective or actually detrimental when supplemented chronically, we have to focus on another pathway of the complex arginine metabolism, which is controlled by the ubiquitous enzymes arginases (types, 1 and 2), which compete with NOS for arginine as a substrate. The effect of both arginases is to catalyze the cleavage of urea from arginine, thus forming ornithine. Urea controls osmolarity and water content of plasma and cells. The other product of the reaction, ornithine, may be recycled to citrulline and also synthesized de novo by the liver from glutamine, and then transformed by the gut into citrulline. The kidneys provide further recycling of citrulline to arginine, which is finally released into the plasma (Van de Poll et al., 2007). Arginase 1 (ARG 1) is a cytosolic enzyme, mostly expressed in the liver and red blood cells. Deletion of the ARG 1 gene, as occasionally found in humans, is incompatible with prolonged life, and the accompanying hyperargininemia is associated with a several-fold increase in the activity of the mitochondrial arginase (ARG 2), which accounts for the persistent ureagenesis in those patients (Grody et al., 1989). ARG 2 is widely expressed, mostly in the kidneys, gut and brain. ARG 2 overexpression plays a critical role in the pathophysiology of cholesterol-mediated endothelial dysfunction (Schulman et al., 2006). Arginases and NOS compete for arginine, and - under any conditions - arginase activity exceeds NOS activity at all NOS/arginase molar ratios (Santhanam et al., 2008). Moreover, although the K_M of arginases is 100-fold higher than that of NOS, the enzymes compete for arginine because the maximal catalytic rate of arginases is more than 1000 times higher than that of NOS (Wu et al., 1998; Topal et al., 2006). Therefore, increased expression and/or activity of ARG have a deep impact on NOS efficiency. As an example, arginase activity is increased in type 2 diabetic subjects with impaired NOS activity, and such impairment correlates with the degree of hyperglycemia and is reduced by insulin (Kashyap et al., 2008). The problem is more interesting because results of experimental and clinical studies are quite controversial and can not definitely support relevant conclusions. There are only few experimental studies dealing with this problem.

The study of Maxwell and coworkers (Maxwell et al., 2001) was performed to determine whether supplementation with L-arginine would prevent the decline in aerobic capacity observed in hypercholesterolemic mice. In one previous study the same authors determined the exercise capacity of wild-type (E1) and apolipoprotein E-deficient mice (E2) at 8 week of age when the cholesterol levels of both strains are low. The observed decline in VO_{2max} is associated with endothelial vasodilator dysfunction and reduced urinary nitrate excretion. This study was designed with the intention of averting the impairment in aerobic capacity associated with hypercholesterolemia through chronic supplementation of L-arginine. Those results clearly showed that administration of L-arginine restores exercise-induced EDNO synthesis and normalizes aerobic capacity in hypercholesterolemic mice. In normal mice, L-arginine enhances exercise-induced EDNO synthesis and aerobic capacity. As we noted above, investigation on humans do not clearly support exact role of L-arginine/NO system in exercise-induced vasodilation in healthy subjects, athletes or patients in some pathophysiological statement. Increased vasodilation and large artery distensibility occur after dynamic aerobic exercise (Kingwell et al., 1997; Naka et al., 2003) probably attributed in part to augment •NO release (Green et al., 2002; Gilligan et al., 1994). Although the contribution of •NO to large artery stiffness, wave reflection (AIx), and pressure amplification during exercise has never been tested, it is an important consideration relevant to cardiac structure and cardiovascular risk (Sharman et al., 2007). If reduced •NO bioavailability underlies abnormal exercise hemodynamics in a fashion similar to that found in men with hypercholesterolemia, we may expect that induction of endothelial dysfunction in healthy men (by blockade of •NO synthase) should lead to increased AIx and reduced PP amplification. Sharman and coworkers (Sharman et al., 2008) compared this hypothesis in young men who performed submaximal cycle exercise after intravenous infusion of L-NMMA with an •NO-independent control vasoconstrictor (noradrenaline [NE]) and placebo (saline). This study supports modulatory role of •NO in large artery stiffness, pressure amplification, and myocardial loading under resting conditions. However, this is the first study to report that •NO does not affect large artery hemodynamics and wave reflection such that ventricular-vascular interaction is substantially altered during low-intensity aerobic exercise. Systemic infusion of L-NAME (an inhibitor of NOS) caused significant increase in MAP, but not in VO_{2max} in healthy volunteers at submaximal exercise. Furthermore, that reduction was significant in maximal exercise (Jones et al., 2004). These finding suggest that muscle blood flow was well preserved during submaximal exercise following NOS inhibition, indicating multifactorial influence on blood flow regulation during exercise.

On the other hand, one other study on healthy subjects showed that L-NAME decreases lactate production during exercise, which is quite logical, taken into consideration well known fact that •NO stimulates glycolysis. These data indicated that skeletal muscle •NO production represents an important limitation to the acceleration of oxidative metabolism following the onset of supramaximal exercise in humans (Wilkerson et al., 2004).

In conclusion, the study by Taddei and coworkers (Taddei et al., 2000) demonstrated that regular physical training protects the vascular endothelium from aging-related alterations. The beneficial effect of exercise is related to preservation of •NO availability by a mechanism probably linked to the prevention of oxidative stress and the consequent •NO breakdown. This beneficial effect could be important in accounting for the positive impact of regular exercise on cardiovascular risk in the elderly population. Arginine and glutamine are two non-essential amino acids than can become "conditionally essential" because of elevated needs during pathological conditions, and metabolism may not be able to maintain their concentrations at sufficient levels to match metabolic requirements. Chronic exogenous arginine supplementation has not proven to exert positive clinical effects in different trials, and sequential articulation of the knowledge of introduction of arginine-driven transcriptional, translational, and epigenetic adaptations may give us a key for interpreting those controversial results. Study by Lerman and coworkers demonstrated that oral L-arginine supplementation for 6 months improves coronary endothelial function in association with improvement in symptoms in patients with non-significant coronary artery stenosis. This study proposes a therapeutic role for L-arginine in patients with chest pain and coronary endothelial dysfunction (Lerman et al., 1998).

The role of L-arginine supplementation in cardiac patients remains controversial. Furthermore, it is also unclear if arginine supplementation in the sedentary population can have the same results. Further research will be needed to assess the interaction of these factors and to determine the effects of prolonged administration of arginine and antioxidants on exercise performance. Supporting that, a recent study in trained athletes, who were without any cardiovascular problems, showed beneficial effect of L-arginine in elderly physically active subjects (Chen et al., 2010). An arginine and antioxidant-containing supplement increased the anaerobic threshold and the work at anaerobic threshold at both week one and week three in elderly cyclists. No effect on VO_{2max} was observed. This study indicates a potential role of L-arginine and antioxidant supplementation in improving exercise performance in elderly.

Our data about different supplementation in different sports are in accordance with those results on elderly cyclist (Cubrilo, 2006). So, both vitamins and combination of vitamins+amino acids (L-arginine+L-citruline, as potential improvement of L-arginine/NO system) increased basal level of •NO in different sports: cycling, rowing and taekwondo. Furthermore, two months supplementation by vitamins+amino acids increased release of •NO and showed antioxidant properties in acute exercise, what can be start point for further strategies in systemic supplementation in elite athletes (Cubrilo, 2006).

7. Conslusions and perspectives

Regarding to all notes listed above, general conclusion about role of oxidative stress in exercise suggest that reactive oxygen species play in part role in oxidative damage during exercise. Taken into consideration controversial results of many studies, as well as our own results, we can not clearly conclude which ROS is responsible for such events in strenuous exercise. Furthermore, therapeutic interventions with different antioxidants and other supplements are, also, controversial. Based on our results, SOD might be important marker for acute adaptation on oxidative damage, but this note need further support with serious studies with large number of well randomized participants. Also, based on recent results only measurement of isoprostans, 8-Br-Deoxyguanosine and allantoin in body fluids can support conclusion about oxidative damage. Anyway, we are in a big open field and usage of redox status as potential new therapeutic target needs further big clinical trials.

8. References

Allen, J.D.; Cobb, F.R. & Gow, A.J. (2005). Regional and Whole-Body Markers of Nitric Oxide Production following Hyperemic Stimuli. *Free Radical Biology & Medicine*, Vol. 38, No. 9, (May 2005), pp. 1164-1069

Allen, J.D.; Cobb, F.R.; Kraus, W.E. & Gow, A.J. (2006). Total Nitrogen Oxide following Exercise Testing Reflects Endothelial Function and Discriminates Health Status. *Free Radical Biology & Medicine*, Vol. 41, No. 5, (September 2006), pp. 740-747

Allen, J.D.; Miller, E.M.; Schwark, E.; Robbins, J.L.; Duscha, B.D. & Annex, B.H. (2009). Plasma Nitrite Response and Arterial Reactivity Differentiate Vascular Health and Performance. *Nitric Oxide*, Vol. 20, No. 4, (June 2009), pp. 231-237

Andersen, HR; Nielsen, JB; Nielsen, F. & Grandjean, P. (1997). Antioxidative Enzyme Activity in Human Erythrocytes. *Clinical Chemistry*, Vol. 43, No. 4, (April 1997), pp. 562-568

Andrade, F.H.; Reid, M.B.; Allen, D.G. & Westerblad, H. (1998). Effect of Hydrogen Peroxide and Dithiothreitol on Contractile Function of Single Skeletal Muscle Fibres from the Mouse. *The Journal of Physiology* (London), Vol. 509, No. 2, (June 1998), pp. 565-575

Armstrong, N.; McManus, A.M. & Welsman, J.R. (2008). Aerobic Fitness. In: *Paediatric Exercise Science and Medicine*, N. Armstrong & W. Van Mechelen (Eds.), 269-282, *Oxford University Press*, Oxford, United Kingdom

Auclair, C. & Voisin, E. (1985). Nitroblue Tetrazolium Reduction. In: CRC *Handbook of methods for oxygen radical research,* R.A. Greenwald (Ed.), 123-132, *CRC Press,* Boca Raton, Florida

Balakrishnan, S.D. & Anuradh, C.V. (1998). Exercise, Depletion of Antioxidants and Antioxidant Manipulation. *Cell Biochemistry and Function,* Vol. 16, No. 4, (December 1998), pp. 269-275

Banfi, G.; Malavazos, A.; Iorio, E.; Dolci, A.; Doneda, L.; Verna, R. & Corsi, M.M. (2006). Plasma Oxidative Stress Biomarkers, Nitric Oxide and Heat Shock Protein 70 in Trained Elite Soccer Players. *European Journal of Applied Physiology,* Vol. 96, No. 5, (March 2006), pp. 483-486

Bangsbo, J. (1994). The Physiology of Soccer with Special Reference to High-Intensity Intermittent Exercise. *Acta Physiologica Scandinavica Suppl,* Vol. 619, pp. 1-155

Beutler, E. (1982). Catalase. In: *Red cell metabolism, a manual of biochemical methods.* E. Beutler (Ed.), 105-115, *Grune and Stratton,* New York

Bhandari, A.K. & Scheinman, M. (1985). The Long Q-T Syndrome. *Modern Concepts of Cardiovascular Disease,* Vol. 54, pp. 45-50

Bloomer, R.J. & Fisher-Wellman, K.H. (2008). Blood Oxidative Stress Biomarkers: Influence of Sex, Training Status, and Dietary Intake. *Gender Medicine,* Vol. 5, No. 3, (September 2008) ,pp. 218-228

Boisseau, N. & Delamarche, P. (2000). Metabolic and Hormonal Responses to Exercise in Children and Adolescents. *Sports Medicine,* Vol. 30, No. 6, (December 2000), pp. 405-422

Borg, G.A. (1982). Psychophysical Bases of Perceived Exertion. *Medicine and Science in Sports and Exerc*ise, Vol. 14, No. 5, pp. 377-381

Brites, F.D.; Evelson, P.A.; Christiansen, M.G.; Nicol, M.F.; Basílico, M.J.; Wikinski, R.W. & Llesuy, S.F. (1999). Soccer Players under Regular Training Show Oxidative Stress but an Improved Plasma Antioxidant Status. *Clinical Science,* Vol. 96, No. 4, (April 1999), pp. 381-385

Caiozzo, V.J.; Davis, J.A.; Ellis, J.F.; Azus, J.L.; Vandagriff, R.; Prietto, C.A. & McMaster, W.C. (1982). A Comparison of Gas Exchange Indices Used to Detect the Anaerobic Threshold. *Journal of Applied Physiology,* Vol. 53, No. 5, (November 1982), pp. 1184-1189

Casajus, J.A. (2001). Seasonal Variation in Fitness Variables in Professional Soccer Players. *J. The Journal of Sports Medicine and Physical Fitness,* Vol. 41, No. 4, (December 2001), pp. 463-469

Cazzola, R.; Russo-Volpe, S.; Cervato, G. & Cestaro, B. (2003). Biochemical Assessments of Oxidative Stress, Erythrocyte Membrane Fluidity and Antioxidant Status in Professional Soccer Players and Sedentary Controls. *European Journal of Clinical Investigation,* Vol. 33, No. 10, (October 2003), pp. 924–930

Chang, C.K.; Tseng, H.F.; Hsuuw, Y.D.; Chan, W.H. & Shieh L.C. (2002). Higher LDL Oxidation at Rest and after a Rugby Game in Weekend Warriors. *Annals of Nutrition & Metabolism,* Vol. 46, No. 3-4, pp. 103-107

Chen, S.S.; Chang, L.S. & Wei, Y.H. (2001). Oxidative Damage to Proteins Anddecrease of Antioxidant Capacity in Patients with Varicocele. *Free Radical Biology & Medicine,* Vol. 30, No. 11, (June 2001), pp. 1328-1334

Chen, S.; Kim, W.; Henning, S.M.; Carpenter, C.L. & Li, Z. (2010). Arginine and Antioxidant Supplement on Performance in Elderly Male Cyclists: a Randomized Controlled Trial. *Journal of the International Society of Sports Nutrition,* Vol. 7, (March 2010), pp. 13

Chevion, S.; Moran, D.S.; Heled, Y., Shani, Y.; Regev, G.; Abbou, B.; Berenshtein, E.; Stadtman, E.R. & Epstein, Y. (2003). Plasma Antioxidant Status and Cell Injury after Severe Physical Exercise. *Proceedings of the National Academy of Sciences of the United States of America,* Vol. 100, No. 9, pp. 5119-5123

Childs, A.; Jacobs, C.; Kaminski, T.; Halliwell, B. & Leeuwenburgh, C. (2001). Supplementation with Vitamin C and N-Acetyl-Cysteine Increases Oxidative Stress in Humans after an Acute Muscle Injury Induced by Eccentric Exercise. *Free Radical Biology & Medicine*, Vol. 31, No. 6, (September 2001), pp. 745-753

Chiueh, C.C. (1999). Neuroprotective Properties of Nitric Oxide. *Annals of the New York Academy of Sciences,* Vol. 890, pp. 301–311

Cooper, D.M.; Nemet, D. & Galassetti, P. (2004). Exercise, Stress, and Infammation in the Growing Child: From the Bench to the Playground. *Current Opinion in Pediatrics,* Vol. 16, No. 3, (June 2004), pp. 286–292

Crisafulli, A.; Melis, F. & Lai, A.C. (2002). Haemodynamics during a Complete Exercise Induced Atrioventricular Block. *British Journal of Sports Medicine,* Vol. 36, No. 1, (February 2002), pp. 69-70

Crouse, S.F.; Rohack, J.J. & Jacobsen, D.J. (1992). Cardiac Structure and Function in Women Basketball Athletes—Seasonal-Variation and Comparisons with Nonathletic Controls. *Research Quarterly for Exercise and Sport,* Vol. 63, No. 4, (December 1992), pp. 393–401

Crouse, S.F.; Meade, T.; Hansen, E.B.; Green, J.S. & Martin, S.E. (2009). Electrocardiograms of Collegiate Football Athletes. *Clinical Cardiology,* Vol. 32, No. 1, (January 2009), pp. 37-42

Cubrilo, D. (2006). The Evaluation of L-Arginine/NO System Function and Oxidative Stress Parameters in Elite Athletes. *Master thesis.* Faculty of Medicine, University of Kragujevac, Kragujevac, Serbia

Cubrilo, D. (2009). Systemic Effects of Redox Disturbance Induced by Intensive Training in Young Football Players. *PhD thesis.* Faculty of Medicine, University of Kragujevac, Kragujevac, Serbia

Cubrilo, D.; Djordjevic, D.; Zivkovic, V.; Djuric, D.; Blagojevic, D.; Spasic, M. & Jakovljevic, V. (2011). Oxidative Stress and Nitrite Dynamics under Maximal Load in Elite Athletes: Relation to Sport Type. *Molecular and Celular Biochem*istry, Vol. 355, No. 1-2, (September 2011), pp. 273-279

DeCoursey, T.E. & Ligeti, E. (2005). Regulation and Termination of NADPH Oxidase Activity. *Cellular and Molecular Life Sciences,* Vol. 62, No. 19-20, (October 2005), pp. 2173–2193

Dillard, C.J.; Litov, R.E. & Savin, W.M. (1978). Effects of Exercise, Vitamin E, and Ozone on Pulmonary Function and Lipid Peroxidation. *Journal of Applied Physiology,* Vol. 45, No. 6, (December 1978), pp. 927 - 932

Djordjevic, D.; Cubrilo, D.; Zivkovic, V.; Barudzic, N.; Vuletic, M. & Jakovljevic, V. (2010a). Pre-Exercise Superoxide Dismutase Activity Affects the Pro/Antioxidant Response to Acute Exercise. *Serbian Journal of Experimental and Clinical Research*, Vol. 11, No. 4, (December 2010), pp. 145–153

Djordjevic, D.; Jakovljevic, V.; Cubrilo, D.; Zlatkovic, M.; Zivkovic ,V. & Djuric, D. (2010b). Coordination between Nitric Oxide and Superoxide Anion Radical during Progressive Exercise in Elite Soccer Players. *The Open Biochemistry Journal*, Vol. 4, (December 2010), pp. 100-106

Djordjevic, D. (2011). The Influence of Training Status on Morphofunctional Characteristics and Redox State of Young Handball Players. *PhD Thesis*. Faculty of Medicine, University of Kragujevac, Kragujevac, Serbia

Douglas, P.S.; O'Toole, M.L.; Katz, S.E.; Ginsburg, G.S.; Hiller, W.D. & Laird, R.H. (1997). Left Ventricular Hypertrophy in Athletes. *American Journal of Cardiology*, Vol. 80, No. 10, (November 1997), pp. 1384–1388.

Dröge W. (2002). Free Radicals in the Physiological Control of Cell Function. *Physiological Review*, Vol. 82, No. 1, (January 2002), pp. 47–95

Ekblom, B. (1986). Applied Physiology of Soccer. *Sports Medicine*, Vol. 3, No. 1, (January-February 1986), pp. 50–60

Evelson, P.; Gambino, G.; Travacio, M.; Jaita, G.; Verona, J.; Maroncelli, C.; Wikinski, R.; Llesuy, S. & Brites, F. (2002). Higher Antioxidant Defenses in Plasma and Low Density Lipoproteins from Rugby Players. *European Journal of Clinical Investigation*, Vol. 32, No. 11, (November 2002), pp. 818-825

Fagard R. (2003). Athlete's Heart. *Heart*, Vol. 89, No. 12, (December 2003), pp. 1455–1461

Faina, M.; Gallozi, C.; Lupo, S.; Colli, S.; Sassi, R. & Marini, C. (1988). Definition of Physiological Profile of the Soccer Players. In: *Science and football*. T. Reilly (Ed.), 158-163, *E&FN Spon*, London, United Kingdom

Ferrer, M.D.; Tauler, P.; Sureda, A.; Tur, J.A. & Pons, A. (2009). Antioxidant Regulatory Mechanisms in Neutrophils and Lymphocytes after Intense Exercise. *Journal of Sports Sciences*, Vol. 27, No. 1, (January 2009), pp. 49–58

Finaud, J.; Lac, G. & Filaire, E. (2006). Oxidative Stress: Relationship with Exercise and Training. *Sports Medicine*, Vol. 36, No. 4, pp. 327-358

Finkel, T. (1998). Oxygen Radicals and Signaling. *Current Opinion in Cell Biology*, Vol. 10, No. 2, (April 1998), pp. 248-253

Fisher-Wellman, K. & Bloomer, R.J. (2009). Acute Exercise and Oxidative Stress: a 30 Year History. *Dynamic Medicine*, Vol. 8, (January 2009), pp. 1-25

Flam, B.R.; Eichler, D.C. & Solomonson, L.P. (2007). Endothelial Nitric Oxide Production is Tightly Coupled to the Citrulline-NO Cycle. *Nitric Oxide*, Vol. 17, No. 3-4, (November-December 2007), pp. 115–121

Ghafourifar, P. & Cadenas, E. (2005). Mitochondrial Nitric Oxide Synthase. *Trends in Pharmacological Sciences*, Vol. 26, No. 4, (April 2005), pp. 190–195

Gilligan, D.M.; Panza, J.A.; Kilcoyne, C.M.; Waclawiw, M.A.; Casino, P.R. & Quyyumi, A.A. (1994). Contribution of Endothelium-Derived Nitric Oxide to Exercise-Induced Vasodilation. *Circulation*, Vol. 90, No. 6, (December 1994), pp. 2853–2858

Goldfarb, A.H.; Bloomer, R.J. & McKenzie, M.J. (2005). Combined Antioxidant Treatment Effects on Blood Oxidative Stress after Eccentric exercise. *Medicine and Science in Sports and Exercise*, Vol. 37, No. 2, (February 2005), pp. 234-239

Gougoura, S.; Nikolaidis, M.G.; Kostaropoulos, I.A.; Jamurtas, A.Z.; Koukoulis, G. & Kouretas, D. (2007). Increased Oxidative Stress Indices in the Blood of Child Swimmers. *European Journal of Applied Physioogy*, Vol. 100, No. 2, (May 2007), pp. 235-239

Green, L.C.; Wagner, D.A.; Glogowski, J.; Skipper, P.L.; Wishnok, J.S. & Tannenbaum, S.R. (1982). Analysis of Nitrate, Nitrite and [15 N] Nitrate in Biological Fluids. *Analytical Biochemistry*, Vol. 126, No. 1, (October 1982), pp. 131-138

Green. D.; Cheetham, C.; Mavaddat, L.; Watts. K.; Best, M.; Taylor. R. & O'Driscoll, G. (2002). Effect of Lower Limb Exercise on Forearm Vascular Function: Contribution of Nitric Oxide. *American Journal of Physiology. Heart and Circulatory Physiology*, Vol. 283, No. 3, (September 2002), pp. H899–H907

Green, D.J.; Maiorana, A.; O'Driscoll, G. & Taylor, R. (2004). Effect of Exercise Training on Endothelium-Derived Nitric Oxide Function in Humans. *The Journal of Physiology*, Vol. 561, No. 1, (November 2004), pp. 1–25

Grody, W.W.; Argyle, C.; Kern, R.M.; Dizikes, D.J.; Spector, E.B.; Strickland, A.D.; Klein, D. & Cederbaum, S.D. (1989). Differential Expression of the Two Arginase Genes in Hyperargininemia. Enzymatic, Pathologic, and Molecular Analysis. *The Journal of Clinical Investigation*, Vol. 83, No. 2, (February 1989), pp. 602–609

Groussard, C.; Rannou-Bekono, F.; Machefer, G.; Chevanne, M.; Vincent, S.; Sergent, O.; Cillard, J. & Gratas-Delamarche, A. (2003). Changes in Blood Lipid Peroxidation Markers and Antioxidants after a Single Sprint Anaerobic Exercise. *European Journal of Applied Physiology*, Vol. 89, No. 1, (March 2003), pp. 14-20

Gubner, R. & Ungerleider, H. (1943). Electrocardiographic Criteria for Left Ventricular Hypertrophy. Factors Determining the Evolution of the Electrocardiographic Patterns in Hypertrophy and Bundle Branch Block. *Archives of Internal Medicine*, Vol. 72, No. 2, (August 1943), pp. 196-209

Halliwell, B. & Gutteridge, J.M.C. (1999). *Free Radicals in Biology and Medicine*, Oxford University Press, New York

Halliwell, B. & Gutteridge, J.M.C. (2007). *Free Radicals in Biology and Medicine*, Oxford University Press, New York

Hambrecht, R.; Fiehn, E.; Weigl, C.; Gielen, S.; Hamann, C.; Kaiser, R.; Yu, J.; Adams, V.; Niebauer, J. & Schuler, G. (1998). Regular Physical Exercise Corrects Endothelial Dysfunction and Improves Exercise Capacity in Patients with Chronic Heart Failure. *Circulation*, Vol. 98, No. 24, (December 1998), pp. 2709–2715

Henschen, S. (1899). Skilanglauf und Skiwettlauf. Eine Medizinische Sport Studie. *Mitt Med Klin Uppsala (Jena)*, Vol.2, pp. 15-18

Herrero A. & Barja, G. (1997). ADP Regulation of Mitochondrial Free Radical Production is Different with Complex I or Complex II Linked Substrates: Implications for the Exercise Paradox and Brain Hypermetabolism. *Journal of Bioenergetics and Biomembranes*, Vol. 29, No. 3, (June 1997), pp. 241-249

Howley, E.T.; Bassett, D.R.Jr. & Welch, H.G. (1995). Criteria for Maximal Oxygen Uptake: Review and Commentary. *Medicine and Science in Sports and Exercise*, Vol. 27, No. 9, (September 1995), pp. 1292-1301

Huston, T.P.; Puffer, J.C. & Rodney, W.M.; (1985). The Athletic Heart Syndrome. *The New England Journal of Medicine,* Vol. 313, No. 1, (July 1985), pp. 24 -32

Ignarro, L.J.; Buga, G.M.; Wood, K.S.; Byrns, R.E. & Chaudhuri, G. (1987). Endothelium-Derived Relaxing Factor Produced and Released from Artery and Vein is Nitric Oxide. *Proceedings of the National Academy of Sciences of the United States of America,* Vol. 84, No. 24, (December 1987), pp. 9265–9269

Jackson, M. J.; Pye, D. & Palomero, J. (2007). The Production of Reactive Oxygen and Nitrogen Species by Skeletal Muscle. *Journal of Applied Physiology,* Vol. 1025, No. 4, (April 2007), pp. 1664-1670

Jakovljevic, V.Lj.; Zlatkovic, M.; Cubrilo, D.; Pantic, I. & Djuric, D.M. (2011). The Effects of Progressive Exercise on Cardiovascular Function in Top Level Athletes: Focus on Oxidative Stress. *Acta Physiologica Hungarica,* Vol. 98, No. 1, (March 2011), pp. 50-57

Ji, L.L. & Fu, R. (1993). Antioxidant Enzyme Response to Exercise and Ageing. *Medicine and Science in Sports and Exercise,* Vol. 25, No. 2, (February 1993), pp. 225-231

Ji, L.L. (1999). Antioxidants and Oxidative Stress in Exercise. *Proceedings of the Society for Experimental Biology and Medicine,* Vol. 222, No. 3, (December 1999), pp. 283-292

Jones, M.A. & Carter H. (2000). The Effect of Endurance Training on Parameters of Aerobic Fitness. *Sports Medicine,* Vol. 29, No. 6, (June 2000), pp. 373-386

Jones, A.M.; Wilkerson, D.P. & Campbell, I.T. (2004). Nitric Oxide Synthase Inhibition with L-NAME Reduces Maximal Oxygen Uptake but not Gas Exchange Threshold during Incremental Cycle Exercise in Man. *The Journal of Physiology,* Vol. 56, No. 1 (October 2004), pp. 329-338

Jungersten, L.; Ambring, A.; Wall, B. & Wennmalm, A. (1997). Both Physical Fitness and Acute Exercise Regulate Nitric Oxide Formation in Healthy Humans. *Journal of Applied Physiology,* Vol. 82, No. 3, (March 1997), pp. 760-764

Kakhlon, O. & Cabantchik, Z.I. (2002). The Labile Iron Pool: Characterization, Measurement, and Participation in Cellular Processes. *Free Radical Biology & Medicine,* Vol. 33, No. 8, (October 2002), pp. 1037–1046

Kashyap, S.R.; Park, Y.M.; Lara, A.; De Fronzo, R.A. & Zhang, R. (2008). Insulin Reduces Plasma Arginase Activity in Type 2 Diabetic Patients. *Diabetes Care,* Vol. 31, No. 1, (January 2008), pp. 134–139

Kayatekin, B.M.; Gonenc, S.; Acikgoz, O.; Uysal, N. & Dayi, A. (2002). Effects of Sprint Exercise on Oxidative Stress in Skeletal Muscle and Liver. *European Journal of Physiology,* Vol. 87, No. 2, (June 2002), pp. 141-144

Kew, S.; Mesa, M.D; Tricon, S.; Buckley, R.; Minihane, A.M. & Yaqoob, P. (2004). Effects of Oils Rich in Eicosapentaenoic and Docosahexaenoic Acids on Immune Cell Composition and Function in Healthy Humans. *The American Journal of Clinical Nutrition,* Vol. 79, No. 4, (April 2004), pp. 674–681

Kingwell, B.A. (2000). Nitric Oxide-Mediated Metabolic Regulation during Exercise: Effects of Training in Health and Cardiovascular Disease. *The FASEB journal: Official Publication of the Federation of American Societies for Experimental Biology,* Vol. 14, No. 12, (September 2000), pp. 1685–1696

Kingwell, B.A.; Berry, K.L.; Cameron, J.D.; Jennings. G.L. & Dart, A.M. (1997). Arterial Compliance Increases after Moderate-Intensity Cycling. *American Journal of*

Physiology. Heart and Circulatory Physiology, Vol. 273, No. 5 Pt 2, (November 1997), pp. H2186–H2191

Kingwell, B.A.; Sherrard, B.; Jennings, G.L. & Dart, A.M. (1997). Four Weeks of Cycle Training Increases Basal Production of Nitric Oxide from the Forearm. *American Journal of Physiology,* Vol. 272, No. 3 Pt 2, (March 1997), pp. H1070-H1077

Kurz, S. & Harrison, D.G. (1997). Insulin and the Arginine Paradox. *The Journal of Clinical Investigation,* Vol. 99, No. 3, (February 1997), pp. 369–370

Laaksonen, D.E.; Atalay, M.; Niskanen, L.; Uusitupa, M., Hänninen, O. & Sen, C.K. (1999). Blood Glutathione Homeostasis as a Determinant of Resting and Exercise-Induced Oxidative Stress in Young Men. *Redox Report: Communications in Free Radical Research,* Vol. 4, No. 1-2, pp. 53-59

Lagranha, C.J.; Alba-Loureiro, T.C.; Martins, E.F.; Pithon-Curi, T.C.; & Curi, R. (2008). Neutrophil Fatty Acid Composition: Effect of a Single Session of Exercise and Glutamine Supplementation. *Amino Acids,* Vol. 35, No. 1, (June 2008), pp. 243–245

Lander H.M. (1997). An Essential Role of Free Radicals and Derived Species in Signal Transduction. *The FASEB journal: Official Publication of the Federation of American Societies for Experimental Biology,* Vol. 11, No. 2, (February 1997), pp. 118-124

Lekhi, C.; Gupta, P.H. & Singh, B. (2007). Influence of Exercise on Oxidant Stress Products in Elite Indian Cyclists. *British Journal of Sports Medicine,* Vol. 41, No. 10, (October 2007), pp. 691-693.

Lerman, A.; Burnett, J.C.; Higano, S.T.Jr.; McKinley, L.J. & Holmes, D.R.Jr. (1998). Long-Term L-Arginine Supplementation Improves Small-Vessel Coronary Endothelial Function in Humans. *Circulation,* Vol. 97, No. 21, (June 1998), pp. 2123-2128

Levine, DB (2008). VO_2max: What do We Know, and What do We Still Need to Know? *The Journal of Physiology,* Vol. 586, No. 1, (January 2008), pp. 25-34

Li, C.; Huang, W.; Harris, M.B.; Goolsky, J.M. & Venema, R.C. (2005). Interaction of Endothelial Nitric Oxide Synthase with the CAT-1 Arginine Transporter Enhances NO Release by a Mechanism not Involving Arginine Transport. *The Biochemical Journal,* Vol. 386, No. 3 (March 2005), pp. 567–574

Lim, J.S.; Hwang, J.S.; Lee, J.A.; Kim, D.H.; Park, K.D.; Jeong, J.S. & Cheon, G.J. (2009). Cross-Calibration of Multi-Frequency Bioelectrical Impedance Analysis with Eight-Point Tactile Electrodes and Dual-Energy X-Ray Absorptiometry for Assessment of Body Composition in Healthy Children Aged 6-18 Years. *Pediatrics International: Official Journal of the Japan Pediatric Society,* Vol. 51, No. 2, (April 2009), pp. 63-68

Liochev, S.I.. & Fridovich, I. (2002). The Haber–Weiss Cycle—70 Years Later: an Alternative View. *Redox Report : Communications in Free Radical Research,* Vol. 7, No. 1, pp. 55–57

Maeda, S.; Miyauchi, T.; Kakiyama, T.; Sugawara, J.; Iemitsu, M.; Irukayama-Tomobe, Y.; Murakami, H.; Kumagai, Y.; Kuno, S. & Matsuda M. (2001). Effects of Exercise Training of 8 Weeks and Detraining on Plasma Levels of Endothelium-Derived Factors, Endothelin-1 and Nitric Oxide in Healthy Young Humans. *Life Sciences,* Vol. 69, No. 9, (July 2001), pp. 1005-1016

Maiorana, A.; O'Driscoll, G.; Taylor, R. & Green, D. (2003). Exercise and the Nitric Oxide Vasodilator System. *Sports Medicine,* Vol. 33, No. 14, pp. 1013-1035

Maron, B.J.; Thompson, P.D.; Puffer, J.C.; McGrew, C.A.; Strong, W.B.; Douglas, P.S.; Clark, L.T.; Mitten, M.J.; Crawford, M.H.; Atkins, D.L.; Driscoll, D.J. & Epstein, A.E.

(1996). American Heart Association Statement for Health Professionals: Cardiovascular Preparticipation Screening of Competitive Athletes. *Circulation,* Vol. 94, No. 4, (August 1996), pp. 850–856

Maron, B.J. (1998). Heart Disease and Other Causes of Sudden Death in Young Athletes. *Current Problems in Cardiology,* Vol. 23, No. 9, (September 2003), pp. 477–529

Maron, B.J. (2003). Sudden Death in Young Athletes. *New England Journal of Medicine,* Vol. 349, No. 11, (September 2003), pp. 1064–1075

Maron, B.J. & Zipes, D.P. (2005). 36th Bethesda Conference: Eligibility Recommendations for Competitive Athletes with Cardiovascular Abnormalities. *Journal of the American College of Cardiology,* Vol. 45, No. 8, pp. 1312–1375

Maron, J.B. & Pelliccia A. (2006). The Heart of Trained Athletes. Cardiac Remodeling and the Risks of Sports, Including Sudden Death. *Circulation,* Vol. 114, No. 15, (October 2006), pp. 1633-1644

Martin, W.H. 3rd.; Coyle, E.F.; Bloomfield, S.A. & Ehsani, A.A. (1986). Effects of Physical Deconditioning after Intense Endurance Training on Left Ventricular Dimensions and Stroke Volume. *Journal of the American College of Cardiology,* Vol. 7, No. 5, (May 1986), pp. 982–989

Marzatiko, F.; Pansarasa, O.; Bertorelli, L.; Somenzini, L. & Della Valle, G. (1997). Blood Free Radical Antioxidant Enzymes and Lipid Peroxides following Long-Distance and Lactacidemic Performances in Highly Trained Aerobic and Sprint Athletes. *The Journal of Sports Medicine and Physical Fitness,* Vol. 37, No. 4, (December 1997), pp. 235-239

Mastaloudis, A.; Leonard, S.W. & Traber, M.G. (2001). Oxidative Stress in Athletes during Extreme Endurance Exercise. *Free Radical Biology & Medicine,* Vol. 31, No. 7, (October 2001), pp. 911-922

Maxwell, A.J.; Ho, H.K.V.; Le, C.Q.; Lin, P.S.; Bernstein. D. & Cooke, J.P. (2001). L-Arginine Enhances Aerobic Exercise Capacity in Association with Augmented Nitric Oxide Production. *Journal of Applied Physiology,* Vol. 90, No. 3, (March 2001), pp. 933-938

McCord, J.M. & Fridovich, I. (1969). The Utility of Superoxide Dismutase in Studying Free Radical Reactions. I. Radicals Generated by the Interaction of Sulfite, Dimethyl Sulfoxide, and Oxygen. *The Journal of Biological Chemistry,* Vol. 244, No. 22, (November 1969), pp. 6056-6063

Metin, G.; Gumustas, M.K.; Uslu, E., Belce, A.; & Kayserilioglu, A. (2003). Effect of Regular Training on Plasma Thiols, Malondialdehyde and Carnitine Concentracions in Young Soccer Players. *Chinese Journal of Physiology,* Vol. 46, No. 1, (March 2001), pp. 35-39

Misra, H.P. & Fridovich, I. (1972). The Role of Superoxide-Anion in the Autooxidation of Epinephrine and a Simple Assay for Superoxide Dismutase. *The Journal of Biological Chemistry,* Vol. 247, No. 10, (May 1972), pp. 3170-3175

Miyazaki, H.; Oh-ishi, S.; Ookawara, T.; Kizaki, T.; Toshinai, K.; Ha, S.; Haga, S.; Ji, L.L. & Ohno, H. (2001). Strenuous Endurance Training in Humans Reduces Oxidative Stress following Exhausting Exercise. *European Journal of Applied Physiology*, Vol. 84, No. 1-2, (January-February 2001), pp. 1-6

Morganroth, J.; Maron, B.J.; Henry, W.L. & Epstein, S.E. (1975). Comparative Left Ventricular Dimensions in Trained Athletes. *Annals of Internal Medicine,* Vol. 82, No. 4, (April 1975), pp. 521–524

Morris, S.M.Jr. (2006): Arginine: beyond Protein. *The American Journal of Clinical Nutrition, Vol.* 83, No. 6, (February 2006), pp. 508S–512S

Muller, F.L., Liu, Y., & Van Remmen, H. (2004). Complex III Releases Superoxide to Both Sides of the Inner Mitochondrial Membrane. *The Journal of Biological Chemistry,* Vol. 279, No. 47, (November 2004), pp. 49064–49073

Naghizadeh, H.; Afzalpour, M.E. & Zarban A. (2009). The Comparison of Antioxidant Status and Lipid Profile of Karate Athletes with Non-Athletes. *Journal of Birjand University of Medical Sciences,* Vol. 16, No. 3, pp. 54-61

Naka, K.K.; Tweddel, A.C.; Parthimos. D.; Henderson. A.; Goodfellow. J. & Frenneaux, M.P. (2003). Arterial Distensibility: Acute Changes following Dynamic Exercise in Normal Subjects. *American Journal of Physiology. Heart and Circulatory Physiology,* Vol. 284, No. 3, (March 2003), pp. H970–H978

Ohkawa, H.; Ohishi, N. & Yagi, K. (1979). Assay for Lipid Peroxides in Animal Tissues by Thiobarbituric Acid Reaction. *Analytical Biochemistry,* Vol. 95, No. 2, (June 1979), pp. 351-358

Ookawara, T.; Haga, S.; Ha S.; Oh-Ishi ,S.; Toshinai, K.; Kizaki, T.; Ji, L.L.; Suzuki, K. &. Ohno H. (2003). Effects of Endurance Training on Three Superoxide Dismutase Isoenzymes in Human Plasma. *Free Radical Research,* Vol. 37, No. 7, (July 2003), pp. 713-719

Ørtenblad, N.; Madsen, K. & Djurhuus, M.S. (1997). Antioxidant Status and Lipid Peroxidation after Short-Term Maximal Exercise in Trained and Untrained Humans. *American Journal of Physiology,* Vol. 272, No. 4 Pt 2, (April 1997), pp. R1258-R1263

Ostojić, MS. (2000). Physical and Physiological Characteristics of Elite Serbian Soccer Players. *Facta Universitatis Series Physical Education andSports,* Vol. 7, No. 7, pp. 23-29

Palazzetti, S.; Richard, M.J.; Favier, A. &. Margaritis, I. (2003). Overload Training Increases Exercise-Induced Oxidative Stress and Damage. *Canadian Journal of Applied Physiol*ogy, Vol. 28, No. 4, (August 2003), pp. 588-604

Pastor, N.; Weinstein, H.; Jamison, E. & Brenowitz, M. (2000). A Detailed Interpretation of OH Radical Footprints in a TBPDNA Complex Reveals the Role of Dynamics in the Mechanism of Sequence-Specific Binding. *Journal of Molecular Biology,* Vol. 304, No. 1, (November 2000), pp. 55–68

Pelliccia, A.; Maron, B.J.; Spataro, A.; Proschan, M.A. & Spirito, P. (1991). The Upper Limit of Physiologic Cardiac Hypertrophy in Highly Trained Elite Athletes. *New England Journal of Medicine,* Vol. 324, No. 5, (January 1991), pp. 295–301

Pelliccia, A.; Maron, B.J.; Culasso, F.; Spataro, A. & Caselli, G. (1996). Athlete's Heart in Women: Echocardiographic Characterization of Highly Trained Elite Female Athletes. *JAMA: the Journal of the American Medical Association,* Vol. 276, No. 3 (July 1996), pp. 211–215

Pelliccia, A.; Maron, B.J.; Culasso, F.; Di Paolo, F.M.; Spataro, A.; Biffi, A.; Caselli, G. & Piovano, P. (2000). Clinical Significance of Abnormal Electrocardiographic Patterns in Trained Athletes. *Circulation,* Vol. 102, No. 3, (July 2000), pp. 278-84

Pelliccia, A.; Maron, B.J.; de Luca, R.; Di Paolo, F.M.; Spataro, A. & Culasso, F. (2002). Remodeling of Left Ventricular Hypertrophy in Elite Athletes after Long-Term Deconditioning. *Circulation,* Vol. 105, No. 8, (February 2002), pp. 944 -949

Pelliccia, A.; Maron, B.J.; Di Paolo, F.M.; Biffi, A.; Quattrini, F.M.; Pisicchio, C.; Roselli, A.; Caselli, S. & Culasso, F. (2005). Prevalence and Clinical Significance of Left Atrial Remodeling in Competitive Athletes. *Journal of the American College of Cardiology,* Vol. 46, No. 4, (August 2005), pp. 690–696

Pelliccia, A.; Culasso, F.; Di Paolo, M.F.; Accettura, D.; Cantore, R., Castagna, W.; Ciacciarelli, A.; Costini, G.; Cuffari, B.; Drago, E.; Federici, V.; Gribaudo, C.G.; Lacovelli, G.; Landolfi, L.; Menichetti, G.; Atzeni, O.U.; Vagnini, A.; Casasco, M. & Di Luigi, L. (2007). Prevalence of Abnormal Electrocardiograms in a Large Unselected Population Undergoing Pre-Participation Cardiovascular Screening. *European Heart Journal.* Vol. 28, No. 16, (August 2007), pp. 2006-2010

Pick, E. & Keisari, Y. (1980). A Simple Colorimetric Method for the Measurment of Hydrogen Peroxide Produced by Cells in Culture. *Journal of Immunological Methods,* Vol. 38, No. 1-2, pp. 161-170

Pincemail, J.; Lecomte, J.; Castiau, J. ; Collard, E. ; Vasankari, T. ; Cheramy-Bien, J. ; Limet, R. & Defraigne, J. (2000). Evaluation of Autoantibodies against Oxidized LDL and Antioxidant Status in Top Soccer and Basketball Players after 4 Months of Competition. *Free Radical Biology & Medicine,* Vol. 28, No. 4, (February 2000), pp. 559-565

Ponorac, N; Matavulj, A; Grujic. N; Rajkovaca, Z; & Kovacevic, P. (2005). Maximal Oxigen Uptake (VO_2max) as the Indicator of Physical Working Capacity In Sportsmen. *Acta Medica Medianae.* Vol. 44, No. 4, pp. 17-20

Poveda, J.J.; Riestra, A.; Salas, E.; Cagigas, M.L.; López-Somoza, C.; Amado, J.A. & Berrazueta, J.R. (1997). Contribution of Nitric Oxide to Exercise-Induced Changes in Healthy Volunteers: Effects of Acute Exercise and Long-Term Physical Training. *European Journal of Clinical Investigation,* Vol. 27, No. 11, (November 1997), pp. 967-971

Pryor, W.A. & Squadrito, G.L. (1995). The Chemistry of Peroxynitrite: a Product from the Reaction of Nitric Oxide with Superoxide. *American Journal of Physiology,* Vol. 268, No. 5 Pt 1, (May 1995), pp. L699-L722

Radak, Z.; Chung, H.Y. & Goto, S. (2008a). Systemic Adaptation to Oxidative Challenge Induced by Regular Exercise. *Free Radical Biology & Medicine,* Vol. 44, No. 2, (January 2008), pp. 153-159

Radak, Z.; Chung, H.Y.; Koltai, E.; Taylor, A.W. & Goto, S. (2008b). Exercise, Oxidative Stress and Hormesis. *Ageing Research Reviews,* Vol. 7, No. 1, (January 2008),pp. 34-42

Ramel, A.; Wagner, K.H. & Elmadfa, I. (2004). Plasma Antioxidants and Lipidoxidation after Submaximal Resistance Exercise in Men. *European Journal of Nutrition,* Vol. 43, No. 1, (February 2004), pp. 2-6

Rassaf, T.; Lauer, T.; Heiss, C.; Balzer, J.; Mangold, S.; Leyendecker, T.; Rottler, J.; Drexhage, C.; Meyer, C. & Kelm, M. (2007). Nitric Oxide Synthase-Derived Plasma Nitrite

Predicts Exercise Capacity. *British Journal of Sports Medicine,* Vol. 41, No. 10, (October 2007), pp. 669-673

Ridnour, L.A.; Thomas D.D.; Mancardi D.; Espey, M.G.; Miranda K.M.; Paolocci N.; Feelisch, M.; Fukuto, J. & Wink, D.A. (2004). The Chemistry of Nitrosative Stress Induced by Nitric Oxide and Reactive Nitrogen Oxide Species. Putting Perspective on Stressful Biological Situations. *Biological Chemistry,* Vol.385, No. 1, (January 2004), pp. 1–10

Ridnour, L.A.; Isenberg, J.S.; Espey, M.G.; Thomas, D.D.; Roberts, D.D. & Wink, D.A. (2005). Nitric Oxide Regulates Angiogenesis Through a Functional Switch Involving Thrombospondin-1. *Proceedings of the National Academy of Sciences of the United States of America,* Vol. 102, No. 37, (September 2005), pp. 13147–13152

Santhanam, L.; Christianson, D.W.; Nyhan, D. & Berkowitz, D.E. (2008). Arginase and Vascular Aging. *Journal of Applied Physiology,* Vol. 105, No. 5, (November 2008), pp. 1632–1642

Schulman, S.P.; Becker L.C.; Kass, D.A.; Champion, H.C.; Terrin, M.L.; Forman, S.; Ernst K.V., Kelemen, M.D.; Townsend, S.N.; Capriotti, A.; Hare, J.M. & Gerstenblith, G. (2006). L-Arginine Therapy in Acute Myocardial Infarction. The Vascular Interaction with Age in Myocardial Infarction (VINTAGE MI) Randomized Trial. *JAMA: the Journal of the American Medical Association,* Vol. 295, No. 1, (January 2006), pp. 58–64

Schwartz, P.J.; Periti, M. & Malliani, A. (1975). The Long QT Syndrome. *American Heart Journal,* Vol. 89, No. 3, (March 1975), pp. 378-390

Selamoglu, S.; Turgay, F.; Kayatekin, B.M.; Gönenc, S. & Yslegen, C. (2000). Aerobic and Anaerobic Training Effects on the Antioxidant Enzymes in the Blood. *Acta Physiologica Hungarica* Vol. 87, No. 3, pp. 267-273

Sharma, S.; Maron, B.J., Whyte, G.; Firoozi, S.; Elliott, P.M. & McKenna, W.J. (2002). Physiologic Limits of Left Ventricular Hypertrophy in Elite Junior Athletes: Relevance to Differential Diagnosis of Athlete's Heart and Hypertrophic Cardiomyopathy. *Journal of the American College of Cardiology,* Vol. 40, No. 8, (October 2002), pp. 1431–1436

Sharman, J.E.; McEniery, C.M.; Campbell, R.; Pusalkar, P.; Wilkinson, I.B.; Coombes, J.S. & Cockcroft, J.R. (2008). Nitric Oxide Does not Significantly Contribute to Changes in Pulse Pressure Amplification during Light Aerobic Exercise. *Hypertension,* Vol. 51, No. 4, (April 2008), pp. 856-861

Shephard, R.J. (1999). Biology and Medicine of Soccer: an Update. *Journal of Sport Sciences,* Vol. 17, No. 10, (October 1999), pp. 757-786

Stolen, T.; Chamari, K. & Castagna, C. (2005). Physiology of Soccer: an Update. *Sports Medicine,* Vol. 35, No. 6, pp. 501-536

Sureda, A.; Ferrer, M.D.; Tauler, P.; Tur, J.A. & Pons, A. (2008). Lymphocyte Antioxidant Response and H_2O_2 Production after a Swimming Session: Gender Differences. *Free Radical Research,* Vol. 42, No. 4, (April 2008), pp. 312–319

Taddei, S.; Galetta, F.; Virdis, A.; Ghiadoni, L.; Salvetti, G.; Franzoni, F.; Giusti, C. & Salvetti, A. (2000). Physical Activity Prevents Age-Related Impairment in Nitric Oxide Availability in Elderly Athletes. *Circulation,* Vol.101, No. 25, (June 2000), pp. 2896-2901

Tanskanen, M.; Atalay, M. & Uusitalo, A. (2010). Altered Oxidative Stress in Overtrained Athletes. *Journal of Sports Sciences,* Vol. 28, No. 3, (February 2010), pp. 309-317

Tappel, A.L. (1953). The Inhibition of Hematin-Catalyzed Oxidations by Alpha-Tocopherol. *Archives of Biochemistry and Biophysics,* Vol. 47, No. 1, (November 1953), pp. 223-225

Tauler, P.; Ferrer, M.D.; Sureda, A.; Pujol, P.; Drobnic, F.; Tur, J.A. & Pons, A. (2008). Supplementation with an Antioxidant Cocktail Containing Coenzyme Q Prevents Plasma Oxidative Damage Induced by Soccer. *European Journal of Applied Physiology,* Vol. 104, No. 5, (November 2008), pp. 777–785

Tinken, T.M.; Thijssen, D.H.J.; Black, M.A.; Cable, N.T. & Green, D.J. (2008). Time Course of Change in Vasodilator Function and Capacity in Response to Exercise Training in Humans. *The Journal of Physiology.* Vol. 586, No. 20, (October 2008), 5003-5012

Topal, G.; Brunet, A.; Walch, L.; Boucher, J.L. & David-Dufilho, M. (2006). Mitochondrial Arginase II Modulates Nitric-Oxide Synthesis through non Freely Exchangeable L - Arginine Pools in Human Endothelial Cells. *The Journal of Pharmacology and Experimental Therapeutics,* Vol. 318, No. 3, (September 2006), pp. 1368–1374

Valko, M.; Izakovic; M.; Mazur, M.; Rhodes, C.J. & Telser, J. (2004). Role of Oxygen Radicals in DNA Damage and Cancer Incidence. *Molecular and Celular Biochemistry,* Vol. 266, No. 1-2, (November 2004), pp. 37–56

Valko M.; Morris H. & Cronin, M.T.D. (2005). Metals, Toxicity and Oxidative Stress. *Current Medicinal Chemistry,* Vol. 12, No. 10, pp. 1161–1208

Valko, M.; Rhodes, C.J; Moncol, J.; Izakovic, M. & Mazur, M. (2006). Free Radicals, Metals and Antioxidants in Oxidative Stress-Induced Cancer. *Chemico-Biological Interactions,* Vol. 160, No. 1, (March 2006), pp. 1–40

Van de Poll, M.C.G.; Siroen, M.P.C.; van Leeuwen, P.A.M.; Soeters, P.B.; Melis, G.C.; Boelens, P.G.; Deutz, N.E.P. & Dejong, C.H.C. (2007). Interorgan Amino Acid Exchange in Humans: Consequences for Arginine and Citrulline Metabolism. *The American Journal of Clinical Nutrition,* Vol. 85, No. 1, (January 2007), pp. 167–172

Vollaard, N.B.; Shearman, J.P. & Cooper, C.E. (2005). Exercise-Induced Oxidative Stress: Myths, Realities and Physiological Relevance. *Sports Medicine,* Vol. 35, No. 12, pp. 1045-1062

Vorbach, C.; Harrison, R. & Capecchi, M.R. (2003). Xanthine Oxidoreductase is Central to the Evolution and Function of the Innate Immune System. *Trends in Immunology,* Vol. 24, No. 9, (September 2003), pp. 512–517

Wasserman, K.; Whipp, B.J.; Koyal, S.N. & Beaver, W.L. (1973). Anaerobic Threshold and Respiratory Gas Exchange during Exercise. *Journal of Applied Physiology,* Vol. 35, No. 2, (August 1973), pp. 236-243

Wilkerson, D.P.; Campbell, I.T. & Jones, A.M. (2004). Influence of Nitric Oxide Synthase Inhibition on Pulmonary O_2 Uptake Kinetics during Supra-Maximal Exercise in Humans. *The Journal of Physiology,* Vol. 561, No. 2, (December 2004), pp. 623-635

Wolin, M.S. (2000). Interactions of Oxidants with Vascular Signaling Systems. *Arteriosclerosis, Thrombosis, and Vascular Biology,* Vol. 20, No. 6, (June 2000), pp. 1430-1442

Wu, G. & Morris, S.M.Jr. (1998). Arginine Metabolism: Nitric Oxide and Beyond. *The Biochemical Journal,* Vol. 336, No. 1, (November 1998), pp. 1–17

Yamaner, F. (2010). Oxidative Predictors and Lipoproteins in Male Soccer Players. *Turkish Journal of Medical Sciences,* Vol. 40, No. 3, pp. 1-8

4

Transient Cold Shock Induces Oxidative Stress Events in Antarctic Fungi

Nedelina Kostadinova, Ekaterina Krumova, Tzvetanka Stefanova, Vladislava Dishliyska and Maria Angelova
The Stephan Angeloff Institute of Microbiology, Bulgarian Academy of Sciences, Bulgaria

1. Introduction

The Antarctic biota has evolved under the influence of a suite of geological and climatic factors, including the geographic isolation of the landmass and the continental shelves, extremely low temperatures and intense seasonality (Russo et al., 2010). The isolation and environmental history of Antarctica have led to a unique biota. Many groups of organisms became extinct in Antarctica as a result of the extremely cold conditions. Although this continent is the coldest, highest, windiest, driest, wildest and most pristine of all of the continents, it is full of life. In addition to its well-known inhabitants, such as penguins and seals, it also has a diverse and unique range of microbial diversity (Nichols et al., 1999; Vincent, 2000). Microorganisms successfully colonise cold habitats and play a major role in the processes of nutrient turnover at low temperatures. In recent years, a growing attention in research has been devoted to cold-adapted microorganisms. This interest in Antarctic microorganisms stems from several reasons. Antarctica's environmental extremes present conditions in which microorganisms have evolved unique characteristics for survival, which are of great scientific interest. Moreover, the availability of novel Antarctic species, which are generally isolated from extreme environments, opens the door for biotechnological exploration. Investigations of psychrotolerant and psychrophilic microorganisms are also important for human health because microorganisms can cause food spoilage and food-borne diseases. Research on cold shock raises a number of questions: which cellular function is affected most upon cold shock, what makes cell growth stop, and are there well-conserved or common cold shock proteins as in the case of heat-shock proteins? These questions are no less important than those in the case of heat shock (Inouye, 1999).

Cold-adapted microorganisms include both psychrophilic (organisms with an optimal growth temperature at or below 15°C and a maximum growth temperature below 20°C) and psychrotrophic or psychrotolerant (organisms exhibiting the ability to grow at temperatures below 15°C but exhibiting maximum growth rates at temperature optima above 18°C) organisms, and they are often subjected to other extreme environmental parameters (Morgan-Kiss et al., 2006). Antarctic microflora is also represented by mesophilic and thermophilic bacteria that are isolated from geothermal soils and by mesophilic fungi, which are present as viable propagules that are able to grow actively, at least under Antarctic summer conditions (Pepi et al., 2005; Ruisi, 2007).

Cold-adapted microorganisms have developed several strategies to adapt to low temperatures (Onofri et al., 2007). This adaptation includes the production of cold-active enzymes (Feller & Gerday, 2003; Collins et al., 2008; Gatti-Lafranconi et al., 2010), the modulation of lipid compositions to maintain the fluidity of the cell membrane (Chintalapati et al., 2004; Russell, 2008), the production of RNA chaperones to suppress the formation of undesired secondary structures of RNA (Kwak et al., 2011) and the synthesis of antifreeze (García-Arribas et al., 2007) and cold shock proteins (Horn et al., 2007).

Additionally, low temperatures can induce oxidative stress due to the enhanced generation of reactive oxygen species (ROS), such as the superoxide anion ($O_2{\bullet}^-$), hydrogen peroxide H_2O_2, and the hydroxyl radical •OH (Chattopadhyay, 2002; Gocheva et al., 2009). These ROS are highly damaging to cellular components, including DNA, lipids and proteins (Sies, 1993). To scavenge ROS and prevent damage, all aerobic cells have evolved a complex defence system consisting of both low molecular mass scavengers and high molecular mass antioxidants, particularly, antioxidant enzymes, such as superoxide dismutase (SOD) and catalase (CAT) (Fridovich, 1998). Thus, antioxidant defence may play a significant role in the microbial survival mechanism under extremely cold conditions (Chattopadhyay, 2002; Chattopadhyay et al., 2011).

Terrestrial microfungal communities in the Antarctic are rarely investigated (Ruisi et al., 2007). There are even fewer studies of the relationship between low temperatures and the induction of oxidative stress events. The decrease in temperature gives signal to organism for evoking cold-shock response generating molecules that are required for growing at the lower range of growth temperature. These cold-shock response are also critical for survival and growth at lower temperatures (Ray, 2006). Our previous studies indicated that growth at a low temperature induced oxidative stress in fungal strains isolated from soil samples of three regions of Antarctica: Casey Station, Terra Nova Bay and South Georgia (Gocheva et al., 2005, 2006, 2009). In recent years, we have also isolated filamentous fungi from samples taken from another Antarctic region – the permanent Bulgarian Antarctic base "St. Kl. Ohridski" on Livingston Island in the South Shetland Islands (Maritime Antarctica) – during the Bulgarian Antarctic expedition 2006/07 (Tosi et al., 2010). Published data about the filamentous and larger fungi of Livingston Island are scarce (Gray & Smith 1984; Wirtz et al., 2003). Furthermore, the filamentous fungi from the Bulgarian area on Livingston Island have only been investigated by our research team. Moreover, the studies on the specificity of the cell response to cold-induced oxidative stress between two different thermal classes of Antarctic fungi, psychrophilic (psychrotolerant) and mesophilic, are scant (Gocheva et al., 2009).

This paper is an attempt to enlarge our previous investigation focusing on two Antarctic fungi (psychrotolerant and mesophilic strains) that were isolated from the Bulgarian Antarctic area on Livingston Island. This study was designed to compare the effect of short-term cold shock on the growth of mycelia, glucose consumption and the level of oxidative stress biomarkers, including ROS, oxidatively damaged proteins, reserve carbohydrates and trehalose metabolising enzymes. Moreover, the Antarctic strains were used to obtain information concerning changes in the activities of two antioxidant enzymes, SOD and CAT, under a short-term cold treatment.

2. Materials and methods

2.1 Fungal strains, culture media and cultivation

The fungal strains, *Penicillium sp.* 161 and *Aspergillus glaucus* 363 (with optimal growth temperatures of 20°C and 25°C, respectively), isolated from Livingston Island (South Shetlands archipelago, Antarctica) (Tosi et al., 2010) were used for the experiments. The strains belong to the Mycological collection at the Institute of Microbiology, Sofia, and they were maintained at 4°C on beer agar at pH 6.3.

The composition of the seed and production media has been described previously (Angelova et al., 1996). The cultivation was performed in a 3 L bioreactor, which was ABR-09-developed and constructed by the former Central Laboratory for Bioinstrumentation and Automatisation (CLBA) of the Bulgarian Academy of Sciences. The bioreactor was equipped with temperature, pH and automatic dissolved oxygen (DO) monitoring equipment and a control system.

For the submerged cultivation, 74 ml of seed medium was inoculated with 6 ml of spore suspension at a concentration of 2×10^8 spores/ml in 500-ml Erlenmeyer flasks. The cultivation was performed at 20°C for 48 h for the psychrotolerant strain and at 25°C for 24 h for mesophilic strain on a rotary shaker (220 rpm). For the bioreactor cultures, 200 ml of the seed culture was brought into the 3-L bioreactor, which contained 1800 ml of the production medium. The cultures were grown at an optimal temperature with a stirrer speed of 400 rpm and an air flow of 0.5 vvm. During the middle of the exponential phase (24 h for *Penicillium sp.* 161 and 18 h for *A. glaucus* 363), the temperature was reduced to 4 or 10°C. This downshift was reached in approximately 40 min. After an incubation of 6 h under cold stress conditions, the temperature was shifted up to the optimal value. The control variants were grown at their optimal temperature during the entire period.

2.2 Cell-free extract preparation and isolation of the cytosolic and mitochondrial fractions

The cell-free extract, in addition to the cytosolic and mitochondrial fractions, were prepared as previously described (Krumova et al., 2008). All of the steps were performed at 0–4°C.

2.3 Determination of ROS

For the measurement of the $O_2^{\bullet-}$ production rate, the method of superoxide dismutase-inhibitable reduction of cytochrome *c* was used (Hassan & Fridovich, 1979) with some modifications. Briefly, the cell suspensions or mitochondrial fractions, which were taken from the control and cold-stressed cultures, were incubated for 60 min at 30°C on a water bath rotary shaker at 150 rpm. The reaction mixtures contained 50 μM cytochrome *c*, 2% non-autoclaved glucose, 20 mM NADPH in the presence and absence of 50 μg ml^{-1} of the superoxide dismutase from bovine erythrocytes in a 0.05 M potassium phosphate buffer with a pH of 7.8. The reaction was stopped by cooling in an ice-cold water bath. The cells were removed by centrifugation before measuring absorbance at 550 nm to determine the extent of cytochrome *c* reduction. A molar extinction coefficient of 2.11×10^4 was used to calculate the concentration of reduced cytochrome C. For measurement of the hydrogen peroxide production, the method of Pick & Mizel (1981) was used. Briefly, fungal cells treated with temperature were suspended in a 0.05 M potassium phosphate buffer with a pH of 7.8 and containing 50 μg ml^{-1} of horseradish peroxidase type VI-A. After incubation at 30°C for 45 min, the reaction was stopped by the addition of 1N NaOH, and the absorbance was read at 620 nm. For the calculations, a standard curve with H_2O_2 concentrations (from 5 to 50 μM) was used.

2.4 Enzyme activity determination

SOD activity was measured by the nitro-blue tetrazolium (NBT) reduction method of Beauchamp & Fridovich (1971). The reaction mixture contained 56 μM (NBT), 0.01 M methionine, 1.17 μM riboflavin, 20 μM KCN and 0.05 M phosphate buffer with a pH of 7.8. Superoxide was measured by the increase in absorbance at 560 nm at 30° C after 6 min of incubation from the beginning of the illumination. One unit of SOD activity was defined as the amount of enzyme required for the inhibition reduction of NBT by 50% (A_{560}) and was expressed as units per mg protein [U/mg protein]. Cyanide (5 mM) was used to distinguish between the cyanide-sensitive isoenzyme Cu/Zn-SOD and the cyanide-resistant Mn SOD. The Cu/Zn-SOD activity was obtained as the total activity minus the activity in the presence of 5 mM cyanide.

The catalase activity was determined by monitoring the decomposition of 18 mM H_2O_2 at 240 nm (Beers & Sizer, 1952). One unit of activity was that which decomposes 1 μmol of H_2O_2 min^{-1} mg $protein^{-1}$ at 25°C and a pH of 7.0. Specific activity is given as U/mg protein.

The neutral trehalase (NT) and trehalose-6-phosphate synthase (TPS) enzyme activities were assayed in cell-free extracts by the methods of Müller et al. (1992) and Vandercammen et al. (1989), respectively, as described by El-Bashiti et al. (2005).

2.5 Analytical methods

The glycogen and trehalose contents were determined following the procedure of Becker (1978) and Vandecamen et al. (1989) and modified by Parrou et al. (1997). The soluble reducing sugars were determined by the Somogy-Nelson method (Somogy 1952).

The protein oxidative damage was measured spectrophotometrically as the protein carbonyl content using the 2,4-dinitrophenylhydrazine (DNPH) binding assay (Hart et al., 1999) slightly modified by Adachi & Ishii (2000). The cell-free extracts were incubated with DNPH for 1 h at 37°C; the proteins were precipitated in 10% cold TCA and washed with ethanol:ethylacetate (1:1) to remove any excess of DNPH and finally dissolved in 6 M guanidine chloride with a pH of 2. The optical density was measured at 380 nm, and the carbonyl content was calculated using a molar extinction coefficient of 21 mM^{-1} cm^{-1} as nanomoles of DNPH incorporated (protein carbonyls) per mg of protein.

Protein was estimated by the Lowry procedure (Lowry, 1951) using a solution of bovine serum albumin as a standard.

The dry weight determination was performed on samples of mycelia harvested throughout the culture period. The culture fluid was filtered through a Whatman (Clifton, USA) No. 4 filter The separated mycelia were washed twice with distilled water and dried to a constant weight at 105°C.

2.6 PAGE analyses

Non-denaturing polyacrylamide gel electrophoresis of the extract was followed by activity staining for each of the enzymes tested. A forty μg of total protein was applied to a 10% nondenaturing PAGE. Staining for SOD and CAT was performed as described by Beauchamp and Fridovich (1971) and Woodbury et al. (1971), respectively.

2.7 Statistical evaluation of the results

The results obtained in this investigation were evaluated from at least three repeated experiments using three or five parallel runs. The statistical comparison between the controls and the treated cultures was determined by the Student's *t*-test for MIE (mean interval estimation) and by a one-way analysis of variance (ANOVA) followed by Dunnet's post-test, with a significance level of 0.05.

3. Results

Our previous study indicated that the optimal growth temperatures for both Antarctic strains, the psychrotolerant *Penicillium sp.* 161 and the mesophilic *A. glaucus* 363, are 20°C and 25°C, respectively (Tosi et al., 2010). The duration of the temperature shift (6 h) was chosen because this length of time was found to be sufficient to give a clear contrast between the control and the stressed cultures.

3.1 Response of the fungal growth and glucose consumption to temperature shifts

Mycelia of the Antarctic strains, which were grown until the middle of the exponential phase at optimal temperature, were shifted to colder temperatures, i.e., 4°C and 10°C. Figure 1 shows the growth curves of the model strains after the temperature downshift exposure and the subsequent restoration of the normal conditions.

Within the first 4 h of the beginning of the stress, the growth of the psychrotolerant strain *Penicillium sp.* 161 ceased and the biomass as measured by dry weight decreased sharply in comparison to the control (Fig. 1A). In the next 2 h, growth resumed and the return to the optimal 20°C after 6 h allowed the biomass to increase relative to the control levels. A similar trend was demonstrated for the Antarctic mesophilic strain, *A. glaucus* 363, after a shift from 25°C to 10°C or 4°C (Fig. 1B), but the difference in biomass production between the control and treated mycelia was more significant compared to the psychrotolerant strain. Moreover, after 4 h of recovery from either temperature treatment, biomass production was restored and the dry weight reached its pre-stress level.

The concentration of glucose in the culture medium was measured throughout the experiment (Fig. 2). The maximum glucose consumption occurred in cultures incubated at an optimal growth temperature was comparable to the stressed cultures.

A comparison of the model strains also shows that the consumption of glucose by the mesophilic strain (Fig. 2B) was faster than that of the psychrotolerant fungus (Fig. 2A). Comparing the curves in Fig. 2A and Fig. 2B, it was possible to verify that the downshift of temperature from the optimal temperature to 10°C or 4°C caused a significant decrease in glucose consumption. This tendency continued even after the return to the optimal temperature.

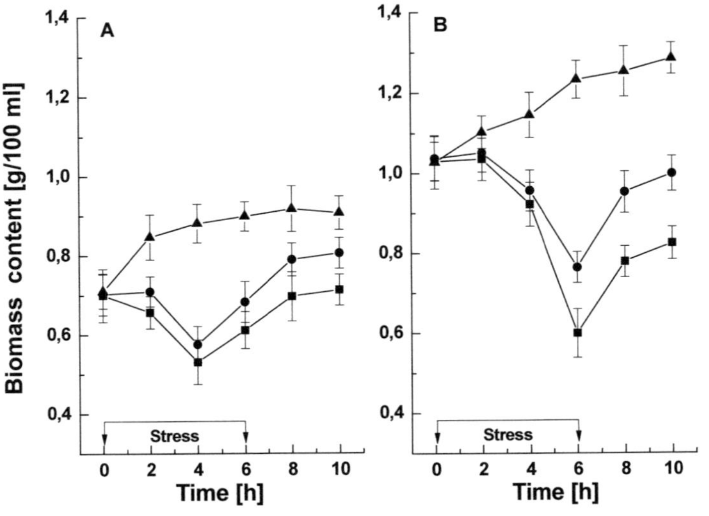

Fig. 1. Effect of cold shock on biomass production by *Penicillium sp.* 161 (A) and *A. glaucus* 363 (B) ▲ - growth at optimal temperature; ■ - downshift from optimal temperature to 4°C; ● - downshift from optimal temperature to 10°C. Bars represent SD of means. The effect of treatment was significant for the temperature treatment ($P \leq 0.05$)

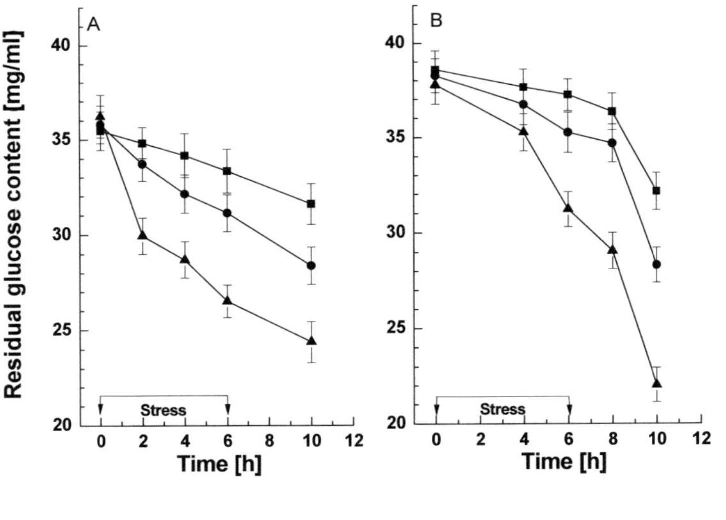

Fig. 2. Glucose consumption in *Penicillium sp.* 161 (A) and *A. glaucus* 363 (B) during cold shock and following recovery period. (▲) – growth at optimal temperature; (■)- downshift from optimal temperature to 4°C; (●) - downshift from optimal temperature to 10°C. Bars represent SD of means. The effect of treatment was significant for the temperature treatment ($P \leq 0.05$)

3.2 Effect of cold stress on ROS generation

To further characterise the fungal cell response to cold stress, the generation of ROS in both intact cells and mitochondrial fractions was determined. Table 1 shows the effect of decreased temperatures on the superoxide anion level in the model Antarctic strains. Elevated $O_2\bullet^-$ content was found for all of the temperatures tested. In the psychrotolerant strain, the temperature downshift to 10°C resulted in an increase in the $O_2\bullet^-$ level (approximately 116% and 124% for the intact cells and mitochondria, respectively), compared with the control.

The next diminution in the temperature to 4°C caused a more significant increase in the $O_2\bullet^-$ content in the intact cells (196%) and the mitochondrial fraction (143%). Furthermore, the mesophilic strain that was treated by cold stress to 10°C and 4°C showed extremely high $O_2\bullet^-$ levels, especially in the mitochondria (410% and 660% of the control, respectively). In both of the strains, the exposure to the above-mentioned low temperatures had a less pronounced induction in the H_2O_2 levels compared to the superoxide anion content (Table 2).

Variant	$O_2\bullet^-$ [µM/mg d.w. per h]			
	Penicillium sp. 161		*A. glaucus* 363	
	Intact cells	Mitochondrial fraction	Intact cells	Mitochondrial fraction
Control	3.59±0.03	5.24±0.03	4.70±0.02	0.55±0.01
10°C	4.15±0.04	6.50±0.01	6.78±0.01	2.25±0.01
4°C	7.05±0.18	7.47±0.05	11.1±0.07	3.63±0.02

Table 1. Increase in $O_2\bullet^-$ generation in the intact cells or mitochondrial fractions of *Penicillium* sp. 161 and *A. glaucus* 363 treated by low temperatures

Variant	H_2O_2 [mM/mg d.w. per h]			
	Penicillium sp. 161		*A. glaucus* 363	
	Intact cells	Mitochondrial fraction	Intact cells	Mitochondrial fraction
Control	5.68±0.18	12.11±0.10	11.39±1.10	6.40±0.28
10°C	6.22±0.40	15.00±0.37	13.72±1.09	9.79±0.33
4°C	9.18±0.11	21.57±0.44	14.61±0.56	11.58±0.43

Table 2. Increase in H_2O_2 generation in the intact cells or mitochondrial fractions of *Penicillium* sp. 161 and *A. glaucus* 363 treated by low temperatures

These results also demonstrated a more active process of ROS generation in the mitochondrial fractions than in the intact cells, and in the mesophilic *A. glaucus* 363 compared to the psychrotolerant *Penicillium* sp. 161.

3.3 Cold shock causes protein oxidation

The reaction of proteins with oxygen radicals leads to the appearance of carbonyl groups in polypeptide chains (Davies and Goldberg 1987). Thus, carbonyl formation is a marker for protein oxidation. We investigated whether the exposure of Antarctic strains to a temperature downshift causes oxidative damage to proteins during the stress period and in the post-stress recovery phase (Fig. 3). The level of protein carbonyl content in the control variants of both Antarctic strains did not change notably during the 6 h of stress. However, after the temperature downshift, the amount of carbonyl groups in the cell proteins increased with the duration of the exposure compared to the control. The increase was dependent on the grade of the temperature shift used in the experiment and less on the temperature requirements of the strains. Both of the strains demonstrated a sharp elevation in oxidatively damaged proteins compared to the control immediately after the start of the treatment.

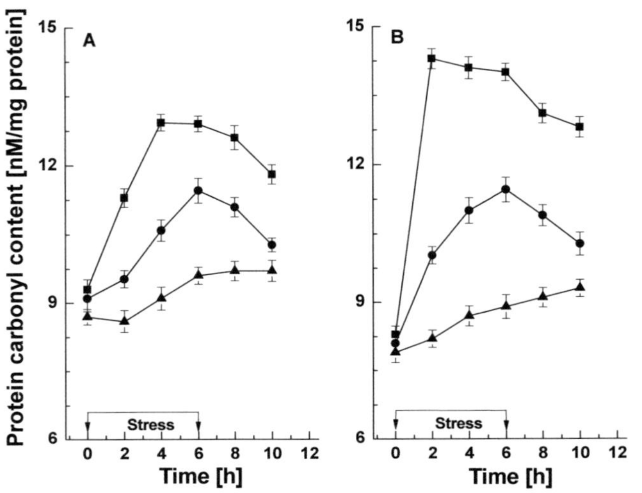

Fig. 3. Oxidative damage of proteins in *Penicillium* sp. 161 (A) and *A. glaucus* 363 (B), upon optimal temperature (▲) and upon temperature downshift from optimal to 4°C (■) or 10°C (●). Bars represent SD of means. The effect of treatment was significant for the temperature treatment ($P \leq 0.05$)

As seen in Fig. 3, the maximum values of the carbonyl content in the mesophilic strain during the downshift to 10ºC and 4ºC were 1.3- and 1.6-fold higher than in the control, respectively. A similar trend was found for the psychrotolerant strain (1.2- and 1.3-fold, respectively), but it showed a comparably lower sensitivity to cold temperatures. Although the content of the damaged proteins decreased significantly during the recovery phase, its value remained considerably higher than in the control variants.

3.4 Accumulation of reserve carbohydrates under conditions of cold shock

Another physiological consequence of low temperature exposure in both of the Antarctic fungi was the accumulation of reserve carbohydrates through stress. The production of glycogen and trehalose after a temperature downshift to 4°C or 10°C was detected. As is shown in Fig. 4, there was a sharp increase in the glycogen and trehalose content after the beginning of the stress. In the experiments with the psychrotolerant strain, *Penicillium* sp. 161, glycogen was accumulated up to 1.2 - 1.75 times the basal level and was accompanied by a 1.7-fold increase in trehalose content (Fig. 4A and 4B). It is noteworthy that higher initial levels of both of the reserve carbohydrates were detected in the mesophilic strain, *A. glaucus* 363. The stress conditions (4°C or 10°C) caused an additional sizeable rise in glycogen (1.4- and 1.3-fold, respectively) and trehalose (2.2- and 1.9-fold, respectively) content (Fig. 4C and 4D). The data demonstrated a clear dependence on both the degree of cold shock and the thermal characteristics of the model strain. Four hours after the removal of the temperature stress, the levels of trehalose and glycogen returned to the control values with the exception of glycogen at 4°C. In the recovery phase, both strains maintained the high glycogen amount that had accumulated during the stress conditions.

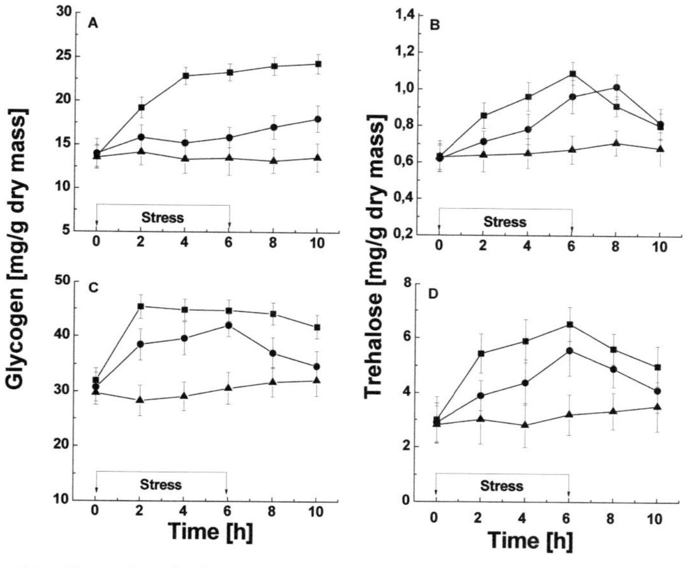

Fig. 4. Glycogen (A, C) and trehalose (B, D) content in *Penicillium* sp. 161 (A, B) and *A. glaucus* 363 (C, D) upon optimal temperature (▲) and upon temperature downshift from optimal to 4°C (■) or 10°C (●). Bars represent SD of means. The effect of treatment was significant for the temperature treatment ($P \leq 0.05$)

3.5 Changes in TPS and NT activities during cold stress conditions

Our results indicated that trehalose metabolism plays an important role in the response to cold stress (Fig. 5). In both strains used, the trehalose accumulation (see Fig. 4) paralleled a significant increase in the activity of TPS for up to 4 h, after which time the activity decreased (Fig. 5A and 5C). When *A. glaucus* 363 cells were exposed to a temperature shift to 4°C or 10°C, the TPS activity rapidly increased approximately 4- and 2.5-fold compared to the control culture, respectively. At the same conditions, *Penicillium* sp. 161 cultures demonstrated 2.4-fold higher activity than the control. The mesophilic strain demonstrated strong temperature shift-dependent changes in TPS activity, while no temperature dependence was observed in the psychrotolerant strain. Curiously, the enhanced levels of TPS in both of the Antarctic fungi coincided with an increase in NT activity (Fig. 5B and 5D). These increases were approximately 1.6- and 1.3-fold in the psychrotolerant fungus and 2.2-

and 1.7-fold in the mesophilic strain after cold shock at 4°C or 10°C, respectively, over the control value. We also found that the exposure to cold stress increased the NT levels in a time- and temperature shift-dependent manner. During the recovery phase, both strains showed a significant reduction in TPS activity, and the enzyme values reached those of the control cultures. In contrast, NT activity was maintained 4 h after the return to the optimal temperature.

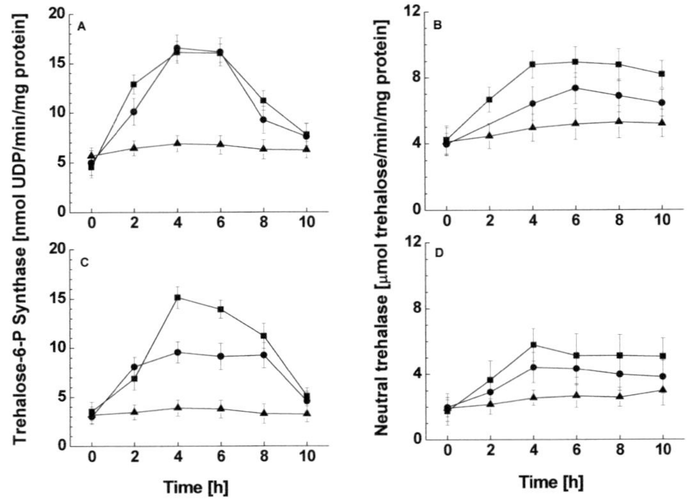

Fig. 5. Cold stress induced trehalose metabolism. TPS (A, C) and NT (B, D) accumulation in *Penicillium* sp. 161 (A, B) and *A. glaucus* 363 (C, D) upon optimal temperature (▲) and upon temperature downshift from optimal to 4°C (■) or 10°C (●). Bars represent SD of means. The effect of treatment was significant for the temperature treatment ($P \leq 0.05$)

3.6 Activation of antioxidant enzyme defence

When the mycelia, taken from middle of the exponential phase at the optimal temperature, were subjected for 6 h to 4°C or 10°C, the antioxidant enzyme defence was activated. The time courses of SOD activities for both of the strains during the cold shock treatment are shown in Fig. 6.

The results showed that transient exposure to 4°C and 10°C resulted in elevated total superoxide scavenging activity compared with the control variant, and the increase was in a dose- and time-dependent manner until 10 h, i.e., 4 h after the temperature recovery. The mesophilic strain had an approximately 7-fold higher SOD activity than the psychrotolerant strain. Moreover, the antioxidant cell response of *Penicillium* sp. 161 included an approximately 1.4- and 1.2- fold increase in SOD activity compared with the control culture at 4°C and 10°C, respectively (Fig. 6A). The fungal cultures of *A. glaucus* 363 also demonstrated a temperature-dependent response that was higher at 4°C compared with the condition at 10°C. The maximum increase was approximately 1.6- and 1.3-fold in the variants that were exposed to 4°C and 10 °C, respectively (Fig. 6B).

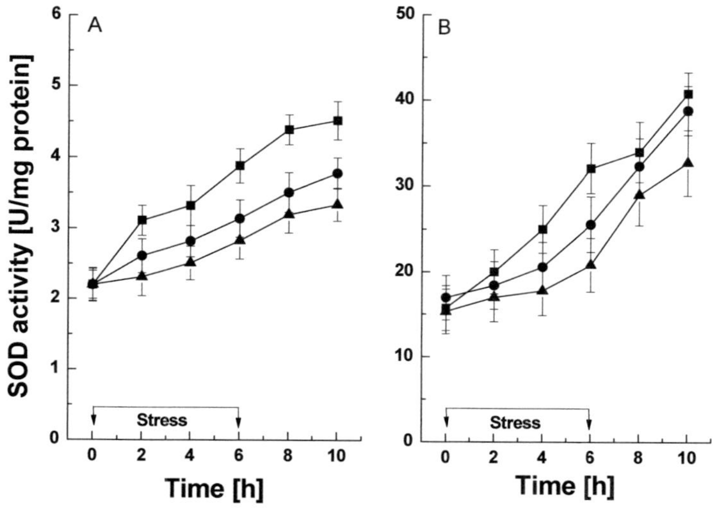

Fig. 6. SOD activities in cultures of *Penicillium* sp. 161 (A) and *A. glaucus* 363 (B) upon optimal temperature (▲) and upon temperature downshift from optimal to 4°C (■) or 10°C (●). Bars represent SD of means. The effect of treatment was significant for the temperature treatment ($P \leq 0.05$)

There were significant differences in the expression of Mn-SOD and Cu/Zn-SOD among strains of both the psychrotolerant and mesophilic thermal classes (Fig. 7). Under optimal conditions, the psychrotolerant strain produced mostly Mn-SOD and less Cu/Zn-SOD (Fig. 7A), while in the mesophilic fungus, the Cu/Zn-SOD content was found to be significantly higher than Mn-containing enzyme (Fig. 7B). The cold shock treatment caused an enhanced level of both isoenzymes, but the increased total SOD activity was due to the Cu/Zn-SOD isoform for both strains.

To confirm the levels of Mn-SOD and Cu/Zn-SOD activity in the fungal cultures, the enzymatic activity of these antioxidant enzymes was determined using the native gel electrophoresis technique (Fig. 8). The non-denaturing PAGE showed that the cell-free extracts of the psychrophilic strain contained two distinct bands with SOD activity, Cu/Zn- and Mn-SOD. At least three clear SOD isoforms were observed after cell-free extract proteins of the mesophilic strain were electrophoresed on a 10% native PAGE.

These isoforms were identified as one Mn-SOD and two Cu/Zn-SOD, which were named I and II in order of increasing migration. The temperature treatment did not change the isoenzyme profile of either strain. At the same time, there was a temperature degree-dependent increase in Cu/Zn- and Mn-containing SOD (Figs. 8B and 8C).

Figure 9 presents the data on the changes in CAT activity. In contrast to SOD, the psychrophilic strain showed a higher CAT level (1.5-fold) than that of the mesophilic strain.

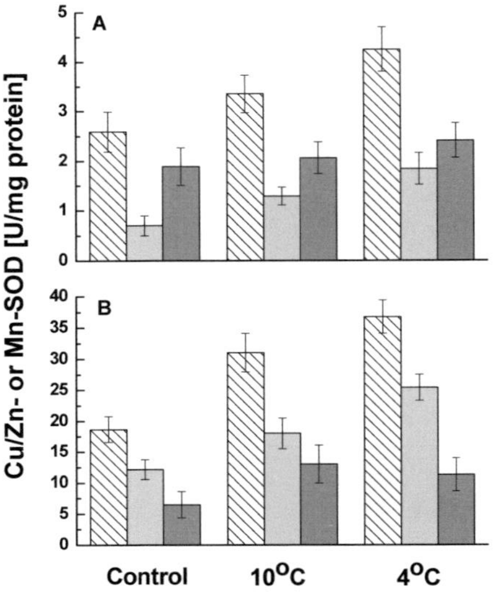

Fig. 7. Effect of low temperatures on isoenzyme profiles of *Penicillium* sp. 161 (A) and *A. glaucus* 363 (B) 4 h after the treatment: total SOD (hatchet bars); Cu/Zn-SOD (light grey bars); Mn-SOD (dark grey bars). Bars represent SD of means. The effect of treatment was significant for the temperature treatment (P ≤ 0.05).

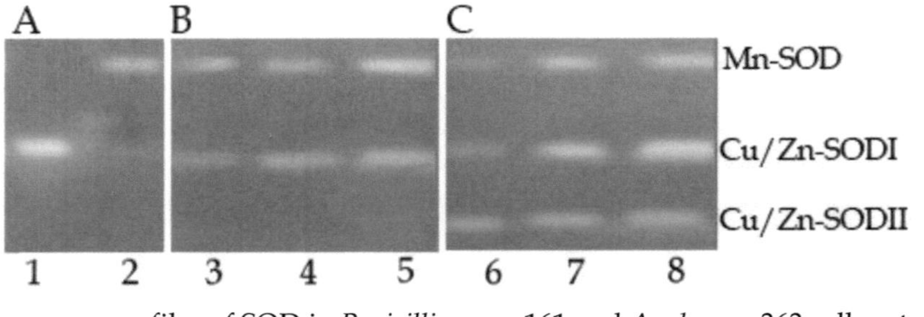

Fig. 8. Isoenzyme profiles of SOD in *Penicillium* sp. 161 and *A. glaucus* 363 cells : standards (A); SOD activity in *Penicillium* sp. 161 (B) and in *A. glaucus* 363 (C) cells evaluated by polyacrylamide gel electrophoresis (10% gel): lane 1, standard Cu/Zn-SOD from bovine erythrocytes; lane 2, standard Mn-SOD from *Escherichia coli*; lanes 3, 4, and 5, SOD in the *Penicillium* sp. 161 sp cells; lanes 6, 7 and 8, SOD in the *A. glaucus* 363 cells, cultivated at control temperatures (lanes 3 and 6), at downshift to 10°C (lanes 4 and 7) and downshift to 4°C.

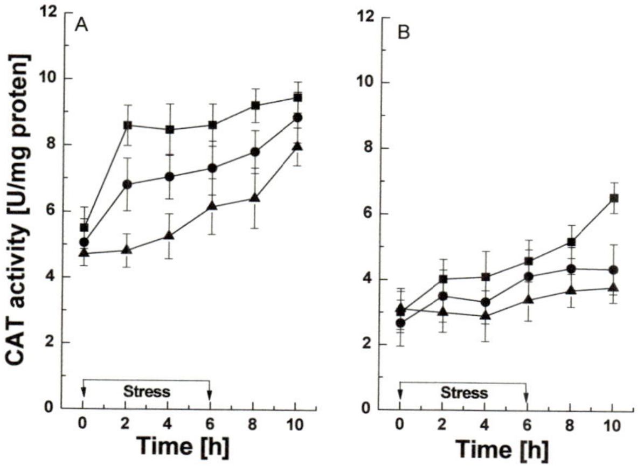

Fig. 9. CAT activities in cultures of *Penicillium* sp. 161 (A) and *A. glaucus* 363 (B) upon optimal temperature (▲) and upon temperature downshift from optimal to 4°C (■) or 10°C (●). Bars represent SD of means. The effect of treatment was significant for the temperature treatment ($P \leq 0.05$)

The low-temperature treatment resulted in a significant increase in CAT activity. The trend of the change was similar to that for SOD (a dose- and time-dependent manner). In all of the variants, the enzyme level exceeded the control value. The highest percentage of increased CAT activity (approximately 2-fold) was found in the psychrotolerant strain after the temperature downshift from 20°C to 4°C. The overexpression of CAT under the cold shock conditions was confirmed by PAGE analysis (Fig. 10). When the protein extracts were separated by native electrophoresis, one CAT isoenzyme was observed in both *Penicillium* sp. 161 and *A. glaucus* 363.

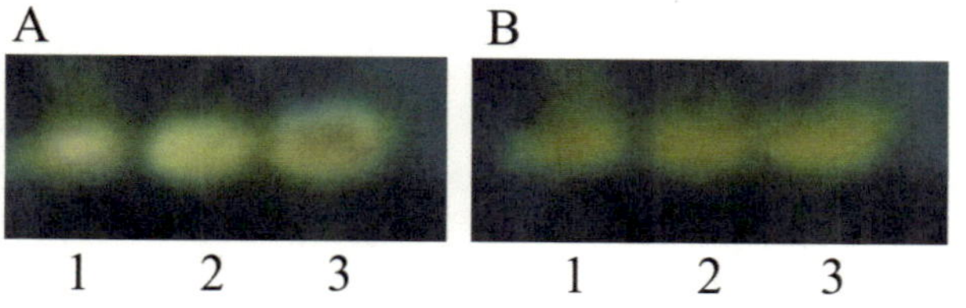

Fig. 10. CAT activity in cultures of *Penicillium* sp. 161 (A) and *A. glaucus* 363 (B) evaluated by polyacrylamide gel electrophoresis (10% gel): lane 1, Control; lane 2, cultures treated with 10°C; lane 3, cultures treated with 4°C

It is noteworthy that the tendency of increased SOD and CAT activities compared to the control variant continued even after the return to the optimal temperature.

4. Discussion

The evaluation of oxidative stress events during low-temperature treatments provides information on the relationship between oxidative stress and cold shock in lower eukaryotes such as filamentous fungi. Very little information is available on the oxidative cell response of microorganisms from cold habitats (Gocheva et al., 2006, 2009; Tosi et al., 2010; Chattopadhyay et al., 2011; Nichols et al., 1999). Even less is known concerning filamentous fungi from the Bulgarian Antarctic area. We choose to compare the cell responses between fungi belonging to two different thermal classes, psychrotolerant and mesophilic, that are able to grow in a wide range of temperatures. According to many reports on fungal biodiversity, these fungi are prevalent in Antarctica (Onofri et al., 2007, 2008; Arenz & Blanchette, 2011). Our results confirmed that the downshift of temperature induces typical events of oxidative stress. The exposure of both fungal cultures to 4°C or 10°C caused a statistically significant reduction in biomass production and glucose consumption compared to the control. Despite their genetic adaptation to survive in the harsh Antarctic environment, the sharp decrease in temperature resulted in a cessation of growth, which was more pronounced in the mesophilic fungus than in the psychrophilic fungus. A possible explanation or cause for this discrepancy is that most of the Antarctic mycoflora is metabolically active whenever a combination of favourable abiotic conditions occurs during the short growing summer season (Ruisi et al., 2007). However, Hébraud M. & Potier (1999) assumed that, as opposed to mesophiles, one or more regulatory factors exists in cold-adapted microorganisms prior to cold shock that allows for the maintenance of functional translational machinery at low temperatures. However, despite this continuous protein synthesis, the growth of psychrotrophic strains ceases transiently after a cold shock, and additional regulatory mechanisms may exist that allow for growth resumption at low temperatures.

Additionally, the data for all cold shock variants reveal enhanced ROS generation. These ROS can be generated via a variety of extra-environmental conditions, including low temperature exposure (Belozerskaya & Gessler 2007; Gessler et al., 2007; Gómez-Toribio et al., 2009; Gocheva et al., 2009; Krumova et al., 2009). As is known, the unstressed fungal cells produce $O_2{\bullet}^-$ and H_2O_2, presumably due to a single electron reduction of 2% of the consumed oxygen as has been previously suggested (Joseph-Horne et al., 2001). Our direct assay showed a significant increase in $O_2{\bullet}^-$ under cold exposure to 4°C and 10°C in dose-dependent manner. A similar but less-pronounced tendency was found for H_2O_2 levels. This increase was more extreme in experiments with mitochondria than in those with intact cells. Thus, cold stress likely imposes an oxidative burden of which $O_2{\bullet}^-$ is a major component. A similar direct analysis of ROS content in fungal cells has not often been reported. Chattopadhyay et al. (2011) reported that the production of free radicals in the Antarctic bacterium *Pseudomonas fluorescens* MTCC 667 was higher at low temperature compared to that at 22°C. When cells are exposed to low temperature, the rate of enzymatic reactions decreases, leading to a decrease in the demand for ATP and an accumulation of electrons at certain points in the respiratory chain. The situation promotes a sudden increase in the production of a number of ROS, which remove the burden of the excess reducing potential (Chattopadhyay, 2002). According to Zhang et al. (2003), the growth temperature downshift increases ROS levels in *Saccharomyces cerevisiae.* The authors reported that H_2O_2 is a major component in low temperature-exposed yeast cells. In contrast to heterotrophic organisms, most plants possess photosynthetic systems that, when out of control, may produce a lot of ROS and other reactive species (Lushchak, 2011). According to Hu et al.(2008), cold-stress inhibited net photosynthetic rate (P_N) and cytochrome respiratory pathway but enhanced the photosynthetic electron flux to O_2 and over-reduction of respiratory electron transport chain, resulting in ROS accumulation in cucumber leaves. For fish living at low temperatures, the increased polyunsaturation of mitochondrial membranes should raise rates of mitochondrial respiration which would in turn enhance the formation of ROS, increase proton leak and favour peroxidation of these membranes (Guderley, 2004).

Consistent with previous reports (Şahin & Gümüşlü, 2004; Manfredini et al., 2005; Nyström, 2005; Li et al., 2009), the present study demonstrated the coexistence of enhanced ROS activity and oxidatively damaged proteins in Antarctic fungal models. ROS can modify proteins leading to the formation of additional carbonyl groups caused by the oxidation of specific amino acid residues or the protein backbone (Davies & Goldberg, 1987). The level of proteins with formed additional carbonyl groups, which is thought to be an excellent marker for protein oxidation, has been shown to be related to oxidative stress in various studies. Low temperature stress was shown to induce ROS accumulation in cells (Suzuki & Mittler, 2006; Ouellet, 2007), which leads to a production of oxidised proteins. A similar response was observed in hyper-oxygenated cultures of white-rot fungus *Phanerochaete chrysosporium* (Belinky et al., 2003), heavy metal treatment (Krumova et al., 2009; Belozerskaya et al., 2010), addition of oxidant species (Angelova et al., 2005) and temperature exposure (Kim et al., 2006; Li et al., 2008; Gocheva et al., 2009), for example. In the temperature-induced protein carbonylation, Li et al. (2008) speculated that respiratory enzymes and those within mitochondria may be especially vulnerable because oxygen-derived free radicals from the respiratory chain are generally held to be the main source of the oxidative damage seen during stress.

The psychrotolerant strain *Penicillium* sp. 161 demonstrated a lower level of oxidatively damaged proteins than that in the mesophilic strain *A. glaucus* 363 during the exposure to 4°C or 10°C. At the same time, our unpublished data demonstrate the availability of cold-acclimation proteins (Caps) in *Penicillium* sp. 161 and cold shock proteins (Csps) in *A. glaucus* 363. According to D'Amico et al. (2006), the existing distinctions between the mesophilic and the psychrophilic cold shock response include the lack of repression of housekeeping protein synthesis and the presence of Caps in psychrophiles. The authors considered that many of the Csps involved in various cellular processes, such as transcription, translation, protein folding and the regulation of membrane fluidity and observed in mesophiles (D'Amico et al., 2006, as cited in Phadtare, 1994), act as Caps in psychrophiles, and are constitutively rather than transiently expressed at low temperatures. Furthermore, this differential regulation of expression indicates that a temperature-sensory system exists in psychrophiles, and thermosensors at the cell membrane level, which sense changes in fluidity, have been reported (D'Amico et al., 2006, as cited in Ray et al., 1994).

Trehalose and glycogen accumulation in microbial cells when they are subjected to unfavourable growth conditions could also be used as stress indicators (Parrou et al., 1997; Türkel, 2006; Ocón et al., 2007; Iturriaga et al., 2009; Gonçalves et al., 2011). Both carbohydrates are important storage compounds in fungal vegetative cells and spores (Robinson, 2001; Rúa et al., 2008). In Antarctic fungi, *Penicillium* sp. 161 and *A. glaucus* 363, that were exposed to cold stress, a reduction in glucose consumption coincided with a significant increase in the trehalose and glycogen content (Fig. 4). These results agree with earlier studies regarding the microbial response against different types of abiotic stress (Ostrovskii et al., 2003; Jules et al., 2008; Robinson, 2001). Our previous studies have confirmed the role of these reserve carbohydrates in the survival strategies of filamentous fungi under stress conditions such as heavy metal toxicity (Krumova et al., 2009), heat shock (Abrashev et al., 2008), oxidative stress-agent treatment (Angelova et al., 2005) and low temperature exposure (Gocheva et al., 2009). In contrast, heat stress led to a significant decrease in the glycogen and trehalose content of *Pichia anomala* (Parrou et al., 1997) and *P. angusta* cells (Türkel, 2006), respectively. However, the nitrogen starvation-induced biosynthesis and the accumulation of trehalose and glycogen in the same strain, *P. angusta*, suggested that both stress responses do not have an overlapping signal transduction pathway in the activation of trehalose and glycogen biosynthesis in this *Pichia* species (Türkel, 2006). Jules et al. (2008) confirmed an interconnection between trehalose and glycogen in yeast cell carbon storage management. They found that glycogen was accumulated earlier and faster, which indicated the presence of a fine-tuning control during periods of stress. In contrast, our results showed that both reserve carbohydrates started to increase simultaneously, immediately after the temperature downshift (Fig. 4), and glycogen maintained its higher level after the recovery to the optimal temperature longer than trehalose did.

Considering that almost all of the genes encoding the enzymes involved in the metabolism of these two reserve carbohydrates contain between one and several copies of the stress-responsive element (STRE), Parrou et al. (1997) suggested the possibility of a link between the potential transcriptional induction of these genes and the accumulation of glycogen and trehalose under different stress conditions. In this study, we evaluated the activities of the trehalose-metabolising enzymes TPS and NT during cold stress conditions. In the used Antarctic strains, cold shock elicits a complex response that involves the activation of both biosynthesis and biodegradation pathways, but not to the same extent. The increase in TPS activity significantly exceeds that of NT activity. This situation leads to a high level of accumulation of trehalose during low temperature exposure. Thus, the intracellular level of trehalose is the result of a well-regulated balance between enzymatic synthesis and degradation. In *Saccharomyces cerevisiae*, an enhanced level of trehalose was the result of the stimulation of trehalose synthase and the inhibition of trehalase by high temperature (Parrou et al., 1997). Moreover, after the stress had ceased, trehalose concentrations in our experiments returned to basal levels, indicating that neutral trehalase activity could have a role in cold shock recovery.

It is well known that the antioxidant systems of microorganisms, including fungi, act as important tolerance mechanisms against abiotic stress. Enhanced SOD and CAT activities have been associated with the induced resistance of fungi to different stress factors (Manfredini et al., 2005; Li et al., 2009; Belozerskaya et al., 2010; Chattopadhyay et al., 2011). Both activities are involved in superoxide radical and hydrogen peroxide scavenging. Our results demonstrated that the exposure to cold shock for 6 h induced the expression of SOD and CAT in a psychrotolerant strain and in a mesophilic strain.

Furthermore, the increase in SOD activity was more pronounced for the mesophilic fungus, while the activation of antioxidant enzyme defence in the psychrotolerant *Penicillium* was primarily due to an increase in CAT activity. One possible explanation for this finding is a much higher level of $O_2{\bullet}^-$ generated upon low temperature stress in the mesophilic strain and higher levels of H_2O_2 in psychrotolerant cultures (Table 1 and Table 2). To antagonise ROS, *A. glaucus* 363 induces SOD synthesis to a much higher degree, while *Penicillium* sp. 161 enhances CAT levels.

The observations by PAGE analysis suggest that in *A. glaucus* 363, Cu/Zn-SODI is primarily involved in the fungal line of defence under the conditions of temperature downshift. In contrast, Mn-SOD activity was considerably higher than Cu/Zn-SODI in *Penicillium* sp. 161, possibly because of a higher level of $O_2{\bullet}^-$, which was generated in mitochondria during the cold stress (Table 1). Moreover, the PAGE results confirmed the significant role of CAT induction in the cold-stress response, especially in the psychrotolerant strain (Fig. 10).

5. Conclusions

Our results provide further evidence indicating a relationship between cold shock and oxidative stress. Growth at low temperature clearly induced oxidative stress events in both of the Antarctic fungal strains tested, which consisted of enhanced levels of oxidatively damaged proteins, the accumulation of reserve carbohydrates and increased activity of the antioxidant enzyme defence. Despite the significant induction of antioxidant enzymes, exposure to low temperature is damaging and appears to exceed the antioxidant defence. Although the tested strains were isolated from Antarctic soil samples, they both demonstrated a different degree of oxidative stress damage and strategy of antioxidant defence. The psychrotolerant strain *Penicillium* sp. 161. more successfully confronted the challenge of low temperature, which slows metabolite flux, than the mesophilic *A. glaucus* 363. As shown by Methé et al. (2005), comparative genome analyses suggests that the psychrophilic lifestyle is most likely conferred not by a unique set of genes, but by a collection of synergistic changes in the overall genome content and amino acid composition.

Analysis of the relationship between oxidative stress and cold stress in different thermal classes of Antarctic microorganisms will improve our understanding, at the molecular level, of the strategies and mechanisms that facilitate fungal cell survival in harsh environments, and the necessary foundation for practical applications. Future study using comparative molecular analysis of genes coding for proteins with antioxidant enzyme activity in psychrotolerant and mesophilic strains by the bioinformatics approach will enhance the knowledge of cold stress signaling mechanisms in eukaryotic microorganisms. From a biotechnological point of view, new information of cold stress response of Antarctic fungi can help further our efforts to develop an effective technology for production of cold-active enzyme SOD. Such SOD can be very important antioxidant in cryopreservation processes (Rossi et al., 2001), cryosurgery and organ transplantation (Sun et al., 2006), *in vitro* fertilization (Agarwal et al., 2006) etc. Cold-active SOD can be used in all cosmetic formulations, to promote, at least, younger looking skin (Lods et al., 2000; Diehl, 2009).

6. Acknowledgment

This work was supported by the NCSI of the Ministry of Education and Science, Bulgaria (grant DO02-172/08 and grant VU-B-205/06) to which sincere thanks are owed.

7. References

Abrashev, R.I., Pashova, S. B., Stefanova, L.N., Vassilev, S.V., Dolashka-Angelova, P.A. & Angelova, M.B. (2008). Heat-shock-induced oxidative stress and antioxidant response in *Aspergillus niger* 26. *Canadian Journal of Microbiology*, Vol. 54, No. 12, (December 2008), pp. 977-983, ISSN 1480-3275

Adachi, H. & Ishii, N. (2000). Effects of tocotrienols on life span and protein carbonylation in Caenorhabditis elegans. *Journals of Gerontology Series A, Biological Sciences and Medical Sciences*, Vol. 55, No. 6, (June 2000), B280–B285, ISSN 1079-5006

Angelova, M., Genova, L., Pashova, S., Slokoska, L. & Dolashka, P. (1996). Effect of cultural conditions on the synthesis of superoxide dismutase by *Humicola lutea* 110. *Journal of Fermentation and Bioengineering*, Vol. 82, No. 5, (July 1996), pp. 464-468, ISSN 0922-338X

Angelova, M.B., Pashova, S.B., Spasova, B.K., Vassilev, S.V. & Slokoska, L.S. (2005). Oxidative stress response of filamentous fungi induced by hydrogen peroxide and paraquat. Mycological Research, Vol. 109, No. 2, (February 2005), pp. 150–158, ISSN 0953-7562

Agarwal A., Gupta S. & Sikka S. (2006). The role of free radicals and antioxidants in reproduction. *Current Opinion in Obstetrics and Gynecology*, 2006, Vol. 18, № 3 (June 2006) pp. 325–332, ISSN 1040872X - 1040-872X

Arenz, B.E. & Blanchette, R.A. (2011). Distribution and abundance of soil fungi in Antarctica at sites on the Peninsula, Ross Sea Region and McMurdo Dry Valleys. *Soil Biology and Biotechnology*, Vol. 43, № 2, (February 2011), pp. 308-315, ISSN 0038-0717

Beauchamp, C. & Fridovich, I. (1971). Superoxide dismutase: Improved assay and an assay applicable to polyacrylamide gels. *Analytical Biochemistry*, Vol. 44, No. 1, (November 1971), pp. 276-287, ISSN 0003-2697

Becker, J. (1978). A method of glycogen determination in whole yeast cells. *Analytical Biochemistry*, Vol. 86, No. 1, (May 1978), pp. 56-64, ISSN 0003-2697

Beers, R.F. & Sizer, I.W. (1952). A spectrophotometric method for measuring the breakdown of hydrogen peroxide by catalase. *Journal of Biological Chemistry*, Vol. 195, No. 1, (March 1952), pp. 133-140, ISSN 0021-9258

Belinky, P.A., Flikshtein, N., Lechenko, S., Gepstein, S. & Dosoretz, C. G. (2003). Reactive oxygen species and induction of lignin peroxidase in *Phanerochaete chrysosporium*. Applied and Environmental Microbiology, Vol. 69, No. 11, (November 2003), pp. 6500–6506, ISSN 0099-2240

Belozerskaya, T.A., & Gessler, N.N. (2007). Reactive Oxygen Species and the Strategy of Antioxidant Defense in Fungi: *A Review. Applied and Biochemistry and. Microbiology*, Vol. 43, No. 5, (September 2007), pp. 506–515, ISSN 0003-6838

Belozerskaya, T., Aslanidi, K., Ivanova, A., Gessler, N., Egorova, A., Karpenko, Yu. & Olishevskaya, S. (2010). Characteristics of Extremophylic Fungi from Chernobyl Nuclear Power Plant. In: *Current Research, Technology and Education Topics in Applied Microbiology and Microbial Biotechnology*, Mendez-Vilas A. (*ed.*), 88-94, Vol. 1, Formatex Research Center, *ISBN* 978-84-614-6194-3, Badajoz, Spain

Chattopadhyay, M. K. (2002). Low temperature and oxidative stress. *Current Science*, Vol. 83, No. 2, (July 2002), pp. 109, ISSN 0011-3891

Chattopadhyay, M. K., Raghu, G., Sharma, Y. V. R. K., Biju, A.R., Rajasekharan, M.V. & Shivaji, S. (2011). Increase in Oxidative Stress at Low Temperature in an Antarctic Bacterium. *Current Microbiology*, Vol. 62, No. 2, (February 2011), pp. 544–546, ISSN 1432-0991

Chintalapati, S., Kiran, M.D. & Shivaji, S. (2004). Role of membrane lipid fatty acids in cold adaptation. *Cellular and Molecular Biology*, Vol. 50, No. 5, (July 2004), pp. 631-642, ISSN 1165-158X

Collins, T., Roulling, F., Piette, F., Marx, J. C., Feller, G., Gerday, C. & D'Amico, S. (2008). Fundamentals of Cold-Adapted Enzymes. In: *Psychrophiles: from Biodiversity to Biotechnology*, Margesin R., Schinner F., Gerday C., Marx J.C. (Eds), 211-227, Springer-Verlag, ISBN 978-3-540-74334-7, Berlin, Heidelberg

D'Amico, S., Collins, T., Marx, J., Feller, G., & Gerday, C. (2006). Psychrophilic microorganisms: challenges for life. *European Molecular Biology Organization Reports*, Vol. 7, No. 4, (April 2006), pp. 385 - 389, ISSN 1469-221X

Davies, K.J.A. & Goldberg, A.L. (1987). Proteins damaged by oxygen radicals are rapidly degraded in extracts of red blood cells. *Journal of Biological Chemistry*, Vol. 262, No. 17, (June 1987), pp. 8227-8234, ISSN 0021-9258

Diehl C. (2009). Use of a Mixture of Superoxide Dismutase and Catalase for Treating Inflammatory Skin Lesions. United States Patent Pub. № US 2009/0092591

El-Bashiti, T., Hamamcı, H., Öktem, H.A. & Yücel, M. (2005). Biochemical analysis of trehalose and its metabolizing enzymes in wheat under abiotic stress conditions. *Plant Science*, Vol. 169, No. 1, (July 2005), pp. 47–54, ISSN 0168-9452

Feller, G. & Gerday, C. (2003). Psychrophilic enzymes: Hot topics in cold adaptation. *Nature reviews. Microbiology*, Vol. 1, No. 3, (December 2003), pp. 200-208, ISSN 1740-1526

Fridovich, I. (1998). Oxygen toxicity, a radical explanation. *The Journal of Experimental Biology.*, Vol. 201, No. 8, (April 1998), pp. 1203-1209, ISSN 0022-0949

García-Arribas, O., Mateo, R., Tomczak, M.M., Davies, P.L. & Mateu, M.G. (February 2007). Thermodynamic stability of a cold-adapted protein, type III antifreeze protein, and energetic contribution of salt bridges. In: *Protein Science*, Vol. 16, No. 2, (February 2007), pp. 227–238, ISSN 1469-896x

Gatti-Lafranconi, P., Natalello, A., Rehm, S., Doglia, S.M., Pleiss, J. & Lotti, M. (2010). Evolution of stability in a cold-active enzyme elicits specificity relaxation and highlights substrate-related effects on temperature adaptation. Journal of Molecular Biology, Vol. 395, No. 1, pp. 155-166, ISSN 0022-2836

Gessler, N.N., Averýanov, A.A. & Belozerskaya, T.A. (2007). Reactive oxygen species in regulation of fungal development. *Biochemistry (Mosc)*, Vol. 72, No. 10, (October 2007), pp. 1091-109, ISSN 0006-2979

Gocheva, Y., Krumova, E., Slokoska, L., Gesheva, V. & Angelova, M. (2005). Isolation of filamentous fungi from Antarctica. *Comptes-rendus de l'Académie Bulgare des Sciences*, Vol. 58, No. 4, pp. 403-408, ISSN 1310-1331

Gocheva, Y., Krumova, E., Slokoska, L., Miteva, J. & Angelova, M. (2006). Cell response of Antarctic and temperate strains of *Penicillium* spp. to different growth temperature. *Mycological Research*, Vol. 110, No. 11, (November 2006), pp. 1347-1354, ISSN 0953-7562

Gocheva, Y., Tosi, S., Krumova, E., Slokoska, L., Miteva, J., Vassilev, S. & Angelova, M. (2009). Temperature downshift induced antioxidant response in fungi isolated from Antarctica. *Extremophiles*, Vol. 13, No. 2, (March 2009), pp. 273-281, ISSN 1431-0651

Gómez-Toribio, V., García-Martín, A.B., Martínez, M.J., Martínez, Á.T. & Guillén, F. (2009). Induction of Extracellular Hydroxyl Radical Production by White-Rot Fungi through Quinone Redox Cycling. *Applied and Environmental Microbiology,* Vol. 75, No. 12, (June 2009), pp. 3944–3953, ISSN 0099-2240

Gonçalves, R.D., Cupertino, F.B., Freitas, F.Z., Luchessi, A.D. & Bertolini, M.C. (2011). A genome-wide screen for Neurospora crassa transcription factors regulating glycogen metabolism. *Molecular and Cellular Proteomics,* (July 2011) ISSN: 1535-9484 MCP Papers in Press. Published on July 18, 2011 as Manuscript M111.007963 DOI: 10.1074/mcp.M111.007963

Gray, N.F., & Lewis-Smith, R.I. (1984). The distribution of nematophagous fungi in the maritime Antarctic. *Mycopathologia,* Vol. 85, No. 1-2, (March 1984), pp. 81-92, ISSN 0301-486X

Guderley H. (2004). Metabolic responses to low temperature in fish muscle. *Biological Reviews,* Vol. 79, № 2, (May 2004), pp. 409-427, ISSN 1464-7931

Hart, P.J., Belbirine, M.M., Ogihara, N.L., Nersissian, A.M., Weiss, M.S., Valentine, J.S. & Eisenberg, D (1999). A structure-base mechanism for cooper-zinc superoxide dismutase. Biochemistry, 38 (7), (February 1999), pp. 2167-2178, ISSN: 0006-2960.

Hassan, H.M. & Fridovich, I. (1979). Intracellular production of superoxide radical and of hydrogen peroxide by redox active compounds. *Archives of Biochemistry and Biophysics,* Vol. 196, No. 2, (September 1979), pp. 385-395, ISSN 0003-9861

Hébraud, M. & Potier, P. (1999). Cold Shock Response and Low Temperature Adaptation in Psychrotrophic Bacteria. *Journal of Molecular Microbiology and Biotechnology,* Vol. 1, No. 2, (November 1999), pp. 211-219, ISSN 1464-1801

Horn, G., Hofweber, R., Kremer, W. & Kalbitzer, H.R. (2007). Structure and function of bacterial cold shock proteins. *Cellular and Molecular Life Sciences,* Vol. 64, No. 12, (April 2007), pp. 1457–1470, ISSN 1420-682X

Hu W.H., Song X.S., Shi K., Xia X.J., Zhou Y.H. & Yu J.Q. (2008). Changes in electron transport, superoxide dismutase and ascorbate peroxidase isoenzymes in chloroplasts and mitochondria of cucumber leaves as influenced by chilling. *Photosynthetica,* Vol. *46,* No. 4, (December 2008), pp. 581-588, ISSN 0300-3604

Inouye, M. (1999). Cold-Shock Response and Adaptation. *Journal of Molecular Microbiology and Biotechnology,* Vol. 1, No. 2, (November *1999*), pp. 191, ISSN 1464-1801

Iturriaga, G., Suárez, R. & Nova-Franco, B. (2009). Trehalose Metabolism: From Osmoprotection to Signaling. *International Journal of Molecular Sciences,* Vol. 10, No. 9, (September 2009), pp. 3793-3810, ISSN 1422-0067

Joseph-Horne, T., Hollomon, D.W., & Wood, P.M. (2001). Fungal respiration: a fusion of standard and alternative components. *Biochimica et biophysica acta,* Vol. 1504, No. 2-3, (April 2001), pp. 179-195, ISSN 0006-3002

Jules, M., Beltran, G., François, J. & Parrou, J.L. (2008). New Insights into Trehalose Metabolism by *Saccharomyces cerevisiae*: *NTH2* Encodes a Functional Cytosolic Trehalase, and Deletion of *TPS1* Reveals Ath1p-Dependent Trehalose Mobilization. *Applied and Environmental Microbiology,* Vol. 74, No. 3, (February 2008), pp. 605–614, ISSN 0099-2240

Kim, I. S., Moon, H. Y., Yun ,H. S. & Jin, I. (2006). Heat shock causes oxidative stress and induces a variety of cell rescue proteins in Saccharomyces cerevisiae KNU5377. *Journal of microbiology*, Vol. 44, No. 5, (October 2006), pp. 492-501, ISSN 1225-8873

Krumova, E., Dolashki, A., Pashova, S., Dolashka-Angelova, P., Stevanovic, S., Hristova, R., Stefanova, L., Voelter, W. & Angelova, M. (2008). Unusual location and characterization of Cu/Zn-containing superoxide dismutase from filamentous fungus *Humicola lutea*. *Archives of Microbiology*, Vol. 189, No. 2, (February 2008), pp. 121-30, ISSN 0302-8933

Krumova, E.Z., Pashova, S.B., Dolashka-Angelova, P.A., Stefanova, T. & Angelova, M.B. (2009). Biomarkers of oxidative stress in the fungal strain *Humicola lutea* under copper exposure. *Process Biochemistry*, Vol. 44 No. 3, (March 2009), pp. 288-295, ISSN 1359-5113

Kwak, K.J., Park, S.J., Han, J.H., Kim, M.K., Han O.S., Han, Y.S. & Kang, H. (2011). Structural determinants crucial to the RNA chaperone activity of glycine-rich RNA-binding proteins 4 and 7 in Arabidopsis thaliana during the cold adaptation process. *Journal of Experimental Botany*, Vol. 62, No. 11, (April 2011), pp. 4003-4011, ISSN 0022-0957

Li, Q., Harvey, L.M. & McNeil, B. (2008). The effects of elevated process temperature on the protein carbonyls in the filamentous fungus, *Aspergillus niger* B1-D. *Process Biochemistry*, Vol. 43, No. 8, (August 2008), pp. 877-881, ISSN 1359-5113

Li, Q., Harvey, L.M. & McNeil, B. (2009). Oxidative stress in industrial fungi. *Critical Review in Biotechnology*, Vol. 29, No. 3, (September 2009), pp. 199-213, ISSN 0738-8551

Lods L.M., Dres C., Johson C., Scholz D.B. & Brooks G.J. (2000). The future of enzymes in cosmetics *International Journal of Cosmetic Science*, Vol. 22, № 2, (April 2000), pp. 85-94, ISSN 0142-5463

Lowry, O.H., Rosenbrough, H.J., Faar, A.L. & Randall, R.J. (1951). Protein measurement with the Folin phenol reagent. *Journal of Biological Chemistry*, Vol. 193, No. 1, (November 1951), pp. 265-275, ISSN 0021-9258

Lushchak V.I. (2011). Adaptive response to oxidative stress: Bacteria, fungi, plants and animals. *Comparative Biochemistry and Physiology, Part C* Vol. 153, No. 2, (March 2011), pp. 175–190, ISSN 0742-8413

Manfredini, V., Martins, V.D., do Carmo, M., Peralba, R. & Benfato, M.S. (2005). Adaptative response to enhanced basal oxidative damage in *sod* mutants from *Saccharomyces cerevisiae*. *Molecular and Cellular Biochemistry*, Vol. 276, No. 1-2, (August 2005), pp. 175–181, ISSN 0300-8177

Methé, B.A., Nelson, K.E., Deming, J.W., Momen, B., Melamud, E., Zhang, X., Moult, J., Madupu, R., Nelson, W.C., Dodson, R.J., Brinkac, L.M., Daugherty, S.C., Durkin, A.S., DeBoy, R.T., Kolonay, J.F., Sullivan, S.A., Zhou, L., Davidsen, T.M., Wu, M., Huston, L.M., Weaver, B., Weidman, J.F., Khouri, H., Utterback, T.R., Feldblyum, T.V. & Fraser, C.M. (2005). The psychrophilic lifestyle as revealed by the genome sequence of *Colwellia psychrerythraea* 34H through genomic and proteomic analyses. *Proceedings of the National Academy of Sciences of the United States of America*, Vol. 102, No. 31, (August 2005), pp. 10913-10918, ISSN 0027-8424

Morgan-Kiss, R.M., Priscu, J.C., Pocock, T., Gudynaite-Savitch, L. & Huner, N.P.A. (2006). Adaptation and Acclimation of Photosynthetic Microorganisms to Permanently

Cold Environments. *Microbiology and Molecular Biology Reviews*, Vol. 70, No. 1, (March 2006), pp. 222-252, ISSN 1092-2172

Müller, J., Staehelin, C., Mellor, R., Boller, T. & Wiemken, A. (1992). Partial purification and characterization of trehalose from soybean nodules. *Journal of Plant Physiology*, Vol. 140, No.1, (January 1992), pp. 8–13, ISSN 0176-1617

Nichols, D., Bowman, J., Sanderson, K., Nichols, C., Lewis, T., Mcmeekin, T. & Nichols, P.D. (1999). Development with Antarctic microorganisms: culture collections, bioactivity screening, taxonomy, PUFA production and cold-adapted enzymes. *Current Opinion in Biotechnology*, Vol. 10, No. 3, (June 1999), pp. 240-246, ISSN 0958-1669

Nyström, T. (2005). Role of oxidative carbonylation in protein quality control and senescence, *The EMBO Journal*, Vol. 24, No. 7, (April 2005), pp. 1311–1317 ISSN 0261-4189

Ocón, A., Hampp, R. & Requena, N. (2007). Trehalose turnover during abiotic stress in arbuscular mycorrhizal fungi. *New Phytologist*, Vol. 174, No. 4, pp. 879–891, (June 2007), ISSN 0028- 646X

Onofri, S., Barreca, D., Selbmann, L., Isola, D., Rabbow, E., Horneck, G., de Vera, J.P.P., Hatton, J. & Zucconi, L. (2008). Resistance of Antarctic black fungi and cryptoendolithic communities to simulated space and Martian conditions. *Studies in Mycology*, Vol. 61, No. 1, (December 2008), pp. 99–109, ISSN 0166-0616

Onofri, S., Selbmann, L., de Hoog, G.S., Grube, M., Barreca, D., Ruisi, S. & Zucconi, L. (2007). Evolution and adaptation of fungi at boundaries of life. *Advances in Space Research*, Vol. 40, No. 11, (December 2007), pp. 1657–1664, ISSN 0273-1177

Ostrovskii, D.N., Demina, G.R., Biniukov, V.I., Shashkov, A.S. & Schloter, M. (2003). Free radicals in mercury-resistant bacteria indicate a novel metabolic pathway. *Mikrobilogiia (In Russian)*, Vol. 72, No. 5, (September-October 2003), pp. 594-599, ISSN 0026-3656

Ouellet, F. (2007). Cold Acclimation and Freezing Tolerance in Plants. In: *Encyclopedia of life sciences*, (September 2007), John Wiley & Sons, Ltd., http://onlinelibrary.wiley.com/doi/10.1002/9780470015902.a0020093/pdf

Parrou, J.L., Teste, M.A. & François, J. (1997). Effects of various types of stress on the metabolism of reserve carbohydrates in *Saccharomyces cerevisiae*: genetic evidence for a stress-induced recycling of glycogen and trehalose. *Microbiology*, Vol. 143, No. 6, (June 1997) pp. 1891-1900, ISSN 0026-2617

Pepi, M., Agnorelli, C. & Bargagli, R. (2005). Iron Demand by Thermophilic and Mesophilic Bacteria Isolated from an Antarctic Geothermal Soil. *Biometals*, Vol. 18, (October 2005), pp. 529–536, ISSN 0966-0844

Pick, E. & Mizel, D. (1981). Rapid microassays for the measurement of superoxide and hydrogen peroxide production by macrophages in culture using an autoenzyme immunoassay reader. *Journal of Immunological Methods*, Vol. 46, No. 2, (January 1981), pp. 211–226, ISSN 0022-1759

Ray M. K. (2006). Cold-stress response of low temperature adapted bacteria. In: Stress Response: A Molecular Biology Approach, A.S. Sreedhar & U.K. Srinivas (Eds.), pp. 1-23, Research Signpost, ISBN: 81-308-0109-4, Kerala, India

Robinson, C.H. (2001). Cold adaptation in Arctic and Antarctic fungi. *New Phytologist*, Vol. 151, No. 2, (August 2001), pp. 341-353, ISSN 0028- 646X

Rossi T., Mazzilli F., Delfino M. & Dondero F. (2001). Improved Human Sperm Recovery Using Superoxide Dismutase and Catalase Supplementation in Semen Cryopreservation Procedure. *Cell and Tissue Banking,* Vol. 2, № 12, (March 2001), pp. 9-13, ASSN 1389-9333

Rúa, J., de Cima, S., del Valle, P., Gutiérrez-Larraínzar, M., Busto, F., & de Arriaga, D. (2008) Glycogen and trehalose mobilization by acetic acid in *Phycomyces blakesleeanus*: dependence on the anion form. *Research in Microbiology,* Vol. 159, No. 3, (April 2008), pp. 200-206, ISSN 0923-2508

Ruisi, S., Barreca, D., Selbmann, L., Zucconi, L., & Onofri, S. (2007). Fungi in Antarctica. *Reviews in Environmental Science and Biotechnology,* Vol. 6, No. 1-3, (August 2007), pp. 127- 141, ISSN 1569-1705

Russell, N.J. (2008). Membrane components and cold sensing. Fundamentals of Cold-Adapted Enzymes. In: *Psychrophiles: from biodiversity to biotechnology,* Margesin R., Schinner F., Gerday C., Marx J. C. (Eds), 177-190, Springer-Verlag, ISBN 978-3-540-74334-7, Heidelberg

Russo, R., Riccio, A., di Prisco, G., Verde, C. & Giordano, D. (2010). Molecular adaptations in Antarctic fish and bacteria. *Polar Science,* Vol. 4, No. 2, (August 2010), pp. 245-256

Şahin, E., & Gümüşlü, S. (2004). Cold-stress-induced modulation of antioxidant defense: role of stressed conditions in tissue injury followed by protein oxidation and lipid peroxidation. *International journal of biometeorology,* Vol. 48, No. 4, (May 2004), pp. 165-171, ISSN 0020-7128

Sies, H. (1993). Strategies of antioxidant defence. *European Journal of Biochemistry,* Vol. 215, No. 2, (July 1993), pp. 213-219, ISSN 0014-2956

Somogy, M. (1952). Notes on sugar determination. *Journal of Biological Chemistry,* Vol. 195, No. 1, (March 1952), pp. 19-23, ISSN 0021-9258

Sun, H-W., Shen F. & Zhou Y-M. (2006). Influence of perfusion by gaseous oxygen persufflation on rat donor liver. *Hepatobiliary & Pancreatic Diseases International,* Vol. 5, *№ 2, (*May 2006), pp. 195-198, ISSN 1499-3872

Suzuki, N., & Mittler, R. (2006). Reactive oxygen species and temperature stresses: A delicate balance between signaling and destruction. *Physiologia Plantarum,* Vol. 126, No. 1, (January 2006), pp. 45–51, ISSN 0031-9317

Tosi, S., Kostadinova, N., Krumova, E., Pashova, S., Dishliiska, V., Spassova, B., Vassilev, S. & Angelova, M. (2010). Antioxidant enzyme activity of filamentous fungi isolated from Livingston Island, Maritime Antarctica. *Polar Biology,* Vol. 33, No. 9, (May 2010), pp. 1227-1237, ISSN 0722-4060

Türkel, S. (2006). Comparative Analysis of Glycogen and Trehalose Accumulation in Methylotrophic and Nonmethylotrophic Yeast. *Microbiology,* Vol. 75, No. 6, (December, 2006), pp. 639–643, ISSN 0026-2617

Vandercammen, A., Francois, J. & Her, H. (1989). Characterization of trehalose-6-phosphate synthase and trehalose-6-phosphate phosphatase of *Saccharomyces cerevisiae. European Journal of Biochemistry,* Vol. 182, No. 5, (July 1989), pp. 613–620, ISSN 0014-2956

Vincent, W.F. (2000). Evolutionary origins of Antarctic microbiota: invasion, selection and endemism. *Antarctic Science*, Vol. 12, No. 3, pp. 374-385, ISSN 0954-1020

Wirtz, N., Thorsten Lumbsch, H., Allan Green, T. G., Türk, R., Pintado, A., Sancho, L. & Schroeter, B. (2003). Lichen fungi have low cyanobiont selectivity in maritime Antarctica. *New Phytologist*, Vol. 160, No. 1, (October 2003), pp. 177-183, ISSN 0028646X - 0028-646X

Woodbury W., Spencer, A.K. & Stahmann, M.A. (1971). Animproved procedure using ferricyanide for detecting catalase isozymes. *Analitical Biochemistry*, Vol. 44, № 1, (November 1971), pp. 301-305, ISSN 0003-2697

Zhang, L., Onda, K., Imai, R., Fukuda, R., Horiuchi, H. & Ohta, A. (2003). Growth temperature downshift induces antioxidant response in *Saccharomyces cerevisiae*. *Biochemical and Biophysical Research Communications*, Vol. 307, No. 2, (July 2003), pp. 308-314, ISSN 0006-291X

5

Changes in Hydrogen Peroxide Levels and Catalase Isoforms Expression are Induced With Freezing Tolerance by Abscisic Acid in Potato Microplants

Martha E. Mora-Herrera[1,2], Humberto López-Delgado[1], Ernestina Valadez-Moctezuma[3] and Ian M. Scott[4]

[1]Programa Nacional de Papa, Instituto Nacional de Investigaciones Forestales Agrícolas y Pecuarias, (INIFAP), Metepec
[2]Centro Universitario Tenancingo, Universidad Autónoma del Estado de México, Carr. Tenancingo-Villa Guerrero Km 1.5 Tenancingo,
[3]Departamento de Fitotecnia, Universidad Autónoma Chapingo, Chapingo,
[4]Institute of Biological, Environmental and Rural Sciences, Aberystwyth University, Ceredigion,
[1,2,3]México
[4]UK

1. Introduction

There is evidence that abscisic acid (ABA) has a protective signaling role in freezing stress in plants (Kobayashi et al., 2008), including mosses (Minami et al., 2003). ABA signaling networks and their actions are not totally understood, but H_2O_2 has been implicated as an intermediary in several ABA responses, where its roles include induction of the antioxidant system (Cho et al., 2009). Mora-Herrera & López-Delgado (2007), using *in vitro* microplants as employed in potato production programs, found freezing tolerance was enhanced by culture on ABA-containing medium. This ABA treatment tripled survival of a -6°C incubation in the cold-sensitive cv. Atlantic, while in the more cold-tolerant cv. Alpha, survival improved by two-thirds. In the ABA-treated microplants, they found the H_2O_2-scavenging enzyme ascorbate peroxidase increased in activity.

Stress tolerance in potato is growing in importance, as increases in potato production by developing countries greatly exceed other major crops (FAO, 2008). The present study used the microplant system to investigate effects of prolonged ABA treatment on catalase, another enzyme important in controlling cellular H_2O_2. Catalases are tetrameric, heme-containing oxidoreductases that dismutate H_2O_2 to water and oxygen. In plants, their peroxisomal location coincides with the cellular site of H_2O_2 generation by photorespiration or fatty acid β-oxidation (Feierabend, 2005)(Scheme 1). Evidence for catalase involvement in these processes includes susceptibility of catalase mutants to photorespiration-promoting conditions (Queval et al., 2007), and catalase induction in nutrient stress conditions promoting fatty acid catabolism (Contento & Bassham, 2010). Catalases respond to a wide range of stresses (Du et al., 2008) and, most relevantly here, have been functionally implicated in low-temperature tolerance by transgenic experiments on rice (Matsumura et al., 2002). Moreover, there is evidence that catalase is an integral component of ABA-activated stress protection mechanisms (Xing et al., 2008).

Plant catalases occur in small gene families, whose differential expression reflects different roles (Feierabend, 2005). In *Arabidopsis*, *CAT2*, expressed in photosynthetic tissues (Du et al., 2008), is needed to cope with photorespiration (Queval et al., 2007). *Arabidopsis CAT1* is induced by treatments including cold and ABA (Du et al., 2008). Pharmacological and mutant studies have revealed that *CAT1* induction by ABA involves mitogen-activated protein kinase (MAPK) cascades, in which H_2O_2 is involved (Xing et al., 2008). Among maize catalases, *CAT3* is a chilling-acclimation responsive gene in seedlings, and a long-standing example of regulation by H_2O_2(Prasad et al., 1994). Maize *CAT1* is highly expressed as seeds dehydrate in late embryogenesis, and its promoter has an ABRE (ABA-responsive) element, while H_2O_2 was also implicated as a signal by Guan et al. (2000) and Zhang et al. (2006) showed *CAT1* induction by ABA in maize leaves involved MAPK cascades and H_2O_2.

In potato, previous studies have identified two, differentially expressed catalase genes (Santos et al., 2006). In photosynthesizing tissues, where photorespiration occurs, the principal one expressed was *CAT1*. Phylogenetic comparisons by Santos et al. (Santos et al., 2006) found potato *CAT1* was less similar to potato *CAT2* than to *Nicotiana CAT1* genes. Potato *CAT2* shares high identity with *N. plumbaginifolia CAT2*, characteristics of which include inducibility by stressful exposure to ultraviolet light, ozone or SO_2(Willekens et al., 1994). Consistent with an analogous role in stress conditions, potato *CAT2* was induced in plants suffering nematode or bacterial infection (Niebel et al., 1995). More recently, *CAT2* was found to be induced in potato leaves treated with H_2O_2, while *CAT1* was not (Almeida et al., 2005).

This study was undertaken with the hypotheses of catalase and H_2O_2 involvement in ABA-induced freezing tolerance in potato microplants. Moreover, catalase isoforms were predicted to show differential patterns of expression and activity in this process.

2. Materials and methods

2.1 Microplant material

Virus-free microplants of *Solanum tuberosum* L. cv. Alpha and cv. Atlantic, from the Germplasm Bank of the National Potato Program of the National Institute for Forestry Agriculture and Livestock Research (INIFAP), Toluca, México, were micropropagated as nodal cuttings*in vitro* at 20 °C, following previous protocols (Mora-Herrera et al., 2005). In every experiment, 24 microplants were cultured per treatment, and pooled into samples to achieve the weight required for the particular measurement.

2.2 Freezing treatments

Microplants cultivated 28 d on medium with 10µM (±)-*cis, trans*-ABA (Sigma, USA), or as controls without ABA, were transferred to peat moss (in 3 × 5 cm pots) pre-sterilized for 1 h at 120 °C. These transplanted microplants were kept for 24 h at 20 °C under a 16 h photoperiod (fluorescent lights, 35 μmol m^2 s^{-1}, 400-700 nm), to allow recovery from the stress of transplantation, prior to exposure to -6 ±1 °C in darkness for 4 h, as previously (Mora-Herrera et al., 2005). H_2O_2 and catalase measurements were performed immediately after this freezing incubation.

2.3 Determination of H_2O_2 content

H_2O_2 was measured by luminol-dependent chemiluminescence, as in Mora-Herrera et al. (2005), in 3 experiments, with 3 samples per treatment, and each assay replicated 6 times.

2.4 Quantification of catalase activity

Frozen shoot tissue (0.5 g) was powdered under liquid N_2, and extracted in 2 mL 50 mM potassium phosphate buffer (pH 7.2) containing 5 mM dithiothreitol, 1 mM ethylenediamine tetraacetic acid, and 1% polyvinylpyrrolidone. After clarification by centrifugation (11,000 *g*, 15 min, 4 °C), catalase activity (EC 1.11.1.6) was determined according to Aebi (1984). The total reaction mixture (3 mL) contained 20 μL extract (100 μg protein) and 30 mM H_2O_2 in 50 mM sodium/potassium phosphate buffer (pH 7.0). The reaction was initiated by H_2O_2 addition and followed by absorbance decrease at 240 nm (extinction coefficient 39.4 mM^{-1} cm^{-1}) every 20 s for 3 min, at 26 °C. Protein was determined using Bradford reagent. Catalase was measured in 3 experiments, each with 3 samples (assayed in triplicate) per treatment.

2.5 Catalase zymograms

Enzymes were extracted by a similar method to Cruz-Ortega et al. (2002). Frozen tissue (0.1 g) was powdered in liquid N_2, then extracted in 100 μL potassium phosphate buffer (pH 7.8, 1 mM ethylenediaminetetraacetate, 1 mM phenylmethanesulfonyl fluoride, 10 mM dithiothreitol, 2% polyvinylpyrrolidone). The extracts were clarified at 11,000 *g* (10 min, 4 °C). Non-denaturing polyacrylamide gel electrophoresis, as described by Ougham (1987), was performed for 18-20 h at 4 °C. Catalase activity staining used the ferricyanide method of Woodbury et al., 1971. The GE Healthcare Life Sciences HMW Native Marker Kit containing bovine liver catalase (232 kDa) was used as a gel marker. Results shown are representative of 6 experiments.

2.6 Analysis of transcripts by RT-PCR

Total RNA extractions used TRIzol isolation reagent, and treatment with DNase I (Invitrogen, USA). cDNAs were synthesized with Oligo$(dT)_{12-18}$ primer and SuperScript II reverse transcriptase (Invitrogen). PCR amplications of potato *CAT1* and *CAT2* used the primers of Santos et al. (2006). As an internal quantitative control, potato actin (NCBI accession X55751) primers (forward, 5'-AGACGCCTATGTGGGAGATG-3'; reverse, 5'-GCGAGCTTTTCTTTCACGTC-3') were used. After 40 cycles at 52 °C, PCR products were electrophoresed in 1% agarose and visualized with ethidium bromide. Images were acquired by a gel documentation system (UVItec, UK), and relative transcipt levels estimated with Quantity One v.4.6.5 software (Bio-Rad, USA). Results shown are representative of 4 experiments.

2.7 Statistical analysis

Statgraphics Plus v.5.0 (StatPoint Technologies, USA) was used for *t*-tests, and ANOVA with Tukey *post-hoc* tests ($P < 0.05$).

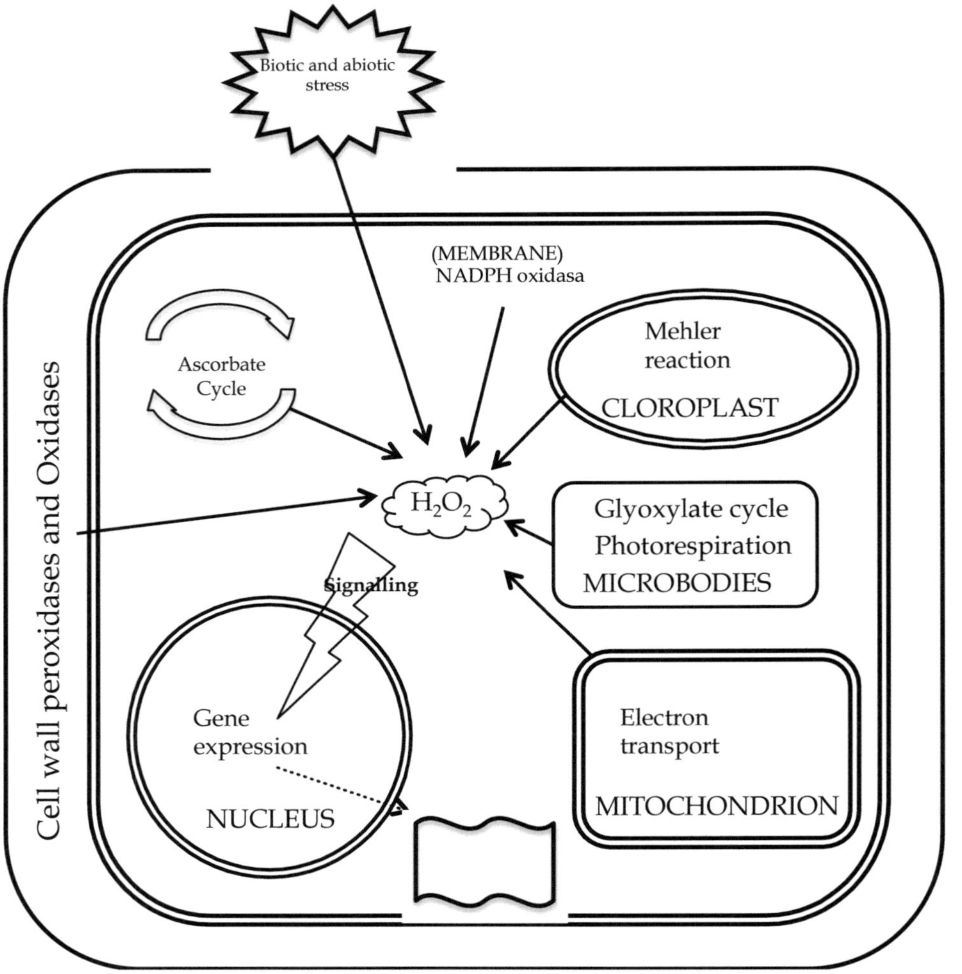

Scheme 1. H_2O_2 is produced in chloroplasts via the Mehler reaction, photorespiration in peroxisomes , glyoxylate cycle, and via electron transport in mitochondria. Cell wall peroxidases and NADPH oxidases in the plasma membrane also can increase the H_2O_2 production when the plant is under biotic or abiotic stress. The signaling role of H_2O_2 is mediated by enzymatic antioxidants one of them is catalase.

3. Results

3.1 Effects of ABA on H_2O_2 content of potato microplants

In vitro microplants were cultured for 28 d on MS medium supplemented with 10 µM ABA. In the study of Mora-Herrera & López-Delgado (2007), 10 µM was the highest ABA concentration used, which gave greater improvements in freezing tolerance than lower concentrations. It also caused growth inhibition, but this did not detract from eventual growth and tuber yield of microplants transplanted to compost and glasshouse conditions (Mora-Herrera & López-Delgado, 2007). We investigated H_2O_2 and catalase in cv. Alpha and Atlantic microplants at three stages: (a) after 28 d of culture in the presence (or absence) of ABA, (b) 24 h after transplanting from *in vitro* culture to compost, and (c) after 4 h of freezing (-6 °C) in microplants transplanted 24 h previously to compost.

Shoot H_2O_2 contents were on average 24% higher in microplants (of either cv.) that had been cultured for 28 d on ABA-containing medium (Fig. 1). This ABA-induced elevation of H_2O_2 contents persisted in microplants transplanted for 24 h to compost, and also after these transplanted microplants had been subjected to freezing (Fig. 1). It was also notable that freezing treatment also increased H_2O_2, by 23% on average in the transplanted microplants (Fig. 1).

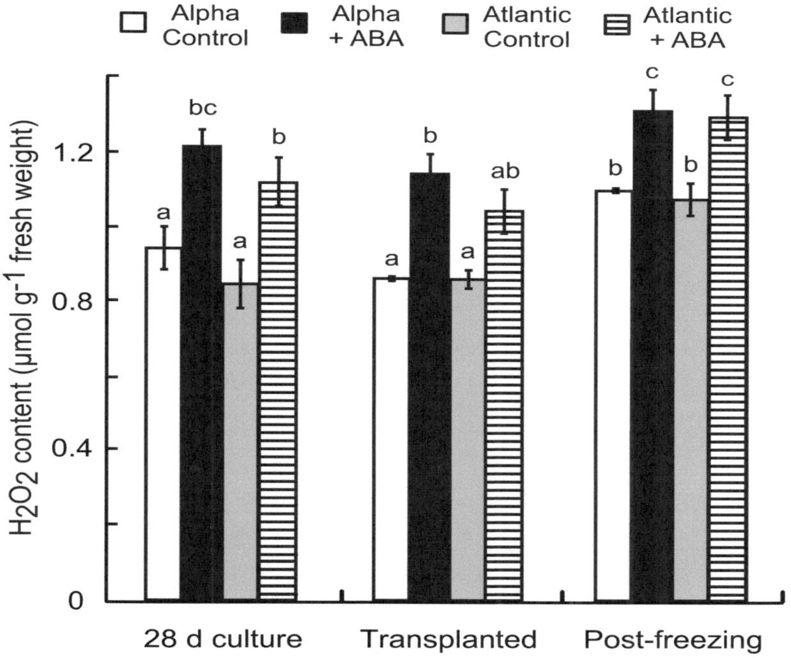

Fig. 1. H_2O_2 content of microplants (cvs. Alpha and Atlantic) grown in the presence of ABA (10 µM), or its absence (controls), assayed at three stages. '28 d culture', after 28 days of *in vitro* culture. 'Transplanted', 24 h after transfer to compost. 'Post-freezing', immediately after 4 h of freezing (-6 °C). Bars show means (n = 3) ± SE, those with different letters differing significantly (ANOVA, P< 0.05).

3.2 Effects of ABA on CAT1 and CAT2 transcripts

RT-PCR with primers specific to *CAT1* or *CAT2* was used to compare the abundance of their transcripts in response to ABA treatment (Fig. 2). The results showed*CAT1* and *CAT2* were differentially regulated by ABA, in both cvs. Relative abundance of *CAT1* transcripts was lower by 25% on average, while *CAT2* transcripts increased up to 4-fold, in ABA-treated microplants (Fig. 2). In consequence, *CAT2* was the gene predominantly expressed in ABA-treated microplants.

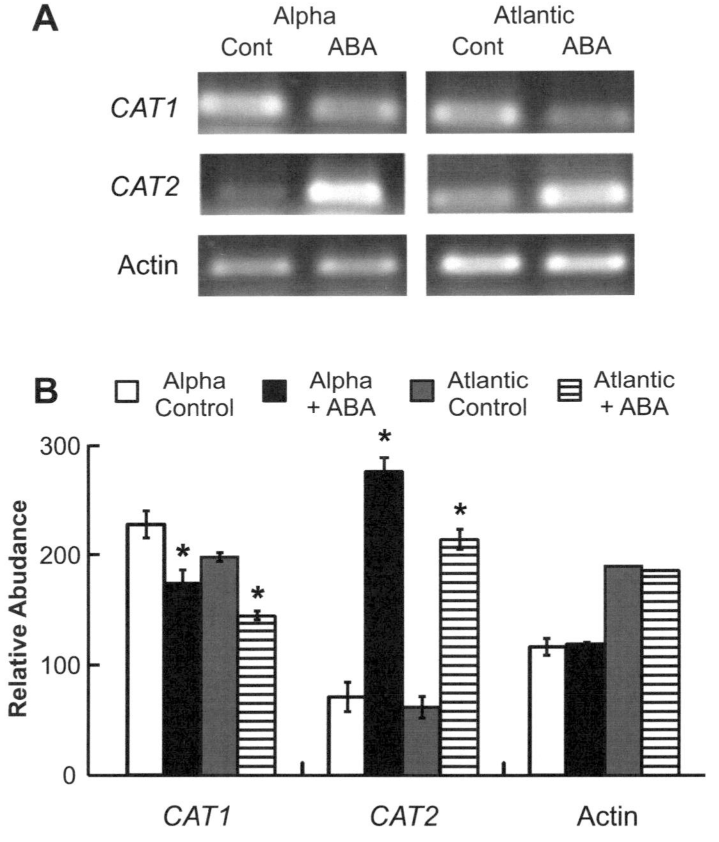

Fig. 2. Effects of ABA (10 μM) on *CAT1* and *CAT2* transcripts in microplants (cvs. Alpha and Atlantic) grown *in vitro* for 28 d. (A) RT-PCR products in agarose gels typical of 4 experiments. (B) Mean relative abundance (± SE) of RT-PCR products in 4 experiments. Actin was the internal control. *ABA treatments significantly different to controls (*t*-tests, $P < 0.05$).

3.3 Effects of ABA on catalase activities

Native gels stained for enzyme activity ('zymograms') confirmed the occurrence of catalase isozymes (Fig. 3), as would be expected from the expression of more than one gene. The faster-migrating native isozyme was greatly increased in ABA-treated microplants of both cvs (Fig. 3), and was attributed to the CAT2 protein, based on the similar effects of ABA on *CAT2* transcripts (Fig. 2) and the immunological evidence of Santos et al. (2006). This isozyme showed similar migration to a 232-kDa standard of bovine liver catalase (Fig. 3). Less expected was the occurrence of more than one slower-migrating isozyme (Fig. 3), since Santos et al. (2006) reported only one, which they assigned as CAT1. The two slower-migrating bands were apparently absent in zymograms of ABA treatments, which represented a more dramatic difference in CAT1 activity than the 25% reduction in *CAT1* transcripts seen in RT-PCR.

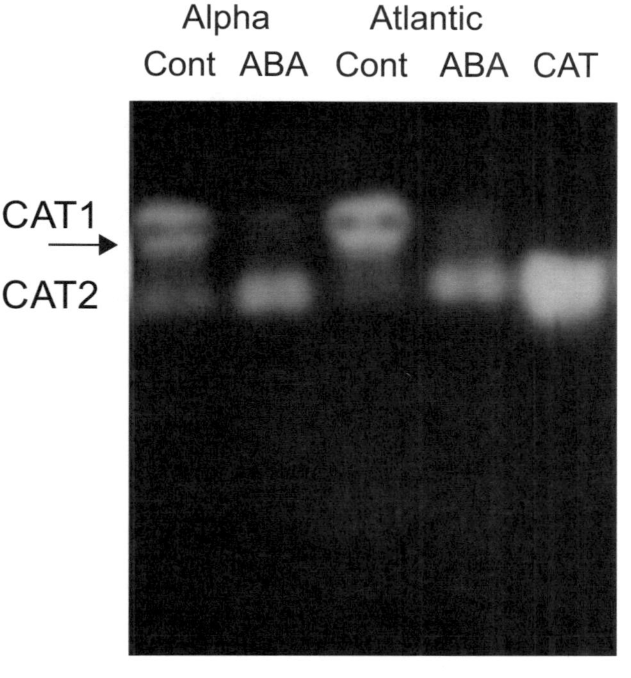

Fig. 3. Catalase zymograms of microplants (cvs. Alpha and Atlantic) grown in vitro in the presence of ABA (10 μM), or its absence (controls) for 28 d. 'CAT' lane: bovine liver catalase (232 kDa). Labels on left: attribution of bands to CAT1 or CAT2 isoforms. Arrow indicates possible heterotetrameric form.

The isozymes had different distributions in microplant shoot tissues. In zymograms of leaves, the two slower-migrating bands dominated, though a faint CAT2 band was visible (Fig. 4). Stem zymograms, in contrast, showed the CAT2 band only (Fig. 4).

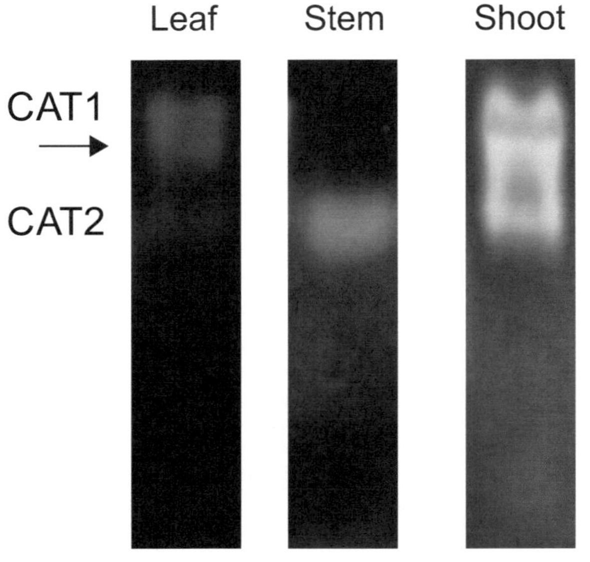

Fig. 4. Catalase zymograms of leaves, stems or whole-shoots of microplants (cv. Alpha) cultured *in vitro* (without ABA). Labels on left: attribution of bands to CAT1 or CAT2 isoforms. Arrow indicates possible heterotetrameric form.

Quantifications of catalase activity indicated the changed isozyme profiles induced by growth on ABA medium resulted in a net decrease, at least in the enzymic assay conditions used. Significant reductions (of 22% on average) were observed in ABA-treated microplants of either cv., relative to untreated controls, both before and after transplantation from *in vitro* culture to compost (Fig. 5).

Catalase activities in ABA-treated and control microplants showed differential responses to freezing. Post-freezing catalase activities in ABA-treated microplants were not significantly different to pre-freezing levels (Fig. 5). In controls, by contrast, catalase activities were lower after freezing, by 33% on average. The net result was that post-freezing catalase activity was not significantly different in ABA-treated and control microplants (Fig. 5).

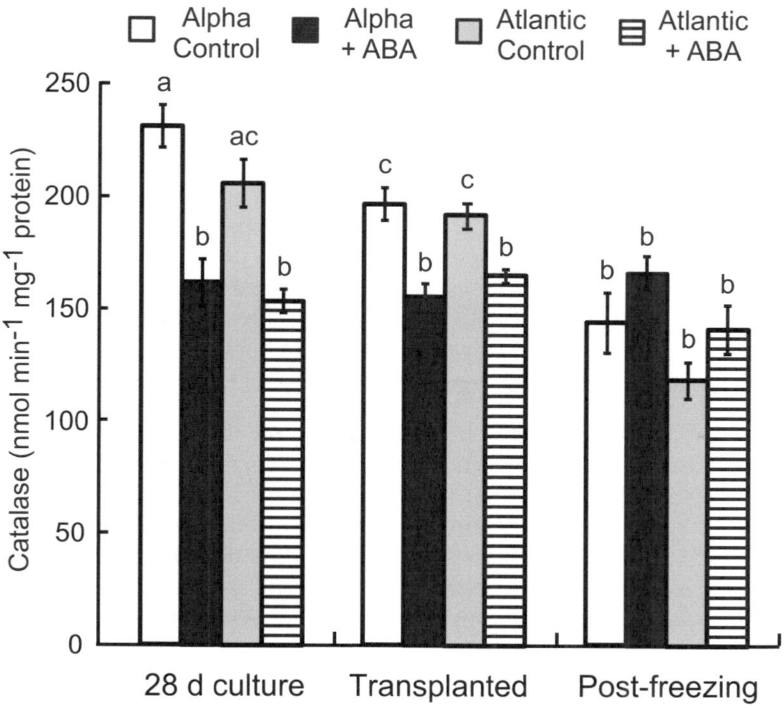

Fig. 5. Catalase activity of microplants (cvs. Alpha and Atlantic), grown in the presence of ABA (10 µM), or its absence (controls), assayed at three stages. '28 d culture', after 28 days of *in vitro* culture. 'Transplanted', 24 h after transfer to compost. 'Post-freezing', immediately after 4 h of freezing (-6 °C). Bars are means (n = 5 - 6) ± SE, those with different letters differing significantly (ANOVA, P< 0.05).

4. Discussion

This paper belongs to a series on protection by growth regulators against freezing stress in potato microplants (Mora-Herrera et al., 2005, Mora-Herrera & López-Delgado 2007). One finding was that freezing treatment increased H_2O_2 levels. Despite recognition that abiotic stress is likely to promote formation of reactive oxygen species (Jaspers & Kangasjärvi, 2010), direct studies of the effects of sub-zero temperatures on tissue H_2O_2 are surprisingly sparse. It is therefore worth aligning our results with the only comparable recent study (Yang et al., 2007), especially since concerns have been expressed about variability of H_2O_2 literature data (Queval et al., 2008).

Yang et al. (2007) subjected wheat plants to -6 °C for 6 h, the temperature being changed from, and back to, 20 °C over 6 h periods. H_2O_2 (measured spectrophotometrically after reaction with KI)increased from ca. 1.2 to 2.1 μmol g^{-1} in this treatment (Yang et al., 2007). These values are comparable to H_2O_2 in potato microplants in this and previous papers (López-Delgado et al., 1998; Mora-Herrera et al., 2005). The increase from ca. 0.87 to 1.1 μmol $g^{-1}H_2O_2$ in our (ABA-untreated) microplants resulted from a treatment of similar severity (-6 °C for 4 h), but was measured without any post-freezing period.

The present study was prompted by the finding that 28 d culture with ABA protected microplants in freezing (Mora-Herrera & López-Delgado, 2007). In these prolonged exposures to ABA, H_2O_2 levels were higher by an average (± SD) of 24 ± 7.3% across cvs. and experimental stages (n = 6). This was also seen in treatments with another class of protective growth regulators, the salicylates (López-Delgado et al., 1998; Mora-Herrera et al., 2005). The H_2O_2 increment in culture with these growth regulators was notably consistent. H_2O_2 was 27% higher on 100 μM salicylate (Mora-Herrera et al., 2005), and 24% on 1 μM acetylsalicylate (López-Delgado et al., 1998). This may reflect a tight control of maximal H_2O_2 in healthy tissues to avoid toxic concentrations (Queval et al., 2008).

Despite the increase in H_2O_2 induced by freezing treatment, the difference between ABA-treated and untreated microplants was maintained. Therefore, cellular mechanisms for H_2O_2 generation were not saturated by either treatment. The origin of H_2O_2 induced by ABA has been identified as superoxide generation by plasma membrane NADPH oxidases, encoded by *Rboh* (*r*espiratory *b*urst *o*xidase *h*omolog) genes (Cho et al., 2009). Recent work in maize indicates that the ABA-induced expression and activity of NADPH oxidases is further stimulated by the resultant H_2O_2 in a MAPK-regulated positive feedback (Lin et al., 2009).

The *Arabidopsis* RbohD NADPH oxidase was recently also implicated in a systemic reactive oxygen signal in plants subjected to stresses including ice-water cooling (Miller et al., 2009). This class of enzymes, which have now been characterized in potato tubers (Kobayashi et al., 2007), are therefore candidates for H_2O_2 production in both ABA and freezing treatments of the microplants. While it is obviously probable that freezing resulted in H_2O_2 generation by cellular processes under stress (Jaspers & Kangasjärvi, 2010), cellular signaling may also have been involved.

The redox state adjustment indicated by higher H_2O_2 levels may have been a factor in the growth retardation that was another shared effect of ABA (Mora-Herrera & López-Delgado, 2007) and acetylsalicylate (López-Delgado et al., 1998), since a direct pre-treatment with H_2O_2 can itself inhibit microplant growth in culture (López-Delgado et al., 1998). If NADPH oxidases were responsible for the ABA-induced H_2O_2, it could be relevant that certain *Arabidopsisatrboh* mutants are defective in ABA inhibition of root growth (Kwak et al., 2003).

We investigated catalase, as a principal H_2O_2 scavenger, in ABA-treated microplants. RT-PCR and zymogram analyses revealed contrasting ABA responses for different catalase forms. *CAT2* transcripts and the relevant isozyme were strongly ABA-inducible. Given the increased H_2O_2 levels in ABA-treated microplants, and the H_2O_2-inducibility of potato *CAT2* (Almeida et al., 2005), this gene may have an ABA-induction mechanism like *Arabidopsis CAT1* (Xing et al., 2008) and maize *CAT1*(Lin et al., 2009). Our data are consistent with potato *CAT2* as the ortholog of the stress-inducible *N. plumbaginifolia CAT2* (Willekens et al., 1994; Santos et al., 2006).

CAT1 transcripts, in contrast, showed a 25% reduction in abundance in ABA-treated microplants. Zymograms showed more dramatic difference, with the putative CAT1 band absent in ABA treatments. Almeida et al. (Almeida et al., 2005) found H_2O_2 treatment reduced CAT1 in immunoblots and zymograms, whereas *CAT1* in RNA gel blots did not show the same decline. As in our study, therefore, there was a disparity between the RNA and protein findings, which suggested post-transcriptional effects of ABA and H_2O_2 on CAT1 expression. Spectrophotometric assays showed a consistent net reduction in catalase activity in ABA-treated microplants at standard temperature. This suggests the zymograms, where the decline in CAT1 appeared more dramatic, were better indicators of enzymic activity than the RT-PCR.

In zymograms of field-grown plants, Almeida et al. (Almeida et al., 2005) saw only one slower-migrating band, attributed to CAT1, whereas our *in vitro* microplants yielded two slower-migrating bands. An extra isozyme could reflect a third, uncharacterized catalase potato gene, since at least three occur in confamilial species (Santos et al., 2006). On the other hand, the coincidental expression patterns (Figs. 3-4) of the two slower-migrating bands suggested at least one (presumably the faster-migrating) may have been a heterotetramer of CAT1 and CAT2 proteins, analogous to those in other species (Feierabend, 2005).

Heterotetrameric isoforms probably depend on the different loci being co-expressed in a given cell type (Feierabend, 2005), and in some respect the distribution of *CAT1* and *CAT2* expression may have differed *in vitro* and in the field. In microplants under standard conditions, the isoforms did have different tissue distributions. In stem zymograms only the CAT2 band was visible, while leaf zymograms were dominated by the two bands that putatively included CAT1, consistent with an association of CAT1 with photorespiration (Santos et al., 2006). It is furthermore possible that catalase could be differentially distributed in different types of leaf cells, as has been observed for H_2O_2 and ascorbate peroxidase (Galvez-Valdivieso et al., 2009).

In theory, the reduced catalase activity seen in spectrophotometric assays could have facilitated a controlled H_2O_2 increase to adjust growth and prime defenses against abiotic stress. Our data suggest the leaf would be the critical site of these events, since it was the leaf-localized isoforms whose decline was evidenced by isozyme results.

Tsai & Kao (2004) also saw a decrease in catalase activity in ABA treatment, of rice roots. On the other hand, other studies have found increased activity in response to ABA (Agarwal et al., 2005, Zhang et al., 2006, Du et al., 2008, Kumar et al., 2008). Our model system was different in that the microplants experienced prolonged growth on ABA-containing medium. This may have brought a different physiological adjustment to those seen in single, brief treatments, whose effects are transient (Du et al., 2008). It may be more pertinent that, on exposure to freezing, catalase activity levels were maintained in ABA-treated microplants, but not in controls.

The potential value for food security of ABA and catalase lies in their association with coping mechanisms for stresses that challenge crop production (Cho et al., 2009). Crop species in which the effects of ABA on catalase had previously been investigated were cereals (Tsai & Kao 2004, Agarwal et al., 2005, Zhang et al., 2006) or legumes (Kumar et al., 2008). We have added potato to this list. Morever, we suggest that experimental systems like the cultured microplants may have particular biotechnological relevance, because exploitation of stress tolerance mechanisms are likely to involve the stable changes in physiology seen in prolonged treatments.

5. Conclusion

Freezing tolerance-enhancing treatments with ABA caused differential changes in catalase isoforms and activities, in concert with changes in H_2O_2 levels. At least one isoform may have been a heterotetramer of CAT1 and CAT2 proteins. This may reflect a tight control of maximal H_2O_2 in healthy tissues to avoid toxic concentrations. Knowledge of stress tolerance mechanisms involve stable changes in physiology during prolonged treatments.

6. Acknowledgments

This research was supported by CONACYT project SEP/CONACYT/2003/CO2/45016. The first author acknowledges postgraduate and posdoctoral CONACYT scholarships. We extend our sincere thanks to PhD Silvia Ivonne Mora-Herrera for technical advice and MPhil Ricardo Martinez-Gutierrez and Martha Alvarado-Ordoñez for technical support.

7. References

Aebi, H. (1984). Catalase *in vitro*. *Method. Enzymol*, 105, pp. 121-126.

Agarwal, S., Sairam, R. K., Srivastava, G.C., Tyagi A. & Meena, R.C. (2005). Role of ABA, salicylic acid, calcium and hydrogen peroxide in antioxidant enzyme induction in wheat seedlings. *Plant Sci*, 169, 3, (September 2005), pp. 559-570.

Almeida, J. M., Fidalgo F., Confraria, A., Santos A., Pires H. & Santos, I.(2005). Effect of hydrogen peroxide on catalase gene expression, isoform activities and levels in leaves of potato sprayed with homobrassinolide and ultrastructural chages in mesophyll cells. *Funct. Plant Biol*, 32, 8, pp. 707-720.

Cho, D., Shin, D., Jeon B. W. & Kwak, J.M. (2009). ROS-mediated ABA signaling. *J. Plant Biol*, 52, pp. 102–113.

Contento, A. L. & Bassham, D. C. (2010). Increase in catalase-3 activity as a response to use of alternative catabolic substrates during sucrose starvation. *Plant Physiol. Biochem*, 48,4, (April 2010), pp. 232-238.

Cruz-Ortega, R., Ayala-Cordero G. & Anaya, A.L. (2002). Allelochemical stress produced by the aqueous leachate of *Callicarpa acuminata*: effects on roots of bean, maize, and tomato. *Physiol. Plant*, 116, 1, (September 2002), pp. 20-27.

Du, Y. Y., Wang, P. C., Chen J. & Song, C. P. (2008). Comprehensive functional analysis of the catalase gene family in *Arabidopsis thaliana*. *J. Integr. Plant Biol*, 50, 10, pp. 1318–1326.

FAO. (2008). International Year of the Potato, *International Year of the Potato Secretariat, Food and Agriculture Organization of the United Nations*.ISBN 978-92-5-106142-8, Rome.

Feierabend, J. (2005). Catalases in plants: molecular and functional properties and role in stress defence. In: *Antioxidants and reactive oxygen in plants*. N. Smirnoff, pp. 101–140. Blackwell Publishing, ISBN: 9781405171465, Oxford.

Galvez-Valdivieso, G., Fryer, M. J., Lawson, T., Slattery, K., Truman, W., Smirnoff, N., Asami,T., Davies,W. J., Jones,A. M., Baker N. R. & Mullineaux, P. M. (2009). The high light response in *Arabidopsis* involves ABA signaling between vascular and bundle sheath cells. *Plant Cell,* 21, 7, (July 2009) , pp. 2143–2162.

Guan, L. M., ZhaoJ. & Scandalios, J.G. (2000). *Cis*-elements and trans-factors that regulate expression of the maize *Cat1* antioxidant gene in response to ABA and osmotic stress: H_2O_2 is the likely intermediary signaling molecule for the response. *Plant J.,* 22, pp. 87-95.

Jaspers, P. & Kangasjärvi, J. (2010): Reactive oxygen species in abiotic stress signaling. *Physiol. Plant.* 138, 4, (April 2010), pp. 405–413.

Kobayashi, F., Takumi S. & Nakamura, C. (2008). Increased freezing tolerance in an ABA-hypersensitive mutant of common wheat. *J. of Plant Physiol.* 165, pp. 224-232.

Kobayashi, M., Ohura, I., Kawakita, K., Yokota N., Fujiwara, M., Shimamoto, K., Doke N. & Yoshioka, H. (2007). Calcium-dependent protein kinases regulate the production of reactive oxygen species by potato NADPH oxidase. *Plant Cell,* 19, 3, (March 2007), pp. 1065–1080.

Kumar, S., Kaur G., & Nayyar, H. (2008). Exogenous application of abscisic acid improves cold tolerance in chickpea (*Cicer arietinum* L.). *J. Agron. Crop Science* 194, 6, (December 2008), pp. 449-456.

Kwak, J. M., Mori, I. C., Pei, Z. M., Leonhardt, N., Torres, M.A., Dangl, J. L., Bloom, R. E., Bodde, S., Jones J. D. & Schroeder, J. I. (2003). NADPH oxidase *AtrbohD* and *AtrbohF* genes function in ROS-dependent ABA signaling in *Arabidopsis. EMBO J.* 22, 11, (June 2003), pp. 2623–2633.

Lin, F., Ding, H., Wang, J., Zhang, H., Zhang, A., Zhang, Y., Tan, M., Dong W. & Jiang, M. (2009). Positive feedback regulation of maize NADPH oxidase by mitogen-activated protein kinase cascade in abscisic acid signalling. *J. Exp. Bot.* 60, 11, (July 2009), pp. 3221–3238.

López-Delgado, H., Dat, J.F., Foyer C. H. &Scott, I.M. (1998). Induction of thermotolerance in potato microplants by acetylsalicylic acid and H_2O_2. *J. Exp. Bot.* 49, 321, (April 1998), pp. 713-720.

Matsumura, T., Tabayashi, N., Kamagata, Y., Souma C. & Saruyama, H. (2002). Wheat catalase expressed in transgenic rice can improve tolerance against low temperature stress. *Physiol. Plant.* 116, 3, (November 2002), pp. 317-327, ISSN 0031-9317.

Miller, G., Schlauch,K., Tam, R., Cortes, D., Torres, M.A., Shulaev, V., Dangl J.L., & R. Mittler R. (2009). The plant NADPH oxidase RBOHD mediates rapid systemic signaling in response to diverse stimuli, In: *Sci. Signal.* 2, ra45.

Minami, A., Nagao,M., Arakawa, K., Fujikawa S. & Takezawa, D. (2003). Abscisic acid-induced freezing tolerance in the moss *Physcomitrella patens* is accompanied by increased expression of stress-related genes. *J. Plant Physiol.* 160, 5, pp. 475-483.

Mora-Herrera, M. E., López-Delgado, H., Castillo-Morales A. & Foyer, C.H. (2005). Salicylic acid and H_2O_2 function by independent pathways in the induction of freezing tolerance in potato. *Physiol. Plant.* 125, 4, (December 2005), pp. 430-440.

Mora-Herrera, M. E. & López-Delgado, H.A.(2007). Freezing tolerance and antioxidant activity in potato microplants induced by abscisic acid treatment. *Am. J. Potato Res.* 84, 6, pp. 467-475.

Niebel, A., Heungens, K., Barthels, N., Inzé, D., M. Van Montagu & Gheysen, G. (1995). Characterization of a pathogen-induced potato catalase and its systemic expression upon nematode and bacterial infection. *Mol. Plant Microbe In.* 8, 3, pp. 371-378.

Ougham, H. J., (1987). Gene expression during leaf development in *Lolium temulentum*: patterns of protein synthesis in response to heat-shock and cold-shock. *Physiol. Plant.* 70, 3, (April 2006), pp. 479-484.

Prasad, T. K., Anderson, M. D., Martin B.A. & Stewart, C.R. (1994). Evidence for chilling-induced oxidative stress in maize seedlings and a regulatory role for hydrogen peroxide. *Plant Cell* 6, pp. 65-67.

Queval, G., Hager, J., Gakière B. & Noctor, G.(2008). Why are literature data for H_2O_2 contents so variable? A discussion of potential difficulties in the quantitative assay of leaf extracts. *J. Exp. Bot.* 59, pp. 135-146.

Queval, G., Issakidis-Bourguet, E., Hoeberichts, F.A.,Vandorpe, M., Gakière, B., Vanacker, H., Miginiac-Maslow, M., Van Breusegem F. & Noctor, G. (2007). Conditional oxidative stress responses in the Arabidopsis photorespiratory mutant *cat2* demonstrate that redox state is a key modulator of daylength-dependent gene expression, and define photoperiod as a crucial factor in the regulation of H_2O_2-induced cell death. *Plant J.* 52, 4, (September 2007) pp. 640–657.

Santos, I., Pires, H., Almeida, J.M., Fidalgo, F., Confraria, A., Duarte, M., Borlido J., & Salema, R. (2006). Phylogenetic relationship of potato *CAT1* and *CAT2* genes, their differential expression in non-photosynthetic organs and during leaf development, and their association with different cellular processes. *Funct. Plant Biol.* 33, 7, pp. 639-651, ISSN 1445-4408.

Tsai, Y. C. & Kao, C.H.(2004). The involvement of hydrogen peroxide in abscisic acid-induced activities of ascorbate peroxidase and glutathione reductase in rice roots. *Plant Growth Regul.* 43, 3, pp. 207-212.

Willekens, H., Van Camp, W., Van Montagu, M., Inzé, D., Langebartels C. & Sandermann, H. (1994). Ozone, sulfur dioxide, and ultraviolet B have similar effects on mRNA accumulation of antioxidant genes in *Nicotiana plumbaginifolia* L. *Plant Physiol.* 106, 3, (November 1994), pp. 1007-1014.

Woodbury, W., Spencer A.K. and Stahmann, M.A. (1971). An improved procedure using ferricyanide for detecting catalase isozymes. *Anal. Biochem.* 44, pp. 301-305.

Xing, Y., Jia W. & Zhang, J. (2008). AtMKK1 mediates ABA-induced *CAT1* expression and H_2O_2 production via AtMPK6-coupled signaling in Arabidopsis. *Plant J.* 54, 3, (May 2008), pp. 440–451.

Yang, S. H., Wang L. J. & Li, S.H.(2007). Ultraviolet-B irradiation-induced freezing tolerance in relation to antioxidant system in winter wheat (*Triticum aestivum* L.) leaves. *Environ. Exp. Bot.* 60, pp. 300–307.

Zhang, A., Jiang, M., Zhang, J., Tan M. and Hu, X. (2006). Mitogen-activated protein kinase is involved in abscisic acid-induced antioxidant defense and acts downstream of reactive oxygen species production in leaves of maize plants. *Plant Physiol.* 141, 2, (June 2006), pp. 475–487.

Section 3

Chemical Factors

6

Oxidative Stress Induced by the 2,4-Dichlorophenoxyacetic Herbicide

Tayeb Wafa, Nakbi Amel, Chaieb Ikbal and Hammami Mohamed
Laboratory of Biochemistry, UR03/ES08 'Human Nutrition & Metabolic Disorders', USCR Mass Spectrometry, Faculty of Medicine Monastir, Tunisia

1. Introduction

Phenoxyacetic herbicides constitute one of the largest groups of herbicides sold in the world. Among them, since 1946, 2,4-Dichlorophenoxyacetic acid (2,4-D), whose structural formula is shown in Fig. 1, has been the most used. Nowadays, new formulations of 2,4-D are continuously made available. In fact, there are over 600 2,4-D products currently on the market. For over 60 years, 2,4-D has been the most commonly and widely used herbicide throughout the world (Tayeb et al., 2011 a).

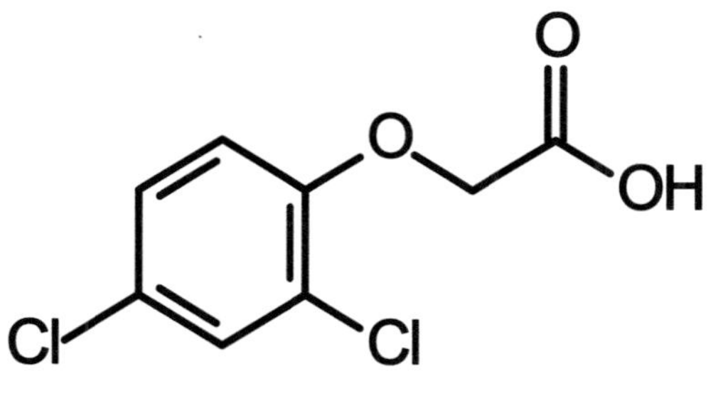

Fig. 1. Chemical structure of 2,4-D

2,4-D is primary used as a weed control in agriculture, forestry, and lawn care practices. It is marked as a selective systemic herbicide. Its herbicidal activity is mediated by an auxin-like capacity to alter normal protein synthesis and cell division in plant meristems and leaves (Stevens and Breckenridge, 2001). While at low concentrations 2,4-D acts as an auxin analogue promoting plant growth, at high concentrations it is lethal and used as herbicide against broad-leafed and woody plants (Mullison, 1987). Yet, Romero-Puertas et al. (2004) have recently suggested that the 2,4-D- herbicidal activity may also be due to an increase in the production of oxygen reactive species (ROS). The latter's lead to the generation of oxidative stress in the weed.

Upon application, 2,4-D is distributed into various compartments of the environment in accordance with its physical/chemical properties and local environmental conditions (Tayeb et al., 2011a). The toxicity of 2,4-D and other related compounds was attributed to the free acid form of the chemicals (Munro et al., 1992). It is known that it disturbs metabolism (Palmeira et al., 1995). Moreover, immunosuppressive (Pistl et al., 2003), neurotoxic (Bortolozzi et al., 2004) and hepatotoxic effects have been well documented (Tuschl and Schawb, 2003; Tayeb et al., 2010). As a phenoxyherbicide, 2,4-D may cause an array of adverse effects to the nervous system such as myotonia, disruption of the activity of nervous system and behavioral changes (Bortolozzi et al., 2004). In addition, it is known that 2,4-D provokes changes in the animal nervous system due to interaction with acetylcholinesterase (AChE) activity (Sarikaya and Yılmaz, 2003; Caglan et al., 2008; Cattaneo et al., 2008).

A review of the toxicology and mechanism of toxicity of 2,4-D is necessary to assess the potential risk for animal and human health. Literature on the induction of oxidative stress and involvement of lipid peroxidation after 2,4-D in vitro and in vivo exposure will be reviewed here to provide an updated scientific basis to derive future research studies on this compound. Included in the review was information on the possible implication of 2,4-D exposure in the pathogenesis of health problems; and, a survey of current studies dealing with the antioxidant properties of some substances to decrease the oxidative stress induced by 2,4-D.

2. The 2,4-D herbicide and oxidative stress

Although the exact mechanisms by which this herbicide is incorporated into cells are not totally understood, 2,4-D has been reported to be a peroxisome proliferator (Bradberry et al., 2000). In plant cells 2,4-D induces mitotic and meiotic irregularities both in vivo and in vitro (Khalatkar and Bhargava, 1982). In mammalian cells in vitro, 2,4-D inhibits cell growth, protein and DNA synthesis, and arrests cells in the G/S phase of the cell cycle (Rivarola et al., 1985). Later, Maire et al. (2007), in their study on mammalian cells, showed that DNA damage detected by the comet assay could be related to oxidative stress. 2,4-D was found to induce oxidative stress, a mechanism responsible for DNA damage measured by the comet assay in fish (Martinez-Tabche et al., 2004). The induction of oxidative stress leading to secondary genotoxicity was proposed as a possible mechanism for carcinogenicity (Beddowes et al., 2003).

Pesticide exposure can lead to oxidative stress through unregulated generation of reactive oxygen species (ROS) such as superoxide anion, hydrogen peroxide, hydroxyl radical, peroxyl radicals and singlet oxygen. ROS are produced during normal process in the cell. Under normal conditions antioxidant systems of the cell minimize damage caused by ROS. When ROS generation increases to an extent that it overcomes the cellular antioxidant systems, the result is oxidative stress. It is known that pesticides can cause oxidative stress, resulting in the generation of free radicals (Banerjee et al., 1999). It is suspected that pesticides induce alterations in antioxidants or free oxygen radical scavenging enzyme systems. In addition, it is generally believed that lipid peroxidation is one of the molecular mechanisms involved in pesticide induced toxicity (Akhgari et al., 2003). Indeed, Phenoxyherbicides stimulate generation/production of ROS. Selassie et al., (1998) suggests that this is related to two properties, one being the formation of free radicals from them, and the second being a direct attack of these phenoxyl radicals on biochemical processes in a number of sensitive metabolic pathways.

Herbicide 2,4-D has been suggested as a potential environmental endocrine disruptor and oxidative damage inducer (Munro et al., 1992; Mi et al., 2007). Several studies have shown that 2,4-D produces oxidative stress and/or depletes antioxidants both in vitro and in vivo. In vitro reports have looked, especially, at the effects of 2,4-D on hepatocytes and red blood cells (Palmeira et al., 1995; Bukowska, 2003). In vivo oxidative activity has been studied in many species including yeast, plants, fish and rats (Romero-Puerats et al., 2004; Teixeira et al., 2004; Oruc and Uner, 1999; Celik et al., 2006). Lipid peroxidation has been suggested as one of the molecular mechanisms involved in pesticide-induced toxicity; as a consequence such pesticides can disturb the biochemical and physiological functions of some organs.

2.1 *In vitro* studies

To study the in vitro effects of pesticide exposure, several researchers use biological lipid membranes model like erythrocyte ghosts as they are sensitive to the peroxidative process; since they are rich in polyunsaturated fatty acids in their membranes, a class of compounds highly susceptible to lipid peroxidation. The majority of works dealing with the effects of 2,4-D and its metabolites on erythrocytes were summarized in table 1. Toxic influence of 2,4-D may provoke disturbances in bilayer phospholipid structure that plays an important role in the correct function of cell membrane. Phenoxyherbicides interact with proteins and lipids of erythrocyte membrane (Suwalsky and Berites, 1996). Indeed, Janik and Wolf (1992) have demonstrated the inhibitory effect of chlorinated compounds on the Ca-ATPase which indicates a toxic effect to human erythrocytes functions. Bukowska et al. (1998) have found the increase in the level of methemoglobin (metHgb) and the change of the oxygen affinity of haemoglobin under the influence of 2,4-D. Later, Bukowska (2003) reported that treatment of human erythrocytes in vitro with 2,4-D at 250 and 500ppm resulted in decreased levels of reduced glutathione, decreased activity of superoxide dismutase, and increased levels of glutathione peroxidase. These significant changes in antioxidant enzyme activities and evidence of oxidative stress indicate that 2,4-D should be taken seriously as a cytotoxic and potentially genotoxic agent. In 2008, Bukowska et al. present the evidence for a direct prooxidant activity of phenoxyherbicides. In fact, the pro-oxidative action of these compounds is strongly dependent on the localization of the substituent in the phenol ring. Indeed, the compounds with chlorine residues in the second and fourth position of phenol ring cause strong damage to antioxidative enzymes and lipid peroxidation (Bukowska, 2003; Bukowska et al., 2000; Duchnowicz et al., 2002). Also, they much more easily penetrate the cell membrane. Bukowska et al. (2008) proposed a metabolic reaction chain that explains the mechanism of action of 2,4-D in vivo. The authors have noted that the prooxidative capability of this herbicide are related with its hydrolysis to 2,4-dichlorophenol that may generate radicals oxidizing H_2DCF, marker of oxidative status of the cells.

Other in vitro studies, dealing with the induction of oxidative stress after 2,4-D exposure, were conducted on hepatocytes. Palmeira et al. (1994) suggested that 2,4-D can decrease ATP, GSH and NADH levels while conversely increasing the levels of AMP, NAD, LDH and GSSG in rat hepatocytes. This herbicide at (1- 10 mM) may induce cell death by decreasing cellular GSH/GSSG ratio, promoting loss of protein thiol contents and inducing lipid peroxidation (Palmeira et al., 1995). In fact, it is suggested that membrane protein thiols can be attacked by radicals, resulting in a membrane protein thiol loss which in turn may also be associated with the development of hepatocellular injury.

Parameter	2,4-D	References
Decreased activity of CAT	- Effects observed at 1000 ppm (24 hours)	Bukowska, 2000
Induction of hemolysis Lipid peroxidation	- Effects observed at 1mM	Duchnowicz et al., 2002
Oxidation of haemoglobin	- Effects observed at 0.5 mM	Duchnowicz et al., 2002
Increase in membrane fluidity	- Effects observed at 1 mM	Duchnowicz and Koter, 2003
Glutathione peroxidise activity	- Effects observed at 1.13 mM	Bukowska et al., 2003
Depletion of GSH level Increased activity of GSH-Px Decreased activity of SOD	- Effects observed at 500 ppm - Effects observed at 250 ppm - Effects observed at 250 ppm (1hour)	Bukowska, 2003
W/S parameter that reflect denaturation or protein conformational in membrane	- Effects observed at 2 mM	Duchnowicz et al., 2005
Carbonyl group content Oxidation of H_2DCF	- Effects observed at 1.13 mM - Effects observed at 1.13 mM (3 h incubation)	Bukowska et al., 2008

Table 1. Effects of 2,4-D on human erythrocytes

2.2 *In vivo* studies

Studies in order to understand the toxic mechanism of 2,4-D in living cells have been performed using, e.g., Saccharomyces cerevisiae (Teixeira and Sa-Correia, 2002). Indeed, yeast has proved to be a useful experimental model for the study of basic molecular mechanisms underlying the toxicological effects of the important agrochemical 2,4-D and the associated adaptive responses (Papaefthimiou et al., 2004).

2.2.2 Yeast

At low pH (e.g. acidic soils, the alimentary canal of animals), the highly lipophilic weak acid 2,4-D exists in its undissociated lipophilic toxic form (RCOOH), which can readily cross the plasma membrane by passive diffusion. In the neutral cytosol, the undissociated form of 2,4-D dissociates, leading to internal acidification (Simoes et al., 2003; Fernandes et al., 2003) and to accumulation of the toxic counter-ion ($RCOO^-$), which cannot easily cross the plasma membrane lipid bilayer (Figure 3). Therefore, at low pH the toxic potential of the herbicide increases dramatically (Cabral et al., 2003).

Herbicide accumulation in the yeast cell leads to a dose-dependent increase in the level of hydroxyl radicals, as detected using in vivo electron paramagnetic resonance (EPR) spectroscopy (Teixeira et al., 2004). A coordinate transient increase in hydroxyl radical and lipid peroxidation levels was registered as a consequence of acute 2,4-D stress (Figure 2).

Results from of the early yeast response to 2,4-D provided additional clues to its possible mode of pro-oxidant action (Table 2). The response to 2,4-D includes the upregulation of genes involved in peroxisomal beta oxidation and mitochondrial oxidative phosphorylation, two metabolic processes leading to the endogenous generation of reactive oxygen species (ROS). Electron leakage from the mitochondrial respiratory chain might further increase the level of ROS generated during short-term cell exposure to 2,4-D (Figure 2).

The global yeast response to 2,4-D, revealed by microarray and proteomic analyses, indicates the upregulation of a large number of genes involved in alternative carbon and nitrogen source metabolism (Teixeira et al., 2006a) and in the uptake and biosynthesis of amino acids (Teixeira et al., 2006a; Teixeira et al., 2005), correlating with a dramatic reduction of the intracellular concentration of amino acids (Teixeira et al., 2005). These adaptive mechanisms might be a response to the deleterious effects exerted by 2,4-D on plasma membrane lipid organization and permeability, leading to nutrient import inhibition (Bradberry et al., 2000) (Figure 2).

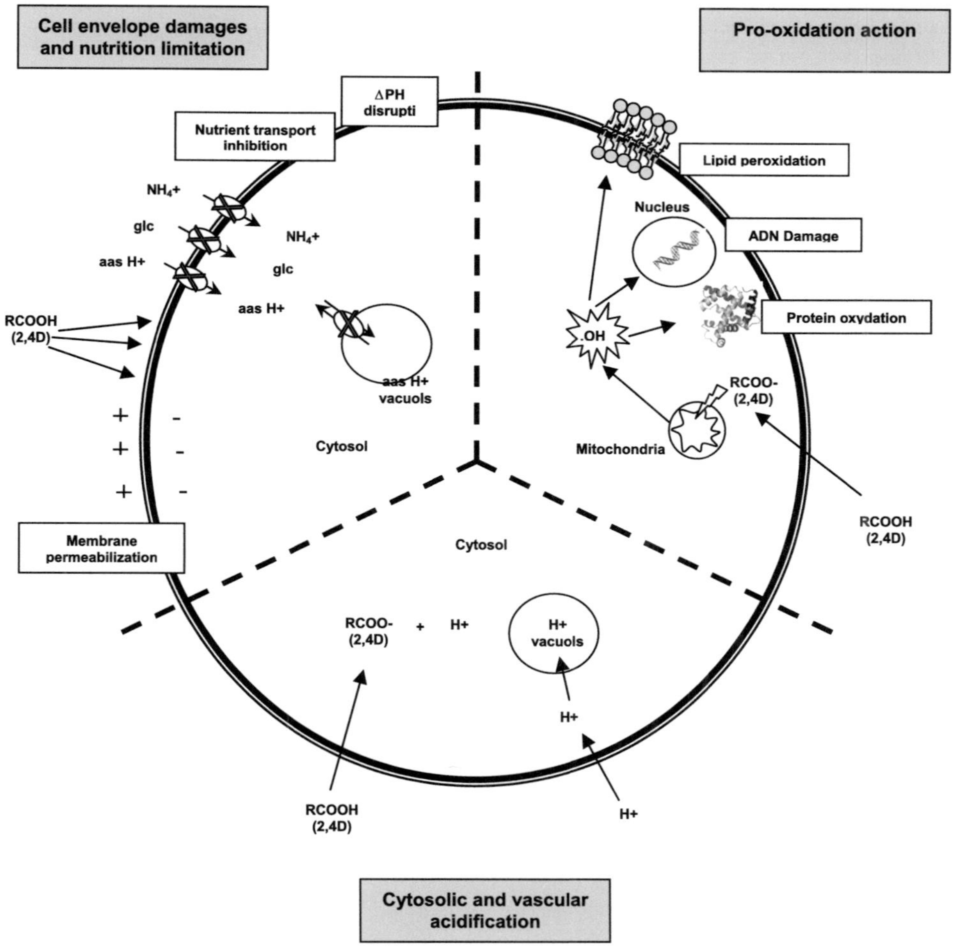

Fig. 2. Model for the mechanisms underlying the toxicity of 2,4-D in the yeast

Moreover studies on S. cerevisiae, Benndorf et al. (2006) studied the responses of Pseudomonas putida to chlorophenoxy herbicides. They described the induction of Stress proteins by antioxidant enzymes in the organisms P. putida KT2440 (cells) when stressed with 2,4-D. Also, 2,4-D induced oxidative stress in spermatogonial cells (Mi et al., 2007). In fact, exposure to 2,4-D elicited TBARS products of lipid peroxidation, and decreased glutathione content and SOD activity in embryonic chickens. In 2010, Park et al. studied the Biological and molecular responses of Chironomus riparius (Diptera, Chironomidae) to herbicide 2,4-D. Their results showed that the responses of HSPs and GST in C. riparius exposed to 2,4-D suggest that it can induce oxidative damage and changes in the endocrine system. The authors found that the upregulation of ferritin genes in C. riparius exposed to 2,4-D may lead to protection responses against 2,4-D induced oxidative damage.

Effects observed after 2,4-D exposure	**Reference**
- Production of many stress and heat-shock proteins during the adaptation period, and that eventually cell division occurred in the presence of 2,4-D.	Teixeira and Sa-Correia, 2002
- Coordinated stimulation of vacuolar and plasma membrane H^+-ATPase activities, to counteract the dissipation of the physiological H^+-gradients across vacuolar and plasma membranes occurring under 2,4-D stress.	Fernandes et al., 2003
- The transient increase in free radical (hydroxyl radicals) generation and lipid peroxidation in the yeast cell challenged with 2,4-D correlates with the stimulation of the activity of antioxidant enzymes (Ctt1p, Sod1p, Grx1p and Grx2p).	Teixeira et al., 2004
- Increased content of Vma1p and Vma2p (two submits of vacuolar H^+-ATPase).	Teixeira et al., 2005
- Changes at the level of membrane lipid composition in yeast cells adapted to 2,4-D, including an increase in the saturation degree of membrane fatty acids.	Viegas et al., 2005
- Upregulation of genes involved in peroxisomal β-oxidation and mitochondrial oxidative phosphorylation, two metabolic processes leading to the endogenous generation of ROS	Teixeira et al., 2006a
- Identification of Msn2p and Msn4p as the putative transcriptional regulators of 20% of the 2,4-D-activated genes. These target genes encode heat shock proteins, molecular chaperones and antioxidant enzymes	Teixeira et al., 2006b

Table 2. Effects of 2,4-D on yeast

2.2.2 Fish

On fish, at tissue level, the 2,4-D toxicity follow the common route through the gills and external tegumenta and by the digestive tract to a small extent. The absorbed chemical has been shown to bind proteins of plasma so as to be transported throughout the organisms (Arnold and Beasley, 1989). Structural abnormalities like vacuolation of erythrocytes is a regular feature of 2,4-D (Ateeq et al., 2002) like many other chemicals.

Several studies showed that the antioxidants of fish may be useful biomarkers of exposure to aquatic pollutants as 2,4-D herbicide (Ahmad et al., 2000). Table 3 present the most important works done in this field. The study of Ozcan Oruc and Uner (2000) aims to investigate the effects of the herbicide 2,4-D and the insecticide azinphosmethyl on hepatic antioxidant enzyme activities and lipid peroxidation in tilapia. Fish were exposed to 27 ppm 2,4-D, 0.03 ppm azinphosmethyl and to a mixture of both for 24, 48, 72 and 96 h. It was concluded that the metabolism of pesticide-exposed *O. niloticus* resisted the oxidative stress using the antioxidant mechanism and prevented the increase of lipid peroxidation. Later on 2004, these authors studied the tissue-specific oxidative stress responses in fish exposed to 2,4-D and azinphosmethyl. Results indicate that the toxicities of azinphosmethyl and 2,4-D may be related to oxidative stress. In fact, this last study revealed that fish exposed to pesticides develop tissue-specific adaptive responses to protect cells against oxidative stress. Moreover, according to our results, the elevations in gill SOD activity and kidney GST activity serve as biomarkers of oxidative stress and may be helpful in assessing the risk of environmental contaminants. Also, Zhang et al. (2004) explored the hepatic antioxidant responses of fish *Carassius auratus* to long-term exposure of 2,4- dichlorophenol. They concluded that SOD and Se-GPx may be potential early biomarkers of 2,4-DCP contamination in aquatic ecosystems.

Specie	**Organ**	**Dose**	**Results**	**References**
O. niloticus	liver	27 ppm for 24, 48, 72 and 96 h	Increase in GPx (for 96h) and GR activities	Ozcan Oruc and Uner, 2000
O. niloticus	kidney	87 ppm for 96h	Increase in GPx and GST activities	Ozcan Oruc and Uner, 2004
	Gill		Increase in SOD activity	
	Brain		Decrease in GPx activity	
C.carpio	kidney	87 ppm for 96h	Increase in CAT, GPx and GST activities	Ozcan Oruc and Uner, 2004
	Gill		Increase in SOD activity	
C. auratus	Liver	0.005 - 1.0 mg/l for 40 days	Alteration in CAT, SOD and GPx activities	Zhang et al., 2004

Table 3. Effects of 2,4-D on fish

2.2.3 Rats

On view of the data concerning rats, it can be concluded that exposure to 2,4-D induced oxidative stress and lipid Peroxidation (table 4). The first important study was done by Celik et al. (2006), who studied the effects of 2,4-D on serum marker enzymes, erythrocyte and tissue antioxidant defense and lipid peroxidation in rats. The authors found that the administration of 1.5 and 3mg/day of 2,4-D during 25 days induced in vivo oxidation. Recently, from the series of experiments described by Tayeb et al. (2010; 2011 b; 2011c) and

Nakbi et al. (2010; 2011a; 2011b): it is quite evident that subacute exposure of rats to 5, 15, 75 and 150 mg/kg BW during 28 days caused significant negative changes in the erythrocyte, liver and kidney functions. The fatty acid composition of the erythrocyte membranes also of hepatocytes has been altered by the 2,4-D exposure with 2,4-D exposure increased levels of SFA and the decreased level of unsaturated fatty acids (UFA); increase in the index of fatty acid unsaturation. These results may explain the higher amounts of MDA observed in 2,4-D treated group. Furthermore, the antioxidant enzyme activities, in liver, erythrocytes and kidneys were significantly affected.

Thus, results described in table 4 indicated the potential effects of 2,4-D to cause oxidative stress in rat. These results are partly in accordance despite the differences between studies in their settings, materials and experimental designs. While, in an investigation done by Dinamarca et al. (2007) who looked for the effects of 2,4-D on the generation of oxidative stress during early pregnancy in mice, they proved that 2,4-D in the concentrations usually found in blood can not provoke oxidative stress. Indeed, these negative results agree with others in that 2,4-D seems to induce in vivo oxidation only with high doses and with increasing length of administration period. It is known that phenoxyacetic acid herbicides are eliminated by a renal anion transport system which is saturated as plasma concentration increases. Since saturation of the rodent renal transporter is reported to occur at doses in excess of 50 mg/kg/day, then the rise in blood concentration as dose of herbicide increases may lead to the distribution of the compound into cells and tissues which then become susceptible to oxidative stress. So, this may account for observations of oxidation being induced in rats.

3. The 2,4-D exposure and human diseases

Works done by Mountassif et al. (2008) Nakbi et al. (2011) and Tayeb et al., (2011 b; 2011c) clearly demonstrated that subacute exposure to 2,4-D significantly modified lipidic status, disrupt lipid metabolism, also, we have noted an increase in the LDL/HDL and TC/HDL ratios, which are pertinent indices of the incidence of cardiovascular risk. All these findings support the hypothesis that high doses of 2,4-D might contribute to development of vascular and cardiac pathologies.

Indeed, some pesticides have been implicated in the pathogenesis of cardiovascular disorders, hypertension and other health related problems (Singh et al., 2007). Kang et al. (2006) have noted that there are long-term health consequences of Agent orange herbicide (a mixture of 2,4-D and 2,4,5-trichlorophenoxyacetic acid (2,4,5-T)) exposure among army Vietnam veterans who were exposed to this herbicide. The study group showed significantly higher risk of diabetes, heart diseases, and circulatory diseases such as hypertension. Also, England (1981) reported that prolonged exposure to herbicides such as 2,4-D has been associated with Coronary Artery Ectasia (CAE). It was reported that more than 50% of CAE were caused by atherosclerosis (Lin et al. 2008). Recently, Schreinemachers (2010) indicate that human exposure to 2,4-D was associated with changes in biomarkers that, have been linked to risk factors related to the pathogenesis of acute myocardial infarction and type-2 diabetes, such as dyslipidemia and impaired glucose metabolism.

Matrices	Doses	Results	References
- Erythrocytes, brain, liver, kidney, heart	1.5, 3mg/kg/B.W/day during 25 days	- Induction of in vivo oxidation: changes in the GSH, GST, GR, SOD, CAT activities and MDA levels.	Celik et al., 2006
- Liver	3 mg/kg/B.W/day for 4 weeks	- Moderated oxidative stress in liver cells: increase of lipide peroxidation (MDA) and decrease in CAT activity	Mountassif et al., 2008
- Blood	0.01, 0.1 and 100 mg/kg/B.W/day during gestation days 0-9	- Catalase activity and TBARs were not modified -TAC (Total antioxidant capacity) was significantly decreased at 100 mg/kg/d of 2,4-D	Dinamarca et al., 2007
- Erythrocytes	5 mg/kg/B.W/day for 4 weeks	- Significant decrease of SOD, GPx, GR and CAT activities - Changes of the fatty acid profile in erythrocyte membranes	Nakbi et al., 2010 a
- Liver	5 mg/kg/B.W/day for 4 weeks	- Increased hepatic lipid peroxidation (MDA, conjugated dienes) and decreased hepatic antioxidant enzyme activities (SOD, CAT, GPx, GR) - Modification of liver's fatty acid composition	Nakbi et al., 2010 b
- Erythrocytes	15, 75 and 150 mg/kg/B.W/day for 4 weeks	- The MDA level was significantly increased in 2,4-D treated groups. - Fatty acid composition of the erythrocytes was also significantly changed with 2,4-D exposure, in favor of the peroxidation of polyunsaturated fatty acids. - Antioxidant enzyme (SOD, CAT, GPx, and GR) activities were significantly decreased	Tayeb et al., 2011 b
- Liver	15, 75 and 150 mg/kg/B.W/day for 4 weeks	- Significantly increase in The MDA and conjugated dienes level - Fatty acid composition of the liver was significantly changed - Hepatic antioxidant enzyme (SOD, CAT, GPx, and GR) activities were significantly affected.	Tayeb et al., 2011c Tayeb et al. 2010
- Kidney	15, 75 and 150 mg/kg/B.W/day for 4 weeks	- Increase in kidney MDA - The activities of CAT, SOD, GPx, GR were significantly affected due to 2,4-D exposure.	Tayeb et al., 2011 c
- Kidney	600 mg/L from the 14th day of pregnancy until day 14 after delivery.	- Increase in TBARs and protein carbonyl levels - Decrease in antioxidant enzyme activities (CAT, SOD, GPx) in the kidneys of suckling pups and their mothers. - Significant decline in kidney glutathione, non-protein thiol and vitamin C levels.	Troudi et al., 2011

Table 4. Effects of 2,4-D on rats

4. Protective effects of some antioxidants on 2,4-D toxicity

Several scavenging agents and antagonists are established to reduce pesticides toxicity (Kalender et al., 2004; Grajeda-Cota et al., 2004). The prevention of the peroxidation processes by using antioxidants and free radical sweepers of plant origin becomes an important issue of clinical nature. So, in their *in vitro* investigations, Bors et al. (2009; 2011) have evaluated the impact of extracts of *Uncaria tomentosa* leaves and bark on human erythrocytes as well as the antioxidant properties of U. *tomentosa* extracts against oxidative stress induced by 2,4-D and its environmental transformation products 2,4-DCP and catechol. Their studies showed that U. *tomentosa* extracts protected against the induction of hemolysis, haemoglobin oxidation and ROS increase in human erythrocytes incubated with 2,4-D (Bors et al., 2009).

Other in vivo studies have been investigated in the exploitation of the anti-inflammatory and vascular protective properties of olive oil polyphenols. Recently, Nakbi et al. (2010a; 2010b; 2011) confirmed the beneficial effects of extra virgin olive oil and its hydrophilic and lipophilic fractions for their lipid-lowering, antioxidative, and protective effects against oxidative damage induced by 2,4-D. In fact, extra virgin olive oil and its extracts administered to 2,4-D-treated rats protected tissues and erythrocyte membranes against oxidative damage by means of preventing excessive lipid peroxidation to increase the monounsaturated fatty acid composition and by maintaining serum marker enzymes and antioxidants enzymes at near normal concentrations. So including olive oil in the diet may offer benefits in decreasing tissue damage and the atherosclerotic process during 2,4-D exposure in rats.

5. Conclusions

Until nowadays, acute toxicological tests have been conducted on the various forms of 2,4-D. The precise mechanism of 2,4-D acute toxicity may involve disruption of plasma and intracellular membranes or uncoupling of oxidative phosphorylation; this last mechanism was involved in the generation of oxidative stress. In fact, Several lines of evidence indicate that oxidative stress and ROS formed in the presence of 2,4-D could be responsible for its toxic effects in many settings *in vitro* and *in vivo*. Consequently, increased tissue oxidative stress can lead to cell damage. Given the implications of oxidative stress in several human genetic diseases, ageing, inflammation and cancer development, these results are of concern in situations of eventual massive or repeated exposure to the herbicide. In recent years, scientists have focused on the preventive effects of some natural antioxidant against degenerative diseases mediated by the ROS. Our recent finding suggests that including olive oil in the diet may offer benefits in decreasing tissue damage and the atherosclerotic process during 2,4-D exposure in rats. Further experimental evidence, mechanism-oriented studies and clinical trials are needed to understand and to further characterize the toxic effects of 2,4-D herbicide.

6. References

Ahmad, I.; Hamid, T.; Fatima, M.; Chand, H.; Jain, S.; Athar, M. & Raisuddin, S. (2000). Induction of hepatic antioxidants in freshwater catfish (Channa punctatus Bloch) is a biomarker of paper mill effluent exposure. *Biochimica et Biophysica Acta (BBA) - General Subjects*, Vol. 1523, 37-48.

Akhgari, M.; Abdollahi, M.; Kebryaeezadeh, A.; Hosseini, R. & Sabzevari, O. (2003). Biochemical evidence for free radical induced lipid peroxidation as a mechanism for subchronic toxicity of malathion in blood and liver of rats. *Human Experimental Toxicology,* Vol . 22, 205–211.

Arnold, E. & Beasley, V. (1989). The pharmacokinetics of chlorinated phenoxy acid herbicides: a literature review. *Veterinary and Human Toxicology,* Vol. 31, 121– 125.

Ateeq, B.; Farah, M.; Ali, M. & Ahmad, W. (2002). Induction of micronuclei and erythrocyte alteration in catfish Clarias batrachus by 2,4-dichlorophenoxy acetic acid and butachlor. *Mutation Research,* Vol. 518, 135– 144.

Banerjee, B.; Seth, V.; Bhattacharya, A.; Pasha, S. & Chakraborty, A. (1999). Biochemical effects of some pesticides on lipid peroxidation and free radical scavengers, *Toxicology Letters,* Vol. 107, 33–47.

Beddowes, E.; Faux, S. & Chipman, J. (2003). Chloroform, carbon tetrachloride and glutathione depletion induce secondary genotoxicity in liver cells via oxidative stress, *Toxicology,* Vol. 187, 101–115.

Benndorf, D.; Thiersch, M.; Loffhagen, N.; Kunath, C. & Harms, H. (2006). Pseudomonas putida KT2440 responds specifically to chlorophenoxy herbicides and their initial metabolites. *Proteomics,* Vol. 6, 3319–3329.

Bors, M.; Baczynska, J.; Bukowska, B.; Duda, D.; Michałowicz, J.; Pilarski, R. & Gulewicz, K., 2009. The protective role of extracts from leaves and bark of Uncaria tomentosa (Willd.) DC in oxidative damage of human erythrocytes provoked by 2,4-D herbicide. In: Garbacz, J.K. (Ed.), *Diagnosing of State of the Environment, Research Methods – Prognosis,* vol. 3. Bydgoszcz, Poland, pp. 231–240.

Bors, M.; Bukowska, B.; Pilarski, R.; Gulewicz, K.; ski , J.; Michałowicz, J. & Koter-Michalak, M. (2011). Protective activity of the Uncaria tomentosa extracts on human erythrocytes in oxidative stress induced by 2,4-dichlorophenol (2,4-DCP) and catechol. *Food and Chemical Toxicology,* Vol. 49, 2202–2211.

Bortolozzi, A.; Evangelista, A.; De Duffard, R. & Duffard, M. (2004). Antonelli, Effects of 2,4-dichlorophenoxyacetic acid exposure on dopamine D2-like receptors in rat brain, *Neurotoxicology Teratology,* Vol. 26, 599–605.

Bradberry, S. et al. (2000) Mechanisms of toxicity, clinical features, and management of acute chlorophenoxy herbicide poisoning: a review. *Journal of Toxicology and Clininical Toxicology,* Vol. 38, 111–122.

Bukowska, B.; Rychlik, B. & Krokosz, A. (2008). Michalowicz, Phenoxyherbicides induce production of free radicals in human erythrocytes: oxidation of dichlorofluorescine and dihydrorhodamine 123 by 2,4-D-Na and MCPA-Na, *Food and Chemical Toxicology,* Vol. 46, 359–367.

Bukowska, B. (2003). Effects of 2,4-D and its metabolite 2,4-dichlorophenol on antioxidant enzymes and level of glutathione in human erythrocytes. *Comparative Biochemistry and Physiology Part C,* Vol.135, 435–441.

Bukowska, B.; Chajdys, A.; Duda, W. & Duchnowicz, P. (2000). Catalase activity in human erythrocytes: effect of phenoxyherbicides and their metabolites. *Cell Biology* International, Vol. 24, 705–711.

Bukowska, B.; Reszka, E. & Duda, W. (1998). Influence of phenoxyherbicides and their metabolities on the form of oxy- and deoxyhemoglobin of vertebrates, *Biochemistry and Molecular Biology International,* Vol. 4, 47–59.

Cabral, M. et al. (2003). Toxicity of chlorinated phenoxyacetic acid herbicides in the experimental eukaryotic model Saccharomyces cerevisiae: role of pH and of growth phase and size of the yeast cell population. *Chemosphere,* Vol. 51, 47–54.

Caglan, A.; Karasu Benli, Sarıkaya, R.; Sepici-Dincel, A.; Selvi, M.; Sahin, D. & Erko, F. (2007). Investigation of acute toxicity of (2,4-dichlorophenoxy) acetic acid (2,4-D) herbicide on crayfish (Astacus leptodactylus Esch. 1823). *Pesticide Biochemistry and Physiology,* Vol. 88, 296–299.

Cattaneo, R.; Loro, V.; Spanevello, R.; Silveira, F.; Luz, L.; Miron, D.; Fonseca, M.; Moraes, B. & Clasen, B. (2008). Metabolic and histological parameters of silver catfish (Rhamdia quelen) exposed to commercial formulation of 2,4- dichlorophenoxiacetic acid (2,4-D) herbicide. *Pesticide Biochemistry and Physiology,* Vol. 92, 133–137.

Celik, I.; Tuluce, Y. & Isik, I. (2006). Influence of subacute treatment of some plant growth regulators on serum marker enzymes and erythrocyte and tissue antioxidant defense and lipid peroxidation in rats. *Journal of Biochemistry and Molecular Toxicology,* Vol. 20, 174–182.

Dinamarca, V.; Hidalgo, M. & Cavieres, M. (2007). Lack of effects of 2,4-Dichlorophenoxyacetic acid administration on markers of oxidative stress during early pregnancy in mice. *Toxicology,* Vol. 237, 104–110.

Duchnowicz, P. & Koter, M. (2003). Damage to the erythrocyte membrane caused by chlorophenoxyacetic herbicides. *Cellular Molecular Biology Letters,* Vol. 8, 25–30.

Duchnowicz, P.; Koter, M. & Duda, W. (2002). Damage of erythrocyte by phenoxyacetic herbicides and their metabolites. *Pesticide Biochemistry and Physiology,* Vol. 74, 1-7.

Duchnowicz, P.; Szczepaniak, P. & Koter, M. (2005). Erythrocyte membrane protein damage by phenoxyacetic herbicides and their metabolites. *Pesticide Biochemistry and Physiology,* Vol. 82, 59–65.

England, J. (1981). Herbicides and coronary ectasia.*The Medical Journal of Australia,* Vol. 1, 371.

Fernandes, A.; et al. (2003). Activation and significance of vacuolar H+- ATPase in Saccharomyces cerevisiae adaptation and resistance to the herbicide 2,4-dichlorophenoxyacetic acid. *Biochemical Biophysical Research Communications,* Vol. 312, 1317–1324.

Grajeda-Cota, P.; Ramirez-Mares, M. & Mejia, E. (2004). Vitamin C protects against in vitro cytotoxicity of cypermethrin in rat hepatocytes. *Toxicology In Vitro,* Vol. 18, 13–19.

Janik, F. & Wolf, H. (1992). The Ca-ATPase of human erythrocytes as an in vitro toxicity test system - acute effects of some chlorinated compounds. *Journal of Applied Toxicology,* Vol. 12, 351–358.

Kalender, S.; Kalender, Y.; Ogutcu, A.; Uzunhisarcikli, M.; Durak, D. & Acikgoz, F. (2004). Endosulfan-induced cardiotoxicity and free radical metabolism in rats: the protective effect of vitamin E. *Toxicology,* Vol. 202, 227–35.

Kang, H.; Dalager, N.; Needham, L.; Patterson, D.; Lees, P.; Yates, K. & Matanoski, G. (2006). Health status of army chemical corps Vietnam veterans who sprayed defoliant in Vietnam. *American Journal of Industrial Medicine,* Vol. 49, 875–884.

Khalatkar, A. & Bhargava, Y. (1982). 2,4-dichlorophenoxyacetic acid—A new environmental mutagen. *Mutation Research Letters,* Vol. 103, 111-114.

Lin, C.; Chen, C.; Lin, T. & Lin, C. (2008). Coronary arteria ectasia, *Tzu Chi Medical Journal,* Vol. 20, 270–274.

Maire, M.; Rast, C. ; Landkocz, Y. & Vasseur, P. (2007). 2,4-Dichlorophenoxyacetic acid: Effects on Syrian hamster embryo (SHE) cell transformation, c-*Myc* expression, DNA damage and apoptosis. *Mutation Research,* Vol. 631, 124–136.

Martinez-Tabche, L.; Madrigal-Bujaidar, E. & Negrete, T. (2004). Genotoxicity and lipoperoxidation produced by paraquat and 2,4-dichlorophenoxyacetic acid in the gills of rainbow trout (*Oncorhynchus mykiss*), *Bulletin Environnemental Contamination Toxicology.* Vol. 73, 146–152.

Mi,Y.; Zhang, C. & Taya, K. (2007). Quercetin protects spermatogonial cells from2,4-d-induced oxidative damage in embryonic chickens. *Journal of Reproduction and Development,* Vol. 53, 749–754.

Mountassif, D.; Kabine, M.; Mounchid, K.; Mounaji, K.; Latruffe, N. & El Kebbaj, M. (2008). Biochemical and histological alterations of cellular metabolism from jerboa (*Jaculus orientalis*) by 2,4-Dichlorophenoxyacetic acid: Effects on D-3-hydroxybutyrate dehydrogenase. *Pesticide Biochemistry and Physiology,* Vol. 90, 87-96.

Mullison, W.R. "Environmental Fate of Phenoxy Herbicides. Fate of Pesticides in the Environment" *California Agricultural Experiment Station Publication 1987.* 3320, pp. 121- 131 .

Munro, I.; Carlo, G.; Orr, J.; Sund, K.; Wilson, R.; Kennepohl, E.; Lynch, B. & Jablinske, M., (1992). A comprehensive, integrated review and evaluation of the scientific evidence relating to the safety of the herbicide 2,4-D, *International Journal of Toxicology,* Vol. 11, 559–664.

Nakbi, A.; Tayeb, W.; Dabbou, S; Chargui, I.; Issaoui, M.; Zakhama, A.; Miled, A. & Hammami, M., (2011). Hypolipidimic and antioxidant activities of Virgin Olive Oil and its fractions in 2,4-D- treated rats. *Nutrition,* doi:10.1016/j.nut.2011.02.009

Nakbi, A.; Tayeb, W.; Dabbou, S.; Issaoui, M.; Grissa, A.; Attia, N. & Hammami, M. (2010 a). Dietary olive oil effect on antioxidant status and fatty acid profile in the erythrocyte of 2,4-D exposed rats. *Lipids in Health and Disease,* Vol. 9:89.

Nakbi, A.; Tayeb, W.; Grissa, A.; Issaoui, M.; Dabbou, S.; Chargui, I.; Ellouz, M.; Miled, A. & Hammami, M., (2010 b). Effects of olive oil and its fractions on oxidative stress and the liver's fatty acid composition in 2,4- Dichlorophenoxyacetic acid- treated rats, *Nutrition and metabolism,* Vol. 7:80.

Ozcan Oruc, E. & Üner, N. (1999). Effects of 2,4-Diamin on some parameters of protein and carbohydrate metabolisms in the serum, muscle and liver of Cyprinus carpio, *Environmental Pollution,* Vol.105, 267–272.

Ozcan Oruc, E. & Uner, N. (2004). Tissue-specific oxidative stress responses in fish exposed to 2,4-D and azinphosmethyl. *Comparative Biochemistry and Physiology Part C,* Vol. 137, 43–51.

Ozcan Oruc, E. & Uner, N. (2000). Combined effects of 2,4-D and azinphosmethyl on antioxidant enzymes and lipid peroxidation in liver of *Oreochromis niloticus Comparative Biochemistry and Physiology Part C,* Vol. 127, 291–296.

Palmeira, C.; Moreno, A. & Madeira, V. (1994). Interactions of herbicides 2,4-D and dinoseb with liver mitochondrial bioenergetics, *Toxicological Applied. Pharmacology,* Vol. 127, 50–57.

Palmeira, C.; Moreno, A. & Madeira, V. (1995). Thiols metabolism is altered by the herbicides paraquat, dinoseb and 2,4-D: a study in isolated hepatocytes, *Toxicology Letters,* Vol. 81, 115–123.

Papaefthimiou, C. et al. (2004). Comparison of two screening bioassays, based on the frog sciatic nerve and yeast cells, for the assessment of herbicide toxicity. *Environ. Toxicology Chemistry,* Vol. 23, 1211– 1218.

Park, K.; Park, J.; Kim, J. & Kwak., I. (2010). Biological and molecular responses of Chironomus riparius (Diptera, Chironomidae) to herbicide 2,4-D (2,4-dichlorophenoxyacetic acid). *Comparative Biochemistry and Physiology Part C,* Vol. 151, 439–446.

Pistl, J.; Kovalkovicova, N.; Holovska, V.; Legath, J. & Mikula, I. (2003). Determination of the immunotoxic potential of pesticides on functional activity of sheep leukocytes in vitro, *Toxicology,* Vol. 188, 73–81.

Rivarola, V.; Bergesse, J. & Balegno, H. (1985). DNA and protein synthesis inhibition in chinese hamster ovary cells by dichlorophenoxyacetic acid . *Toxicology Letters,* Vol. 29, 137-144.

Romero-Puertas, M.; Mccarthy, I.; Gomez, M.; Sandalio, L.; Corpas, F.; Del Rio, L. & Palama, J. (2004). Reactive oxygen species-mediated enzymatic systems involved in the oxidative action of 2,4-dichlorophenoxyacetic acid, *Plant, Cell and Environement.* Vol.27, 1135–1148.

Sarikaya, R. & Yılmaz, M. (2003). Investigation of acute toxicity and the effect of 2,4-D (2,4-dichlorophenoxyacetic acid) herbicide on the behavior of the common carp (Cyprinus carpio L., 1758; Pisces, Cyprinidae), *Chemosphere,* Vol. 52,195–201.

Schreinemachers, D. (2010). Perturbation of lipids and glucose metabolism associated with previous 2,4-D exposure: a cross-sectional study of NHANES III data, 1988–1994, *Environmental Health,* Vol. 26, 9–11.

Selassie, C.; De Soya, T.; Rosario, M.; Gao, H. & Hansch, C. (1998). Phenol toxicity in leukemia cells: a radical process? *Chemico-Biological Interactions,* Vol. 113, 175–190.

Simoes, T. et al. (2003). Adaptation of Saccharomyces cerevisiae to the herbicide 2,4-dichlorophenoxyacetic acid, mediated by Msn2p- and Msn4p-regulated genes: important role of SPI1. *Applied and Environmental Microbiology,* Vol. 69, 4019–4028

Singh, V.; Jyoti Krishna Reddy, M.; Kesavachandran, C.; Rastogi, S. & Siddiqui, M. (2007). Biomonitoring of organochlorines, glutathione, lipid peroxidation and cholinesterase activity among pesticide sprayers in mango orchards, *Clinica Chimica Acta,* Vol. 377, 268–272.

Stevens, J. & Breckenridge, C. Crop protection chemicals, in: W.A. Hayes (Ed.), *Principles and Methods of Toxicology,* Taylor & Francis, Philadelphia, 2001, pp. 718–816.

Suwalsky, M. & Berites, M. (1996). Interaction of 2,4-dichlorophenoxyacetic acid (2,4-D) with cell and modal membranes, *Biochimica et Biophysica Acta,* Vol. 1285, 267- 276.

Tayeb, W.; Nakbi, A.; Trabelsi, M.; Attia, N.; Miled, A. & Hammami, M., 2010. Hepatotoxicity induced by sub-acute exposure of rats to 2,4-dichlorophenoxyacetic acid based herbicide ''Désormone lourd'', *Journal of Hazardous Materials,* Vol. 180, 225–233

Tayeb, W. (2011 c). "Risk assessment of the occupational exposure to pesticides: Study of the in vivo and in vitro toxicity of an herbicide considered at risk: the 2,4-D ». Monastir, Tunisia. Thesis, 350 p.

Tayeb, W.; Chaieb, I. & Hammami, M. (2011 a). Environmental fate and effects of 2,4-Dichlorophenoxyacetic herbicide. Chap de livre «*Herbicides: Properties, crop protection and Environnemental hazards* ». Book Editor: PIOTROWSKI, K.P., Nova Sciences publisher, NY. in press.

Tayeb, W.; Nakbi, A.; Chargui, I.; Cheraief, I.; Miled, A. & Hammami, M. (2011 b). Subacute effects of 2,4-Diclorophenoxyacetic herbicide on antioxidant defense system and lipid peroxidation in rat erythrocytes. *Pesticide Biochemistry Physiology,* Vol. 99, 256 - 264.

Teixeira, M. et al. (2005). A proteome analysis of the yeast response to the herbicide 2,4-dichlorophenoxyacetic acid. *Proteomics,* Vol. 5, 1889–1901.

Teixeira, M. et al. (2006 a). Early transcriptional response of Saccharomyces cerevisiae to stress imposed by the herbicide 2,4-dichlorophenoxyacetic acid. *FEMS Yeast Research,* Vol.6, 230–248.

Teixeira, M.C. et al. (2006 b). The YEASTRACT database: a tool for the analysis of transcription regulatory associations in Saccharomyces cerevisiae. *Nucleic Acids Research,* Vol. 34, D446–D451

Teixeira, M.; Duque, P. & Sa´-Correia, I. (2007). Environmental genomics: mechanistic insights into toxicity of and resistance to the herbicide 2,4-D. *TRENDS in Biotechnology* , Vol.25 , 363- 370.

Teixeira, M. & Sa-Correia, I. (2002). Saccharomyces cerevisiae resistance to chlorinated phenoxyacetic acid herbicides involves Pdr1p-mediated transcriptional activation of TPO1 and PDR5 genes. *Biochemical and Biophysical Research Communications,* Vol. 292, 530–537.

Teixeira, M.; Telo, J.; Duarte, N. & Sá-Correia, I. (2004). The herbicide 2,4 dichlorophenoxyacetic acid induces the generation of free-radicals and associated oxidative stress responses in yeast, *Biochemical and Biophysical Research Communications,* Vol. 324, 1101–1107.

Troudi, A.; Soudani, N.; Mahjoubi Samet, A.; Ben Amara, I. & Zeghal, N., (2011). 2,4-Dichlorophenoxyacetic acid effects on nephrotoxicity in rats during late pregnancy and early postnatal periods. *Ecotoxicology and Environemental Safety,* doi:10.1016/j.ecoenv.2011.07.032.

Tuschl, H. & Schwab, C. (2003). Cytotoxic effects of the herbicide 2,4-dichlorophenoxyacetic acid in HepG2 cells, *Food Chemical Toxicology,* Vol. 41, 385–393.

Viegas, C. et al. (2005). Yeast adaptation to 2,4-dichlorophenoxyacetic acid involves increased membrane fatty acid saturation degree and decreased OLE1 transcription. *Biochemical and Biophysical Research Communications,* Vol. 330, 271–278.

Zhang, J.; Shen, H.; Wang, X.; Wu, J. & Xue, Y. (2004). Effects of chronic exposure of 2,4-dichlorophenol on the antioxidant system in liver of freshwater fish Carassius auratus. *Chemosphere, Vol.* 55, 167–174

7

Environmental Pollution and Oxidative Stress in Fish

Oksana B. Stoliar[1] and Volodymyr I. Lushchak[2]
[1]Ternopil National Pedagogical University Named after Volodymyr Hnatiuk
[2]Precarpathian National University Named after Vassyl Stefanyk
Ukraine

1. Introduction

Living systems encounter a variety of stresses during their continuous interaction with environment. Environmentally-induced stresses frequently activate the endogenous production of reactive oxygen species (ROS), most of which are generated as side products of tissue respiration. Hence, constant exposure to stressors may enhance ROS-mediated oxidative damage. Increased number of agricultural and industrial wastes enter aquatic environment and being taken up by aquatic organisms induce plural changes. Some of them directly enhance ROS formation whereas others act indirectly, for example, by binding with cellular thiols and reducing antioxidant potential. Fish are particularly threatened by water pollution. The use of sentinel species in biomonitoring needs to be discussed due to different level of their vulnerability by environmental toxicants.

Oxidative stress is defined as a situation when steady-state ROS concentration is transiently or chronically enhanced, disturbing cellular metabolism and its regulation and damaging cellular constituents (Lushchak, 2011). The activation of oxidative manifestations leads to the response of antioxidants, activation of expression of genes encoding antioxidant enzymes, elevation of the concentration of ROS scavengers. Nevertheless, there are considerable gaps in our knowledge on response to oxidative stress, particularly in the feral animals. Indeed, in field studies, wide spectrum of inter-site differences (higher, equal or lower activities of various antioxidant enzymes with tissue peculiarities and disbalance) have been observed in polluted compared to clean areas reflecting both mild stress conditions of the location or strong oxidative damage. Different models of the aquatic animal response, therefore, need to be analysed before conclusions can be drawn. In any case, the integrated approach with the appreciation of balance between prooxidant manifestations and antioxidant defence (enzymatic and nonenzymatic) in biological systems needs to be a control point to assess toxic effects under stressful environmental conditions.

In field investigations, there are problems on selection of reference sites even in the cases when these sites were selected by generally appreciated criteria. It is clear that, given the spontaneous human activities, we will not be able to find a true reference site in some areas.

The practical use of oxidative stress markers in fish is also connected to significant difficulties, because of their considerable seasonal variation. Moreover, animals can adapt to low pollution conditions and, under these circumstances, seasonal factors might affect biomarker responses to a greater extent than pollution variations. Therefore, with the aim to standardize the results and avoid the effect of adaptation to chronically polluted environment, caged organisms, including fish are used for biomonitoring. However, for fish, the responses of caged specimens are studied sparsely.

This chapter summarizes current knowledge on oxidative stress responses of fish in field conditions and their potential for environmental toxicology studies and biomonitoring.

2. Peculiarities of field pollution as a stressful factor

Unlike model oxidative stress that is usually caused by singular substance acting under controlled laboratory conditions (concentration, period) (Bagnyukova et al., 2005, 2006, 2007; Kubrak et al., 2010; Lushchak et al., 2007, 2008, 2009a,b,c; Yi et al., 2007; Sun, et al., 2008), environmental impact is usually developed according to multiple stressor effects. Indeed, ecosystems are under the pressure of complex mixtures of contaminants released in the environment due to various human activities. They may originate from miscellaneous sources such as chemical and drug manufacture, domestic sewage, polymer and petrochemical-based industries, oil refineries, mining, glass blowing, battery manufacture and many others. Hydrological changes, hydromorphological degradation and invasive species also can contribute to the set of stressing factors (Amado et al., 2006; Sureda et al., 2006; dos Anjos et al., 2011). According to origin, two primary routes of pollution can be selected: (1) point-source pollution and (2) non-point-source pollution. **Point-source pollution** originates from discrete sources whose inputs into aquatic systems can often be defined in a spatially explicit manner. Examples of point-source pollution include industrial effluents (pulp and paper mills, steel plants, food processing plants), municipal sewage treatment plants and combined sewage-storm-water overflows, resource extraction (mining), and land disposal sites (landfill sites, industrial impoundments). In opposit, **non-point-source pollution** originates from diverse poorly defined, diffuse sources that typically occur over broad geographical scales. Examples of non-point-source pollution include agricultural runoff (pesticides, pathogens, and fertilizers), storm-water and urban runoff, and atmospheric deposition (wet and dry deposition of persistent organic pollutants such as polychlorinated biphenyls (PCBs) and mercury) (Ritter et al., 2002). Basically the most prevalent xenobiotics arising out of agricultural and industrial activities are pesticides and trace metal ions.

Two examples will illustrate the complexity of natural water pollution that gives very poor prediction of its impact on biota. The first one is connected with industrial area, in the estuary near the city Göteborg, at the Swedish Western coast. The analyses of the sediments in this area showed high concentrations of anthropogenic compounds, such as polycyclic aaromatic hydrocarbons (PAHs), PCBs, tributyl tin, and dioxin, as well as transition metal ions. Toxicological analyses of the sediments in the Göteborg harbor area indicated that the levels of pollutants high enough to exert harmful effects on the ecosystem. To make the harbor more accessible and to secure future oil imports to Sweden, the dredging of the fairways Göteborg harbor was completed during 2003 (Sturve et al., 2005). In result, biomarker responses in the eelpout (*Zoarces viviparus*) sampled both before and during the dredging indicated that fish were chronically affected by pollutants compared to those in a reference area. However, the results during the dredging activities clearly show that fish were even more affected by remobilized pollutants. The second example illustrates the composition of three aquatic bodies in generally low industrially disturbed area in Western Ukraine during three seasons. However, spontaneous pollution together with the use of collapse of water purification systems constitutes further pressures for the aquatic environment.

The represented example of physico-chemical analyses of the water from three typical field sites in Western Ukraine, forestry site F near the spring of the river, agricultural site A in the lower part of the river, and forestry site N on the bank of the cooling pond of Nuclear Power Plant, showed that run-off and sewage discharges, industrial processes could be important sources of phosphates, phenol, nitrites, ammonium to surface waters at sites F and A and trace metal ions, particularly Cd and Cu, are typical pollutants at site N. While site A is proved as the most polluted river site in the region due to information from the Public Administration of the Environmental Protection, the high level of anthropogenic impact at site F near the municipal water inlet of the city was unexpected. High pollution caused by Cu and Cd at site N, where agricultural activity is low, may be explained by specific composition of the sewages from Nuclear Power Plant. The results indicate that the levels of Cu, Cd, nitrite, and phosphate even exceed environmental quality levels (EU Council Directive 98/83/EC) (http://www.emwis.org/IFP/law_EU.htm) (Table 1). In any case, studied chemical parameters in general terms confirm a plurality of compounds which mutual effect could probably induce toxic effects to aquatic organisms. Moreover, the relativity of the concept of the reference site is clear, even though the reference site was selected by generally accepted criteria. Page: 133
This needs to be taken into consideration at examination of environmental impacts.

Hence, aquatic environment is a sink for many environmental contaminants which can be absorbed by aquatic organisms leading to disturbing of antioxidant/prooxidant balance in fish (Lackner, 1998; Livingstone, 2001, Lushchak, 2011). That may cause oxidative stress, determined as a state when antioxidant defenses are overcomed by prooxidant forces (Livingstone, 2001, 1991; Livingstone, 2001,). Moreover, dependently on the source of pollutant, steady-state ROS concentration can be enchanced transiently or chronically, disturbing cellular metabolism and its regulation and damaging cellular constituents (Lushchak, 2011). Synergistic or antagonistic effects of mixtures of pollutants are hardly interpreted and predicted exclusively from the chemical analyses; some contaminants are substantially accumulated in specific tissues without recorded toxic effects (Viarengo & Nott, 1993), while others demonstrate high toxicity even at low levels. So, oxidative stress response of fish cannot be predicted using data on the level of certain pollutants in their tissues. In some cases, the correspondence between these characteristics was reported, for example, the 22-fold increase in PCB concentrations in white muscle of brown bullhead (*Ameriurus nebulosus*) was accompanied by disturbance of antioxidant defence in the tissues of this fish compared to fish from the nonpolluted site (Otto & Moon, 1996). On the other

hand, Machala and colleagues (1997; 2001) did not find any correlation between markers of oxidative stress in liver of chub (*Leuciscus cephalus*) and concentrations of specific contaminants, namely organochlorine compounds, PAHs, and metals in several sampling sites of a river with various pollution types and rates. So only the direct determining of the stress response, namely oxidative stress markers and may be some others, in biological systems has become the most adequate tool for early warning in environmental toxicology studies (Valavanidis et al., 2006).

Parameter	Site	Spring	Summer	Autumn
Phosphates, μM	F	16.7±2.1*	5.1±0.4[b*]	20.7±2.2[b*]
	A	15.1±1.7*	10.1±0.9[a,b*]	24.5±2.5[b*]
	N	1.0±0.1[a]	3.9±0.3[a,b]	1.9±0.1[a,b]
Nitrites, mg $N\text{-}NO_2^-\cdot L^{-1}$	F	2.2±0.2*	6.4±0.5[b*]	4.1±0.4[b*]
	A	0.4±0.04[a]	1.4±0.2[a,b*]	1.4±0.1[a,b*]
	N	0.4±0.04[a]	1.4±0.1[a*]	0.9±0.1[a,b*]
Nitrates, mg $N\text{-}NO_2^-\cdot L^{-1}$	F	2.0±0.2	0.3±0.0[b]	0.1±0.0[b]
	A	2.9±0.3[a]	12.8±1.2[a,b]	3.0±0.4[a]
	N	0.9±0.1[a]	0.1±0.0[a,b]	0.1±0.0[b]
NH_4^+, $mg\cdot L^{-1}$	F	6.4±0.8*	2.9±0.3[b*]	3.6±0.3*
	A	2.8±0.3[a*]	1.0±0.1[a.b*]	0.1±0.0[a,b]
	N	1.6±0.2[a*]	1.9±0.2[a*]	0.1±0.0[a,b]
Oxidisability, mg $O_2\cdot L^{-1}$	F	47.8±2.1*	24.8±2.2[b*]	17.7±2.1[b*]
	A	30.1±3.1[a*]	8.5±0.5[a,b*]	8.9±0.8[a,b*]
	N	11.4±1.2[a*]	41.2±3.2[a,b*]	17.0±1.6[b*]
Hardness, mM $CaCO_3$	F	1.1±0.1	1.1±0.1	0.5±0.1[b]
	A	1.3±0.1[a*]	1.7±0.2[a,b*]	1.0±0.1[a,b]
	N	0.5±0.1[a]	0.8±0.1[a,b]	0.9±0.1[a,b]
Phenol, $\mu g\cdot L^{-1}$	F	3.6±0.4*	0.9±0.1[b]	1.5±0.2[b*]
	A	0.8±0.1[a]	1.7±0.2[a,b*]	4.9±0.5*
	N	0.6±0.1[a]	0.7±0.1[a]	0.8±0.1
Cu, $\mu g\cdot L^{-1}$	F	2.5±0.2	1.3±0.1[b]	2.1±0.2[b]
	A	3.5±0.3[a*]	1.9±0.2[a,b]	6.3±0.6[a,b*]
	N	7.5±0.8[a*]	5.3±0.5[a,b*]	5.1±0.5[a,b*]
Cd, $\mu g\cdot L^{-1}$	F	2.2±0.2	4.3±0.4[b]	3.6±0.4[b]
	A	2.9±0.3[a]	4.1±0.4[b]	3.4±0.3
	N	2.8±0.3[a]	8.9±0.7[a,b*]	6.3±0.6[a,b*]

Table 1. Physico-chemical parameters of water in three seasons, M±SD, n=3 (from Falfushynska et al., 2010c with permission): *Exceeding of maximum permitted concentration allowed for the protection of freshwater aquatic life. The values are expressed as the mean ± SD; [a]Significantly different from spring value at the same site with $P < 0.05$; [b]significantly different from site F value in the same season with $P < 0.05$.

Xenobiotic-induced stress responses can be broadly categorized as intoxication and detoxication signals. Intoxication signals manifest debilitating phenomena while the detoxication signals are adaptive in nature and provide protection to the biological systems when affronted with toxic xenobiotics (Bhattachary, 2001). Dependently on the intensity and duration of toxicant effect and resistance of the studied organism, different manifestations of the oxidative stress can be expected. However, there are considerable gaps in our understanding of oxidative stress response mechanisms in the feral animals (Valavanidis et al., 2006). The long-term effect of pollutants, typical for chronically and heavily polluted areas, the enhancement of ROS level and perturbation of antioxidant efficiency often prelude the onset of significant alterations like protein and DNA damage, lipid peroxidation (LPO) and enzyme inhibition (Winston & Di Giulio, 1991). Fish are particularly threatened by aquatic pollution, and the environmental stress they face may help to shape their ecology, evolution, or biological systems (Padmini, 2010).

3. Oxidative manifestations: Reactive oxygen species, damage to lipids, proteins and DNA

When the effect of environmental pollution on the antioxidant defence is elucidated, the exceeding of the resiliency of this system, and consequently, oxidative stress could be approved only basing on the elevations of the rate of oxidative manifestations. The expression of specific lesions known to arise specifically at oxidative stress, e.g. lipid peroxidation (membrane damage), oxidized bases in DNA and accumulation of lipofuscin pigments were found in many aquatic animals exposed to contaminants (Winston, 1991). However, whilst in the laboratory a wide spectrum of these indices is measured, only single parameter is often explored in Environmental Risk Assessment (ERA). In any case, the rate of oxidative damage is the control point of the effective adaptation to oxidative stress.

Lipid peroxidation or oxidation of polyunsaturated fatty acids, measured usually as a level of thiobarbituric acid reactive substances (TBARS), has been used most frequently to analyse the effect of pollutants (Livingstone, 2001; Lushchak et al., 2007, 2008, 2009 a, b, c, 2011). The elevated LPO in fish from heavily polluted field sites was observed (Ferreira et al., 2005; Farombi et al., 2007; Sanchez et al., 2007). For example, in the African catfish (*Clarias gariepinus*) from the Ogun River located close to major industries in the South Western part of Nigeria, TBARS levels of *C. gariepinus* were significantly higher in the liver, kidney, gills and heart by 177%, 102%, 168% and 71% respectively compared to that from fish farm which was considered as a reference site (Farombi et al., 2007). Elevated levels of LPO products were indicated in the blood of three cichlid fish species (*Oreochromis niloticus*, *Tilapia rendalli*, and *Geophagus brasiliensis*) from metal-contaminated site (Bonafé et al., 2008). Dorval et al. (2005) demonstrated higher level of hepatic LPO products in white sucker (*Catostomus commersoni*) from the river sites in Québec (Canada), impacted by agricultural chemicals. The killifish (*Fundulus heteroclitus*) inhabiting a creosote-polluted inlet of the Elizabeth River also exhibited higher LPO as compared to the reference population (Bacanskas et al., 2004). Differences of the level of TBARS in a liver of common carp (*Cyprinus carpio*) were also detected between fish from rural and industrial sites in relatively low polluted area in Western Ukraine (Falfushynska & Stoliar, 2009). In this study, the gills demonstrated significantly lower level of TBARS than the liver. Moreover, the correlation between TBARS levels and $O_2^{\bullet -}$ production was detected, confirming the conclusion on potential mechanisms of oxidative damage in fish. In crucian carp (*Carassius carassius)* from the similar two areas of comparison, in Western Ukraine (basin of the river Dnister) the differences in TBARS concentration were also observed. Especially high level of TBARS was observed in fish from industrial site in summer (Falfushynska et al., 2010).

Other examples demonstrate the absence of differences in TBARS concentration between fish from polluted and clean areas. In the study of Pandey et al. (2003), the differences of a broad set of antioxidants in gills, kidney and liver tissues in the Indian freshwater fish Wallago attu (Bl. & Schn.) from clean and polluted river sites were showed. But LPO intensity assessed as TBARS level did not differ between two sites. Similar results were obtained by Huang et al. (2007) in the hepatopancreas of carp from polluted site, unlike the responses of other studied tissues, kidney and intestine. Despite differences in the activities of superoxide dismutase (SOD), glutathione transferase (GST) and glutathione peroxidase (GPx), the level of LPO was the same in the fish from two sites, indicating a stronger antioxidant capacity of this organ. In the series of materials devoted to the consequences of a dredging campaign in Göteborg harbor, Sweden, to eelpout (*Z. viviparous*), as a sentinel species, TBARS did not show inter-site differences (Almroth et al., 2005). Similarly, in the liver of labrid fish (*Coris julis*) despite the variations in the antioxidant enzyme activities, there was no significant difference in TBARS concentrations (Sureda et al., 2006).

The end-products of LPO can be accumulated in lysosomes as insoluble granules containing autofluorescent pigments and are usually referred as lipofuscins. The indication of these pigments in the lysosome vacuolar system of fish hepatocytes also can be used for the assessment of the level of membrane LPO (Viarengo et al., 2007). The authors even recommend evaluation of lipofuscin levels as more valid characteristic of damage to lipids than TBARS. However, the corresponding studies with feral fish are scant and connected solely histological studies that do not permit to assess the oxidative stress response accurately. For example, histopathologic biomarkers in feral freshwater fish populations, namely redbreast sunfish (*Lepomis auratus*) and largemouth bass (*Micropterus salmoides*), showed the signs of lipofuscin accumulation only in polluted sites (Teh et al., 1997). The comparison of fish, barbels (*Barbus graellsii*) and bleaks (*Alburnus alburnus*) from areas located upstream and downstream of a mercury cell chlor-alkali plant on the Cinca River (NE Spain), demonstrated that the prominent elevation of the concentration of mercury in the tissues of fish sampled downstream of the plant (10- and 30-times higher in the muscle and liver of barbels downstream of the factory) was accompanied by significantly higher prevalence of liver pathologies consistent with the prooxidant effect of trace metals (Raldúa et al., 2007). Fifty paddlefish (*Polyodon spathula*) collected from two sites on the Ohio River, USA, demonstrated significantly higher organochlorine concentrations that even exceeded the Food and Drug Administration's action limit for chlordane (0.30 μg/g) than the fish from Cumberland River as a reference site. That was accompanied by the presence of hepatic hemosiderosis (Gundersen et al., 2000). However, concerning the signs of oxidative stress, these results represent only initial stage of study.

The formed free radicals cause various kinds of genotoxicity, particularly modifications to DNA bases. Most of the analytical assays have been focused on measuring of products of guanosine hydroxylation, namely 8-OHdG or 8-oxodG, and its free base 8-hydroxyguanine, in urine as an indirect method for oxidative damage by free radicals (Shigenaga & Ames, 1991). In the studies of feral fish these methods are presented scanty. A study with the fish (*Sparus aurata*) found that 8-oxodG determination in chromosomal DNA was a potentially useful biomarker of oxidative stress caused by urban and industrial environmental pollution (Rodriguez-Ariza et al., 1999). However, the proof of oxidative stress as a reason for genotoxicity is usually explored only in model studies, but not in ERA.

Proteins are considered to be important targets of free radical attack in cells (Eustace & Jay 2004; Almroth et al., 2008b; Lushchak, 2011) and thus compromise antioxidant defense, cellular function, and survival (Padmini, 2010). Therefore, protein oxidation, often under investigation in proteomic studies, has been recently proposed as a biomarker of oxidative stress (Sheehan, 2006; Lushchak, 2011). In flounders, living in contaminated waters with xenobiotics, increased levels of oxidised proteins were reported (Fessard & Livingstone, 1998). Studies on dynamics showed that proteins can be oxidized before lipids or DNA in ROS-exposed cells (Du & Gebicki, 2004). At the same time, many other factors can influence cell cycle and correspondingly, injury of proteins, related particularly to their oxidative damage. In any case, protein carbonyls (PC), so successfully explored in the studies of model oxidative stress in short-term laboratory experiments (Parvez & Raisuddin, 2005; Kubrak et al., 2010; Lushchak, 2011), are very seldomly used in the field studies for the assessment of environmental effects on fish. In the set of studies devoted to the consequences of a dredging campaign in Göteborg harbor, Sweden, to fish on the example of eelpout (*Z. viviparous*), as a sentinel species, monitor the impact of these events, the formation of additional carbonyl groups in proteins was studied (Almroth et al., 2005; 2008a). They confirmed that unlike LPO, PC, measured using an ELISA method, show differences between the reference and polluted sites in the field, as well as differences between time periods (before and during dredging and following the oil spill detected in this area were found. Particular results were reported for the fish from spontaneously polluted area. In the study with *C. carassius* from two field sites, significant differences of PC were indicated. However, lower level, particularly in the gills was found in fish from industrial site. The inter-site differences were opposite to that of the concentrations of GSH and metallothionein-related thiols (Falfushynska et al., 2010b). These data were interpreted from high tolerance of fish of genus Carassius to adverse conditions in the industrial site. On the other hand, for *C. carpio* higher levels of PC were detected in the liver and gills in two seasons in industrial site and only in summer the inter-site difference was opposite (Falfushynska et al., 2009).

Direct studies of intensity of ROS production in the field works are limited. The measurement of superoxide anion radical ($O_2^{\bullet-}$) production in the liver and gills of *C. carpio* from rural and industrial areas in Western Ukraine showed that the $O_2^{\bullet-}$ production was elevated at the industrial site in the majority of samples (Falfushynska & Stolyar, 2009). In this study, the negative correlation between Mn-SOD activity and $O_2^{\bullet-}$ production was observed and production of $O_2^{\bullet-}$ and TBARS correlated positively. In the compared groups, coherent changes of PC and $O_2^{\bullet-}$ levels were also detected in the liver and gills. In the study of *C. carassius* from two field sites, the significant difference of PC corresponded to variations in $O_2^{\bullet-}$ production, particularly in gills (Falfushynska et al., 2009b).

4. Non-enzymatic antioxidants: Glutathione and other scavengers

Non-enzymatic antioxidants are represented by ROS scavengers (both hydrophilic such as low-molecular mass thiols, glutathione (GSH), metallothioneins (MTs), ascorbic and uric acids, as well as lipophilic ones such as vitamin E and carotenoids (Viarengo et al., 2007). In the field studies, GSH is the most frequently studied scavenger. The hepatic ratio of oxidized to reduced glutathione (GSSG/GSH), a value used as an indicator of the "redox status" of the cell, may be appropriate biomarker for oxidative stress. However both GSH and GSSG levels have only been measured in a limited number of field studies (Van der Oost et al., 2003). It should be noted, that GSH can be involved in diverse processes different from related to free radical metabolism, within the cell and its variability can not be considered entirely in connection to oxidative stress.

In moiety of field studies, the elevated level of GSH was indicated in fish from polluted areas (Van der Oost et al., 2003). English sole (*Pleuronectes vetulus*) sampled from the Duwamish Waterway, a contaminated urban site in Puget Sound, Washington, showed increased GSH concentrations. The findings also indicated that induction of GSH synthesis from L-*Cys* was not a major factor in the increase of hepatic GSH in contaminant-exposed fish whereas it was not accompanied by changes in either L-Cys concentrations or gamma-glutamylcysteine synthetase activity (Nishimoto et al., 1995). A population of killifish (*F. heteroclitus*) inhabiting a creosote-polluted inlet of the Elizabeth River demonstrated higher total glutathione concentrations in adult hepatic tissue as compared to the reference population (Bacanskas et al., 2004). Study on the Indian freshwater fish *Wallago attu* (Bl. & Schn.) collected from two sites along the river Yamuna demonstrated that GSH in liver, kidney and gills was found to be substantially higher in the fish collected from more polluted site (Pandey, 2003). In the African catfish (*C. gariepinus*) from the Ogun River located close to major industries in the South Western part of Nigeria, GSH concentration was higher by 81%, 83% and 53% in the liver, kidney and heart, respectively, compared to that from the reference site (Farombi et al., 2007). Opposite response of GSH in the gills (lower by 44% in the fish from polluted area) was indicated in this study. At the same time, some field studies of contaminated sites did not detect differences in GSH level with the fish from the reference site (Jenner et al., 1990; Eufemia et al., 1997) or detected decreases (Otto & Moon, 1996; McFarland et al., 1999), and one study found an initial decrease followed by a sustained elevation (Steadman et al., 1991). Dorval et al. (2005) demonstrated that in hepatic and adrenal tissues of white sucker (*C. commersoni*) from a river that drains an agricultural region, GSH level was higher in reference site compared to fish from contaminated sites. Also, the three-spined stickleback (*Gasterosteus aculeatus* L.) sampled from heavily contaminated stream in France exhibited decreased liver GSH levels (Sanchez et al., 2007).

The increase of the ratio of GSSG/GSH in fish due to either direct scavenging of radicals or increased peroxidase activity can be expected. However, increases in total glutathione without increases in the percent of GSSG have been observed in channel catfish under the effects of sediments from polluted site (Di Giulio et al., 1993), in larvae of killifish (*F. heteroclitus*) inhabiting a Superfund site on the Elizabeth River (VA, USA) (Meyer et al., 2003). The levels of both, GSH and GSSG, in *C. carpio* from the river in Western Ukraine were higher in industrial site than in rural site and were more sensitive to spatial peculiarities in liver than in gills (Falfushynska & Stolyar, 2009). In the gills, GSH redox status was in the range 0.77–0.97, but in summer, it decreased to 0.54, with no difference between the sites. On the other hand, in *C. carassius* tissues high GSH levels and redox state of GSSG/GSH couple (particularly in the gills), were indicated in polluted site, which was expected to confer some advantages to this higly tolerant to environmental stresses fish (Falfushynska et al., 2010). Nevertheless, after 21 days in captivity, the fish from these two sites demonstrated opposite difference in GSH and GSSG levels (Falfushynska et al., 2011).

Some general relationships between Redox Index of GSH (RI GSH) calculated as the ratio of content GSH/(GSH+2GSSG) and other markers of oxidative stress were observed in field studies of fish (Falfushynska et al., 2010a), that can be explained by versatility of glutathione functions. The examples of discrepancies between RI GSH, LPO products and the activities of antioxidant enzymes in the field studies on fish were analysed in a review by Kelly et al. (1998). However, in any case, the elevation of GSH level and RI GSH can possess benefit to fish in its ability to survive in polluted environment, while glutathione depletion is usually associated to enhancing of peroxidation processes in the cell membrane and leads to stress and can prominently contribute in hepatotoxicity (Viarengo et al., 2007).

Metallothioneins (MTs) are low molecular mass intracellular cysteine-enriched proteins that are suggested to be related to oxidative stress response. They constitute a diverse family of thermostable intracellular low molecular mass proteins, which are enriched in cysteines and bind metal ions in metal-thiolate clusters. Now they are considered to participate in the storage and detoxification of metal ions such as zinc, copper, and cadmium, and in the scavenging of ROS in diverse living organisms (Viarengo et al., 2007; Fernandes et al., 2008). However till now, despite a lot of publications devoted to MTs, their biological functions, relationship and necessity for the organism are discussed and adjusted. Some recent data demonstrate MTs induction in fish by other than metal pollutants, particularly in connection with oxidative stress (Paris-Palacios et al., 2000; 2003). However, several metals which are not essential for MTs (ferrum and nickel, for example) and also endocrine-disrupting chemicals have been known to be inhibitors of the MT gene transcription (Rhee et al., 2009; Lee et al., 2010).

Since their low redox potential, the metal-thiolate clusters of MTs can be easily reduced or oxidized *in vitro* and *in vivo* with concomitant binding/release of metal ions (Maret & Valee, 1998). It was found that MT levels in mammalian tissues under physiological conditions could be rather high to harbour important implications for MTs operation in Zn and redox metabolism (Capdevila et al., 1997; Capasso et al., 2005; Kelly et al., 2006). Different substances besides metals, such as fungicides fenhexamid, mancozeb, and also hydrogen peroxide, induce the elevation of MT content in fish (Viarengo et al., 1999; Cavaletto et al., 2002; Paris-Palacios et al., 2003; Mosleh et al., 2005; Kang, 2006). Inter-relation of elevated MT level with other stress proteins (catalase, GST) and negative relation to LPO products was confirmed for *C. carassius* over three seasons in a mixed polluted area characterized by spontaneous agricultural activities (Falfushynska et al., 2010a). On the other hand, studies on the *C. carpio,* have demonstrated the inability of their MTs to maintain high level of antioxidant defense, but elevated metal-binding capacity at the industrial site polluted by metals (Falfushynska & Stolyar, 2009; Falfushynska & Stoliar, 2009).

The participation of MTs in antioxidant defense can be explained by the high content of thiols and the particular metal binding/release dynamics intrinsic to these proteins (Atif et al., 2006; Monserrat et al., 2007; Viarengo et al., 2007). However, the effect of pollution on the relations between the metal binding and antioxidant functions of the MTs in aquatic animals has not been clarified (Chesman et al., 2007).

The participation of MTs in antioxidant defence can be also indirectly connected to the distribution of metal ions within the cell in deposited form and unbound, potentially toxic form. Complex field pollution can decrease the metal-binding function of MTs and promote the metal-related generation of ROS. With a view to include MTs in biomonitoring programs, simultaneous studies of the response of their expression, metal-binding capacity and thiol concentration in the field conditions must be undertaken. Metal-keeping function and possible participation in the antioxidant defense, expressed by concentrations of complexes of MT with metal ions (MT-Me) and total MTs (MT-SH), can be alternative/complementary characteristics of MTs in aquatic animals in complex field pollution.

Main discovery in the study of MTs expression is the distinguishing of basal and related emergency gene products both in vertebrate and invertebrate aquatic animals. In general, it seems that constitutive MT isoforms represented a primary action under 'less stressful' or 'sublethal' conditions whereas the activation of other isoforms became important under 'more stressful' or 'lethal' circumstances (Bargelloni et al., 1999; Lee et al., 2010). However, some limitations are still evident demonstrating that MTs are regulated at translational and transcriptional levels. The concentration MT proteins and their multiplicity do not appear to correlate always with constitutive expression of MTs. For example, two icefish species show the same number of MT genes despite a lack of expressed MTs at the protein level. In brown trout (*Salmo trutta*) from some polluted areas, the MT content was not elevated even when transcription of MT genes was enhanced (Hansen et al., 2006a, b). Therefore, with the aim to understand the importance of MTs in the response under oxidative stress, the study needs to combine the determination of gene expression at all levels including their properties.

There is limited evidence of induction of MTs due to exposure to environmental trace organic contaminants, and thus they usually are not discussed in corresponding literature. Moreover, they respond not only to anthropogenic pollution but also to physical stress and other natural factors. This makes them extremely difficult to be used as "stand-alone" biomarkers (Lam & Wu, 2003).

Studies addressed other potential ROS scavengers in the feral fish under stress conditions are very scarce. In a large-scale field study in Sweden, perch inhabiting water bodies contaminated with bleached kraft pulp mill effluents consistently displayed higher ascorbate concentrations than fish from the reference site (Andersson et al., 1988). α-Tocopherol (vitamin E), a lipid-soluble antioxidant, that is synthesized by plants, but required in the diets of animals, appears to play a major role in protecting of cell membranes from LPO (Stegeman et al., 1992). Measurements of these substances in tissues of feral fish are expected to be useful for ERA, especially within the set of oxidative stress indices.

5. Antioxidant enzymes

Antioxidant enzymes are included in the environmental pollution assessment because of their inducibility under conditions of mild oxidative stress and their potential role in adaptation to pollutant-induced stress. It is expected that they may be more sensitive at detecting of initial insults than such markers as histopathologies, changes in growth rates, etc. (Adams & Greeley, 2000). Laboratory studies confirmed that the measurement of changes in the expression of a large number of specific genes or activities of certain enzymes of antioxidant defence can be explored in an early warning system of toxicant exposure (Livingstone, 2001; Lushchak, 2011). However, if in the model studies, the enzyme response to disposable effect of toxic chemicals can depend on duration of pollutant influence, showing a bell-shaped relationship, in the field studies the results often indicate that antioxidant enzyme responses are transient and variable for different species, enzymes and chemicals (Viarengo et al., 2007). Obviously, the early warning can be used when temporal effect of pollution is expected. In the field studies, fish is frequently subjected to long-term exposure of number of factors. Therefore, the observed difference in the activity of antioxidant enzymes between two sites may be attributed both to their activation under mild stress conditions of the location or to their suppression due to strong oxidative damage. Different models of the aquatic animal response, therefore, need to be analysed before conclusions can be drawn. Changes in gene expression may occur relatively quickly during an exposure, but the effect of long-term exposure on the expression may differ (Bagnyukova et al., 2005, 2006, 2007; Kubrak et al., 2010; Lushchak et al., 2007, 2008, 2009a, b, c). Additionally, contaminant-independent reference expression patterns should include natural fluctuations of the level. Indeed, in field studies, higher, equal or lower activities of various antioxidant enzymes have been observed in polluted compared to cleaner areas (Narbonne et al., 1999; Bonafé et al., 2008).

Typically, the battery of oxidative stress parameters in feral fish includes usually the activities of either SOD, or catalase. At least one of GSH-related enzymes is also often included in the study. Due to easily carried out and low-cost enzymatic tests, the assessment of catalase and GST activities has most often been used in biomonitoring programmes for fish (Romeo et al., 2000; Viarengo et al., 2007).

Whereas the field studies mostly belong to biomarker-type studies, the specification of antioxidant enzymes in them is limited. In according to this, it should be noted that including in the study the assessment of only one antioxidant enzyme and (desirable) its measurement only in single tissue/organ facilitates nominally the discussion of obtained data. In this case, the final conclusion concerning the indication of adverse effect is derived from the initial knowledge concerning the relative level of pollution (site in the lower stream of the river or situated close to the certain point of pollution). In any case, indication of changing enzyme activity comparing to selected reference site is considered as a sign of pollution effect. For example, significant differences in SOD, GPx, and catalase activities in the blood of three cichlid fish (*O. niloticus, T. rendalli,* and *G. brasiliensis*) taken during two seasons from site polluted by industrial effluents compare to reference site was used as an evidence of pollution in the area (Bonafé et al., 2008).

It is more difficult to give the multiple explanations of obtained results when a set of biomarkers includes several enzymes and they are studied in more than one tissue. In this case, discrepancy between different activities is usual attribute of oxidative stress in fish. In general, the elevated enzyme activity in fish from polluted areas is considered to be the main feature of compensatory response within its tolerance range and the lower its activity witness about the exceed of the resilience of this response.

Many biomarker-type studies have identified increases in antioxidant defenses in aquatic organisms (Collier & Varanasi, 1991; Stein et al., 1992; Rodriguez-Ariza et al., 1993; Livingstone et al., 1995; van der Oost et al., 1996; Eufemia et al., 1997; Stephensen et al., 2000). That is probably the result of both, physiological acclimation and/or genetic adaptation in the populations (Meyer et al., 2003). This activation is mainly connected to SOD, particularly in fish from sites contaminated with persistent organic substances (Buet et al., 2006). A field study at the Elizabeth River polluted by creosot demonstrated elevated Cu,Zn-SOD activity in feral spot (*Leiostomus xanthurus*) (Roberts et al., 1987).

The elevated SOD activity can be combined with the decrease or stable activity of the second main antioxidant enzyme, catalase. For example, in the field study *C. carpio* were collected in two sites of the upper Yellow River, and the results showed that in polluted site, SOD and GST activities were higher and catalase and also GPx activities were lower in almost each case of comparison (activities were determined in hepatopancreas, kidney and intestine) (Huang et al., 2007). The concerted elevation of SOD and GPx activities was indicated in liver of sterlet (*Acipenser ruthenus* L.) collected from the Danube-oil refinery site compared to that from the reference site, while no differences were found in other studied enzymes (catalase, GST, the same as the enzymes of biotransformation in liver, aspartate aminotransferase, alanine aminotransferase, and gamma-glutamyl transferase in serum) (Stanic et al., 2006). In the study of three populations of brown trout (*S. trutta*) exposed to elevated Cd and Zn or Cu levels in their natural environment, both metal-exposed groups had higher activities of SOD in liver compared to unexposed trout from reference site, and catalase activity in the liver was the same in all tree populations (Hansen et al., 2006a).

Disbalansed antioxidant activities were shown in the various oxidative stress biomarkers in the Indian freshwater fish *Wallago attu* (Bl. & Schn.) (Pandey et al., 2003). In this study, the fish from polluted river site possessed higher activities of SOD and xanthine oxidase (in liver, kidney and gills), glutathione reductase (GR) in liver and gills whilst catalase activity in both liver and kidney was found to be significantly lower when compared with values in tissues of fish collected from clean site.

In the African catfish (*C. gariepinus*) from the Ogun River located close to major industries in the South Western part of Nigeria, SOD activity was higher by 61% in the liver, 50% in the kidney and in the heart by 28 % compared to that from Agodi fish farm. The levels of GST activities in the liver, kidney and heart of this fish was higher by 62%, 72% and 37%, respectively. Only in the gills of fish from polluted area, a significantly lower SOD (44%) and GST (41%) activities were observed. On the contrary, there was 46%, 41%, 50% and 19% lower catalase activity in the liver, kidney, gills and heart, respectively (Farombi et al., 2007).

Low intensity, but prolonged effect of spontaneous sources of pollution can deplete SOD activity in fish tissues. That is examplified by very few cases stydied to date (Bacanskas et al., 2004, Pandey et al., 2008; Falfushynska & Stolyar, 2009; Falfushynska et al., 2010a). The decreased SOD activity along with increased $O_2^{\bullet-}$ levels suggests the weakness of antioxidant defences in common carp *C. carpio* at the chronically polluted industrial site (Falfushynska & Stolyar, 2009). Mn-SOD in *C. carpio* was both more abundant and more sensitive to local influence than Cu,Zn-SOD. Moreover, the importance of Mn-SOD was supported in this study by the showing the negative correlation between Mn-SOD activity and $O_2^{\bullet-}$ production. Similar site-related difference was obtained for SOD (total activity) of relatively more tolerant fish, *C. carassius*, from the same area (Falfushynska et al., 2010a).

The increase in catalase activity is often observed in the model experiments and also can occurre without relation to SOD responses and due to high pollutant impact (Üner et al., 2005; Moraes et al., 2007; Lushchak, 2011). The higher catalase activity was reported for the liver subcellular fractions of red mullet *Mullus barbatus* collected along the Western Mediterranean coast (the Northern Iberian Shelf). Moreover, only catalase activity was well related to pollution in the area and showed about doubled activities in four most contaminated sites in comparison with the reference sites. For SOD activity, significant difference among sampling sites were found in this study, but they had no clear relationship to the levels of studied pollutants. Additionally, no pollution- or site-related difference was observed for GPx activities (Se-dependent and total) (Porte et al., 2002). On the other hand, the study of fish *M. barbatus* from a coastal marine area of Salento Peninsula (Italy) indicated that catalase activity did not show any significant variation between animals sampled from urbanized and conditionally uncontaminated sites, whilst acetylcholinesterase indicated the neurotoxicity of environment (Lionetto et al., 2003).

The depletion of catalase activity or its stability along with increment of SOD activity were reported (Pandey et al., 2003; Stanic et al., 2006; Huang et al., 2007), and even although catalase mRNA levels were higher in the exposed fish (Hansen et al., 2006a). Dorval and colleagues (2005) in hepatic and adrenal tissues of white sucker (*C. commersoni*) from a river Yamaska that drains an agricultural region in Québec (Canada) found that in fish from the contaminated sites catalase and GPx activities were lower than those in the fish from reference site.

In recent study (Falfushynska & Stolyar, 2009), the low catalase activity in *C. carpio* was attributed to high production of $O_2^{\bullet -}$, which has been reported to inhibit catalase in the case of excess of production (Kono & Fridovich, 1982). However, the negative correlations between them were not regular in three seasons. Two-fold higher catalase activity was found in the liver and gills of carps from industrially polluted site as compare to reference site. Principal component analysis showed that catalase activity was not included in the significantly important set of markers unlike other oxidative stress indexes (Fig. 1).

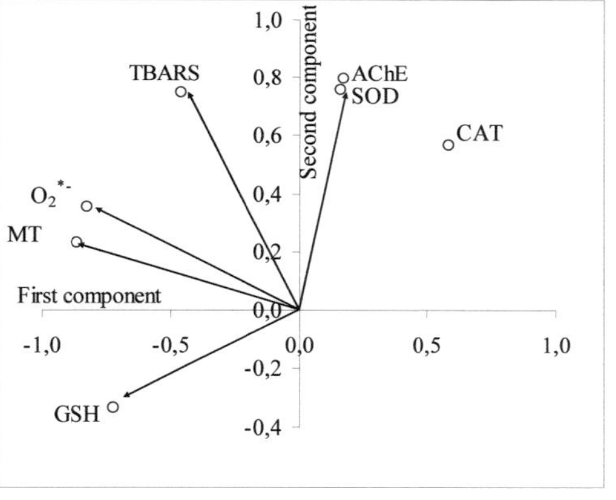

Fig. 1. Principal component analysis of the parameter data set in the gills of common carp from rural (and industrial sites from the river Seret in Western Ukraine. Parameters: activity of SOD, catalase; AChE, level of MT, GSH, TBARS, and $O_2^{\bullet -}$ production. The arrows indicate biomarkers having significant factorial weights > 0.7 (from Falfushynska & Stolyar, 2009 with pemission).

It seems that catalase activation can be considered as a last refuge of antioxidant defense in the feral fish. The particular catalase role in the antioxidant defense of feral fish was grounded by Porte et al. (2002), basing on the information on its activation by H_2O_2 at high concentrations. They suggested that catalase normally plays a relatively minor role in H_2O_2 catabolism at low rates of peroxide generation, but it becomes indispensable when the rate of H_2O_2 production is enhanced, for example, at oxidative stress. Comparatively higher stability of GPX activity reported in this and other studies can result from its dependence on both H_2O_2 and lipid peroxides, as substrates of decomposition. Therefore, it is possible that GPX activity would maintain normal cell functions, whereas catalase would form part of a stress-response mechanism (Janssens et al., 2000). Godin & Garnett (1992) found compensatory relationship between catalase and GPX activities: low GPX activity was combined with high catalase activity.

Coordinated activation of antioxidant enzymes in reported sparse can be explained by the possible genetic adaptation to specific aquatic environment. Higher activities of GST, GPx and GR were detected in liver of teleost *C. julis* from two stations with seagrass Caulerpa species that produce toxic methabolite Caulerpenyne, as compare to fish from the area where Caulerpa species were absent. At that, no statistical difference was found in catalase activity between the groups (Sureda et al., 2006). However, in the case of worsening of environmental conditions caused by dredging of contaminated sediments, high activities of catalase, GR and GPx in red mullet *M. barbatus* sampled at a disposal site for dredged sediments was demonstrated by Regoli and colleagues (2002).

Including of different forms of enzymes in the study demonstrated variability of their responses to the environmental pollution. This is mostly connected with SOD (Roberts et al., 1987) and GST (Porte et al., 2002) and is obvious in relation to different genetic origin, location within the cell and function of individual/specific enzyme isoforms. Meyer et al. (2003) stressed that the study of spot (*L. xanthurus*) (Roberts et al., 1987) and killifish (larvae of *F. heteroclitus*) from the same polluted area demonstrated different species-specific sensitivity of Mn-SOD and Cu,Zn-SOD activation. Connecting tissue specificity of enzymes studied in field experiments, the enzymes of liver are considered to be less sensitive to some kinds of pollution than other tissues, because of the best antioxidant adaptation in the liver of fish, since despite the variations in the antioxidant enzyme activities, there was no significant difference in malondialdehyde concentration (Sureda et al., 2006).

Only a few field studies described the modulation of fish GR activity by the chemical stress. In the study of Machala et al. (1997), the GR activity was significantly higher and appeared to be a relevant biochemical marker of exposure to persistent chlorinated contaminants. However, Bainy and colleagues (1996) reported a decrease in hepatic GR activity in Nile tilapia collected at a PCB- and hexachlorocyclohexane-contaminated sampling site.

Sometimes, the attempts to find the relation between the accumulation of toxic substances in the tissues and antioxidant enzyme activities are successful. For example, muscle concentrations of PCB compounds as well as biliary levels of PAH metabolites showed that catalase activity, but not other ones, was well related to PCB body burden (Porte et al., 2002). Relatinship between the level of PCB in white muscle of brown bullhead (*A. nebulosus*) and cytosolic SOD activity in the kidney of fish from the polluted site was reported (Otto & Moon, 1996). However, catalase activity in the kidney, GPx activity in the red and white muscle, and total glutathione in the liver, kidney, and white muscle were decreased relative to fish from the nonpolluted site. The measuring of CAT activity in liver subcellular fractions together with markers of biotransformation, namely, 7-ethoxyresorufin *O*-deethylase (EROD) and UDP-glucuronosyltransferase (UGT), in two different fish species, the four-spotted megrim (*Lepidorhombus boscii*) and the pouting (*Trisopterus luscus*) collected along the Northern Iberian coast, showed a good positive correlation with the amount of alkylphenols and 1-naphthol accumulated in the tissues for EROD and UGT but not for catalase activities (Fernandes et al., 2008). In the study of Machala et al. (1997) the activities of a set of GSH-related enzymes in common carp from several field sites was analysed in concert with the chemical analysis of organic contamination in five sampling sites in ponds.

In the ponds polluted mainly by PAHs and PCBs, the activation of glutathione-dependent enzymes, namely cytosolic GR and GST toward 1-chloro-2,4-dinitrobenzene, ethacrynic acid and 1,2-epoxy-3-(p-nitrophenoxy) propane, and microsomal GST was detected even when these substances were not revealed in the tissues of fish.

Special attention needs to be paid to GST, the GSH-dependent enzyme that promotes the reactions of conjugation in the II phase of the biotransformation of toxic substances, and also participates in the antioxidant defence due to the dependence on GSH and reduction of some peroxides. It is recommended as the biomarker of oxidative stress in fish (Viarengo et al., 2007). Indeed, among all connected antioxidant defence enzymes, GST is frequently activated at pollution. In the study of three species of cyprinids, barbel (*B. plebejus*), chub (*L. cephalus*), and Italian nase (*Chondrostoma söetta*), from two sites of the River Po, located upstream and downstream from the confluence of one of its middle-reach polluted tributaries, the River Lambro, with the exception of a higher GST enzyme activity of barbel from the downstream site, no significant modification was evident in GR, and GPx activities, despite the difference in specific markers of pollution by PAHs and PCB (Vigano et al., 1998). In the study of Machala et al. (1997) GST toward 1-chloro-2,4-dinitrobenzene, ethacrynic acid and 1,2-epoxy-3-(p-nitrophenoxy) propane, and microsomal GST demonstated higher activity in common carp from several field ponds polluted even to small extent (registered in sediments but not in muscle tissue) by PAHs and PCBs. However it can indicate its involvement in the processes of biotransformation more than in the detoxification of oxygen radicals. The toxicity of many exogenous compounds can be modulated by increased activity of GSTs. Effects of inducing agents on total hepatic GST activity have been observed in several fish species (Armstrong, 1990; George, 1994; Commandeur et al., 1997).

However, in several studies no significant differences or decrease of its activity were observed in fish from polluted sites. The effects of the extensive dredging in Göteborg harbor, Sweden on eelpout (*Z. viviparus*) sampled along a gradient, both before and during the dredging, indicated that eelpout were exposed to increased levels of pollutants indicated by elevated EROD activities, cytochrome P4501A levels and MTs gene expression. The prominent increase in GR activity in eelpout from the inner harbour during dredging was indicated, but no difference was observed in GST activities between the sites (Sturve et al., 2005).

The direct measurement of changes in the expression of a large number of genes related to the markers of specific kinds of pollution, namely vitellogenin, cytochrome P450, hsp-related genes began to be explored for early warning of pollution, for example, by xeno-estrogenes or oil in recent years (Tom & Auslander, 2005; dos Anjos et al., 2011). However, at the moment, among piscine genes which expression is increasingly utilized as environmental biomarkers, very few literature data are devoted to antioxidant defence enzymes in feral fish (Nikinmaa & Rees, 2005; Hansen et al., 2006a, b) despite in laboratory studies a successes in the study of corresponding genes were achieved (Cho et al., 2008; Woo et al., 2009; Lee et al., 2010).

The comparison of expression of specific genes (mRNA level) and the activities or protein levels of corresponding antioxidant enzymes did not confirm the relation between them. When three populations of brown trout (*S. trutta*) exposed to different metal ion levels in the natural environments were compared, the data indicated that chronic exposures to Cd, Zn and/or Cu did not involve maintenance of high activities of SOD and CAT in gills, although SOD mRNA levels were higher in the Cd/Zn-exposed trout (Hansen et al., 2006a, b). Further, in livers, mRNA levels of SOD, CAT and GPx were higher in the metal-exposed trout, but only for SOD enzyme activity was higher in liver compared to the unexposed reference trout. That could result from posttranscriptional modifications. Based on these observations, the necessity to combine studies on transcript and protein levels in the evaluation of antioxidant response in feral fish seems to be desirable approach for field studies. To date, only the highest concentrations of Cd, Zn, Cr, corresponding to LC_{50} 96h (1000 ppb) for Japanese medaka (*Oryzias javanicus*), provoked the activation of transcription of SOD whereas Cu (0.1, 10, 100 ppb) did not demonstrate this effect after exposure for 24 h (Woo et al., 2009).

6. Effect of seasons and abiotic factors on the biomarkers of oxidative stress

There are many factors that may influence the response of fish antioxidant system to exposure to field contaminants. The main criticisms that have been presented against the biomarker approach is connected to high seasonal variability that is frequently found in field studies based on biomarkers, particularly of oxidative stress markers. Season-related biotic regularities include biotic factors, such as reproductive and metabolic status of fish and environmental conditions, such as food availability, oxygen level, temperature of water, salinity, photoperiod, etc. (Parihar et al., 1997; Buet et al., 2006; Da Rocha et al., 2009). Inter-species differences in antioxidant responses depend on the quantitative distribution of antioxidant defenses in the different tissues and sub-cellular compartments. Toxic and organ-specific ROS responses can be related to the anatomical localization, exposure routes and distribution of pollutants, as well as to defense capacity (Ahmad et al., 2006; Ruas et al., 2007; Da Rocha et al., 2009). Fasting conditions (Ferreira et al., 2005), location in the trophic chain affect bioaccumulation of toxic substances (Solé et al., 2009). Fish species that develop different mechanisms of tolerancy to environmental conditions are of particular interest. Some benefits of gills related to high level of GSH known for fish of genus Carassius were expected to supply enhanced antioxidant capacity of fish from chronically polluted site (Falfushynska et al., 2010a). The effect of these factors must be considered in the field studies including this phenomenon in diverse aquatic animals (Winston & Di Giulio, 1991; Lushchak, 2011). Species differences in the efficiency of antioxidant defenses may partly explain prevalence of pathological lesions observed with certain fish species (Vigano et al., 1998). In experiments with hepatocytes of male and female flounder, it was demonstrated that many responses to oxidative stress were sex-related (Winzer et al., 2001). Particularly, increased LPO was showed to be related to a variety of insults other than exposure to xenobiotics causing oxidative stress (Kappus, 1987).

Besides all pointed peculiarities, the combinations of site and season related dependences in aquatic animals have been studied rather extensively. Several studies with fish aimed to distinguish between the local pollution, site-related effects of abiotic factors, and common seasonal regularities. Sometimes they demonstrated that seasonal variation was stronger than relation to site (Niyogi et al., 2001; Almroth et al., 2005; Gorbi et al., 2005). For example, the measuring of a set of biomarkers of antioxidant defense, including catalase, GPx, GR, and GST activities, total glutathione concentration and Total Oxyradical Scavenging

Capacity (TOSC-assay) in the field study of the European eel (*Anguilla anguilla*) and the striped mullet (*M. cephalus*) in Mediterranean lagoons on a seasonal basis suggest that natural variation of responses were associated with seasone variation of both environmental and biological factors, mainly temperature and reproductive cycle which, however, differently affected these two species (Gorbi et al., 2005). Striped mullets exhibited the strongest variation in October at spawning, whereas eels were not influenced by a seasonal sexual maturation and showed more marked changes during summer, likely related to the elevated seawater temperature and light irradiance in the lagoon. Obviously, fish can adapt to low pollution conditions and, under these circumstances, seasonal factors might affect biomarker responses to a greater extent than induced by pollution.

Common seasonal regularities of GSH-related parameters (decreased GSH total concentration and increased GR activities from early to late summer, as well as after maintaining in the laboratory) was demonstrated for killifish (*F. heteroclitus*) from two populations, wild caught in reference and polluted sites of Elizabeth River, USA (Bacanskas et al., 2004). A clear seasonality was found for gill GSH levels of all studied species (*Micropogonias furnieri, Pimelodus pintado, Loricariichthys anus* and *Parapimelodus nigribarbis*) from Southern Brazil, with higher concentration during spring (Da Rocha et al., 2009).

Seasonal variations of the wide set of indices in digestive tissue of barnacle, *Balanus balanoides*, from polluted and non-polluted populations have been evaluated by Niyogi and colleagues (2001). As a general trend in barnacles from polluted and non-polluted populations, maximum antioxidant enzyme, including GST, activities were detected in summer followed by a gradual decrease during the autumn with a minimum in the winter. Microsomal LPO exhibited an almost reverse trend of seasonal variation to that of antioxidant enzyme activities indicating an enhanced susceptibility of barnacle to oxidative stress. Among the environmental parameters, only water temperature seemed to have a significant effect on observed variations of the activities of antioxidant enzymes and GST. However, this pattern was similar to tissue concentrations of PAHs, resulting in significant positive correlation with the activities of antioxidant enzymes, mainly catalase and SOD. So, the seasonal dependence was not clear different from the seasonal level of pollution.

Comparison of flounder (*Platichthys flesus*) collected from nine stations once a month over whole year at Sobieszewo (Gulf of Gdańsk) demonstrated strong month (attributed mainly to spawning, and less to pollution) and geographical (attributed to pollution) variations in biomarker activities, as well as gender difference (Kopecka & Pempkowiak, 2008). In this work correlations between GST and catalase activities with abiotic properties of the environment were less important.

In the study of Da Rocha et al. (2009), the biomarkers of freshwater and estuarine fish species from Southern Brazil were compared in terms of seasonal variation and in three organs (muscle, liver and gills) for the four fish species (*M. furnieri, P. pintado, L. anus* and *P. nigribarbis*) in order to perform an environmental diagnosis. Obtained results showed that liver of *L. anus* and gills of *M. furnieri* presented higher total antioxidant capacity against peroxyl radicals during autumn. In terms of oxidative damage (TBARS), liver of *M. furnieri* and gills of *P. nigribarbis* showed higher TBARS levels during fall, whereas *P. pintado* showed the lowest TBARS value. Finally, a conspicuous seasonal effect was observed for purified enzymes of the GST family as well as for non-purified GST, where minimum values were registered during fall, pointing to this season as one where fish species were less competent to perform detoxifying reactions (Da Rocha et al., 2009).

However, majority of season-related studies confirms that the continuous environmental press may affect the seasonal dynamics of biomarker states (Gorbi et al., 2005). When spatial and temporal dependences are compared, statistical approaches need to be applied to distinguish these determinants. For example, in the study of common carp (*C. carpio*) from two sites, despite of the priority of the season effect for the majority of markers, a similarity in the seasonal patterns between two sites was evident only for $O_2^{\bullet -}$, MT and TBARS (Falfushynska & Stoliar, 2009). This disparity as well as Two-factor interaction and the high classification rate between sites according to Discriminant Analysis reflect the impact of the effects of pollution on the seasonal regularity. However, in the areas that are characterised by spontaneous pollution, temporary inter-site differences can occur. Spatial effects could have seasone-related manner due to the specificity of agricultural activity, for example, or oxygen consumption and temperature that can be particularly important for the animals from cooling ponds of nuclear power plants. Therefore, the native "reference site" can be considered as the relative concept even if this site seems to be nearby unaffected. In this study, a battery comprising SOD, GSH and MTs in the liver as well as SOD, catalase and $O_2^{\bullet -}$ in the gills among several examined biomarkers was found to discriminate adequately fish from areas with varied water quality during three seasons.

In the study of more tolerant that *C. carpio* species, *C. carassius*, from the similar two field sites, Two variant ANOVA and the centroid grouping analysis allowed to distinguish the fish according to season more than to site, confirming the effect of season on all studied indices of oxidative stress except catalase activity in liver (Falfushynska et al., 2010a). The centroid grouping analysis of the separate specimens indicated clearly the general temporal dependence for fish from both sites (Fig. 2). However, in each season, especially in summer for gills and in autumn for liver, distinct spatial separation was found. Regarding the separate indices, the typical temporal difference in enzyme activities involved the elevation of values from spring to summer/autumn, probably due to increase in metabolic rates and accumulation of energy reserves was found (Chellappa et al., 1995). In some cases, different temporal dependence (for SOD and catalase activities in the liver) was observed in the groups from industrially polluted and reference sites. Obviously, this distortion of the general patterns could result from local adverse effects. In each season, especially in summer for gills and autumn for liver, distinct spatial separation was found. Temporal dependence with an elevation in summer was shown for the activities of SOD, catalase and concentration of GSSG in both tissues at reference site and only for the activities of SOD in the gills and GSSG in liver and gills at polluted site. The decrease from spring to autumn was typical for TBARS. The inter-site difference was especially high in summer, when lower actitivities of SOD (in the liver) and catalase and elevated TBARS levels were observed in the tissues of fish from polluted site. In autumn, the higher catalase activity in both tissues, and the higher SOD activity and lower TBARS levels in the gills indicated more efficient antioxidant defense in this group of fish. Only glutathione system showed relatively constant differences between fish from the two sites with higher GSH and Redox Index of GSH levels (particularly in the gills) and lower GSSG levels for the fish from industrial site in comparison to the other site.

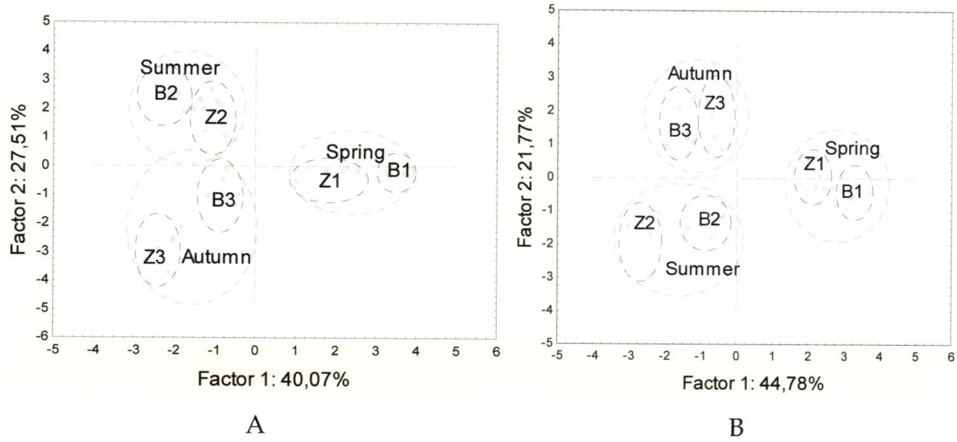

Fig. 2. Centroid grouping analysis of *C. carassius* from two sites, industrial (B) and rural (Z) parameter data set in the liver (A) and gills (B) in spring (1), summer (2) and autumn (3) (from Falfushynska et al., 2010a with permission).

To summarize, there is no doubts in the seasonal variation of antioxidant defence in fish. Different oxidative manifestations may have different dynamics. Environmental pollution, particularly spontaneous pollution, can interfere with seasonal dynamics and disturb it. However, fish can adapt to pollution in the areas with constant impact and do not show particular response compare to clean areas. So, the evidence of common seasonal difference in fish from polluted and clean areas can not be estimated as an absence of the effects of pollution. Integral statistic approach is effective in the discrimination between seasonal physiological regularities and spatial effects, particularly casual effects of non-pointed sources of pollution on fish.

7. Poisoning and adaptation: Caged fish

The main failure in application of markers of oxidative stress with feral fish is expected adaptation to the environmental conditions that leads to compensatory homeostasis occuring in antioxidant system (Barja de Quiroga et al., 1990; Regoli & Principato, 1995; Reynders et al., 2008). The choice of end points in the field studies of fish may be complicated by a history of exposure to xenobiotics causing oxidative stress. For instance, grey mullet (*Mugil* sp.) collected from an estuary polluted with metal ions, PAHs, PCBs, and pesticides demonstrated evidence of oxidative stress as indicated by Redox Index of GSH. However, these fish did not show elevated levels of LPO products while showing elevated activities of antioxidant enzymes (GPx, SOD, catalase, and GR). It is possible that an adaptive response occurred and repair of LPO mignt take place (Rodriquez-Ariza et al., 1993). Therefore, with the aim to standardize the results and avoid the effects of adaptation to chronically polluted environment, transplantation of caged organisms, including fish, to the sites of interest on the time suitable for the response, is recommended as an adequate step for the ERA. Caging studies often utilize farmed fish with known age and nutritional background, though wild captured fish can also be used. Caging allows the exposure of individual fish to conditions at a certain site, for known time (Almroth, 2008). In this case, the early warning of toxic effect can be expected. The transferring of fish in new conditions can critically change their antioxidant profile. However, unlike with molluscs (Viarengo et al., 2007), for fish, the studies on the response of caged specimens are scarce.

For the appreciation of measurable oxidative stress response, 48 h exposure was recommended (Ahmad et al., 2004). This exposure strategy was adapted for different aquatic ecosystems biomonitoring using *A. anguilla* and other fish species. In the study of Ahmad and colleagues (2006), *A. anguilla* was plunged at five study sites located at increasing distances from the entrance point of the main source of contamination, originated from the introduction of agricultural chemicals, trace metal ions, domestic wastes, as well as eutrophication and incorrect utility of resources resulted in an increased water pollution. Inducing trend for total antioxidant enzyme activites (catalase, GPx and GST) was observed in gills of fish caged in the polluted area. In liver and kidney, the exposure typically induced significant decrease in the activity of the abovementioned enzymes. However, each studied parameter displayed a particular pattern in each site. Hepatic GSH concentration was increased, whereas LPO was decreased in individuals from polluted sites (Ahmad et al., 2006). Despite the authors assured that these findings provided a rational use of oxidative stress biomarkers in pollution biomonitoring of freshwater ecosystems, the discrepancy and variability of changes in each polluted site, did not allow to confirm the self-sufficience of this approach. No clear relationship could be established between gill oxidative stress responses and the distance to the main source of pollution in this study. Additionally, no gill LPO induction in this study may be explained by an effective antioxidant action. In general, in this study all organs studied revealed a similar resistance to peroxidative damage, suggesting that the antioxidants are more responsive biomarkers than LPO for short-term exposure. Besides the activation of antioxidant enzymes (as observed in gills), their inhibition (as observed in kidney and liver) should also be considered as a clear marker of pollutant presence and environmental degradation (Ahmad et al., 2006).

The utility of model studies was demonstrated in English sole (*P. vetulus*) exposed in laboratory and in their natural environment to an organic-solvent extract of sediment. Exposure provoked significant increase of hepatic GSH concentrations with a dose-dependent increase. Similarly, fish sampled directly from the polluted site showed higher GSH concentrations compared to fish from a reference site (Nishimoto et al., 1995).

Another period of time, usually used for caging experiments, is about 14-15 days. It is suitable for the indication of the effect of pollution related to the accumulation of specific pollutant/s. The transplantation of brown trout from a river with low levels of metal ions (the Stribekken River) to a river with high levels of Cd and Zn (the Naustebekken River) for up to 15 days allowed to demonstrate difference in the transcription and activities of central antioxidant enzymes and proteins in an environmental setting. This time was sufficient for significant uptake of both Cd and Zn in gills. Moreover, Cd levels were found to correlate significantly with transcript levels of MT, Cu,Zn-SOD, GPx, and GR. The activities of SOD and catalase increased in gills after transfer, but MT protein levels decreased. In liver, SOD activity and MT protein levels increased, while in kidney only MT protein concentrations were elevated after transfer. The detection of a general lack of consistency between mRNA transcription and specific enzyme activities, indicating that these proteins and enzymes are not solely under transcriptional control was very important result of this study (Hansen et al., 2006a).

The transfer of brown trout to a Cu-contaminated river in the Roros region in Central Norway provoked significant increase of MT-A, SOD and GR transcription along with uptake of Cu in gills, while only transcription of MT-A was found to respond in liver and kidney during the exposure. At that, no increase in MT protein levels were observed in gills. The levels of SOD and catalase enzymes were affected in all tissues during the exposure. A negative correlation between SOD and catalase activities was observed in gills indicating that the activities of these enzymes were influenced not only through transcription. The transcript levels of GPx and GR transcript levels correlated positively with each other in gills and liver, indicating their shared function in GSH-turnover (Hansen et al., 2006b).

The response of antioxidant system after depuration of feral fish, mullet, *M. cephalus*, and flounder, *P. flesus*, from the polluted site was expressed as the decrease of the activities of SOD and catalase only in mullet liver (Ferreira et al., 2005). Oxidative damage in liver, evaluated by estimating LPO and PC, increased in both species in most cases. This effect was explained by the decreased antioxidant defence after oxidative stress insults in natural environment.

The initial health status of fish can significantly affect the ability to form stress-related response. Some results in this direction were obtained when animals from two sites were compared under the modelling effect in laboratory (Hasspieler et al., 1994; Meyer et al., 2003; Falfushynska et al., 2011). Up-regulated stress-related parameters in the animals from chronically polluted sites at additional loading was revealed in the set of comparative studies of Di Gulio and colleagues on killifish *F. heteroclitus* inhabiting site polluted by creosote (Meyer et al., 2003; Bacanskas et al., 2004), and in the study of cadmium-acclimated rainbow trout (Chowdhury et al., 2004), whereas the fish from reference site were not able to activate the enzymes of antioxidant defense under the exposure. The comparison of the effects of prooxidant copper (Cu^{2+}, 0.005 and 0.050 mg $\cdot L^{-1}$) or manganese (Mn^{2+}, 0.17 and 1.7 mg $\cdot L^{-1}$) on *C. auratus gibelio* from polluted and unpolluted sites after exposure for 14 days indicated that fish from the polluted site showed lower activities of SOD (Cu,Zn- and Mn-SOD) and GST in the liver and gills. The oxidative stress response was more efficient in fish from the polluted site (Falfushynska et al., 2011) due to the activation of Cu,Zn- and Mn-SOD. The interference of the ability of the antioxidative defence and the origin of fish (pre-exposure to chronic pollution) was found in this study. In general, gibel carp from the polluted site demonstrated a highly effective response of the antioxidant system particularly with SOD activity in liver and gills. In this study site-related differences in the level of LPO and GSH between the two groups were mainly maintained in all groups.

Vega-López and colleagues (2008) studied the responses of antioxidant system (LPO, SOD and catalase activities) in fish *Girardinichthys viviparus* after exposure to water from PCBs contaminated habitants and from other site, expected to be suitable place for the re-introduction of this endangered species. Water enriched by PCB was also inspected in that study during 1, 2, 4, 8 and 16 days of exposure. Four types of responses were observed dependently on the composition of water and sex of fish: (1) increased lipid peroxidation intensity, depressed SOD and increased catalase activities; (2) an increase in all three biomarkers; (3) decreased LPO product levels, unchanged SOD and increased catalase activities; (4) increased LPO intensity and depressed SOD and catalase activities. At that, the only PCB addition to the natural water of fish resulted in decreased LPO, whilst the exposure no native water depressed both studied enzymes in concert with the intensification of LPO. Evaluation of these responses is rather complicated and cannot provide a clear conclusion on the consequence of each response for fish health status. These results also let to conclude that in the field investigations, the selection of the reference site is important to understand the response of fish. The relativity of the concept of the reference site is sometimes indicated even though this site was selected by generally accepted criteria.

8. Biomarkers of oxidative stress in the multi-marker approach: Integrated data analysis

Performed analysis has shown that each of separate markers of oxidative defense alone cannot provide the conclusion on the nature of observed difference between compared groups. Indeed, in the field studies, wide spectrum of inter-site differences (higher, equal or lower activities of various antioxidant enzymes with tissues peculiarities and disbalance) have been observed in polluted compared to clean areas reflecting both mild stress conditions of the location or strong oxidative damage. Therefore, different models of the aquatic animal response need to be analysed before conclusions can be drawn. Only the evaluation of the final effect of oxidative impact can provide the conclusion on the response of organism. All these possibilities and their combinations have been reported (Winston & Di Guilio, 1991; Lushchak, 2011) and this complexity of antioxidant responses to pollutants often leads to a controversy on the use of oxidative stress markers in ecotoxicological studies.

Two approaches for the appreciation of the severity of stress are proposed. The first one is connected to the measuring of the integrated state of the antioxidant capacity which was successfully applied by Regoli and colleagues (Regoli, 2000; Regoli et al., 2002a, b). The total oxyradical scavenging capacity (TOSC) quantifies the capability of the whole antioxidant system to neutralize oxyradicals, allowing to discriminate between different forms of ROS, thus providing useful indications to predict oxyradical-mediated adverse effects under certain physiological conditions of organisms (Regoli, 2000). Appropriate assay conditions have been standardized in which different ROS induced a comparable prooxidant force quantified by the oxidation of the substrate α-keto-γ-methiolbutyric acid. Thus, the efficiency of antioxidants toward various reactive species can be better compared by their ability to inhibit an oxidative pressure induced by specific oxidants. The validity of this approach was confirmed in different field studies. Particullarly, integration of measurement of individual antioxidants with TOSC analysis increased the evaluation of oxidative responses to pollutants in ecotoxicological studies (Regoli et al., 2005). The field study of an Antarctic silverfish (*Pleuragramma antarcticum*), that is developed in the environment with strong pro-oxidant characteristics, revealed particularly prompt responses for GSH metabolism which, however, did not prevent high intensity of LPO. From the analysis of TOSC, the overall efficiency to neutralize peroxyl radicals remained almost constant while slightly lower TOSC values were obtained toward hydroxyl radicals at the end of sampling period (Gorbi & Regoli, 2003; Regoli et al., 2005). The analysis of TOSC revealed that the overall capacity of specific tissues of red mullet (*M. barbatus*) in the area of dredging to absorb various oxidants was not substantially compromised when challenged with increased prooxidant pressures (Regoli, 2002). In the study of Meyer et al. (2003), offsprings of killfish from polluted site showed higher basal TOSC value that was balanced by higher GSH concentrations, and Mn-SOD protein levels.

In general, it is considered that individual antioxidants are useful as "response biomarkers" indicating a varied prooxidant challenges and potentially important early warning signals. Variations of individual antioxidants are useful for understanding the mode of action of a chemical stressor and the possible molecular targets with specific responses. However, their value is more limited for understanding of the biological effects in terms of health conditions of the organisms. The value of TOSC is a less sensitive marker than individual characteristics of antioxidant defence, but it provides more holistic picture of susceptibility to oxidative stress. It is an "biomarker" effect with predictive value since varied capability to counteract oxyradical efects can induce alterations at other levels of biological organization. In conclusion, the combination of measurement of individual antioxidants with TOSC analysis seems to improve the evaluation of oxidative responses to pollutants in ecotoxicological studies (Regoli, 2000).

The second approach to give an integrated assessment of antioxidant response is based on the multi-marker approach and calculation of the balance between prooxidant endogenous and exogenous factors (i.e., environmental pollutants) and antioxidant defenses (enzymatic and nonenzymatic) in biological systems. It gives so-called coefficient - the Integrated Oxidative Stress (IOS) index, calculated as the ratio of antioxidant factors, A, and prooxidant manifestations, O, after data standardization (Falfushynska et al., 2008). The basis of standardization of each factor taking into consideration mean values and standard deviations for each group was described by Leiniö and Lehtonen (2005). The antioxidant factors (A) may include the activities of SOD and catalase, GSH level, and oxidative damage manifestations (O) represented by TBARS and $O_2^{\bullet-}$ levels. They also can include PC and GSSG if they were detected. Lower IOS value in the counterpart implies the weakness of antioxidant defence and higher one – its suppression. In the tissues of common carp from two field sites, all measured markers of oxidative stress contributed to the integrated index. For example, in the polluted area, season-related difference between two groups can be detected (Fig. 3). In most cases, the balance between A and O was found, particularly in gills. However, in spring, the IOS in the liver suggested the unbalance, or presence of oxidative stress in the fish at the reference site. The particularly high overbalancing appeared in the groups for the polluted site for the antioxidant systems of liver and gills in summer and autumn, respectively. In the gills, the highest antioxidant capacity is reflected in autumn according to the calculated IOS. Indeed, the elevation of TBARS level at the R-site in spring and at the I-site in autumn reveals that the variations in activities of different antioxidant enzymes and levels of nonenzymatic antioxidants were unable to prevent oxidative damage in fish liver. In contrast, the gills were significantly more tolerant to oxidative damage. Additionally, this integrative approach applied in different seasons confirmed that the fish in relatively low polluted area can form adaptive response in the particular cases, especially in autumn, and possesses the weakness of this response in spring after the winter metabolic depression.

Generally, it needs to be pointed out that the measuring of oxidative stress markers in the feral fish is almost obligate constituent of the evaluation of its health status. Obviously, as it was shown above, their response are higly variable. Based on this variability, correspondent articles are traditionally finalized by the suggestion that selected parameters are suitable biomarkers of pollution, and studied certain species is suggested to be potential bioindicator organism against environmental pollution. However, the classification of the interrelation of the changes in different tissues, of separate characteristics and their relation to the ability of fish to withstand pollution have not been represented yet. The intercalibration and generalization of results and based on this issue classification of oxidative stress response remains to be very actual technology in ERA.

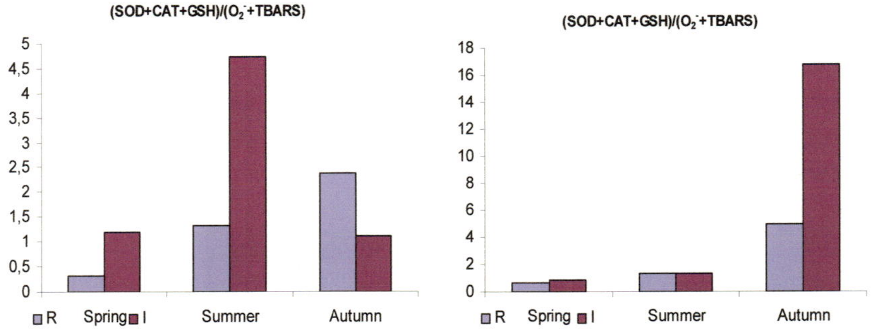

A B

Fig. 3. Integrated Oxidative Stress index of liver (A) and gills (B) of common carp from relatively clean (R) and chronically polluted (I) field sites in three seasons, calculated from their SOD and catalase activities, GSH level, superoxide anion production and TBARS levels (from Falfushynska et al., 2009 with permission).

9. Conclusions and perspectives

Oxidative stress in fish is a general consequence of the environmental pollution. Information on antioxidant defense in fish is meagre despite that fish are constantly exposed to a myriad of environmental stresses including oxidant-induced ones. In the feral organisms, deleterious effects of environment are often difficult to evaluate since many of these effects tend to manifest only after longer periods of time and organisms tend to adapt to them (van der Oost et al., 2003). Early warning of the toxic effects of pollutants, particularly in spontaneously polluted areas can be predicted only using a biomarker approach, including oxidative stress manifestations and adaptive responses.

Antioxidant defense system in fish is very sensitive to environmental conditions. However, in different studies with fish from mixed polluted field sites, the enzymes of antioxidant defence demonstrated case-dependent difference. Whereas, apparently, the study of the frame of resilience of antioxidant enzymes in the fish from polluted and pristine sites needs to be the object of specific study. When several enzymes were studied, their responses were frequently imbalanced, or their transcriptional level and enzyme activity trends might be different. In some studies, the absence of the changes of their activities was observed when biomarkers of specific pollution confirmed toxicity of environment. Studies addressed potential ROS scavengers in the feral fish under stress conditions are very scarce. Free radical-related processes and ROS production were found to be responsible for a variety of oxidative damages leading to adverse health effects and diseases in the feral fish even from comparatively indisturbed areas. Moreover, it is clear that, given spontaneous human activities, nobody will be able to find a true reference site in some areas.

The integrated approach with the appreciation of balance between prooxidant manifestations and antioxidant defences in biological systems needs to be a control point to assess toxic effects under stressful environmental conditions.

10. Acknowledgment

The authors wish to thank Dr. H. Falfushynska for comments on the manuscript and Ph.D. student L. Gnatyshyna for technical assistance with the preparation of the manuscript.

11. References

Adams, S.M. & Greeley, M.S. (2000). Ecotoxicological Indicators of Water Quality: Using Multi-response Indicators to Assess the Health of Aquatic Ecosystems. *Water, Air and Soil pollution*, Vol.123, pp. 103–115, ISSN 0049-6979

Ahmad, I.; Pacheco, M. & Santos M.A. (2004). Enzymatic and Non-enzymatic Antioxidants as an Adaptation to Phagocyte-induced Damage in *Anguilla anguilla* L. Following in Situ Harbor Water Exposure. *Ecotoxicology and Environmental Safety*, Vol.3, No.57, pp. 290–302, ISSN 0147-6513

Ahmad, I.; Pacheco, M. & Santos, M.A. (2006). *Anguilla anguilla* L. Oxidative Stress Biomarkers: An in Situ Study of Freshwater Wetland Ecosystem (Pateira de Fermentelos, Portugal). *Chemosphere*, Vol.65, Issue 6, (November 2006), pp. 952–962, ISSN 0045-6535.

Almroth, B.C. (2008). Oxidative Damage in Fish Used as Biomarkers in Field and Laboratory Studies. Department of Zoology/Zoophysiology, University of Gothenburg, Sweden, ISBN 978-91-628-7420-9

Almroth, B.C.; Albertsson, E., Sturve, J. & Förlin, L. (2008a). Oxidative Stress, Evident in Antioxidant Defences and Damage Products, in Rainbow Trout Caged Outside a Sewage Treatment Plant. *Ecotoxicology and Environmental Safety*, Vol.70, No.3, pp. 370–378, ISSN 0147-6513

Almroth, B.C.; Sturve, J., Berglund, Å. & Förlin, L. (2005). Oxidative Damage in Eelpout (*Zoarces viviparus*), Measured as Protein Carbonyls and TBARS, as Biomarkers. *Aquatic Toxicology*, Vol.73, pp. 171–180, ISSN 0166-445X

Almroth, B.C.; Sturve, J., Stephensen, E., Holth, T.F. & Förlin, L. (2008b). Protein Carbonyls and Antioxidant Defences in Corkwing Wrasse (*Symphodus melops*) from a Heavy Metal Polluted and a PAH Polluted Site. *Marine Environmental Research*. Vol.66, No.2, (April 2008), pp.271–277, ISSN 0141-1136

Amado, L.L.; Robaldo, R.B., Geracitano, L., Monserrat, J.M. & Bianchini, A. (2006). Biomarkers of Exposure and Effect in the Brazilian Flounder *Paralichthys Orbignyanus* (Teleostei: Paralichthyidae) from the Patos Lagoon Estuary (Southern Brazil). *Marine Pollution Bulletin*, Vol.52, No.2, (December 2005), pp. 207–213, ISSN 0025-326X.

Andersson, T.; Forlin, L., Hardig, J. & Larsson, A. (1988). Physiological Disturbances in Fish Living in Coastal Waters Polluted with Bleached Kraft Pulp Mill Effluents. *Canadian Journal of Fisheries and Aquatic Sciences*, Vol. 45, pp. 1525–1536, ISSN 0706-652X.

Armstrong, D.T. (1986). Environmental Stress and Ovarian Function. *Biology of Reproduction*. Vol.34, No.1, (February 1986), pp. 29–39, ISSN 0006-3363.

Atif, F.; Kaur, M., Yousuf, S. & Raisuddin, S. (2006). *In vitro* Free Radical Scavenging Activity of Hepatic Metallothionein Induced in an Indian Freshwater Fish, *Channa punctata* Bloch. *Chemico-Biological Interactions*, Vol.162, (June 2006), pp. 172–180, ISSN 0009-2797.

Bacanskas, L.R.; Whitaker, J. & Di Giulio, R.T. (2004). Oxidative Stress in Two Populations of Killifish (*Fundulus heteroclitus*) with Differing Contaminant Exposure Histories. Marine Environmental Research, Vol.58, No.2-5, (August-December 2004), pp. 597-601, ISSN 0141-1136.

Bagnyukova, T.V.; Chahrak, O.I. & Lushchak, V.I. (2006). Coordinated Response of Goldfish Antioxidant Defenses to Environmental Stress. *Aquatic Toxicology*, Vol. 78, pp. 325-331, ISSN 0166-445X.

Bagnyukova, T.V.; Luzhna, L.I., Pogribny, I.P. & Lushchak, V.I. (2007). Oxidative Stress and Antioxidant Defenses in Goldfish Liver in Response to Short-term Exposure to Arsenite. *Environmental and Molecular Mutagenesis*, Vol. 48, pp. 658–665, ISSN 0893-6692.

Bagnyukova, T.V.; Storey, K.B. & Lushchak, V.I. (2005). Adaptive Response of Antioxidant Enzymes to Catalase Inhibition by Aminotriazole in Goldfish Liver and Kidney. *Comparative Biochemistry and Physiology*, Vol. 142 B, pp. 335–341, ISSN 0305-0491.

Bainy, A.C.D.; Saito, E., Carvalho, P.S.M. & Junquieira, V.B.C. (1996). Oxidative Stress in Gill, Erythrocytes, Liver and Kidney of Nile Tilapia (*Oreochromis Niloticus*) from a Polluted Site. *Aquatic Toxicology*, Vol.34, (February 1996), pp. 151–/162, ISSN 0166-445X.

Bargelloni, L.; Scudiero, R., Parisi, E., Carginale, V., Capasso, C. & Patarnello, T. (1999). Metallothioneins in Antarctic fish: Evidence for Independent Duplication and Gene Conversion. *Molecular Biology and Evolution*, Vol.16, (July 1999), pp. 885–897, ISSN 0737-4038

Barja de Quiroga, G.; Pérez-Campo, R. & López-Torres, M. (1990). Changes on Cerebral Antioxidant Enzymes, Peroxidation, and the Glutathione System of Frogs after Aging and Catalase Inhibition. *The Journal of Neuroscience*. Vol.26, No.3, (July 1990), pp. 370-376, ISSN 0270-6474

Bhattacharya, S. (2001). Stress Response to Pesticides and Heavy Metals in Fish and Other Vertebrates. *Proceedings of the Indian Academy of Sciences*. B67, No.5, (May 2001), pp. 215–246, ISSN 0073-6600.

Buet, A.; Banas, D., Vollaire, Y., Coulet, E. & Roche, H. (2006). Biomarker Responses in European Eel (*Anguilla anguilla*) Exposed to Persistent Organic Pollutants. A Field Study in the Vaccarès Lagoon (Camargue, France). *Chemosphere*. Vol.65, No.10, (December 2006), pp. 1846–1858, ISSN 0045-6535.

Capasso, C.; Carginale, V., Crescenzi, O., Di Maro, D., Spadaccini, R., Temussi, P.-A. & Parisi, E. (2005). Structural and Functional Studies of Vertebrate Metallothioneins: Cross-talk between Domains in the Absence of Physical Contact. *Biochemical Journal*, Vol.391, pp. 95–103, ISSN 0264-6021.

Capdevila, M.; Cols, N., Romero-Isart, N., Gonzalez-Duarte, R., Atrian, S. & Gonzalez-Duarte, P. (1997). Recombinant Synthesis of Mouse Zn3-beta and Zn4-alpha Metallothionein 1 Domains and Characterization of Their Cadmium (II) Binding Capacity. *Cellular and Molecular Life Sciences*, Vol.53, pp. 681–688, ISSN 1421-682X

Cavaletto, M.; Ghezzi, A., Burlando, B., Evangelisti, V., Ceratto, N. & Viarengo, A. (2002). Effect of Hydrogen Peroxide on Antioxidant Enzymes and Metallothionein Level in the Digestive Gland of Mytilus galloprovincialis. *Comparative Biochemistry and Physiology*, Vol.131 C, pp. 447–455, ISSN 0742-8413

Chellappa, S.; Huntingford, F.A., Strang, R.H.C. & Thomson, R.Y. (1995). Condition Factor and Hepatosomatic Index as Estimates of Energy Status in Male Three-spined Stickleback. *Journal of Fish Biology*, Vol.47, pp. 775–787, ISSN 0022-111

Chesman, B.S.; O'Hara, S., Burt, G.R. & Langston, W.J. (2007). Hepatic Metallothionein and Total Oxyradical Scavenging Capacity in Atlantic cod *Gadus morhua* Caged in Open Sea Contamination Gradients. *Aquatic Toxicology*, Vol.84, pp. 310–320, ISSN 0166-445X

Cho, Y.S.; Lee, S.Y. Kim, K.-Y., Bang, I.C., Kim, D.S. & Nam, Y.K. (2008). Gene Structure and Expression of Metallothionein During Metal Exposures in *Hemibarbus mylodon*. *Ecotoxicology and Environmental Safety*, Vol. 71, pp.125–137, ISSN 0147-6513

Collier, T.K. & Varanasi, U. (1991). Hepatic Activities of Xenobiotic Metabolizing Enzymes and Biliary Levels of Xenobiotics in English Sole (*Parophrys vetulus*) Exposed to Environmental Contaminants. *Archives of Environmental Contamination and Toxicology*, Vol.20, pp. 462–473, ISSN 0090-4341

Commandeur, J.N.M.; Hermanns, R.C.A., Meerman, J.H.N., Van Baar, B.L.M. & Vermeulen, N.P.E. (1997). Simultaneous Determination of Eight Lipid Peroxidation Degradation Products in Urine of Rats Treated with Carbon Tetrachloride Using Gas Chromatography with Electroncapture Detection. *Journal of Chromatography*, Vol.694, pp. 277–288, ISSN 0021-9673

Da Rocha, A.M.; Salomão de Freitas. D.P., Burns, M., Vieira, J.P., de la Torre, F.R. & Monserrat, J.M. (2009). Seasonal and Organ Variations in Antioxidant Capacity, Detoxifying Competence and Oxidative Damage in Freshwater and Estuarine Fishes from Southern Brazil. *Comparative Biochemistry and Physiology*, Vol.150 C, pp.512–520, ISSN 0742-8413

Di Giulio, R.T.; Habig, C. & Gallagher, E.P. (1993). Effects of Black Rock Harbor Sediments on Indices of Biotransformation, Oxidative Stress, and DNA Integrity in Channel Catfish. *Aquatic Toxicology*, Vol.26, pp.1–22, ISSN 0166-445X

Dorval, J.; Leblond, V., Deblois, C. & Hontela, A. (2005). Oxidative Stress and Endocrine Endpoints in White Sucker (*Catostomus commersoni*) from a River Impacted by Agricultural Chemicals. *Environmental Toxicology and Chemistry*, Vol.24, No.5, pp. 1273–1280, ISSN 0730-7268

dos Anjos, N.A; Schulze, T., Brack, W., Val, A.L., Schirmer, K. & Scholz, S. (2011). Identification and Evaluation of Cyp1a Transcript Expression in Fish as Molecular Biomarker for Petroleum Contamination in Tropical Freshwater Ecosystems. *Aquatic Toxicology*, Vol.103, No.(1-2), (May 2011), pp. 46–52, ISSN 0166-445X

Du, J. & Gebicki, J.M. (2004). Proteins are Major Initial Cell Targets of Hydroxyl Free Radicals. *The International Journal of Biochemistry & Cell Biology*, Vol.36, No.11, (November 2004), pp. 2334–2343, ISSN 1357-2725

EU Council Directive 98/83/EC), Retrieved from <http://www.emwis.org/IFP/law_EU.htm>

Eufemia, N.; Collier, T.K., Stein, J.E., Watson, D.E. & Di Giulio, R.T. (1997). Biochemical Responses to Sediment-associated Contaminants in Brown Bullhead (*Ameriurus nebulosus*) from the Niagara River Ecosystem. *Ecotoxicology*, Vol.6, pp. 13–34, ISSN 0963-9292

Eustace, B.K. & Jay, D.G. (2004). Extracellular Roles for the Molecular Chaperone, Hsp90. *Cell Cycle*, Vol. 3, pp. 1098–1100, ISSN 1538-4101

Falfushinska, H.; Loumbourdis, N., Romanchyk, L. & Stolyar, O. (2008). Validation of Oxidative Stress Responses in Two Populations of Frogs from Western Ukraine. *Chemosphere*, Vol. 73, pp. 1096–1101, ISSN 0045-6535

Falfushinska, H.I.; Goch, I.V., Bugera, L.I. & Stoliar, O.B. (2009). Metallothioneins and the Indices of Oxidative Damage in the Tissues of Carp *Cyprinus carpio* as the Biomarkers of the Environmental Pollution. Studia Biologica, Vol.3, No.2, pp. 99–108, ISSN 1996-4536

Falfushynska, H. & Stolyar, O. (2009). Responses of Biochemical Markers in Carp *Cyprinus carpio* from Two Field Sites in Western Ukraine, *Ecotoxicology and Environmental Safety*, Vol.72. pp. 729–36, ISSN 0147-6513

Falfushynska, H.; Gnatyshyna, L., Priydun, Ch., Stoliar, O. & Nam, Y.K. (2010a). Variability of Responses in the Crucian Carp *Carassius carassius* from Two Ukrainian Ponds Determined by Multi-marker Approach, *Ecotoxicology and Environmental Safety*, Vol.73, pp. 1896–1906, ISSN 0147-6513

Falfushynska, H.; Gnatyshyna, L., Shymkiv, T. & Stoliar, O. (2010b). Reactivity of Metallothioneins and the Efficiency of their Metal-keeping and Antioxidative Functions in the Adaptation of Fish *Carassius carassius* to Field Pollution. *Proceeding of the Shevchenko Scientific Society. Chemistry & Biochemistry*, Vol. 25, pp. 229–242, ISSN 1563-3569.

Falfushynska, H.I. & Stoliar, O.B. (2009). Function of Metallothioneins in Carp *Cyprinus carpio* from Two Field Sites in Western Ukraine. *Ecotoxicology and Environmental Safety*, Vol.72, pp. 1425–1432, ISSN 0147-6513.

Falfushynska, H.I.; Gnatyshyna, L.L., Farkas, A., Vehovszky, Á., Gyori, J. & Stoliar, O.B. (2010c). Vulnerability of Biomarkers in the Indigenous Mollusk *Anodonta cygnea* to Spontaneous Pollution in a Transition Country. *Chemosphere*, Vol. 81, Issue 10, (November 2010), pp. 1342–1351, ISSN 0045-6535

Falfushynska, H.I.; Gnatyshyna, L.L., Stoliar, O.B. & Nam, Y.K. (2011). Various Responses to Copper and Manganese Exposure of *Carassius auratus gibelio* from the Two Populations. *Comparative Biochemistry and Physiology*, Vol. 154, No. 3, pp. 242–253, ISSN 1532-0456

Farombi, E.O.; Adelowo, O.A. & Ajimoko, Y.R. (2007). Biomarkers of Oxidative Stress and Heavy Metal Levels as Indicators of Environmental Pollution in African Catfish (*Clarias gariepinus*) from Nigeria Ogun River. *International Journal of Environmental Research and Public Health*, Vol.4, pp. 158–165, ISSN 1660-4601

Fernandes, D.; Andreu-Sánchez, O., Bebianno, M.J. & Porte, C. (2008). Assessment of Pollution Along the Northern Iberian shelf by the Combined Use of Chemical and Biochemical Markers in Two Representative Fish Species. *Environmental Pollution*, Vol.155, pp. 327–335, ISSN 0269-7491

Fernandes, D.; Bebianno, M.J. & Porte, C. (2008). Hepatic Levels of Metal and Metallothioneins in Two Commercial Fish Species of the Northern Iberian Shelf, Science of the Total Environment, Vol.391, pp.159–167, ISSN 0048-9697

Ferreira, M.; Moradas-Ferreira, P. & Reis-Henriques, M.A. (2005). Oxidative Stress Biomarkers in Two Resident Species, Mullet (*Mugil cephalus*) and Flounder

(*Platichthys flesus*), from a Polluted Site in River Douro Estuary, Portugal. *Aquatic Toxicology*, Vol.71, pp. 39–48, ISSN 0166-445X

Fessard, V. & Livingstone, D.R. (1998). Development of Western Analysis of Oxidized Proteins as a Biomarker of Oxidative Damage in Liver of Fish. *Marine Environmental Research,* Vol. 46, pp. 407–410, ISSN 0141-1136

George, S.G. (1994). Enzymology and Molecular Biology of Phase II Xenobiotic-conjugating Enzymes in Fish. In: Malins, D.C., Ostrander, G.K. (Eds.). *Aquatic Toxicology;* Molecular, Biochemical and Cellular perspectives. Lewis Publishers, CRC press, pp. 37–85, ISSN 0166-445X

Godin, D.V. & Garnett, M.E. (1992). Species-related Variations in Tissue Antioxidant Status- I. Differences in Antioxidant Enzyme Profiles. *Comparative Biochemistry and Physiology*, Vol.103 B, No. 3, pp. 737–742, ISSN 1879-1107

Gorbi, S. & Regoli, F. (2003). Total Oxyradical Scavenging Capacity as an Index of Susceptibility to Oxidative Stress in Marine Organisms. *Comment on Toxicology,* Vol.9, pp 303–322, ISSN 0886-5140

Gorbi, S.; Baldini, C. & Regoli, F. (2005). Seasonal Variability of Metallothioneins, Cytochrome P450, Bile Metabolites and Oxyradical Metabolism in the European eel *Anguilla anguilla* L. (Anguillidae) and Striped mullet Mugil cephalus L. (Mugilidae). *Archives of Environmental Contamination and Toxicology*. Vol.49, pp. 62–70, ISSN 0090-4341

Gundersen, D.T.; Miller, R., Mischler, A., Elpers, K., Mims, S. D., Millar, J. G. & Blazer, V. (2000). Biomarker Response and Health of Polychlorinated Biphenyl- and Chlordane-contaminated Paddlefish from the Ohio River Basin, USA. *Environmental Toxicology and Chemistry*, Vol.19, pp. 2275–2285, ISSN 1552-8618

Hansen, B.H.; Romma, S., Garmo, O.A., Olsvik, P.A. & Andersen, R.A. (2006a). Antioxidative Stress Proteins and Their Gene Expression in Brown Trout (*Salmo trutta*) from Three Rivers with Different Heavy Metal Levels. *Comparative Biochemistry and Phisiology*, Vol.143 C, pp. 263–274, ISSN ISSN 1532-0456.

Hansen, B.H.; Romma, S., Softeland, L.I., Olsvik, P.A. & Andersen, R.A. (2006b). Induction and Activity of Oxidative Stress-related Proteins during Waterborne Cu-exposure in Brown Trout (*Salmo trutta*). *Chemosphere*, Vol.65, pp. 1707–1714, ISSN 0045-6535

Hasspieler, B.M.; Behar, J.V., Carlson, D.B., Watson, D.E. & Di Giulio, R.T. (1994). Susceptibility of Channel Catfish (*Ictalurus punctatus*) and Brown Bullhead (*Ameriurus nebulosus*) to Oxidative Stress: a Comparative Study. *Aquatic Toxicology,* Vol.28, pp. 53–64, ISSN 0166-445X.

Huang, D.J.; Zhang, Y.M., Song, G., Long, J., Liu, J.H. & Ji, W.H. (2007). Contaminants-induced Oxidative Damage on the Carp *Cyprinus carpio* Collected from the Upper Yellow River, China. *Environmental Monitoring and Assessment*. Vol.128, pp. 483–488, ISSN 0167-6369.

Janssens, B.J.; Childress, J.J., Baguet, F. & Rees, J.F. (2000). Reduced Enzymatic Antioxidative Defense in Deep-sea Fish. *Journal of Experimental Biology*, Vol. 203, No.24, pp. 3717-3725, ISSN 0022-0949

Jenner, N.K.; Ostrander, G.K., Kavanagh, T.J., Livesey, J.C., Shen, M.W., Kim, S.C. & Holmes, E.H. (1990). A Flow Cytometric Comparison of DNA Content and Glutathione Levels in Hepatocytes of English sole (*Parophyrs vetulus*) from Areas of Differing Water Quality. *Archives of Environmental Contamination and Toxicology,* Vol.19, pp. 807–815, ISSN 0090-4341

Kang, Y.J. (2006). Metallothionein Redox Cycle and Function. *Experimental Biology and Medicine*, Vol.231, pp. 1459–1467, ISSN 0037-9727

Kappus, H. (1987). A Survey of Chemicals Inducing Lipid Peroxidation in Biological Systems. *Chemistry and Physics of Lipids*, Vol.45, pp. 105–115, ISSN 1873-2941

Kelly, E.; Duncan, R., Ngu, T.T., Chan, J., Salgado, M.T., Merrifield, M.E. & Stillman, M.J. (2006). Peptide Folding, Metal-binding Mechanisms, and Binding Site Structures in Metallothioneins. *Experimental Biology and Medicine,* Vol.231, pp. 1488–1499, ISSN 1535-3702.

Kelly, K.A.; Havrilla, C.M., Brady, T.C., Abramo, K.H. & Levin, E.D. (1998). Oxidative Stress in Toxicology: Established Mammalian and Emerging Piscine Model Systems. Environmental Health Perspectives, Vol. 106, pp. 375–384, ISSN 0091-6765.

Kono, Y. & Fridovich, I. (1982). Superoxide Radical Inhibits Catalase. *The Journal of Biological Chemistry*, Vol. 257, pp. 57–51, ISSN 0021-9258

Kopecka, J.J.P. & Pempkowiak, J. (2008). Temporal and Spatial Variations of Selected Biomarker Activities in Flounder (*Platichthys flesus*) Collected in the Baltic Proper. *Ecotoxicology and Environmental Safety*, Vol.70, Issue 3, (July 2008), pp. 379–391, ISSN 0147-6513

Kubrak, O.I.; Lushchak, O.V., Lushchak, J.V., Torous, I.M., Storey, J.M., Storey, K.B. & Lushchak, V.I. (2010). Chromium Effects on Free Radical Processes in Goldfish Tissues: Comparison of Cr(III) and Cr(VI) Exposures on Oxidative Stress Markers, Glutathione Status and Antioxidant Enzymes. *Comparative Biochemistry and Physiology*, Vol. 152 C, pp. 360–370, ISSN 1532-0456.

Lackner, R. (1998). "Oxidative stress" in Fish by Environmental Pollutants. In: Braunbeck, T., Hinton, D.E., Streit, B. (Eds.), *Fish Ecotoxicology*. Birkhauser Verlag, Basel, pp. 203–224.

Lam, P. K. S. & Wu, R. S. S. (2003). Use of Biomarkers in Environmental Monitoring. *STAP Workshop on the use of bioindicators, biomarkers and analytical methods for the analysis of POPs in developing countries*, (December 2003), pp. 1–77.

Lee, S.Y.; Stoliar, O. & Nam, Y.K. (2010). Transcriptional Alteration of Two Metallothionein Isoforms in Mud Loach *Misgurnus Mizolepis* fry During Acute Heavy Metal Exposure. *Fisheries and Aquatic Science*, Vol.13, pp. 112–117, ISSN 1303-2712

Leiniö, S. & Lehtonen, K.K. (2005). Seasonal Variability in Biomarkers in the Bivalves *Mytilus edulis* and *Macoma balthica* from the Northern Baltic Sea. *Comparative Biochemistry and Physiology*, Vol.140 C, pp. 408–421, ISSN 1532-0456

Lionetto, M.G.; Caricato, R., Giordano, M.E., Pascariello, M.F., Marinosci, L. & Schettino, T. (2003). Integrated Use of Biomarkers (Acetylcholinesterase and Antioxidant Enzymes Activities) in *Mytilus galloprovincialis* and *Mullus barbatus* in an Italian Coastal Marine Area. *Marine Pollution Bulletin*, Vol. 46, No.3, (March 2003), pp. 324–330, ISSN 0025-326X

Livingstone, D.R. (2001). Contaminant-stimulated Reactive Oxygen Species Production and Oxidative Damage in Aquatic Organisms. *Marine Pollution Bulletin,* Vol. 42, pp. 656–666, ISSN 0025-326X

Livingstone, D.R.; Lemaire, P., Matthews, A., Peters, L.D., Porte, C., Fitzpatrick, P.J., Forlin, L., Nasci, C., Fossatto, V., Wootton, N. & Goldfarb, P. (1995). Assessment of the Impact of Organic Pollutants on Goby (*Zosterisessor ophicephalus*) and Mussel

(*Mytilus galloprovincialis*) from the Venice Lagoon, Italy: Biochemical Studies. Marine Environmental Research, Vol.39, pp. 235–240, ISSN 0141-1136

Lushchak, V.I. (2011). Adaptive response to oxidative stress: Bacteria, fungi, plants and animals. *Comparative Biochemistry and Physiology. Toxicology & Pharmacology: CBP*, Vol. 153, No. 2, pp. 175-190. ISSN 1532-0456

Lushchak, O.V.; Kubrak, O.I., Lozinsky, O.V., Storey, J.M., Storey, K.B. & Lushchak, V.I., (2009a). Chromium (III) Induces Oxidative Stress in Goldfish Liver and Kidney. *Aquatic Toxicology*, Vol. 93, pp. 45–52, ISSN 0166-445X

Lushchak, O.V.; Kubrak, O.I., Nykorak, M.Z., Storey, K.B. & Lushchak, V.I. (2008). The Effect of Potassium Dichromate on Free Radical Processes in Goldfish: Possible Protective Role of Glutathione. *Aquatic Toxicology*, Vol. 87, pp. 108–114, ISSN 0166-445X

Lushchak, O.V.; Kubrak, O.I., Storey, J.M., Storey, K.B., Lushchak, V.I. (2009b). Low Toxic Herbicide Roundup Induces Mild Oxidative Stress in Goldfish Tissues. *Chemosphere*, Vol. 76, pp. 932–937, ISSN 0045-6535.

Lushchak, O.V.; Kubrak, O.I., Torous, I.M., Nazarchuk, T.Y., Storey, K.B. & Lushchak, V.I. (2009c). Trivalent Chromium Induces Oxidative Stress in Goldfish Brain. *Chemosphere*, Vol. 75, pp. 56–62, ISSN 0045-6535.

Lushchak, V. (2011). Environmentally Induced Oxidative Stress in Aquatic Animals. *Aquatic Toxicology*, Vol. 101, pp. 13–30, ISSN 0166-445X

Lushchak, V.I.; Bagnyukova, T.V., Lushchak, O.V., Storey, J.M. & Storey, K.B. (2007). Diethyldithiocarbamate Injection Induces Transient Oxidative Stress in Goldfish Tissues. Chemico-Biological Interactions, Vol. 170, pp. 1–8, ISSN 0009-2797.

Machala, M.; Dusek, L., Hilscherova, K., Kubinova, R., Jurajda, P., Neca, J., Ulrich, R., Gelnar, M., Studnickova, Z. & Holoubek, I. (2001). Determination and Multivariate Statistical Analysis of Biochemical Responses to Environmental Contaminants in Feral Freshwater Fish *Leuciscus Cephalus*, L. *Environmental Toxicology and Chemistry*, Vol. 20, No.5, (May 2001), pp. 1141–1148, ISSN 1552-8618

Maret, W. & Vallee, B. (1998). Thiolate Ligands in Metallothionein Confer Redox Activity on Zinc Clusters. *Proceedings of the National Academy of Sciences of the United States of America*, Vol.95, pp. 3478–3482, ISSN 0027-8424

McFarland, V.A.; Inouye, L.S., Lutz, C.H., Jarvis, A.S., Clarkem, J.U. & McCant, D.D. (1999). Biomarkers of Oxidative Stress and Genotoxicity in Livers of Field-collected Brown Bullhead, *Ameriurus nebulosus*. *Archives of Environmental Contamination and Toxicology*, Vol.37, pp. 236–241, ISSN 0090-4341

Meyer, J.N.; Smith, J.D., Winston, G.W. & Di Giulio, R.T. (2003). Antioxidant Defenses in Killifish (*Fundulus heteroclitus*) Exposed to Contaminated Sediments and Model Prooxidants: Short Term and Heritable Responses. *Aquatic Toxicology*, Vol.65, pp. 377–395, ISSN 0166-445X

Monserrat, J.M.; Martínez, P.E., Geracitano, L.A., Amado, L.L., Martins, C.M.G., Pinho, G.L.L., Chaves, I.S., Ferreira-Cravo, M., Ventura-Lima, J. & Bianchini, A. (2007). Pollution Biomarkers in Estuarine Animals: Critical Review and New Perspectives. *Comparative Biochemistry and Physiology*, Vol.146 C, pp. 221–234, ISSN 1532-0456

Moraes, B.S.; Loro, V.L., Glusczak, L., Pretto, A., Menezes, C., Marchezan, E. & de Oliveira Machado, S. (2007). Effects of Four Rice Herbicides on Some Metabolic and Toxicology Parameters of Teleost Fish (*Leporinus obtusidens*). *Chemosphere*, Vol.68, pp.1597–1601, ISSN 0045-6535

Mosleh, Y.Y.; Paris-Palacios, S., Couderchet, M., Biagianti-Risbourg, S. & Vernet, G. (2005). Effects of the Herbicide Isoproturon on Metallothioneins, Growth, and Antioxidative Defenses in the Aquatic Worm *Tubifex tubifex* (Oligochaeta, Tubificidae). *Ecotoxicology*, Vol.14, pp. 559–571, ISSN 0963-9292

Narbonne, J.F.; Daube`ze, M., Cle´randeau, C. & Garrigues, P. (1999). Scale of Classification Based on Biochemical Markers in Mussels: Application to Pollution Monitoring in European Coasts. *Biomarkers*, Vol. 4, pp. 415–424, ISSN 1875-3183.

Nikinmaa, M. & Rees, B. B. (2005). Oxygen-dependent gene expression in fishes. American Journal of Physiology - Regulatory, Integrative and Comparative Physiology, Vol.288, Issue 5, pp. R1079-R1090, ISSN: 0363-6119.

Nishimoto, M.; Le Eberhart, B.T., Sanborn, H.R. Krone, C., Varanasi, U. & Stein, J.E. (1995). Effects of a Complex Mixture of Chemical Contaminants on Hepatic Glutathione, L-cysteine and {gamma}-glutamylcysteine Synthetase in English Sole (*Pleuronectes vetulus*). *Environmental Toxicology and Chemistry*, Vol.14, Issue 3, pp. 461–469, ISSN 1552-8618

Niyogi, S.; Biswas, S., Sarker, S. & Datta, A.G. (2001). Seasonal Variation of Antioxidant and Biotransformation Enzymes in Barnacle, Balanus balanoides, and their Relation with Polyaromatic Hydrocarbons. Marine Environmental Research, Vol.52, pp. 13–26, ISSN 0141-1136

Otto, D.M.E. & Moon, T.W. (1996). Phase and 11 Enzymes and Antioxidant Responses in Different Tissues of Brown Bullheads from Relatively Polluted and Non-polluted Systems. *Archives of Environmental Contamination and Toxicology*, Vol.31, pp.141–147, ISSN 0090-4341

Padmini, E. (2010). Physiological Adaptations of Stressed Fish to Polluted Environments: Role of Heat Shock Proteins. *Reviews of Environmental Contamination and Toxicology*, Vol. 206, pp. 1–27, ISSN 0179-5953

Pandey, S.; Parvez, S., Sayeed, I., Haque, R., Bin-Hafeez, B. & Raisuddin, S. (2003). Biomarkers of oxidative stress: a comparative study of river Yamuna fish *Wallago attu* (Bl. & Schn.). *Science of the Total Environment*, Vol.309, pp. 105–115, ISSN 0048-9697

Parihar, M.S.; Javeri, T., Hemnani, T., Dubey, A.K. & Prakash, P. (1997). Responses of Superoxide Dismutase, Glutathione Peroxidase and Glutathione Antioxidant Defenses in Gills of the Freshwater Catfish (*Heteropneustes fossilis*) to Short-term Elevated Temperature. *Journal of Thermal Biology*, Vol.22, pp. 151–156, ISSN 0306-4565

Paris-Palacios, S.; Biagianti-Risbourg, S. & Vernet, G. (2003). Metallothionein Induction Related to Hepatic Structural Perturbations and Antioxidative Defences in Roach (*Rutilus rutilus*) Exposed to the Fungicide Procymidone. Biomarkers, Vol.8, pp. 128–141, ISSN 1354-750X

Paris-Palacios, S.; Biagianti-Risbourg, S., Fouley, A. & Vernet, G. (2000). Metallothioneins in Liver of Rutilus Rutilus Exposed to $Cu2^{+}$. Analysis by Metal Summation, SH Determination and Spectrofluorimetry. *Comparative Biochemistry and Phisiology*, Vol.126 C, pp. 113–122, ISSN 1532-0456

Parvez, S. & Raisuddin, S. (2005). Protein Carbonyls: Novel Biomarkers of Exposure to Oxidative Stress-Inducing Pesticides in Freshwater Fish *Channa punctata* (Bloch). *Environmental Toxicology and Pharmacology*, Vol.20, No.1, (December 2004), pp. 112–117, ISSN 1382-6689

Porte, C.; Escartín, E., de la Parra, L.M.G., Biosca, X. & Albaigés, J. (2002). Assessment of Coastal Pollution by Combined Determination of Chemical and Biochemical Markers in *Mullus barbatus*. *Marine Ecology Progress Series*, Vol.235, pp. 205–216, ISSN 0171-8630.

Raldúa, D.; Díez, S., Bayona, J.M. & Barceló, D. (2007). Mercury Levels and Liver Pathology in Feral Fish Living in the Vicinity of a Mercury Cell Chlor-Alkali Factory. *Chemosphere*, Vol.66, Issue 7, (January 2007), pp. 1217–1225, ISSN 0045-6535

Regoli, F. & Principato, G. (1995). Glutathione, Glutathione-dependent and Antioxidant Enzymes in Mussel, *Mytilus galloprovincialis*, Exposed to Metals in Different Field and Laboratory Conditions: Implications for a Proper Use of Biochemical Biomarkers. *Aquatic Toxicology*, Vol. 31, pp. 143–164, ISSN 0166-445X.

Regoli, F. (2000). Total Oxyradical Scavenging Capacity (TOSC) in Polluted and Translocated Mussels: a Predictive Biomarker of Oxidative Stress. *Aquatic Toxicology*, Vol.50, pp. 351–361, ISSN 0166-445X.

Regoli, F.; Gorbi, S., Frenzilli, G., Nigro, M., Corsi, I., Focardi, S. & Winston, G.W. (2002a). Oxidative Stress in Ecotoxicology: from the Analysis of Individual Antioxidants to a More Integrated Approach, *Marine Environmental Research*, Vol.54, pp. 419–423, ISSN 0141-1136

Regoli, F.; Nigro, M., Benedetti, M., Fattorini, D. & Gorbi, S. (2005). Antioxidant Efficiency in Early Life Stages of the Antarctic Silverfish, *Pleuragramma antarcticum*: Responsiveness to Pro-Oxidant Conditions of Platelet Ice and Chemical Exposure. *Aquatic Toxicology*, Vol.75, No.1, pp. 43–52, ISSN 0166-445X.

Regoli, F.; Pellegrini, D., Winston, G.W., Gorbi, S., Giuliani, S., Virno-Lamberti, C. & Bompadre, S. (2002b). Application of Biomarkers for Assessing the Biological Impact of Dredged Materials in the Mediterranean: the Relationship Between Antioxidant Responses and Susceptibility to Oxidative Stress in the Red Mullet (*Mullus barbatus*). *Marine Pollution Bulletin*, Vol.44, No.9, (September 2002), pp. 912–922, ISSN 0025-326X.

Reynders, H.; Bervoets, L., Gelders, M., De Coen, W.M. & Blust, R. (2008). Accumulation and Effects of Metals in Caged Carp and Resident Roach Along a Metal Pollution Gradient. *Science of the Total Environment*, Vol.391, No.1, (December 2007), pp. 82–95, ISSN 0048-9697

Rhee, J.S.; Raisuddin, S., Hwang, D.S., Lee, K.W., Kim, I.C. & Lee, J.S. (2009). Differential Expression of Metallothionein (MT) Gene by Trace Metals and Endocrine-Disrupting Chemicals in the Hermaphroditic Mangrove Killifish, Kryptolebias marmoratus. *Ecotoxicology and Environmental Safety*, Vol.72, pp. 206–212, ISSN 0147-6513.

Ritter, L.; Solomon, K., Sibley, P., Hall, K., Keen, P., Mattu, G. & Linton, B. (2002). Sources, Pathways, and Relative Risks of Contaminants in Surface Water and Groundwater: a Perspective Prepared for the Walkerton Inquiry. *Journal of Toxicology and Environmental Health*. Vol. 65, No.1, (January 2002), pp. 1-142, ISSN 1528-7394

Roberts, M.H.; Sved, D.W. & Felton, S.P. (1987). Temporal changes in AHH and SOD Activities in Feral Spot from the Elizabeth River, a Polluted Sub-estuary. *Marine Environmental Research*, Vol.23, pp. 89–101, ISSN 0141-1136.

Rodriguez-Ariza, A.; Alhama, J., Diaz-Mendez, F. M. & Lopez-Barea, J. (1999). Content of 8-oxodG in Chromosomal DNA of *Sparus aurata* Fish as Biomarker of Oxidative Stress and Environmental Pollution. *Mutation Research/Genetic Toxicology and Environmental Mutagenesis*, Vol.438, No.2, (January 1999), pp. 97-107, ISSN 1383-5718

Rodriguez-Ariza, A.; Peinado, J., Pueyo, C. & Lopez-Barea, J. (1993). Biochemical Indicators of Oxidative Stress in Fish from Polluted Littoral Areas. *Canadian Journal of Fisheries and Aquatic Sciences*, Vol.50, pp. 2568–2573, ISSN 0706-652X

Romeo, M.; Bennani, N., Gnassia-Barelli, M., Lafaurie, M. & Girard, J.P. (2000). Cadmium and Copper Display Different Responses Towards Oxidative Stress in the Kidney of the Sea Baas Dicentrarchus Labrax. *Aquatic Toxicology*, Vol.48, pp. 185–194, ISSN 0742-8413

Ruas, C.B.; Carvalho, C.D., de Araújo, H.S., Espíndola, E.L. & Fernandes, M.N. (2007). Oxidative Stress Biomarkers of Exposure in the Blood of Cichlid Species from a Metal-contaminated River. *Ecotoxicology and Environmental Safety*, Vol.71, Issue 1, (September 2008), pp. 86–93, ISSN 0147-6513.

Ruas, C.B.G.; Carvalho, C.S., de Araújo, H.S.S., Espíndola, E.L.G. & Fernandes, M.N. (2006). Oxidative Stress Biomarkers of Exposure in the Blood of Cichlid Species from a Metal-contaminated River. *Ecotoxicology and Environmental Safety*, Vol. 71, Issue 1, (September 2008), pp. 86–93, ISSN 0147-6513.

Sanchez, W.; Aït-Aïssa, S., Palluel, O., Ditche, J.M. & Porcher, J.M. (2007). Preliminary Investigation of Multi-biomarker Responses in Three-spined Stickleback (*Gasterosteus aculeatus* L.) Sampled in Contaminated Streams. *Ecotoxicology*, Vol.16, pp. 279–287, ISSN 0963-9292

Sheehan, D. (2006). Detection of Redox-based Modification in Two-dimensional Electrophoresis Proteomic Separations. *Biochemical and Biophysical Research Communications*, Vol.349, Issue 2, (October 2006), pp.455–462, ISSN 0006-291X.

Shigenaga, M.K. & Ames, B.N. (1991). Assays for 8-hydroxy-20-deoxyguanosine: A Biomarker of *in vivo* Oxidative DNA Damage. *Free Radical Biology and Medicine*. Vol.10, pp. 211–216, ISSN 0891-5849.

Solé, M.; Rodríguez, S., Papiol, V., Maynou, F. & Cartes, J.E. (2009). Xenobiotic Metabolism in Marine Fish with Different Trophic Strategies and their Relationship to Ecological Variables. Comparative Biochemistry and Physiology, Vol. 149 C, pp. 83–89, ISSN 0742-8413

Stanic, B.; Andric, N., Zoric, S., Grubor-Lajsic, G. & Kovacevic, R. (2006). Assessing Pollution in the Danube River Near Novi Sad (Serbia) Using Several Biomarkers in Sterlet (*Acipenser ruthenus* L.). *Ecotoxicology and Environmental Safety*, Vol. 65, pp. 395–402, ISSN 0147-6513.

Steadman, B.L.; Farag, A.M. & Bergman, H.L. (1991). Exposurerelated Patterns of Biochemical Indicators in Rainbow trout Exposed to No. 2 fuel oil. *Environmental Toxicology and Chemistry*, Vol.10, pp. 365–374., ISSN 1552-8618

Stegeman, J.J.; Brouwer, M., Richard, T.D.G., Forlin, L., Fowler, B.A., Sanders, B.M. & van Veld, P.A. (1992). Molecular Responses to Environmental Contamination: Enzyme and Protein Systems as Indicators of Chemical Exposure and Effect. In: Huggett, R.J., Kimerly, R.A., Mehrle, P.M., Jr, Bergman, H.L. (Eds.), *Biomarkers: Biochemical, Physiological and Histological markers of Anthropogenic Stress*. Lewis Publishers, Chelsea, MI, USA, pp. 235–335.

Stein, J.E.; Collier, T.K., Reichert, W.L., Casillas, E., Hom, T. & Varanasi, U. (1992). Bioindicators of Contaminant Exposure and Sublethal Effects: Studies with Benthic

Fish in Puget Sound, Washington. *Environmental Toxicology and Chemistry*, Vol. 11, pp. 701–714, ISSN 1552-8618

Stephensen, E.; Svavarsson, J., Sturve, J., Ericson, G., Adolfsson-Erici, M. & Forlin, L. (2000). Biochemical Indicators of Pollution Exposure in Shorthorn Sculpin (*Myoxocephalus scorpius*), Caught in Four Harbors on the Southwest Coast of Iceland. *Aquatic Toxicology*, Vol.48, pp. 431–442, ISSN 0166-445X

Sturve, J.; Berglund, Å., Balk, L., Broeg, K., Böhmert, B., Massey, S., Savva, D., Parkkonen, J., Stephensen, E., Koehler, A. & Förlin, L. (2005). Effects of Dredging in Göteborg Harbor, Sweden, Assessed by Biomarkers in Eelpout (*Zoarces viviparus*). *Environmental Toxicology and Chemistry*, Vol.24, No.8, pp. 1951–1961, ISSN 1552-8618

Sun, Y.; Yin, Y., Zhang, J., Yu, H., Wang, X., Wu, J. & Xue, Y. (2008). Hydroxyl Radical Generation and Oxidative Stress in *Carassius auratus* Liver, Exposed to Pyrene. *Ecotoxicology and Environmental Safety*, Vol.71, pp. 446–453, ISSN 0147-6513.

Sureda, A.; Box, A., Enseñat, M., Alou, E., Tauler, P., Deudero, S. & Pons, A. (2006). Enzymatic Antioxidant Response of a Labrid Fish (*Coris julis*) Liver to Environmental Caulerpenyne. *Comparative Biochemistry and Physiology*, Vol. 144 C, Issue 2, (October 2006), pp. 191–196, ISSN 0742-8413

Teh, S.J.; Adams, S. M. & Hinton, D.E. (1997). Histopathologic Biomarkers in Feral Freshwater Fish Populations Exposed to Different Types of Contaminant Stress. *Aquatic Toxicology*, Vol.37, Issue 1, (January 1997), pp. 51–70, ISSN 0166-445X

Tom, M. & Auslander, M. (2005). Transcript and Protein Environmental Biomarkers in Fish—a Review. *Chemosphere*, Vol. 59, Issue 2, (April 2005), pp. 155–162, ISSN 0045-6535

Üner, N.; Oruç, E. & Sevgiler, Y. (2005). Oxidative Stress-related and ATPase Effects of Etoxazole in Different Tissues of *Oreochromis niloticus*. *Environmental Toxicology and Pharmacology*, Vol.20, pp. 99–106, ISSN 1382-6689

Valavanidis, A.; Vlahogianni, Th., Dassenakis, M. & Scoullos, M. (2006). Molecular Biomarkers of Oxidative Stress in Aquatic Organisms in Relation to Toxic Environmental Pollutants. *Ecotoxicology and Environmental Safety*, Vol. 64, Issue 2, (June 2006), pp. 178–189, ISSN 0147-6513

van der Oost, R.; Beyer, J. & Vermeulen, N.P.E. (2003). Fish Bioaccumulation and Biomarkers in Environmental Risk assessment: a Review. *Environmental Toxicology and Pharmacology*, Vol. 13, Issue 2, (February 2003), pp. 57–149, ISSN 0926-6917.

van der Oost, R.; Goksoyr, A., Celander, M., Heida, H. & Vermeulen, N.P.E. (1996). Biomonitoring of Aquatic Pollution with Feral Eel (*Anguilla anguilla*). II. Biomarkers: Pollution Induced Biochemical Responses. *Aquatic Toxicology*, Vol.36, pp. 189–222, ISSN 0166-445X.

Vega-López, A.; Jiménez-Orozco, F.A., García-Latorre E.& Domínguez-López, M.L. (2008). Oxidative Stress Response in an Endangered Goodeid Fish (*Girardinichthys viviparus*) by Exposure to Water from its Extant Localities. *Ecotoxicology and Environmental Safety*, Vol.71, Issue 1, (September 2008), pp. 94–103, ISSN 0147-6513

Viarengo, A. & Nott, J.A. (1993). Mechanisms of Heavy Metal Cation Homeostasis in Marine Invertebrates. *Comparative Biochemistry and Physiology*, Vol.104 C, pp. 355–372, ISSN 0742-8413

Viarengo, A.; Burlando, B., Cavaletto, M., Marchi, B., Ponzano, E. & Blasco, J. (1999). Role of Metallothionein Against Oxidative Stress in the Mussel *Mytilus galloprovincialis*.

American Journal of Physiology – Regulatory, Integrative and Comparative Physiology, Vol.277, pp. 1612–1619, ISSN 0363-6119

Viarengo, A.; Lowe, D., Bolognesi, C., Fabbri, E. & Koehler, A. (2007). The Use of Biomarkers in Biomonitoring: a 2-tier Approach Assessing the Level of Pollutant-induced Stress Syndrome in Sentinel Organisms. *Comparative Biochemistry and Physiology*. Vol.146 C, pp. 281–300, ISSN 0742-8413

Vigano, L.; Arillo, A., Melodia, F., Arlanti, P. & Monti, C. (1998). Biomarker Responses in Cyprinids of the Middle Stretch of the Rover Po, Italy. *Environmental Toxicology and Chemistry*, Vol.17, pp. 404–411, ISSN 1552-8618

Winston, G.W. (1991). Oxidants and Antioxidants in Aquatic Animals. *Comparative Biochemistry and Physiology*, Vol.100 C, pp. 173–176, ISSN 0742-8413

Winston, G.W. & Di Giulio, R.T. (1991). Prooxidant and Antioxidant Mechanisms in Aquatic Organisms. *Aquatic Toxicology*, Vol.19, pp. 137–161, ISSN 0166-445X

Winzer, K.; Winston, G.W., Becker, W., Van Noorden, C.J.F. & Kohler, A. (2001). Sex-related Responses to Oxidative Stress in Primary Cultured Hepatocytes of European flounder (*Platichthys flesus* L.). *Aquatic Toxicology*, Vol. 52, pp. 143–155, ISSN 0742-8413

Woo S.; Yum, S., Kim, D.-W. & Park H.-S. (2009). Transcripts Level Responses in a Marine Medaka (*Oryzias javanicus*) Exposed to Organophosphorus Pesticide. *Comparative Biochemistry and Physiology*, Vol.149 C, No.3, pp. 427–443, ISSN 0742-8413

Yi, X.; Ding, H., Lu, Y., Liu, H., Zhang, M. & Jiang, W. (2007). Effects of Long-term Alachlor Exposure on Hepatic Antioxidant Defense and Detoxifying Enzyme Activities in Crucian Carp (*Carassius auratus*). *Chemosphere*. Vol.68, pp.1576–15151, ISSN 0045-6535

Section 4

Biological Factors and Effects

8

Interference of Oxidative Metabolism in Citrus by *Xanthomonas citri* pv *citri*

Robert C. Ebel and Naveen Kumar
Citrus Physiology, Southwest Florida Research and Education Center
University of Florida, IFAS, Immokalee, FL
USA

1. Introduction

Citrus are one of the most important fruit crops grown worldwide. Among the pathogens that cause disease of *Citrus sp.* and closely related genera, *Xanthomonas citri* pv *citri* (*Xcc*) causes citrus canker, a devastating disease that is found in 30 countries worldwide and has caused significant economic loss (Del Campo et al., 2009; Rigano et al., 2010). The principle mode of transmission of *Xcc* is through heavy rain and high wind events and thus the disease is most severe in regions that experience occasional tropical storms and hurricanes (Graham *et al.*, 2004). Citrus canker outbreaks in Florida, for example, have contributed to a decline in acreage of grapefruit to 61 % by 2009 compared to the acreage in 1994 (Anonymous, 2009). Severe canker can cause fruit drop and even tree death (Graham et al., 2004). Further economic losses can be incurred through restricted movement of infected fruits especially to citrus growing regions where canker is not found (Schubert *et al.*, 2001).

The commercial and dietary importance of citrus and the severity of canker have led to extensive research to identify resistant genotypes that would serve as models of study as well as germplasm for crop improvement. Most commercial citrus are within the *Citrus* genus, however closely related genera are capable of hybridizing with *Citrus sp.* and thus have been included in studies to evaluate variation in plant defense to canker. Citrus genotypes can be classified into four broad classes based on susceptibility to canker (Gottwald, 2002). The most highly-susceptible commercial genotypes are 'Key' lime [*C. aurantifolia* (Christm.) Swingle], grapefruit (*C. paradisi* Macfad.), lemon (*C. limon*), and pointed-leaf Hystrix (*C. hystrix*). Susceptible genotypes include limes (*C. latifolia*), sweet oranges (*C. sinensis*), trifoliate orange (*P. trifoliata*) citranges and citrumelos (*P. trifoliata* hybrids), and bitter oranges (*C. aurantium*). Resistant genotypes include citron (*C. medica* L.) and mandarins (*C. reticulata* Blanco). Highly resistant genotypes include Calamondin [*Citrus margarita* (Lour.)] and kumquat [*Fortunella margarita* (Lour.) Swingle]. The high degree of resistance to Asiatic citrus canker by calamondin, kumquat, and Ichang papeda (*C. ichangenesis*) has been noted in the field (Reddy, 1997; Viloria *et al.*, 2004).

Although *Xcc* can cause disease in kumquat, the cankers are normally much smaller than in *Citrus* species indicating greater resistance (Viloria et al., 2004). Kumquat resistance to *Xcc* has been utilized in breeding programs to produce intergeneric hybrids with *Citrus* species that are more canker resistant than the *Citrus sp.* parent (Viloria et al., 2004). Kumquat is also being used as a model system in research programs to determine the underlying resistance mechanism (eg., Khalaf et al., 2007) with the long term goal of identifying specific genes that could be inserted into commercial *Citrus* species and avoid the much greater genetic variability in yield and fruit quality typically introduced through crosses in traditional breeding programs.

Although development of resistant genotypes is a long-term research goal, commercial industries have been forced to implement a variety of management practices to reduce the impact of this devastating disease including the use of resistant species and cultivars, applications of bactericides especially copper, and in extreme cases removal of infected trees in an attempt to eradicate the disease from a particular region. Resistance alone is insufficient for commercial production, eradication in high wind and rain-prone areas have largely proven ineffective and copper sprays are often unreliable, in part because of increased resistance by the pathogen (Graham et al., 2004). Multiple management approaches will be required to maintain commercial production. One approach that has received limited attention is the application of biotic and abiotic agents that would promote systemic acquired resistance and induced systemic resistance (Valad and Goodman, 2008). Advances in the use of systemic acquired resistance and induced systemic resistance will require a working hypothesis of how *Xcc* interferes with citrus defense. The comparison of resistant and susceptible genotypes has revealed new information regarding the deficiencies in susceptible genotypes that can be developed into a working hypothesis as to how *Xcc* interferes with citrus defense, and from that knowledge strategies can be developed to restore the defense mechanism.

2. Pathogenesis of canker in citrus

Metabolic changes in plants to pathogens coincide with the plant parts affected and the development of the disease. Canker affects all above ground parts of the plant including the leaves, stems and fruit (Graham et al., 2004). Only one bacterium is required to cause canker formation, which enters the plant through stomatal apertures or wounds using its flagella (Gottwald and Graham, 1992; Koizumi and Kuhara, 1982; Stall et al., 1982). Once inside, the bacterium multiplies to reach a population density of 1×10^3 to 1×10^4 bacteria per canker lesion, which is sufficient to act as source of inoculum and under specific conditions promote dispersal (Graham *et al.*, 2004).

Cankers are a localized phenomenon such that plant response in an infected area differs from uninfected areas, and thus bulk sampling of tissues would include both areas. To facilitate sampling of only infected tissues, studies have utilized injection of *Xcc* suspensions into leaf tissues (Khalif et al., 2007). Upon injection, an initial water soaked area is observed and subsequent disease symptoms develop in this region. Thus, sampling the original water soaked area allows sampling of only diseased tissues. The advantage of this approach has been demonstrated by changes in H_2O_2 concentrations in *Xcc* infected areas induced through injection (Kumar et al, 2011a), whereas whole leaf sampling of trees sprayed with *Xcc* suspensions demonstrated inconsistent or no differences in H_2O_2 concentrations (Kumar, data unpublished).

Injection of known concentrations of a specific strain of *Xcc* and maintaining plants under consistent environmental conditions allows repetition of a specific sequence of disease events to which plant response can be correlated. Using this approach, a specific sequence of events in the pathogenesis of *Xcc* in citrus has been described (Burnings and Gabriel, 2003). Following artificial inoculation, the bacterial cells occupy intercellular spaces and begin to divide by the end of the first day after inoculation. Once a critical population threshold is reached, which is about 1 x 10^3 to 1 x 10^4 bacteria per canker lesion, a quorum sensing mechanism (da Silva et al., 2002) is likely the impetus that turns on pathogenicity factors (Bassler, 1999) that includes Rpf encoding genes (Slater et al., 2000). Within 2 days after inoculation, *Xcc* attaches to plant cell walls via specialized proteins called "adhesins" (Lee and Schneewind, 2001) by hrp (hypersensitivity response and pathogenicity) pili or by type IV pili as observed during *xanthomonas* pv. *malvaceraum- Gossypium hirsutum* interaction (Burnings and Gabriel, 2003). Once attached, *Xcc* uses it T3S system to turn on additional pathogenicity genes (Pettersson et al., 1996) and inject pathogenicity factors into the cell including Avr, Pop and Pth proteins such as PthA (Brunings and Gabriel, 2003). PthA presumably stimulates plant cell division and enlargement within 3 days after inoculation that reaches a maximum by 7 days after inoculation (Lawson et al., 1989). Cell enlargement, presence of the bacteria in the apoplast, and its production of hydrophilic polymers causes watersoaking symptoms starting 4 days after inoculation (Duan et al., 1999). The maximum bacterial populations occur at 7 days after inoculation (Khalaf et al., 2007) and about 8 days after inoculation the epidermis ruptures allowing bacteria to egress to the surface (Brunings and Gabriel, 2003). By 10-14 days after inoculation, necrosis develops in the infected areas (Duan et al., 1999) and by 21 days after inoculation leaves abscise (Khalaf et al., 2007).

3. Oxidative response of plants to pathogens

The hypersensitive response (HR) involves a rapid, widespread change in plant cell metabolism intended to alter the chemistry of the region within and surrounding the infected area in order to impact the pathogen by deterring its metabolism, isolating it within the infected region, and even directly killing it (Lamb and Dixon, 1997). As part of the response, programmed cell death (PCD) of plant cells within and adjacent to the infected region is often elicited (Lamb and Dixon, 1997). The HR includes alteration of oxidative metabolism to produce reactive oxygen species (ROS) that promote PCD, sicken pathogen metabolism, and promote changes in cell wall chemistry that isolate the pathogen (Azvedo *et al.*, 2008; Kuzniak and Urbanek, 2000; Lamb and Dixon, 1997). In the case of citrus canker, PCD is evident around infection sites by chlorosis, with the chlorotic rings widening as the canker spreads radially from the infection point and along the plane of the leaf blade (Burnings and Gabriel, 2003).

Reactive oxygen species produced during HR and PCD in response to pathogens include superoxide radicals ($O2^{\cdot -}$), hydrogen peroxide (H_2O_2), and hydroxyl radicals ($OH^{\cdot}$) (Chen et *al.*, 2008; Lamb and Dixon, 1997; Wojtaszek, 1997). Production of ROS occur during normal metabolism of uninfected plants and maintained at low concentrations by several enzymatic and non-enzymatic pathways. In response to infection by pathogens, concentrations of ROS are increased and compartmentalized during HR and PCD via several pathways mediated by signals including salicylic acid, nitrous oxide, and the MAP kinase cascade mechanism (Durrant and Dong, 2004; Vlot et al., 2009) to alter the chemistry within and surrounding the infection site (Mittler, 2002).

One important ROS is H_2O_2, the concentration of which has been correlated with disease resistance (Lamb and Dixon, 1997; Mittler *et al.*, 1999). H_2O_2 concentrations can increase very rapidly from 0 to 6 days after inoculation during plant-bacterial pathogen interactions (Wojtaszek, 1997). Early after infection, elevated concentrations of H_2O_2 serve as diffusible signals to induce defense genes in adjoining cells with the later elevated concentrations serving in the direct inhibition of pathogens (Alverez *et al.*, 1998; Dat et al., 2000; Lamb and Dixon, 1997). The role of H_2O_2 in promoting disease resistance has been confirmed in transgenic potato plants that over-expressed a fungal glucose oxidase gene and accumulated sub-lethal concentrations of H_2O_2 (Wu *et al.*, 1997).

A major source of H_2O_2 is by dismutation of $O_2^{\cdot -}$ via the activity of superoxide dismutase (SOD) (Alscher *et al.*, 2002; Voludakis et al., 2006). SODs are regarded as a first step in reducing oxidative stress by converting $O_2^{\cdot -}$ to H_2O_2 during normal metabolism (Babhita et al., 2002). In response to biotic stress, SOD genes and enzyme concentrations are often up-regulated as part of the resistance mechanism against viral, bacterial and fungal diseases (Barna *et al.*, 2003; Bolwell and Wojtaszek, 1997; Buonaurio *et al.*, 1987; Delledonne *et al.*, 2001; Montalbini and Buonaurio, 1986; Tertivanidis *et al.*, 2004; Voludakis *et al.*, 2006). The importance of SOD in the production of H_2O_2 has been demonstrated in rose cells treated with the Cu-Zn-SOD inhibitor N,N-diethyldithiocarbamate and exposed to phytophthora (Auh and Murphy, 1995). Furthermore, pearl millet (*Pennisetum glaucum*) demonstrated higher SOD activity in resistant genotypes compared to susceptible genotypes when challenged with *Sclerospora graminicola* (Babhita et al., 2002). Similarly, SOD activity was higher in *Xanthomonas campestris* pv. *campestris* resistant cabbage (*Brassica oleracea*) varieties (Gay and Tuzun, 2000).

Based on their metal co-factor, SODs can be classified into three categories: iron SOD (Fe-SOD), manganese-SOD (Mn-SOD), and copper-zinc SOD (Cu-Zn-SOD), each of which is specifically compartmentalized in the cell (Alscher *et al.*, 2002). Fe-SOD is located in the chloroplasts, Mn-SODs in the mitochondria and peroxisomes, and Cu-Zn-SOD in the chloroplast, cytosol, and possibly in the apoplast (Alscher *et al.*, 2002). The various SODs play important roles in plant/pathogen interactions. Fe-SOD, for example, appears to be involved in the early signaling with H_2O_2 by plant cells after infection (Mur *et al.*, 2008; Zurbriggen *et al*, 2009). Mn-SOD has been reported to play an important role in early apoptotic events during PCD in *Gossypium hirsutum-Xanthomonas campestris* pv. *malvecearum* interaction (Voludakis et al., 2006). However, Kukavica et al. (2009) showed the existence of a cell wall bound Mn-SOD that generated $OH^{\cdot}$ in pea roots and probably facilitates cell elongation.

Some of the major enzymes involved in H_2O_2 dismutation and that have been shown to change during pathogenesis include catalase (CAT), ascorbate peroxidase (APOD) and class III peroxidase (POD) (Able et al., 2000; Dat et al., 2003; De Pinto et al., 2006; Gonzalez et al., 2010). Catalase and APOD are the most important enzymes involved in maintaining H_2O_2 at low concentrations in the symplast of healthy plants (Mittler, 2002). Catalase is a tetrameric iron porphyrin that converts millions of H_2O_2 to water and oxygen per second and is generally limited to the peroxisomes where H_2O_2 forms rapidly as a by-product of photorespiration (Willekens et al., 1997). The importance of CAT in disease resistance has been shown in transgenic tobacco (*Nicotiana tabacum* cv AS1) that had reduced CAT1 mRNA and protein (AS1) which demonstrated a HR leading to necrotic lesions upon challenge with *Pseudomonas syringae* pv. *tabaci* (Mittler *et al.*, 1999).

Ascorbate peroxidases contain a heme cofactor and use ascorbate as a substrate as part of the glutathione-ascorbate cycle (Foyer et al., 2009). Ascorbate peroxidase is ubiquitous throughout the cell and thus is important in catalyzing H_2O_2 that is produced as a waste product of different metabolic pathways (Mittler, 2002). The importance of APOD in disease resistance has been shown in transgenic tobacco transformed with antisense cAPX (*Nicotiana tabacum* cv Bel W3) that exhibited PCD accompanied by fragmentation of nuclear DNA after being challenged with *Pseudomonas syringae* pv. *tabaci, Pseudomonas syringae* pv. *phaseolicola* NPS3121 and *Pseudomonas syringae* pv. *syringae* (Mittler *et al.*, 1999; Polidoros et al., 2001).

The use of guaiacol as a substrate to test peroxidase activity is limited to the Class III peroxidases (POD) that are characterized by secretion into the apoplast and utilize phenolic compounds as substrates to cross-link cell walls during cell maturation (De Gara, 2004; Liszkay et al., 2003; Sasaki et al., 2004). During infection, the class III PODs promote lignification, suberization, cross-linking of cell wall proteins, and phytoalexin synthesis to sicken metabolism and isolate the pathogen (Sasaki *et al.*, 2004; Quiroga et al., 2000). The peroxidative cycle of POD uses H_2O_2 as an oxidant to convert phenolic compounds to phenoxy radicals that spontaneously combine to form lignin responsible for cell wall stiffening (Liszkay et al., 2003; Martinez *et al.*, 1998).

4. Comparative analysis of oxidative metabolism in *Xcc* resistant and susceptible genotypes

Recent studies on various *Citrus sp.* and closely related genera have increased our understanding of deficiencies in oxidative metabolism in susceptible genotypes. The most commonly studied resistant genotype is kumquat (*Fortunella margarita* (Lour.) Swingle). The kumquats have been characterized as canker resistant based on fewer canker lesions per leaf and reduced internal bacterial populations per lesion compared to susceptible genotypes (Khalaf *et al.*, 2007; Viloria *et al.*, 2004). Resistance of kumquat has been exhibited in hybrids with *Citrus sp.* such as 'Lakeland' limequat, a cross between the highly *Xcc*-susceptible 'Key' lime and kumquat, which has demonstrated greater canker resistance than 'Key' lime alone under field conditions (Viloria *et al.*, 2004). Furthermore, the Asiatic strain of canker (Canker A) has been shown to reach populations densities consistent with a compatible reaction (Stall *et al.*, 1980) and the lower concentrations of *Xcc* in kumquat indicates a disease resistance mechanism (Viloria *et al.*, 2004). Although oxidative metabolism is complex, recent research has focused on comparing kumquat resistant and susceptible *Citrus* genotypes on their H_2O_2 metabolism in part due to its importance in cell signaling and its involvement in cell wall chemistry during growth and plant defense.

The basal antioxidant metabolism has been shown to vary in different citrus genotypes (Kumar et al., 2001a) which relate to their fundamental differences in resistance. Kumquat, for example, was shown to have higher total SOD activity in kumquat than grapefruit and sweet orange, yet H_2O_2 was lower in kumquat in part because of higher CAT activity. These fundamental differences in basal metabolism are the starting point for changes in oxidative metabolism when challenged with *Xcc*.

5. Oxidative metabolism in canker-resistant kumquat

Using an Asiatic strain of canker (Canker A) and infiltration of kumquat leaves, Kumar et al., (2011c) showed that the *Xcc* populations peaked 4 days after inoculation and declined thereafter. Chlorosis was evident the first day after inoculation and persisted throughout the infection process (Fig. 1). Water soaking was delayed until 4 days after inoculation. H_2O_2 concentrations increased rapidly 1 day after inoculation to almost 2x the controls, about 10 ml, from 6 to 8 days after inoculation and declined slightly thereafter but remained above the controls throughout the infection process (Figs. 1 and 2). The pattern of *Xcc* population and H_2O_2 concentrations is consistent with the latter's role in impeding bacterial growth and promoting PCD, which occurred from 10 to 12 days after inoculation. The rapid necrosis in the localized region of the infected kumquat tissue by *Xcc* has been suggested to be consistent with a hypersensitive response (HR) and induced PCD (Khalaf et al., 2007). Lipid peroxidation was shown to increase rapidly and remain several times higher than the controls in kumquat-*Xcc* interaction (Kumar et al., 2011e). Lipid peroxidation generates free radicals, which in turn are toxic to plant and bacterial cells and is consistent with PCD as part of the HR to pathogens (Gobel *et al.*, 2003; Kumar et al., 2011e; Rusterucci et al., 1996). It is interesting that using the injection method, kumquat did not display much swelling of the epidermis, which is required for egress of Xcc to the leaf surface. Kumar et al., (2011c,e) concluded that the retention of bacteria in the leaf coupled with early leaf abscission, which occurred from days 10 through 12, is consistent with a disease avoidance mechanism.

The production of H_2O_2 occurs mainly through SOD activity. Kumar et al. (2011e) showed that total SOD activity demonstrated two peaks during the course of *Xcc* infection of kumquat with peaks at 1-2 days after inoculation and 6-8 days after inoculation, although the total SOD activity was always higher than the uninfected controls. Analysis of the activity and isoforms of the various SODs were shown to be altered indicating compartmentalization of H_2O_2 production (Kumar et al, 2011c,e). The first peak in total SOD activity was associated with a rapid increase in Fe-SOD activity to 2x the controls by 1 day after inoculation, but the activity dropped rapidly near or below the controls thereafter. Fe-SOD is compartmentalized in chloroplasts and studies on other plant-pathogen interactions have shown that chloroplasts are an important source of ROS signals that initiate changes in oxidative metabolism in other cellular compartments (Mur *et al.*, 2008; Zurbriggen *et al*, 2009). Cu-Zn-SOD is also found in the chloroplasts (Alscher *et al.*, 2002), but Kumar et al. (2011e) found no activity of this SOD isoform during the kumquat-*Xcc* interaction. Mitogen-activated protein kinase (MAPK), which respond to external stimuli, are activated in plant-pathogen interactions and promote ROS generation in chloroplasts by inhibiting CO_2 assimilation that serves as a sink for ROS generated by light (Liu *et al.*, 2007; Zurbriggen *et al*, 2009). Evidence that this mechanism functions during kumquat-*Xcc* interaction is supported by differential expression of related genes (Khalaf *et al.*, 2007). Although Fe-SOD activity initially surged, high concentrations of H_2O_2 have been shown to deactivate Fe-SOD (Giannopolitis and Ries, 1977), which is consistent with suppression of Fe-SOD activity after the first day (Kumar et al., 2011e).

Keeping in mind that total SOD activity in kumquat-*Xcc* interaction increased and remained high throughout pathogenesis, the decline in Fe-SOD activity beyond the first day after inoculation had to be replaced by a different form of SOD that would dominate during the second peak of total SOD activity. Kumar et al., (2011e) found that Mn-SOD activity increased from 2x to 3x that of the control starting 2 days after inoculation and reached a maximum during the second peak of total SOD activity from 6 to 8 days after inoculation. The prolonged, elevated Mn-SOD activity indicated that this class of SOD was responsible for the majority of total SOD activity throughout the entire pathogenesis process. Mn-SOD is generally considered to be limited to mitochondria and peroxisomes (Alscher *et al*, 2002) and recent evidence indicates the importance of mitochondria in generating ROS to promote PCD (Mur *et al*, 2008; Yao *et al.*, 2002). Thus, the elevated H_2O_2 concentration during kumquat-*Xcc* interaction is promoted by SOD activity, first in the chloroplast and thereafter in the peroxisome and mitochondria. Thus, the sustained production of H_2O_2 in peroxisomes and mitochondria indicates that these organelles serve as important generators of H_2O_2 during kumquat-*Xcc* interactions.

The fate of H_2O_2 in kumquat-*Xcc* interaction is determined, in part, by enzymes involved in its dismutation. Catalase is considered the major H_2O_2 scavenging enzyme and is located in peroxisomes of plant cells (Kamada et al., 2003; Hu et al., 2010). During kumquat-*Xcc* interaction, total CAT activity remained similar to the controls up to 5 days after inoculation but declined starting 6 days after inoculation to almost half of the controls (Kumar et al., 2011c). Interestingly, CAT demonstrated qualitative and temporal changes in isoforms (Kumar et al., 2011c). Plants have been shown to contain three CAT genes that code for three subunits and generate at least six isoforms that are classified into three classes (Hu et al., 2010). Class I CATs are abundant in tissues that contain chloroplasts, Class II CATs are mainly expressed in vascular tissues, and Class III CATs are generally found in young and senescent tissues. In uninfected kumquat leaves, Kumar et al. (2011c) identified 4 CAT isoforms (CAT 1-4) that appeared to be constitutive and therefore belong in Class I and II. CAT-3 disappeared, CAT-2 declined starting at 4 days after inoculation, and CAT-4 declined starting at 10 days after inoculation, probably due to termination of all metabolic activity because of necrosis. A novel CAT isoform, CAT-5, was expressed 4 days after inoculation, and appears to belong to Class III since senescence as indicated by chlorosis rapidly developed at this time. There was no evidence of CAT-6.

The decline in CAT activity coincided with the highest concentrations of H_2O_2 but during the stationary phase of *Xcc* population growth (Kumar et al., 2011e). *Xcc* during the log phase of growth in kumquats is highly susceptible to H_2O_2 with almost no survival upon exposure to 1 mM H_2O_2 in comparison to stationary phase populations that can resist up to 30 mM of H_2O_2 (Tondo et al., 2010). H_2O_2 increased to almost 10 mM (Kumar et al., 2011c,e), which was high enough to restrict *Xcc* during the log phase but not enough to impact bacterial populations during the stationary phase of growth (Tondo et al., 2010). The *Xcc* stationary phase populations were able to resist higher external H_2O_2 concentrations due to high bacteria CAT activity via the expression of four CAT genes (*katE, catB, srpA, and katG*) (Tondo et al., 2010). Thus, it appears that the reduced plant CAT activity, which occurred during the stationary phase of *Xcc* population growth, was too late to directly impact the pathogen. Perhaps molecular modification that increasing CAT activity earlier in kumquat would suppress *Xcc* concentrations further by allowing H_2O_2 concentrations to increase during the log phase of *Xcc* growth (Chaouch et al., 2010).

Although the decline in CAT activity was too late to have a direct impact on *Xcc* populations, it may be part of the adaptive response of kumquat to promote necrosis and leaf abscission late in the infection process (Foyer et al., 2009). Recently, Yu et al, (2006) showed that selective degeneration of specific CATs in mouse cell lines subsequently caused an increase in ROS concentrations and induced PCD. Similarly, transgenic plants with reduced CAT expression exhibited necrotic lesions and displayed elevated concentrations of pathogenesis-related proteins in tobacco (*Nicotiana tabacum* cv. Bel w3; Mittler et al., 1999).

Because CATs are limited to peroxisomes, it appears that this organelle serves an important role in canker resistance by elevating H_2O_2 concentrations that diffuses to the rest of the cell and thus could become a promising site for resistance enhancement in susceptible citrus by genetic engineering of CAT gene expression or by post-translational modification of CAT proteins (Chaouch et al., 2010).

Ascorbate peroxidases are ubiquitous peroxidases that help maintain low H_2O_2 concentrations during normal metabolism (Mittler, 2002). During kumquat-*Xcc* interaction, APOD activity declined linearly after *Xcc* inoculation to less than half the activity of the controls by 12 days after inoculation (Kumar et al., 2011c). The immediate and increasing decline in APOD activity is an adaptive plant response to help promote elevated H_2O_2 concentrations throughout the sympast and is the principle enzyme that allowed H_2O_2 concentrations to increase in infected kumquat. There is evidence that higher H_2O_2 concentrations inactivate APODs at both the transcriptional and post-transcriptional levels (Zimmermann et al., 2006; Paradiso et al., 2005).

Higher H_2O_2 concentrations rather than $O_2^{\cdot-}$ in the symplast is interesting because it is a less reactive ROS, which may indicate another role for H_2O_2 than promoting senescence alone. *Xcc* are only found in the apoplast and any positive effect of higher H_2O_2 concentrations would require diffusion out of the symplast. H_2O_2 in the apoplast would allow it to serve as a substrate for the Class III PODs. During normal metabolism of uninfected plants, H_2O_2 is utilized by the Class III PODs to promote loosening of cell walls during cell enlargement and to cross-link cell wall polymers during cell maturation (de Gara, 2004). The Class III PODs are also an adaptive defense mechanism against pathogens since the cross linking of cell wall polymers diminishes their ability to enzymatically digest the cell wall and thus isolates the pathogen in a confined area (Bradley et al., 1992; Passardi et al., 2005). Kumquat POD activity tripled 1 day after inoculation with *Xcc* and continued to increase to 8 days after inoculation (Kumar et al., 2011c). No canker development occurred beyond the initial infection zone as evidenced by water soaking upon injection indicating isolation of the bacteria consistent with activity of the Class III PODs. No up-regulation of POD has been shown for kumquat, but transcriptional analysis has shown up-regulation of POD genes in sweet orange leaves 2 days after inoculation with *Xcc* (Cernadas et al., 2008).

In addition to cross linking cell walls using H_2O_2, Class III PODs are capable of catalyzing reactions utilizing other substrates (Passardi et al., 2005). PODs can convert $O_2^{\cdot-}$ and H_2O_2 to $OH^{\cdot}$ (Schweikert et al., 2000; Schopfer et al., 2002; Liszkay et al, 2003), however, apoplastic generation of $O_2^{\cdot-}$ has not been definitively determined in kumquat-*Xcc* interactions. A potential source of $O_2^{\cdot-}$ is by NADPH oxidase activity (Kasai *et al.*, 2006), which is generally regarded as a critical component of plant defense (Lamb and Dixon, 1997), but that enzyme has not been studied in kumquat exposed to *Xcc*. Any apoplastic SOD activity would de-activate $O_2^{\cdot-}$. One SOD reported to be located in plant apoplasts is Cu-Fe-SOD (Alscher et al., 2002) and in kumquat infected with *Xcc*, a putative Cu-Fe-SOD gene was up-regulated 2 to 7 days after inoculation (Khalaf et al., 2007), however activity of this SOD isoform was not detected (Kumar et al., 2011e). Mn-SOD was also suggested to be involved in cell elongation (Kukavica et al., 2009), which is one of the early events during canker development (Khalaf et al., 2007). Kukavica et al. (2009) proposed a novel role for cell wall bound Mn-SOD that assists in POD-mediated cell elongation by producing $OH^{\cdot}$ in the apoplast. Although the formation of OH˙ during kumquat-*Xcc* is not verified, its formation is consistent with plant defense considering its high toxicity to *Xanthomonas spp.* (Vattanaviboon and Mongkolsuk, 1998). Nevertheless, production of $O_2^{\cdot-}$ and conversion of it plus H_2O_2 to OH˙ in kumquat-*Xcc* interactions needs to be determined.

In summary, kumquat respond to *Xcc* by promoting higher concentrations of H_2O_2 through temporal and qualitative changes in enzymes involved in its synthesis and dismutation. H_2O_2 is produced initially through increased chloroplastic SOD 1 day after inoculation and thereafter through increased mitochondrial and peroxisomal SOD activity. Elevated symplastic H_2O_2 concentrations are maintained by declining APOD and later CAT activity. We propose that the elevated concentration of H_2O_2 diffuses from the symplast to the apoplast where it directly inhibits bacterial metabolism and utilized by POD. The higher POD activity presumably utilizes H_2O_2 to cross-link cell walls and perhaps produce highly toxic OH˙.

7. Oxidative metabolism in canker susceptible grapefruit and sweet orange

Using the same strain of Asiatic canker, infiltration method, and under the same growing conditions as in kumquat (Kumar et al., 2011c,e), the bacterial population in grapefruit and sweet orange leaves grew to 1 x 10^9 CFU/cm^2 (Kumar et al., 2011b,d), which was 10x that of kumquat (Kumar et al., 2011e). In general, the responses of grapefruit and sweet orange to *Xcc* were similar. Whereas the *Xcc* population peaked in kumquat 4 days after inoculation, the population peak occurred 8 days after inoculation in grapefruit (Figs. 1 and 3) and 14 days after inoculation in sweet orange. Chlorosis was evident in grapefruit and sweet orange by the first day after inoculation as in kumquat. However water soaking, which didn't occur until 4 days after inoculation in kumquat, occurred by the second day in grapefruit and sweet orange. Furthermore, swelling of the leaves in the inoculated region was evident starting 6 days after inoculation. Necrosis was evident from 16 to leaf abscission, which occurred a week later than kumquat.

Unlike H_2O_2 concentrations in kumquat that increased and remained high until *Xcc* populations declined, H_2O_2 concentrations in grapefruit and sweet orange leaves demonstrated a biphasic pattern. There was an initial surge in H_2O_2 concentration in both susceptible genotypes to that found in kumquat except it was only to 1/3 the concentration and the surge only lasted until 4 days after inoculation (Kumar et al., 2011b,d). H_2O_2 concentrations declined to or below the controls and then surged a second time but only to the same concentrations or to concentrations slightly above the controls from 12-14 days after inoculation. The crash in H_2O_2 concentration occurred very late in the log phase of bacterial growth, the stage most susceptible to H_2O_2 (Tondo et al., 2010), which allowed extension of that phase resulting in the higher bacterial populations compared to kumquat.

The disturbance in H_2O_2 concentration was related to temporal and qualitative changes in enzyme activities related to H_2O_2 metabolism. Total SOD activity in grapefruit and sweet orange generally followed that of H_2O_2 concentration with a peak in activity occurring 4 days after inoculation followed by a rapid decline with concentrations similar to or less than the controls for the rest of the infection process (Kumar et al., 2011b,d). The initial increase in total SOD activity was due to a surge in Fe-SOD activity similar to that of kumquat. Three

Fe-SOD isoforms were detected in both infected and control leaves of grapefruit, but it was Fe-SOD 2 that contributed most of the Fe-SOD activity observed. Down regulation of *Fe-Sod1*transcription were observed in *Botrytis cinerea* infected cultured cells of *Pinus pinaster* (Azevedo et al., 2008), but whether this gene is involved in *Xcc*-susceptible citrus genotypes is unknown.

Manganese superoxide dismutase activity surged in a manner similar to kumquat but then crashed to concentrations similar to the controls by 4 days after inoculation (Kumar et al., 2011b,d). Thus the decline in H_2O_2 concentration in grapefruit and sweet orange was due in part to suppression of Mn-SOD activity. Four Mn-SOD isoforms were observed in grapefruit (Kumar et al., 2011d). Mn-SOD 3 was constitutively active however Mn-SOD 1 and 2 were higher from 2 and 4 days after inoculation but thereafter gradually disappeared. It appears then that the appearance of Mn-SOD 1 and 2 are originally promoted in response to *Xcc* infection, but response dissipates later in the infection process. A weakly stained Mn-SOD 4 was observed at 10 days after inoculation and appeared to be a last attempt by the host to generate more H_2O_2 to suppress *Xcc* or as part of PCD in the infected zone (Vattanaviboon and Mongkolsuk, 1998).

In addition to changes in activities of the various SODs, H_2O_2 degrading enzymes also demonstrated temporal and qualitative changes in activity (Kumar et al., 2011b,d). Catalase activity increased above the control in grapefruit starting 2 days after inoculation and remained up the control peaking 16 days after inoculation, which is opposite of kumquat where CAT activity was suppressed (Kumar et al., 2011b). Four CAT isoforms were detected in controls and six in *Xcc*-infected grapefruit, with CAT 4 and 5 novel in the latter plants and the intensity of the CAT 2 and 4 bands very high compared to the controls. Higher expression of CAT 2 mRNA in roots of potato was found during pathogenesis of *Corynebacterium sepedonicum* NCPPB 2137 and *Erwinia cartovora* spp. *cartovora* NCPPB 312 and provide the first evidence that class II CAT isoforms are also pathogen induced (Niebel et al., 1995). Thus the elevated CAT activity in grapefruit partially explains the decline in H_2O_2 concentrations in grapefruit.

Unlike kumquat where APOD activity was suppressed in *Xcc*-infected plants, APOD activity in grapefruit increased 4 days after inoculation and remained higher than the controls up to 16 days after inoculation (Kumar et al., 2011b). Like CAT, the higher APOD activity contributed to the lower H_2O_2 concentrations.

The class III POD activity levels were higher in *Xcc*-infected grapefruit and sweet orange leaves 1 days after inoculation (Kumar et al., 2011b,d), which was similar to that in kumquat. Three isoforms (POD 1, 2 and 3) were detected in control and infected leaves of both genotypes with higher intensity of all three bands in infected tissues. In a separate study of *Xcc* infected sweet orange, POD genes were shown to be up-regulated as early as 6 hours after inoculation (Cernadas *et al.*, 2008). More than 70 isoforms of PODs have been identified in plants and it is currently difficult to assign a physiological function to each one due to gene redundancy (Sasaki et al., 2004). Nevertheless, it is interesting that unlike CAT and APOD where there was a differential response in susceptible (grapefruit and sweet orange) and resistant (kumquat) genotypes, POD activity in all three genotypes increased in response to *Xcc*.

8. Proposed model of citrus response to canker

A comparison of *Xcc* population, symptom development, H_2O_2, and activities of enzymes involved in H_2O_2 metabolism between the resistant genotype kumquat and a susceptible genotype such as grapefruit can reveal deficiencies in susceptible genotypes. Although similar concentrations of *Xcc* were injected in leaves of both genotypes, the population was 10x less in kumquat than grapefruit by 3 days after inoculation and remained substantially lower. Activity of chloroplastic Fe-SOD, an organelle that is presumed to be involved in pathogen sensing and signaling, increased 1 day after inoculation in kumquat but 2 days after inoculation in grapefruit, which indicates a delayed response in the latter genotype. The reduced *Xcc* population in kumquat compared to grapefruit was due, in part, to lower H_2O_2. Although H_2O_2 increased in both species upon infection, it was only 1/3 the concentration in grapefruit than kumquat at its peak 5 days after inoculation. The sustained H_2O_2 concentration in kumquat was due to higher and sustained Mn-SOD activity and lower CAT and APOD activities. In grapefruit, however, CAT increased 1 day after inoculation, APOD increased 3 days after inoculation, and Mn-SOD declined 5 days after inoculation. There are reports which showed that *Xanthomonas spp.* are naturally very resistant to O_2^{-} but are susceptible to H_2O_2 (Loprasert et al., 1996; Tondo et al., 2010). Thus, although SOD activity was enhanced in grapefruit, the H_2O_2 was subsequently degraded by enhanced activities of CATs and APODs.

Watersoaking developed earlier in grapefruit (2 days after inoculation) than kumquat (4 days after inoculation). Water soaking is a characteristic symptom of *Xcc* infection in citrus that is caused in part by increased uptake of water through capillary action as a consequence of loss of intercellular space between rapidly dividing and enlarging mesophyll cells (Khalaf et al., 2007; Popham et al., 1993). The earlier watersoaking of grapefruit and the higher raised epidermis is indicative of increased cell growth in this genotype, which was reflected in the observed raising of epidermis compared to kumquat. It is interesting that POD activity in both genotypes was elevated upon *Xcc* infection. Peroxidase serves a dual role of promoting cell enlargement by loosening the cell wall but is also involved in cross-linking of cell wall components during cell maturation, a process that inhibits cell enlargement (Passardi et al., 2004). Which process that occurs would be substrate dependent and would vary temporally and spatially. Such a temporal and spatial variation in POD activity has been shown to occur during cell growth of *Arabidopsis thaliana* leaves where cell enlargement was promoted early and cell wall stiffening occurred later (Abarca et al., 2001). The changes in CAT, APOD and Mn-SOD that lowered H_2O_2 concentrations in grapefruit preceded the raised epidermis and thus it is reasonable to assume that the concentrations of H_2O_2 were necessary to promote cell enlargement in this genotype, whereas the higher concentrations of H_2O_2 that occurred in kumquat were excessive and involved in suppression of *Xcc*. Thus, we propose that the lower H_2O_2 concentrations in grapefruit promoted plant cell growth whereas the higher H_2O_2 concentrations in kumquat were involved in cross linking of cell wall polymers and possibly the production of $OH^{\cdot}$. Solutions to solving *Xcc* in susceptible citrus genotypes such as grapefruit and sweet orange will need to include promoting earlier, higher, and sustained H_2O_2 concentrations.

The comparative studies of oxidative metabolism in susceptible and resistant genotypes to *Xcc* have identified deficiencies in susceptible genotypes. Altering their response either through exogenous applications of chemicals that evoke systemic acquired resistance and induced systemic resistance or through genetic modification should be a focus of future research. In particular, stimulation of Mn-SOD activity, which is important for sustained production of H_2O_2, and suppression of CAT and APOD activity to maintain high concentrations of H_2O_2 in susceptible genotypes should improve resistance to *Xcc*. Strategies that improve H_2O_2 metabolism to enhance resistance should provide new cultural management approaches in commercial groves for reducing the economic impact of this disease.

	Xcc			H_2O_2		Enzyme activity[x]													
	Population[z]	Symptom[y]				Total-SOD		Fe-SOD		Mn-SOD		CAT		APOD		POD			
dai	K/G	K	G	K	G	K	G	K	G	K	G	K	G	K	G	K	G		
0	↔	↔	↔	↔	↔	↔	↔	↔	↔	↔	↔	↔	↔	↔	↔	↔	↔		
1	↔	C	C	↑	↑	↑	↑	↑	↔	↔	↑	↔	↔	↓	↔	↑	↑		
2	↔	C	C,W	↑	↑	↑	↑	↔	↑	↑	↑	↔	↑	↓	↔	↑	↑		
3	↓	C	C,W	↑	↑	↑	↑	↔	↑	↑	↑	↔	↑	↓	↑	↑	↑		
4	↓	C,W	C,W	↑	↑	↑	↑	↔	↑	↑	↑	↔	↑	↓	↑	↑	↑		
5	↓	C,W	C,W	↑	↔	↑	↔	↓	↑	↑	↔	↔	↑	↓	↑	↑	↑		
6	↓	C,W	C,W,E	↑	↓	↑	↔	↓	↔	↑	↓	↓	↑	↓	↑	↑	↑		
7	↓	C,W	C,W,E	↑	↓	↑	↓	↓	↓	↑	↓	↓	↑	↓	↑	↑	↑		
8	↓	C,W	C,W,E	↑	↓	↑	↓	↓	↓	↑	↓	↓	↑	↓	↑	↑	↑		
9	↓	C,W	C,W,E	↑	↓	↑	↓	↓	↓	↑	↓	↓	↑	↓	↑	↑	↑		
10	↓	C,W,N	C,W,E	↑	↓	↑	↓	↓	↓	↑	↓	↓	↑	↓	↑	↑	↑		
11	↓	C,W,N	C,W,E	↑	↓	↑	↓	↓	↓	↑	↔	↓	↑	↓	↑	↑	↑		
12	↓	C,W,N	C,W,E	↑	↔	↑	↔	↓	↔	↑	↔	↓	↑	↓	↑	↑	↑		
13			C,E		↔		↔		↔		↔		↑		↑		↑		
14			C,E		↔		↔		↔		↓		↑		↑		↑		
15			C,E		↓		↔		↓		↓		↑		↑		↑		
16			C,E,N		↓		↔		↔		↓		↑		↑		↑		
17			C,E,N		↓		↔		↔		↓		↑		↑		↑		
18			C,E,N		↔		↔		↔		↓		↔		↔		↑		
19			E,N		↔		↔		↔		↓		↔		↔		↑		
20							↓		↓		↓		↔		↔		↑		

z Populatin concentrations are shown as the ratio of kumquat and grapefruit

y Symptom classification: C= chlorosis, W= watersoaking, E= raised epidermis, N= necrosis

x Enzyme classification: SOD= superoxide dismutase and their various forms as indicated by their metal cofactor, CAT= catalase, APOD= ascrobate peroxidase, and POD= the class III peroxidase

xThe arrows indicate the ratio in Xcc population between kumquat and grapefruit

Fig. 1. Comparison of *Xcc* population, canker symptoms, H_2O_2, and activities of enzymes involved in H_2O_2 metabolism for kumquat (K) and grapefruit (G) by days after inoculation (dai). Arrows for H_2O_2 and enzyme activities indicate a comparison of *Xcc*-infected to uninfected leaves. Data were taken from Kumar et al., 2011b,c,d,e.

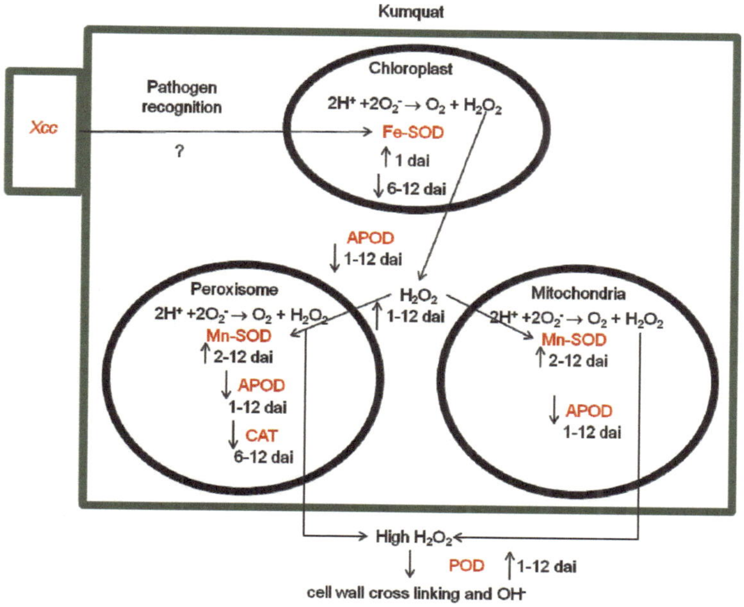

Fig. 2. Proposed mechanism of oxidative metabolism that promotes disease resistance in kumquat. Changes in enzyme activities and H_2O_2 concentration taken from Kumar et al. 2011c,e.

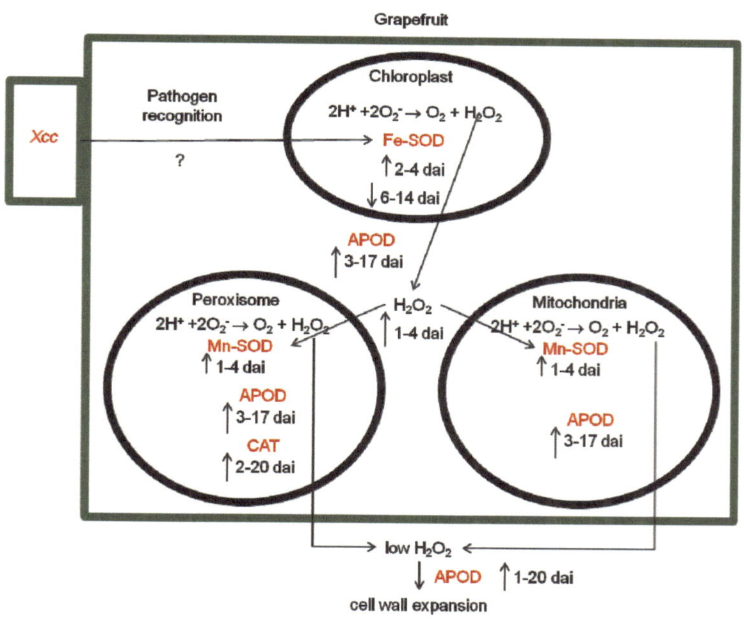

Fig. 3. Proposed mechanism of oxidative metabolism in grapefruit that promotes population growth of *Xcc*. Changes in enzyme activities and H_2O_2 concentration taken from Kumar et al. 2011b,d.

9. References

[1] Abarca, D., Martin, M., Sabater, B., 2001. Differential leaf stress response in young and senescent plants. Physiol. Plant. 113, 409-415.

[2] Able, A.J., Guest, D.I., Sutherland, M.W., 2000. Hydrogen peroxide during the incompatible interaction of tobacco suspension cells inoculated with *Phytophthora nicotiane*. Plant Physiol. 124, 899-910.

[3] Alscher, R.G., Erturk, N., Heath, L.S., 2002. Role of superoxide dismutases (SODs) in controlling oxidative stress in plants. J. Exp. Bot. 53, 1331-1341.

[4] Alvarez, M.E., Penell, R.I., Meijer, P-J., Ishikawa, A., Dixon, R.A. and Lamb, C. 1998. Reactive oxygen intermediates mediate a systemic signal network in the establishment of plant immunity. Cell, 92, 773-784.

[5] Anonymous. 2009. Citrus Summary 2007-2008. Florida Agricultural Statistic Service, USDA NASS.

[6] Auh, C-K., Murphy, T.M., 1995. Plasma membrane redox enzyme is involved in the synthesis of O_2^- and H_2O_2 by *Phytophthora* elicitor-stimulated rose cells. Plant Physiol. 107, 1241–1247

[7] Azevedo, H., Lino-Neto, T., Tavares, R. M. 2008. The necrotroph *Botrytis cinerea* induces a non-host type II resistance mechanism in *Pinus pinaster* suspension-cultured cells. Plant Cell Physiol. 49: 386-395.

[8] Babhitha, M.P., S.G. Bhat, H.S. Prakash, and H.S. Shetty. 2002. Differential induction of superoxide dismutase in downy mildew-resistant and –susceptible genotypes of pearl millet. Plant Path. 51:480-486.

[9] Barna, B., J. Fodor, M. Pogany, and Z. Kiraly. 2003. Role of reactive oxygen species and antioxidants in plant disease resistance. Pest Manag. Sci. 59:459-464.

[10] Bassler, B.L. 1999. How bacteria talk to each other: regulation of gene expression by quorum sensing. Curr. Opin. Microbiol. 2:582-587.

[11] Bolwell, G. P., and Wojtaszek, P. 1997. Mechanisms for the generation of reactive oxygen species in plant defense- a broad perspective. Physiol. Mol. Plant Pathol. 51: 347-366.

[12] Bradley, D., Kjellbom, P., Lamb, C., 1992. Elicitor- and wound-induced oxidative cross-linking of a proline-rich plant cell wall protein: a novel, rapid defense response. Cell 70, 21-30.

[13] Buonaurio, R., Torre, G. D., and Montalbani, P. 1987. Soluble superoxide dismutase (sod) in susceptible and resistant host-parasite complexes of *Phaseolus vulgaris* and *Uromyces phaseoli*. Physiol. Mol. Plant Pathol. 31: 137-184.

[14] Burnings, A.M., Gabriel, D.W., 2003. *Xanthomonas citri*: breaking the surface. Mol. Plant Pathol. 4, 141-157.

[15] del Campo, R., Russi, P., Mara, P., Mara, H., Peyrou, M., de Leon, I.P., Gaggero, C., 2009. *Xanthomonas axonopodis* pv. *citri* enters the VBNC state after copper treatment and retain its virulence. FEMS Microbiol. Lett. 1-6.

[16] Cernadas, R., Camillo, L.R., Benedetti, C.E., 2008. Transcriptional analysis of the sweet orange interaction with the citrus canker pathogen *Xanthomonas axonopodis* pv. *citri* and *Xanthomonas axonopodis* pv. *aurantifolii*. Mol. Plant Pathol. 9, 609-631.

[17] Chaouch, S., Queval, G., Vanderauwera, S., Mhamdi, A., Vandrope, M., Langlois-Meurinne, M., Van Breusegem, F., Saindernan, P., Noctor, G., 2010. Peroxisomal

hydrogen peroxide is coupled to biotic defense responses by isochorismate synthase1 in a daylength-related manner. Plant Physiol. 153, 1692-1705.
[18] Chen, X-R., Wang, X-L., Zhang, Z-G., Wang, Y-C., Zheng, X-B. 2008. Differences in the induction of the oxidative burst in compatible and incompatible interactions of soybean and *Phytophthora sojae*. Physiol Mol Plant Pathol; 73:16-24
[19] Dat, J.F., Pellinen, R., Beckmann, T., Van De Cotte, B., Langebartels, C., Kangasjarvi, J., Inze, D., Van Breusegem, F., 2003. Changes in hydrogen peroxide homeostasis trigger an active cell death process in tobacco plants. Plant J. 33, 621-632.
[20] Dat, J., Vandenabeele, S., Vranova, E., Van Montagu, M., Inze, D., Van Breusegem, F., 2000. Dual action of the active oxygen species during plant stress responses. Cellular and Mol. Life Sci. 57, 779-795.
[21] Delledonne, M., Zeier, J., Marocco, A., and Lamb, C. 2001. Signal interactions between nitric oxide and reactive oxygen intermediates in the plant hypersensitive disease resistance response. Proc. Nat. Acad. Sci. USA 98: 13454-13459.
[22] Duan, Y.P., Castaneda, A.L., Zhao, G., Erdos, G., Gabriel, D.W. 1999. Expression of a single, host-specific, bacterial pathogenicity gene in plant cells elicits division, enlargement and cell death. Mol. Plant-Microbe. Interact. 12, 556-560.
[23] Durrant, W.E. and X. Dong. 2004. Systemic acquired resistance. Ann. Rev. Plant Path., 42:185-209.
[24] Foyer, C.H., Bloom, A.J., Queval, G., Noctor, G., 2009. Photorespiratory metabolism: genes, mutants, energetic, and redox signaling. Annu. Rev. Plant Biol. 60, 455-484.
[25] de Gara, L. (2004) Class III peroxidases and ascorbate metabolism in plants. Phytochem. Rev. 3, 195-205.
[26] Gay, P.A., Tuzun, S., 2000. Temporal and spatial assessment of defense responses in resistant and susceptible cabbage varieties during infection with *Xanthomonas campestris* pv. *campestris*. Physiol. Mol. Plant Pathol. 57, 201-210.
[27] Giannopolitis, C. N., and Ries, S. K. 1977. Superoxide dismutases. I. Occurrence in higher plants. Plant Physiol. 59: 309-314.
[28] Gobel, C., Feussner, I., and Rosahl, S. 2003. Lipid peroxidation during the hypersensitive response in potato in the absence of 9-Lipoxugenases. J. Biol. Chem. 278: 52834-52840.
[29] Gonzalez, A.M., Marcel, T.C., Kohutova, Z., Stam, P., Van Der Linden, C.G., Niks, R.E., 2010. Peroxidase profiling reveals genetic linkage between peroxidase gene clusters and basal host and non-host resistance to rusts and mildew in barley. PLOS ONE 5, e10495.
[30] Gottwald, T. R., Graham, J. H. 1992. A device for precise and nondisruptive stomatal inoculation of leaf tissue with bacterial pathogens. Phytopathology 82: 930-935.
[31] Gottwald, T.R., J.H. Graham, and J.S. Schubert. 2002. Citrus canker: The pathogen and its impact. Plant Health Progress doi:10.1094/PHP-2002-0812-01-RV.
[32] Graham, J. H., Gottwald, T. R., Cubero, J., and Achor, D. 2004. *Xanthomonas axonopodis* pv *citri:* factors affecting successful eradication of citrus canker. Mol. Plant Pathol. 5: 1-15.
[33] Hu, Y-Q., Liu, S., Yuan, H.M., Li, J., Yan, D.W., Zhang, J.F., Lu, Y.T., 2010. Functional comparison of catalase genes in the elimination of photorespiratory H_2O_2 using

promoter and 3′-untranslated region exchange experiments in the *Arabidopsis cat2* photorespiratory mutant. Plant Cell Env. 33:1656-1670.
[34] Kamada, T., Nito, K., Hayashi, H., Mano, S., Hayashi, M., Nishimura, M., 2003. Functional differentiation of peroxisomes revealed by expression profiles of peroxisomal genes in *Arabidopsis thaiana*. Plant Cell Env. 44, 1275-1289.
[35] Kasai, T., Suzuki, T., Ono, K., Ogawa, K., Inagaki, Y., Ichinose, Y., Toyoda, k., Shiraishi, T. 2006. Pea extracellular Cu/Zn-superoxide dismutase responsive to signal molecules from a fungal pathogen. J. Gen. Plant Pathol. 72: 265-272.
[36] Khalaf, A., Moore, G.A., Jones, J.B., Gmitter, F.F. Jr., 2007. New insights into the resistance of 'Nagami' kumquat to canker disease. Physiol. Mol. Plant Pathol. 71, 240-250.
[37] Koizumi, M., and Kuhara, S. 1982. Evaluation of citrus plants for resistance to bacterial canker disease in relation to lesion extension. Bull. Tree Fruit Res. D 4: 73-92.
[38] Kukavica, B., Mojovic, M., Vucnic, Z., Maksimovic, V., Takahama, U., Jovanovic, S.V. 2009. Generation of hydroxyl radical in isolated pea root cell wall, and the role of cell wall-bound peroxidase, Mn-SOD and phenolics in their production. Plant Cell Physiol. 52, 304-317.
[39] Kumar, N., Ebel, R. C. and Roberts, P. D. 2011a. Antioxidant isozyme variability in different genotypes of citrus and kumquat. J. Crop Improv.. 25: 86-100.
[40] Kumar, N., Ebel, R. C. and Roberts, P. 2011b. Antioxidant metabolism of grapefruit infected with *Xanthomonas axonopodis* pv. *citri*. Env. Exp. Bot.. 71: 41-49.
[41] Kumar, N., Ebel, R. C. and Roberts, P. D. 2011c. H_2O_2 degradation is suppressed in kumquat leaves infected with *Xanthomonas axonopodis* pv. *citri*. Sci. Hort.. 130: 241-247.
[42] Kumar, N., Ebel, R. C. and Roberts, P. D. 2011d. H_2O_2 metabolism during sweet orange (*Citrus sinensis* L. Osb.) Hamlin *Xanthomonas axonopodis* pv. *citri*. Sci. Hort.. 128: 465-472.
[43] Kumar, N., Ebel, R. C. and Roberts, P. 2011e. SOD activity in *Xanthomonas axonopodis* pv. *citri* infected leaves of kumquat. J. Hort. Sci. Biotech.. 86: 62-68.
[44] Kuzniak, E. and H. Urbanek. 2000. The involvement of hydrogen peroxide in plant responses to stresses. Acta Physiologia Plant. 22:195-2003.
[45] Lamb, C., and Dixon, R. A. 1997. The oxidative burst in plant disease resistance. Annu. Rev. Plant Physiol. 48: 251-275.
[46] Lawson, R. H., Dienelt, M. M., and Civerlo, E. L. 1989. Histopathology of *Xanthomonas campestris* pv. *citri* from Florida and Mexico in wound-inoculated detached leaves of citrus aurantifolid: light and scanning electron microscopy. Phytopathology 79: 329-335.
[47] Lee, V.T. and O. Schneewind. 2001. Protein secretion and the pathogenesis of bacterial infections. Genes Dev. 15:1725-1752.
[48] Liszkay, A., Kenk, B., Schopfar, P., 2003.Evidence for the involvement of cell wall peroxidase in the generation of hydroxyl radicals mediating extension growth. Planta 217, 658-667.
[49] Liu, Y., Ren, D., Pike, S., Pallardy, S., Gassmann, W. and Zhang, S. (2007). Chloroplast generated reactive oxygen species are involved in hypersensitive response like cell

death mediated by a mitogen-activated protein kinase cascade. The Plant J., 51, 941-954.

[50] Loprasert, S., P. Vattanaviboon, W. Praituan, S. Chamnongpol, and S. Mongkolsuk. 1996. Regulation of the oxidative stress protective enzymes, catalase, and superoxide dismutase in *Xanthomonas* - a review. Gene 179:33-37.

[51] Martinez, C., J.L. Montillet, J.P. Agnel, G.H. Dai, J.F. Daniel, J.P. Geiger, and M. Nicole. 1998. Apoplastic peroxidase generates superoxide anions in cells of cotton cotyledons undergoing the hypersensitive reaction to *Xanthomonas campestris* pv. *Malvacearum* Race 18. Mol. Plant-Microbe Inter. 11:1038-1047.

[52] Mittler, R., 2002. Oxidative stress, antioxidants and stress tolerance. Trends Plant Sci. 7, 405-410.

[53] Mittler, R., Herr, E.H., Orvar, B.L., Van Camp, W., Willekens, H., Inze, D., Ellis, B.E., 1999. Transgenic tobacco plants with reduced capability to detoxify reactive oxygen intermediates are hyperresponsive to pathogen infection. Proc. Nat. Acad. Sci. USA 96, 14165-14170.

[54] Montalbini, P., and Buonaurio, R. 1986. Effect of tobacco mosaic-virus infection on levels of soluble superoxide-dismutase in *Nicotiana tabacum* and *Nicotiana glutinosa* leaves. Plant Sci. 47: 135-143

[55] Mur, L.A., Kenton, P., Lloyd, A.J., Ougham, H. and Prats, F. 2008. The hypersensitive response: the centenary is upon us but how much do we know? J. Exp. Bot., 59, 501-520.

[56] Niebel, A., Heungens, K., Barthels, N., Inze, D., Montagu, M.V., Gheysen, G., 1995. Characterization of a pathogen-induced potato catalase and its systemic expression upon nematode and bacterial infection. Mol. Plant Microbe Inter. 8, 371-378.

[57] Paradiso, A., Tommasi, F., De Gara L., De Pinto, M.C., 2005. Alteration in ascorbate and ascorbate peroxidase in programmed cell death and oxidative stress. BMC Plant Biol. 5 (Suppl I): S28.

[58] Passardi, F., Cosio, C., Penel, C., Dunand, C., 2005. Peroxidases have more functions than a Swiss army knife. Plant Cell Rep. 24, 255-265.

[59] Passardi, F., Penel, C., Dunand, C., 2004. Performing the paradoxical: how plant peroxidases modify the cell wall. Trends Plant Sci. 9, 534-540.

[60] de Pinto, M.C., Paradiso, A., Leonetti, P., De Gara, L. 2006. Hydrogen peroxide, nitric oxide and cytosolic ascorbate peroxidase at the crossroad between defence and cell death. The Plant J. 48, 784-795.

[61] Pettersson, J., R. Nordfelth, E. Dubinina, T. Bergman, M. Gustafsson, D.E. Magnusson, and H. Wolf-Watz. 1996. Modulatioin of virulence factors expression by pathogen target cell contact. Science 273:1231-1233.

[62] Polidoros, A.N., Mylona, P.V., Scandalios, J.G., 2001. Transgenic tobacco plants expressing the maize *Cat2* gene have altered catalase levels that affect plant-pathogen interactions and resistance to oxidative stress. Transgenic Res. 10, 555-569.

[63] Popham, P.L., Pike, S.M., Novacky, A., Pallardy, S.G., 1993. Water relation alterations observed during hypersensitive reaction induced by bacteria. Plant Physiol. 103, 1243-1247.

[64] Quiroga, M., Guerrero, C., Botella, M.A., Barcelo, A., Amaya, I., Medina, M.I., Alonso, F.J., de Forchetti, S.M., Tigier, H., Valpuesta, V., 2000. A tomato peroxidase involved in the synthesis of lignin and suberin. Plant Physiol. 122, 1119-1127.

[65] Reddy, M. R. S. 1997. Sources of resistance to bacterial canker in citrus. J. Mycol. Plant Pathol. 27: 80-81

[66] Rigano, L.A., Marano, M.R., Castagnaro, A.P., Amaral, A.M.D., Vojnov, A.A., 2010. Rapid and sensitive detection of citrus bacterial canker by loop-mediated isothermal amplification combined with simple visual evaluation methods. BMC Microbiol. 10, 176.

[67] Rusterucci, C., Stallaert, V.S., Milat, M-L., Pugin, A., Ricci, P., Blein, J-P., 1996. Relationship between active oxygen species, lipid peroxidation, necrosis, and phytoalexin production induced by elicitin in Nicotiana. Plant Physiol. 111, 885-891.

[68] Sasaki, K., Iwai, T., Hiraga, S., Kuroda, K., Seo, S., Mitsuhara, I., Miyasaka, A., Iwano, M., Ito, H., Matsui, H., Ohashi, Y., 2004. Ten rice peroxidases redundantly respond to multiple stresses including infection with rice blast fungus. Plant Cell Physiol. 45, 1442-1452.

[69] Schopfer, P., Liszkay, A., Bechtold, M., Frahry, G., Wagner, A., 2002. Evidences that hydroxyl radicals mediate auxin-induced extension. Planta 214, 821-828.

[70] Schweikert, C., Liszkay, A., Schopfer, P., 2000. Scission of polysaccharides by peroxidase-generated hydroxyl radicals. Phytochem. 53, 565-570.

[71] Schubert, T. S., Rizvi, S. A., Sun, X., Gottwald, T. R., Graham, J. H., and Dixon, W. N. 2001. Meeting the challenge of eradicating citurs canker in Florida-again. Plant Dis. 85: 340-356.

[72] da Silva, A.C., Ferro, J.A., Reinach, F.C., Farah, C.S., Furlan, L.R., Quaggio, R.B., Monteiro-Vitorello, C.B., van Sluys, M.A., Almeida, N.F., Alves, L.M., do Amaral, A.M., Bertolini, M.C., Carargo, L.E. Camaroote, G., Cannavan, F., Cardozo, J., Chambergo, F., Ciapina, L.P., Cicarelli, R.M., Coutinho, L.L., Cursiono-Santos, J.R., El-Dorry, H., Faria, J.B., Ferreira, A.J., Ferreira, R.C., Ferro, M.I., Formighieri, E.F., Franco, M.C., Greggio, C.C., Gruber, A., Katsuyama, A.M., Kishi, L.T., Leite, R.P., Lemos, E.G., Lemos, M.V., Locali, E.C., Machado, M.A., Madeira, A.M., Martines-Rossi, N.M., Martins, Ec.Ec., Meidanis, J., Mench, C.F., Miyaki, C.Y., Moon, D.H., Moerera, L.,M., Novo, M.T., Okura, V.K., Oliveira, M.C., Oliveira, V.R., Pereira, H.A., Rossi, A., Sena, J.A., Sivla, C., de Souza, R.F., Spinola, L.A., Takita, M.A., Tamura, R.E., Teixeira, E.C., Tessa, R.I., Trindade dos Santos, M., Truffi, D., Tsai, S.M., White, F.F., Setubal, J.C. and Kitahima, J.P. 2002. Comparison of the genomes of two Xanthomona pathogens with differeng host specificities. Nature, 417:459-463.

[73] Slater, H., A. Alvarez-Morales, C.E. Barber, J.J. Danieals, and J.M. Dow. 2000. A two-component system involving an HD-GYP domain protein links cell-cell signaling to pathogenicity gene expression in *Xanthomonas campestris*. Mol. Microbiol. 38:986-1003.

[74] Stall, R.E., Marco, G.M., Canteros de Echenique, B.I. 1982. Importance of mesophyll in mature-leaf resistance to cancrosis of citrus. Phytopath. 72:1097-1100.

[75] Stall, R. E., Miller, J. W., Marco, G. M., and de Echenique, C. 1980. Population dynamics of *Xanthomonas citri* causing cancrosis of citrus in Argentina. Proc. Flo. Hort. Soc. 93: 10-14.

[76] Tertivanidis, K. C. Doudoula, C. Vasilikiotis, E. Hassiotou, R. Perl-Treves, and A. Tsaftaris. 2004. Superoxide dismutase transgenes in sugarbeets confer resistance to oxidative agents and the fungus *C. beticola*. Transgenic Res. 13:225-233.

[77] Tondo, M.L., Petrocelli, S., Ottado, J., Orellano, E.G., 2010. The monofunctional catalase *KatE* of *Xanthomonas axonopodis* pv. *citri* is required for full virulence in citrus plants. Plos One 5, e10803.

[78] Valad, G.E. and R.M. Goodman. 2008. Systemic acquired resistance and induced systemic resistance in conventional agriculture. Crop Sci. 44:1920-1934.

[79] Vattanaviboon, P., Mongkolsuk, S., 1998. Evaluation of the role of hydroxyl radicals and iron play in hydrogen peroxide killing of *Xanthomonas campestris* pv. *phaseoli*. FEMS Microbiol. Lett 169, 255-260.

[80] Viloria, Z., Drouillaerd, D.L., Graham, J.H., Grosser, J.W., 2004. Screening triploid hybrids of Lakeland limequat for resistance to citrus canker. Plant Dis. 88, 1056-1060.

[81] Vlot, A.C., D.A. Dempsey, and D.F. Klessig. 2009. Salicylic acid, a multifaceted hormone to combat disease. Ann. Rev. Plant Phytopath. 47:177-206.

[82] Voludakis, A.E., Marmey, P., Delannoy, E., Jalloul, A., Martinez, C., Nicole, M., 2006. Molecular cloning and characterization of *Gossypium hirsutum* superoxide dismutase gene during cotton-*Xanthomonas campestris* pv. *malvacerum* interaction. Physiol. Mol. Plant Pathol. 68, 119-127.

[83] Willekens, H. S. Chamnongpal, M. Davey, M. Schraudner, C. Langebartels, M. van Montagu, D. Inze, and W. van Camp. 1997. Catalase is a sink for H_2O_2 and is indispensible for stress defence in C_3 plants. The EMBO J. 16:4806-4816.

[84] Voludakis, A.E., Marmey, P., Delannoy, E., Jalloul, A., Martinez, C., Nicole, M., 2006. Molecular cloning and characterization of *Gossypium hirsutum* superoxide dismutase gene during cotton-*Xanthomonas campestris* pv. *malvacerum* interaction. Physiol. Mol. Plant Pathol. 68, 119-127.

[85] Wojtaszek, P. 1997. Oxidative burst: an early plant response to pathogen infection. Biochem. J. 322: 681-692.

[86] Wu, G., Shortt, B. J., Lawrence, E. B., Fitzsimmons, K. C., Levine, E. B., and Shah, D. P. 1997. Activation of host defense mechanisms by elevated production of H_2O_2 in transgenic plants. Plant Physiol. 9: 427-435.

[87] Yao, N., Tada, Y., Sakamoto, M., Nakayashiki, H., Park, P., Tosa, Y., and Mayama, S. 2002. Mitochondrial oxidative burst involved in apoplastic response in oats. Plant J. 30: 567-579.

[88] Yu, L., Wan, F., Dutta, S., Welsh, S., Liu, Z., Freundt, E., Baehrecke, E.H., Lenardo, M., 2006. Autophagic programmed cell death by selective catalase degradation. Proc. Nat. Acad. Sci. USA 103, 4952-4957.

[89] Zimmermann, P., Heinlein, C., Orendi, G., Zentgraf, U., 2006. Senescence-specific regulation of catalases in *Arabidopsis thaliana* (L.) Hyenh. Plant Cell and Env. 29, 1049-1060.

[90] Zurbriggen, M.D., Carrillo, N., Tognetti, V.B., Melzer, M., Peisker, M., Hause, B. and Hajirezaei, M-R. (2009). Chloroplast generated reactive oxygen species play a major role in localized cell death during the non-host interaction between tobacco and *Xanthomonas campestris* pv. *vesicatoria*. The Plant J., 60, 962-973.

9

Effect of Oxidative Stress on Secretory Function in Salivary Gland Cells

Ken Okabayashi, Takanori Narita,
Yu Takahashi and Hiroshi Sugiya
Nihon University College of Bioresource Sciences,
Japan

1. Introduction

Reactive oxygen species (ROS) such as superoxide radical anion, singlet oxygen, hydrogen peroxide and hydroxyl radical are products of oxidative metabolism (Kourie, 1998). Low levels of ROS contribute to important signaling pathways to regulate key biological responses, including cell migration, mitosis and apoptosis (Goldschmidt-Clermont & Moldovan, 1999). For instance, endogenous oxidants protected the vasculature by inhibiting endothelial exocytosis that would otherwise lead to vascular inflammation and thrombosis, because endogenous hydrogen peroxide inhibited thrombin-induced exocytosis of granules from endothelial cells (Matsushita et al., 2005). In rat aortic smooth muscle cells, reduction in the intracellular concentration of hydrogen peroxide by the overexpression of catalase within cellular peroxisomes resulted in suppression of DNA synthesis and cell proliferation, and induction of apoptotic cell death (Brown et al., 1999). On the other hand, ROS are known to be pathogenic factors that induce cellular alterations in different cell types. For example, ROS are considered to be involved in the pathogenesis of postischemic endothelial dysfunction, because hydrogen peroxide induces Ca^{2+} oscillations in human aortic endothelial cells (Hu et al., 1998). In pancreatic β cells, hydrogen peroxide interferences glucose metabolism, which leads to the inhibition of insulin secretion (Krippeit-Drews et al., 1999). In mesangial cells, hydrogen peroxide disturbs Ca^{2+} mobilization, which is considered to be involved in renal injury (Meyer et al., 1996). In neurons, hydrogen peroxide induces apoptotic cell death (Whittemore et al., 1995).

In salivary glands, ROS are involved in alteration of the functions. Oxidative stress demonstrated to induce alteration of secretory function of the rat submandibular gland, because reduction of submandibular saliva components such as protein and calcium was observed in the rat treated with lead acetate (Abdollahi et al., 1997, 2003), which induces oxidative stress (Pande & Flora, 2002). Irradiation, a major treatment modality administered for head and neck cancer, induces hypofunction of the salivary glands and consequent xerostomia (Nagler, 2002; de la Cal et al., 2006), in which ROS are believed to be involved in the hypofunction (Nagler et al., 1997, 2000; Takeda et al., 2003). Regarding Sjögren's syndrome, an autoimmune disease which progressively destroys exocrine glands including the salivary glands, ROS has been suggested to be involved in the onset and pathology of

Sjögren's syndrome (Fox, 2005; Ryo et al., 2006). These findings suggest that oxidative stress from ROS causes salivary gland dysfunction (Vitolo et al., 2004).

Under conditions of oxidative stress, the thiols in cysteine residues within proteins are the most susceptible target among oxidant-sensitive molecules (Biswas et al., 2006; Jacob et al., 2006). There are some thiol-modulating reagents by different mechanisms. Ethacrynic acid, a once commonly used loop diuretic drug, is highly electrophilic and preferentially conjugates with glutathione enzymatically and non-enzymatically, and decreases reduced glutathione (GSH) in the mitochondrial pool (Habig et al., 1974; Meredith & Reed, 1982; Yamamoto et al., 2002). L-buthionine-S,R-sulfoximine (BSO) is an irreversible inhibitor of γ-glutamylcysteine synthetase, a rate-limiting enzyme in GSH biosynthesis (Griffith & Meister, 1985). Such thiol-modulating reagents are useful for the study with effects of thiol-oxidation on cell functions.

In salivary parotid acinar cells, stimulation of β-adrenergic receptors provokes release of amylase, a digestive enzyme. The receptor stimulation by β-adrenergic agonists such as isoproterenol (IPR) activates adenylate cyclase via heterotrimeric GTP-binding protein (G-protein), which leads to an increase in intracellular cAMP levels. The increased cAMP subsequently activates cAMP-dependent protein kinase, which has been well recognized to be essential for consequent exocytotic amylase release (Butcher & Putney, 1980; Quissell et al., 1982; Turner & Sugiya, 2002). In this study, we investigated effects of the thiol-modulating reagents ethacrynic acid on amylase release induced by β-adrenergic receptor activation in rat parotid gland cells.

2. Materials and methods

2.1 Materials

Bovine serum albumin (Fraction V, BSA), collagenase A were obtained from Roche Diagnostics GmbH (Mannheim, Germany). Trypsin (type-I), trypsin inhibitor (type-IS), IPR, N(6),2'-O-dibutyryladenosine 3',5'-cyclic monophosphate (db-cAMP), forskolin, ethacrynic acid, and 3-isobutyl-1-methylxanthine (IBMX) were obtained from Sigma (St. Louis, MO). Mastparan, cysteine, glutathione (reduced form, GSH), BSO, sodium sulfosalicylate (SSA) were obtained from Wako Pure Chemical Industries (Osaka, Japan). Vasoactive intestinal peptide (VIP) was obtained from Peptide Institute (Osaka, Japan). The GSSG/GSH Quantification Kit was obtained from Dojindo (Kumamoto, Japan).

2.2 Preparation of parotid acinar cells

All animal protocols were approved by the Laboratory Animal Committee of the Nihon University. Parotid acinar cells were prepared as previously described (Satoh et al., 2008). Sprague-Dawley rats (male, 200–250 g) were intraperitoneally anesthetized with pentobarbital (50 mg/kg), and the parotid glands were removed and placed in a small volume of Krebs-Ringer-bicarbonate (KRB) solution with the following composition (mM): 116 NaCl, 5.4 KCl, 0.8 $MgSO_4$, 1.8 $CaCl_2$, 0.96 NaH_2PO_4, 25 $NaHCO_3$, 5 Hepes (pH 7.4) and 11.1 glucose. KRB solution was equilibrated with an atmosphere of 95% O_2/5% CO_2. After being minced with a razor, the parotid glands were treated with KRB solution containing 0.5% BSA in the presence or absence of enzyme. First, the glands were incubated with trypsin (0.2 mg/ml) at 37°C for 5 min, after which the trypsin-treated glands were removed by centrifugation at 200 g for 1 min. The glands were subsequently incubated in Ca^{2+}-Mg^{2+}-free KRB solution containing 2 mM EGTA and trypsin inhibitor (0.2 mg/ml) at 37°C for 5 min. After the solution was removed by centrifugation (200 g for 1 min), the glands were incubated in Ca^{2+}-Mg^{2+}-free KRB solution without trypsin inhibitor at 37°C for 5 min. After the solution was removed by centrifugation (200 g for 1 min), the glands were incubated in KRB solution with collagenase A (0.75 mg/ml) at 37°C for 20 min. The suspension was passed through eight layers of nylon mesh to separate the dispersed cells from undigested connective tissue and then was placed on KRB solution containing 4% BSA. After centrifugation (50 g for 5 min), the cells were suspended in appropriate amounts of KRB solution containing 0.5% BSA and 0.02% trypsin inhibitor.

2.3 Amylase release

Parotid acinar cells prepared as described above were stimulated by IPR (1 μM), forskolin (100 μM), mastoparan (50 μM), IBMX (1 mM), db-cAMP (100 μM), carbachol (10 μM), or VIP (10 μM) at 37°C for 20 min. When the effects of the thiol-modulating agents (EA and BSO) were examined, cells were preincubated with the agents for 10 min, and then stimulated. The cell suspensions were passed through a filter paper (Whatmann #1). Amylase activity in the filtrates was measured according to the method described previously (Bernfeld, 1955). Total amylase activity was measured in acinar cells homogenized in 0.01% Triton X-100, and amylase released was described as % of total.

2.4 Total glutathione measurement

Dispersed parotid acinar cells were collected by centrifugation at 10,000 g for 15 s and immediately mixed with 160 μl of 10 mM HCl. The mixture was frozen and thawed three times over, mixed with 40 μl of 5% SSA and then centrifuged at 8,000 g for 10 min. The supernatant was collected and diluted twice for further analysis. Total glutathione was measured by Dojindo GSSG/GSH Quantification Kit. Samples were incubated at 37°C for 10 min and then measured optical density at 405 nm by a micro plate reader (Bio-Rad). Total protein concentrations were determined by the Lowry method (1951).

3. Results

3.1 Effect of ethacrynic acid on IPR-Induced amylase release in parotid acinar cells

We first examined effect of the thiol-modulating reagent ethacrynic acid on amylase release in rat parotid acinar cells. After preincubation in the absence or presence of ethacrynic acid (250 μM) for 10 min, the cells were stimulated with the β-agonist IPR (1 μM) or vehicle (control) for 20 min. As Fig. 1 summarizes, IPR induced amylase release in a time dependent manner in the absence of ethacrynic acid, but the IPR-induced amylase release was partially inhibited in the presence of ethacrynic acid. Ethacrynic acid had no effect on amylase release from the cell non-stimulated. In the cells preincubated with 100, 250 or 500 μM ethacrynic acid and then stimulated with IPR for 20 min, ethacrynic acid inhibited the IPR-induced amylase release in a dose dependent manner, as Fig. 2 shows. These results suggest that the amylase release regulated by β-receptor activation is reduced by thiol-modulation.

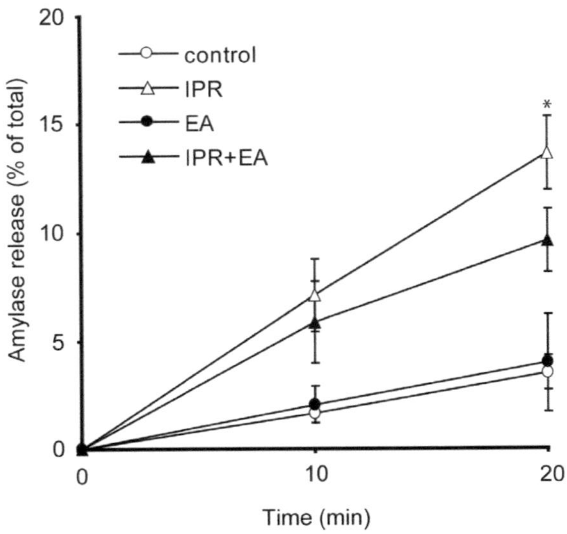

Fig. 1. Inhibition of IPR-induced amylase release by ethacrynic acid in rat parotid acinar cells. After pretreatment of ethacrynic acid (250 μM, EA) or vehicle for 10 min, cells were incubated with (triangles) or without (circles) 1 μM IPR. Value are means ± SE from 5 independent experiments. *P < 0.05

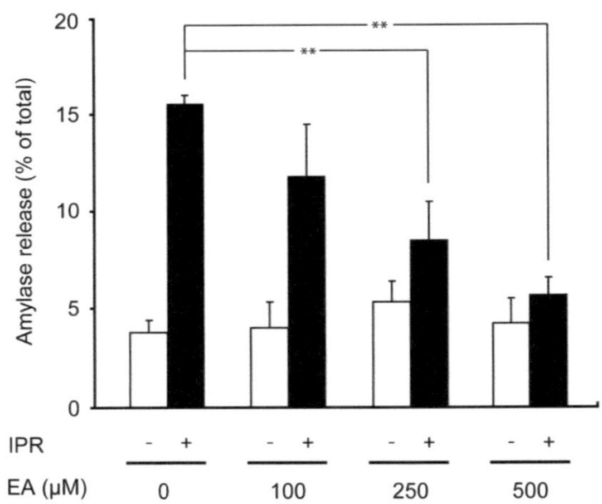

Fig. 2. Dose-dependent effect of ethacrynic acid on IPR-induced amylase release.
After preincubation with 0, 100, 250 or 500 μM ethacrynic acid (EA) for 10 min, rat parotid acinar cells were incubated with (closed columns) or without (open column) 1 uM IPR for 20 min. Values are means ± SE from 3 independent experiments. **P < 0.01

3.2 Relief of the inhibitory effect of ethacrynic acid on IPR-induced amylase release by GSH

To confirm the contribution of thiol-modulation to the inhibition of IPR-induced amylase release by ethacrynic acid, we examined effect of thiol-reducing reagents on the effect of ethacrynic acid. When parotid acinar cells pretreated with ethacrynic acid (250 μM) in absence or presence of GSH (10 mM) or cysteine (10 mM) were stimulated with IPR (1 μM) for 20 min, GSH relieved the inhibitory effect of ethacrynic acid on IPR-induced amylase release, but less cysteine, as Fig. 3 summarizes. These results support that thiol-modulation causes the inhibitory effect of ethacrynic acid on IPR-induced amylase release, although the less effect of cysteine is obscure. GSH and cysteine had no effect on amylase release in the cells non-stimulated (data not shown).

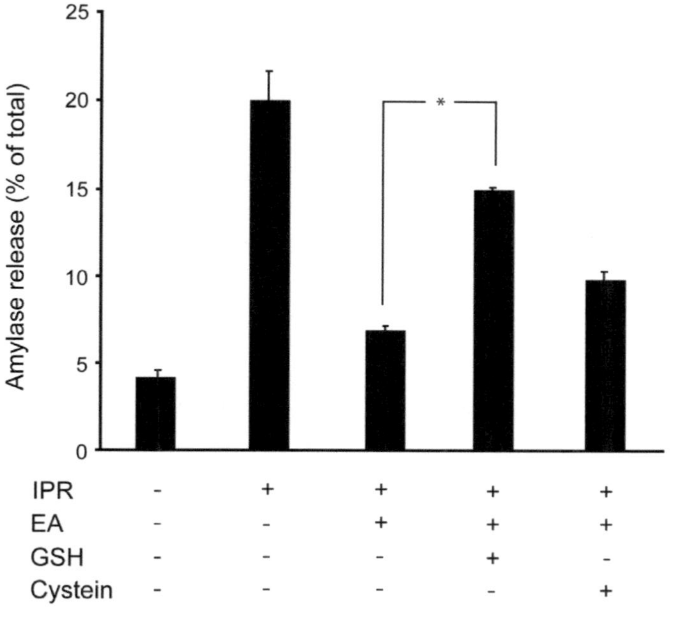

Fig. 3. Relief of the inhibitory effect of ethacrynic acid on the IPR-induced release of amylase by GSH. After pretreatment with 10 mM GSH or 10 mM cysteine in the presence of ethacrynic acid (EA) for 10 min, rat parotid acinar cells were stimulated with 1 μM IPR for 20 min. Values are means ± SE from 3 independent experiments. *$P < 0.05$

3.3 No effect of ethacrynic acid on VIP- and carbachol-induced amylase release

Although β-receptor stimulation dominantly provokes amylase release, stimulation of VIP and muscarinic receptors also evokes amylase release via the increases in intracellular cAMP and Ca^{2+} concentrations, respectively, in rat parotid acinar cells (Scott & Baum, 1985; Yoshimura & Nezu, 1991). Then we next examined the effect of ethacrynic acid on amylase release induced by VIP and carbachol, a muscarinic agonist. When the cells were stimulated with VIP (10 μM) and carbachol (10 μM) for 20 min, amylase release was evoked, although the responses of both secretagogues were lower than that of IPR. However, ethacrynic acid (250 μM) had no effect on VIP- and carbachol-induced amylase release, as shown in Table 1.

Agonist	EA	Amylase release
		(% of total)
(-)	-	3.5 ± 0.4
	+	2.3 ± 0.4
IPR	-	13.6 ± 1.7
	+	9.6 ± 1.5*
VIP	-	5.5 ± 0.4
	+	7.7 ± 1.3
CCh	-	5.0 ± 0.8
	+	4.2 ± 0.4

Table 1. No effect of ethacrynic acid on VIP- and carbachol-induced amylase release in rat parotid acinar cells. After pretreatment of ethacrynic acid (250 μM, EA) or vehicle for 10 min, cells were stimulated with 1 μM IPR, 10 μM VIP or 10 μM carbachol (CCh) for 20 min. Value are means ± SE from 5 independent experiments. *$P < 0.05$

3.4 No effect of ethacrynic acid on amylase release induced by activators of cAMP signaling pathway

It is well known that β-receptor stimulation provokes amylase release via the increase in intracellular cAMP levels in rat parotid acinar cells (Turner & Sugiya, 2002). Then we examined the effect of ethacrynic acid on amylase release induced by activators of cAMP signaling pathway. When parotid acinar cells were incubated with forskolin (100 μM), mastoparan (50 μM), db-cAMP (1 mM) and IBMX (1 mM), a cell-permeable cAMP analogue, an adenylate cyclase activator, a G-protein activator and a cyclic nucleotide phosphodiesterase inhibitor, respectively, for 20 min, amylase release was induced. However, the effects of these drugs on amylase release were not changed even in the cells treated with ethacrynic acid (250 μM), as shown in Table 2. These observations imply that ethacrynic acid has no effect on the cAMP signaling pathway in rat parotid acinar cells.

Reagent	EA	Amylase release
		(% of total)
Forskolin	-	23.6 ± 2.7
	+	21.0 ± 3.8
Mastparan	-	16.6 ± 0.5
	+	15.8 ± 0.2
db-cAMP	-	6.8 ± 0.2
	+	6.3 ± 0.3
IBMX	-	20.7 ± 0.7
	+	16.2 ± 1.5

Table 2. No effect of ethacrynic acid on amylase release induced by cAMP signaling activators. After pretreatment of ethacrynic acid (250 μM, EA) or vehicle for 10 min, rat parotid acinar cells were incubated with forskolin (100 μM), mastoparan (50 μM), db-cAMP (1 mM) or IBMX (1 mM) for 20 min. Value are means ± SE from 5 independent experiments.

3.5 No effect of ethacrynic acid on the intracellular glutathione level

Since EA has been reported to deplete the intracellular glutathione (GSH) (Meredith & Reed, 1982; Dhanbhoora & Babson, 1992), we determined total amount of glutathione in the rat parotid acinar cells treated with ethacrynic acid (250 μM). As Table 3 shows, however, ethacrynic acid had no effect on total amount of glutathione in the cells. Then we next examined effect of the glutathione biosynthesis inhibitor BSO on IPR-induced amylase release. However, BSO (1 mM) had no effect on IPR-induced amylase release, as shown in Fig. 4. These observations suggest that the reduction of glutathione levels is not caused for the inhibitory effect of ethacrynic acid on IPR-induced amylase release.

EA	Total glutathione
	(nmol / mg protein)
(-)	12.67 ± 0.72
(+)	13.97 ± 0.12

Table 3. No effect of ethacrynic acid on total glutathione contents. After treatment of ethacrynic acid (250 μM, EA) or vehicle for 30 min, total glutathione were measured. Values are means ± SE from 3 independent experiments.

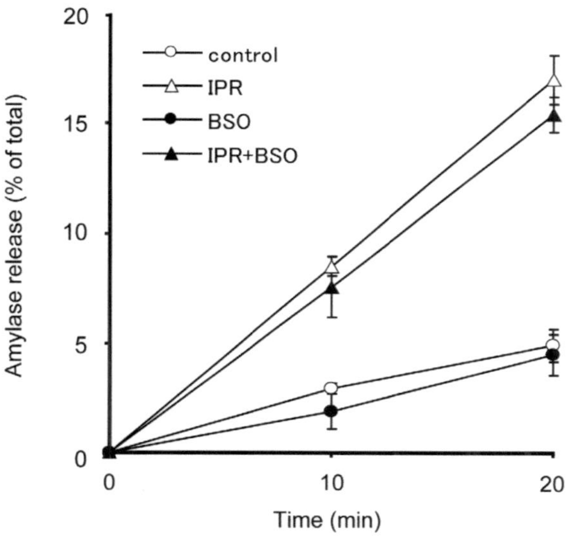

Fig. 4. No effect of BSO on IPR-induced amylase release. After preincubation with 1 mM BSO or vehicle for 10 min, rat parotid acinar cells were incubated with (triangles) or without (circles) 1 μM IPR. Values are means ± SE from 3 independent experiments.

4. Discussion

Amylase release in parotid acinar cells occurs via the two distinct processes, constitutive release and regulatory release (Turner & Sugiya, 2002). The regulatory release is induced by the activation of receptors, whereas the constitutive release is continuously observed without receptor activation. In this study, we demonstrated that the thiol-modulating reagent ethacrynic acid inhibits regulatory amylase release provoked by β-adrenergic receptor stimulation.

Ethacrynic acid has been reported to induce a rapid depletion of glutathione (GSH), subsequent intracellular ROS elevation, and consequent cell injury (Miccadei et al., 1988; Dhanbhoora & Babson, 1992). In fact, deplation of glutathione by treatment with 2-cyclohexene-1-one has been demonstrated to result in inhibition of carbachol-induced amylase release in guinea pig exocrine pancreatic acini (Stenson et al., 1983). In rat pancreatic acinar cells, thiol modulating agents including ethacrynic acid have been reported to reduce the intracellular glutathione levels and inhibition of caerulein-stimulated amylase release (Yu et al., 2002). However, we demonstrated here that ethacrynic acid had no effect on the level of glutathione. Furthermore, the glutathione biosynthesis inhibitor BSO had no effect on IPR-induced amylase release. These observations strongly suggest that the inhibitory effect of ethacrynic acid is not due to depletion of glutathione. Ethacrynic acid had no effect on amylase release induced by cAMP signaling activators and control release and failed to inhibit the effect of IPR in the presence of GSH. Over 90% of cell viability in the cells treated with ethacrynic acid was confirmed by trypan blue extrusion. Therefore, it is also unlikely that cell injury induced by ethacrynic acid causes the inhibition of IPR-induced amylase release.

In the regulatory amylase release, cAMP-dependent signaling pathway is involved. Namely, stimulation of β-adrenergic receptors activates adenylate cyclase via heterotrimeric G-protein, which leads to an increase in intracellular cAMP level. Subsequently, cAMP-dependent protein kinase is activated, which causes exocytotic amylase release (Butcher & Putney, 1980; Quissell et al., 1982; Turner & Sugiya, 2002). However, ethacrynic acid failed to inhibit amylase release induced by the G-protein activator mastparan, the adenylate cyclase activator forskolin, the cyclic nucleotide phosphodiesterase inhibitor IBMX and the cell-permeable cAMP analogue db-cAMP. These results suggest that the cause of the inhibition of IPR-induced amylase release by ethacrynic acid is distinct from the disturbance of cAMP signaling. VIP is another agonist, which induces amylase release via cAMP signaling in rat parotid acinar cells (Scott & Baum, 1985; Inoue et al., 1985). However, ethacrynic acid failed to inhibit VIP-induced amylase release, supporting that EA has no effect on cAMP signaling. Taken together, it is most likely that thiol-modulation of β-adrenergic receptors results in the inhibition of IPR-induced amylase release.

In rat parotid acinar cells, the thiol-oxidizing compound diamide has been demonstrated to reduce the binding affinity of β-adrenergic receptors for ligands and consequently inhibit IPR-induced amylase release (Guo et al., 2010). Diamide had also no effect on mastoparan- or forskolin-induced amylase release and failed to inhibit IPR-induced amylase release in the presence of thiol-reducing reagents, dithiothreitol and GSH, as well as ethacrynic acid described in this paper. Therefore, ethacrynic acid probably leads to thiol-oxidation of β-adrenergic receptors, which results in the reduction of IPR-induced amylase release. Conserved cysteine residues in an extracellular domain of the human β-adrenergic receptor have been suggested to be involved in ligand binding assessed by site-directed mutagenesis (Fraser, 1989). Therefore, it is conceivable that such cysteine residues of β-adrenergic receptor are oxidized by ethacrynic acid. It has been considered that ethacrynic acid is not an oxidant but depletes glutathione by conjugation (Meredith & Reed, 1982). However, currently, independent effects on depletion of intracellular glutathione of ethacrynic acid have been demonstrated (Aizawa et al., 2003; Lu et al., 2009). Therefore, ethacrynic acid appears to have a direct effect as a thiol-oxidating reagent.

Protein thiols are typically maintained in the reduced state. GSH is the most abundant intracellular SH and represents one of the major intracellular defense systems against mediators of oxidative stress (Meister & Tate, 1976). The reducing conditions in cells are primarily maintained by exceedingly large ratio of GSH to GSSG. IPR-induced amylase release inhibited by ethacrynic acid was restored by GSH. Therefore, the antioxidant system by GSH probably plays an important role in maintaining cellular defenses under oxidative stress in rat parotid acinar cells. On the other hand, despite this reducing environment, the formation of mixed disulfides between protein thiols and glutathione has been observed, a process known as S-glutathionylation (Dalle-Donne et al., 2005). S-glutathionylation is considered to occur under physiological conditions and is a reversible cellular response to mild oxidative stress. Involvement of S-glutathionylation in regulating β-adrenergic receptor function under mild oxidative stress in rat parotid acinar cells would be a further study.

5. Conclusion

In this study, we demonstrated that ethacrynic acid, a thiol-modulating reagent, inhibited amylase release induced by β-adrenergic agonist in rat parotid acinar cells. The effect of ethacrynic acid was independent of depletion of glutathione in the cells. Ethacrynic acid failed to inhibit amylase release induced by activators of cAMP signaling pathway, suggesting that the inhibitory effect of ethacrynic acid on amylase release induced by β-adrenergic agonist is caused by the thiol-modulation of β-adrenergic receptors.

6. Acknowledgements

This study was supported in part by a Grant-in-Aid for Scientific Research from the JSPS (#21592375), a Nihon University Multidisciplinary Research Grant for 2011 and a Grant of "Strategic Research Base Development" Program for Private Universities from MEXT, 2010-2014 (S1001024).

7. References

Abdollahi, M.; Dehpour, A.R. & Fooladgar, M. (1997). Alteration of rat submandibulary gland secretion of protein, calcium and N-acetyl-beta-D-glucosaminidase activity by lead. *General Pharmacology*, Vol.29, No.4, pp. 675-680, ISSN 0306-3623

Abdollahi, M.; Fooladian, F.; Emami, B.; Zafari, K. & Bahreini-Moghadam, A. (2003). Protection by sildenafil and theophylline of lead acetate-induced oxidative stress in rat submandibular gland and saliva. *Human & Experimental Toxicology*, Vol.22, No.11, pp. 587-592, ISSN 0960-3271

Aizawa, S.; Ookawa, K.; Kudo, T.; Asano, J.; Hayakari, M. & Tsuchida, S. (2003). Characterization of cell death induced by ethacrynic acid in a human colon cancer cell line DLD-1 and suppression by N-acetyl-L-cysteine. *Cancer Science*, Vol.94, No.10, pp. 886-893, ISSN 1347-9032

Bernfeld P. (1955). Amylase α and β. *Methods in Enzymology*, Vol.1, pp. 149–158, ISSN 0076-6879

Biswas, S.; Chida, A.S. & Rahman, I. (2006). Redox modifications of protein-thiols: emerging roles in cell signaling. *Biochemical Pharmacology*, Vol.71, No.5, pp. 551-564, ISSN 0006-2952

Brown, M.R.; Miller, F.J.Jr.; Li, W.G.; Ellingson, A.N.; Mozena, J.D.; Chatterjee, P.; Engelhardt, J.F.; Zwacka, R.M.; Oberley, L.W.; Fang, X.; Spector, A.A. & Weintraub, N.L. (1999). Overexpression of human catalase inhibits proliferation and promotes apoptosis in vascular smooth muscle cells. *Circulation Research*, Vol.85, No.6, pp. 524-533, ISSN 0009-7330

Butcher, F.R.; Putney, J.W. Jr. (1980). Regulation of parotid gland function by cyclic nucleotides and calcium. *Advances in cyclic nucleotide research*, Vol.13, pp. 215-249, ISSN 0084-5930

Dalle-Donne, I.; Giustarini, D.; Colombo, R.; Milzani, A. & Rossi, R. (2005). S-glutathionylation in human platelets by a thiol-disulfide exchange-independent mechanism. *Free Radical Biology & Medicine*, Vol.38, No.11, pp. 1501-1510 ISSN 0891-5849

de la Cal, C.; Lomniczi, A.; Mohn, C.E.; De Laurentiis, A.; Casal, M.; Chiarenza, A.; Paz, D.; McCann, S.M.; Rettori, V. & Elverdín, J.C. (2006). Decrease in salivary release by radiation mediated by nitric oxide and prostaglandins. *Neuroimmunomodulation*, Vol.13, No.1, pp. 19-27, ISSN 1021-7401

Dhanbhoora, C.M. & Babson, J.R. (1992). Thiol depletion induces lethal cell injury in cultured cardiomyocytes. *Archives of biochemistry and biophysics*, Vol.293, No.1, pp. 130-139, ISSN 0003-9861

Fox, R.I. (2005). Sjögren's syndrome. *Lancet*, Vol.366, No.9482, pp. 321-331, ISSN 0140-6736

Fraser, C.M. (1989). Site-directed mutagenesis of β-adrenergic receptors. Identification of conserved cysteine residues that independently affect ligand binding and receptor activation. *The Journal of Biological Chemistry*, Vol.264, No.16, pp. 9266-9270, ISSN 0021-9258

Goldschmidt-Clermont, P.J. & Moldovan, L. (1999). Stress, superoxide, and signal transduction. *Gene Expression*, Vol.7, No.4-6, pp. 255-260, ISSN 1052-2166

Griffith, O.W. & Meister, A. (1985). Origin and turnover of mitochondrial glutathione. *Proceedings of the National Academy of Sciences of the United States of America*, Vol.82, No.14, pp. 4668-4672, ISSN 0027-8424

Guo, M.Y.; Satoh, K.; Qi, B.; Narita, T.; Katsumata-Kato, O.; Matsuki-Fukushima, M.; Fujita-Yoshigaki, J. & Sugiya, H. (2010). Thiol-oxidation reduces the release of amylase induced by β-adrenergic receptor activation in rat parotid acinar cells. *Biomedical Research*, Vol.31, No.5, pp. 293-299, ISSN 0388-6107

Habig, W.H.; Pabst, M.J. & Jakoby, W.B. (1974). Glutathione S-transferases. The first enzymatic step in mercapturic acid formation. *The Journal of Biological Chemistry*, Vol.249, No.22, pp. 7130-7139, ISSN 0021-9258

Hu, Q.; Corda, S.; Zweier, J.L.; Capogrossi, M.C. & Ziegelstein, R.C. (1998). Hydrogen peroxide induces intracellular calcium oscillations in human aortic endothelial cells. *Circulation*, Vol.97, No.3, pp. 268-275, ISSN 0009-7322

Inoue, Y.; Kaku, K.; Kaneko, T.; Yanaihara, N. & Kanno, T. (1985). Vasoactive intestinal peptide binding to specific receptors on rat parotid acinar cells induces amylase secretion accompanied by intracellular accumulation of cyclic adenosine 3'-5'-monophosphate. *Endocrinology*, Vol.116, No.2, pp. 686-692, ISSN 0013-7227

Jacob, C.; Knight, I. & Winyard, P.G. (2006). Aspects of the biological redox chemistry of cysteine: from simple redox responses to sophisticated signalling pathways. *Biological Chemistry*, Vol.387, No.10-11, pp. 1385-1397, ISSN 1431-6730

Kourie, J.I. (1998). Interaction of reactive oxygen species with ion transport mechanisms. *The American Journal of Physiology*, Vol.275, No.1, pp. C1-C24, ISSN 0002-9513

Krippeit-Drews, P.; Kramer, C.; Welker, S.; Lang, F.; Ammon, H.P. & Drews, G. (1999). Interference of H_2O_2 with stimulus-secretion coupling in mouse pancreatic beta-cells. *The Journal of Physiology*, Vol.514, No.2, pp. 471-481, ISSN 0022-3751

Lowry, O.H.; Rosebrough, N.J.; Farr, A.L. & Randall, R.J. (1951). Protein Measurement with the Folin Phenol Reagent. *The Journal of Biological Chemistry*, Vol.193, No.1, pp. 265-275, ISSN 0021-9258

Lu, D.; Liu, J.X.; Endo, T.; Zhou, H.; Yao, S.; Willert, K.; Schmidt-Wolf, I.G.; Kipps, T.J. & Carson, D.A. (2009). Ethacrynic acid exhibits selective toxicity to chronic lymphocytic leukemia cells by inhibition of the Wnt/β-catenin pathway. *PLoS One*, Vol.4, No.12, e8294, ISSN 1932-6203, Available from http://www.plosone.org/article/info%3Adoi%2F10.1371%2Fjournal.pone.0008294

Matsushita, K.; Morrell, C.N.; Mason, R.J.; Yamakuchi, M.; Khanday, F.A.; Irani, K. & Lowenstein, C.J. (2005). Hydrogen peroxide regulation of endothelial exocytosis by inhibition of N-ethylmaleimide sensitive factor. *The Journal of Cell Biology*, Vol.170, No.1, pp. 73-79, ISSN 0021-9525

Meister, A. & Tate, S.S. (1976). Glutathione and related gamma-glutamyl compounds: biosynthesis and utilization. *Annual Review of Biochemistry*, Vol.45, pp. 559-604, ISSN 0066-4154

Meredith, M.J. & Reed, D.J. (1982). Status of the mitochondrial pool of glutathione in the isolated hepatocyte. *The Journal of Biological Chemistry*, Vol.257, No.7, pp. 3747-3753, ISSN 0021-9258

Meyer, T.N.; Gloy, J.; Hug, M.J.; Greger, R.; Schollmeyer, P. & Pavenstädt, H. (1996). Hydrogen peroxide increases the intracellular calcium activity in rat mesangial cells in primary culture. *Kidney International*, Vol.49, No.2, pp. 388-95, ISSN 0085-2538

Miccadei, S.; Kyle, M.E.; Gilfor, D. & Farber, J.L. (1988). Toxic consequence of the abrupt depletion of glutathione in cultured rat hepatocytes. *Archives of biochemistry and biophysics*, Vol.265, No.2, pp. 311-320, ISSN 0003-9861

Nagler, R.M.; Marmary, Y.; Fox, P.C.; Baum, B.J. & Har-El, R.; Chevion M. (1997). Irradiation-induced damage to the salivary glands: the role of redox-active iron and copper. *Radiation Research*, Vol.147, No.4, pp. 468-476, ISSN 0033-7587

Nagler, R.M.; Reznick, A.Z.; Slavin, S. & Nagler, A. (2000). Partial protection of rat parotid glands from irradiation-induced hyposalivation by manganese superoxide dismutase. *Archives of Oral Biology*, Vol.45, No.9, pp. 741-747, ISSN 0003-9969

Nagler, R.M. (2002). The enigmatic mechanism of irradiation-induced damage to the major salivary glands. *Oral Disease*, Vol.8, No.3, pp. 141-146, ISSN 1354-523X

Pande, M. & Flora, S.J. (2002). Lead induced oxidative damage and its response to combined administration of α-lipoic acid and succimers in rats. *Toxicology*, Vol.177, No.2-3, pp. 187-196, ISSN 0300-483X

Quissell, D.O.; Watson, E. & Dowd, F.J. (1992). Signal transduction mechanisms involved in salivary gland regulated exocytosis. *Critical Reviews in Oral Biology and Medicine*, Vol.3, No.1-2, pp. 83-107, ISSN 1045-4411

Ryo, K.; Yamada, H.; Nakagawa, Y.; Tai, Y.; Obara, K.; Inoue, H.; Mishima, K. & Saito, I. (2006). Possible involvement of oxidative stress in salivary gland of patients with Sjogren's syndrome. *Pathobiology : journal of immunopathology, molecular and cellular biology*, Vol.73, No.5, pp. 252-260, ISSN 1015-2008

Satoh, K.; Matsuki-Fukushima, M.; Qi, B.; Guo, M.Y.; Narita, T.; Fujita-Yoshigaki, J. & Sugiya, H. (2008). Phosphorylation of myristoylated alanine-rich C kinase substrate is involved in the cAMP-dependent amylase release in parotid acinar cells. *American Journal of Physiology. Gastrointestinal and Liver Physiology*, Vol.296, No.6, pp. G1382-G1390, ISSN 0193-1857

Scott, J. & Baum, B.J. (1985). Involvement of cyclic AMP and calcium in exocrine protein secretion induced by vasoactive intestinal polypeptide in rat parotid cells. *Biochimica et Biophysica Acta*, Vol.847, No.2, pp. 255-262, ISSN 0006-3002

Stenson, W.F.; Lobos, E. & Wedner, H.J. (1983). Glutathione depletion inhibits amylase release in guinea pig pancreatic acini. *The American Journal of Physiology*, Vol.244, No.3, pp. G273-G277, ISSN 0002-9513

Takeda, I.; Kizu, Y.; Yoshitaka, O.; Saito, I. & Yamane, G.Y. (2003). Possible role of nitric oxide in radiation-induced salivary gland dysfunction. *Radiation Research*, Vol.159, No.4, pp. 465-470, ISSN 0033-7587

Turner, R.J. & Sugiya, H. (2002). Understanding salivary fluid and protein secretion. *Oral Disease*, Vol.8, No.1, pp. 3-11, ISSN 1354-523X

Vitolo, J.M.; Cotrim, A.P.; Sowers, A.L.; Russo, A.; Wellner, R.B.; Pillemer, S.R.; Mitchell, J.B. & Baum, B.J. (2004). The stable nitroxide tempol facilitates salivary gland protection during head and neck irradiation in a mouse model. *Clinical cancer research : an official journal of the American Association for Cancer Research*, Vol.10, No.5, pp. 1807-1812, ISSN 1078-0432

Whittemore, E.R.; Loo, D.T.; Watt, J.A. & Cotman, C.W. (1995). A detailed analysis of hydrogen peroxide-induced cell death in primary neuronal culture. *Neuroscience*, Vol.67, No.4, pp. 921-932, ISSN 0306-4522

Yamamoto, K.; Masubuchi, Y.; Narimatsu, S.; Kobayashi, S. & Horie, T. (2002). Toxicity of ethacrynic acid in isolated rat hepatocytes. *Toxicology in vitro : an International Journal Published in Association with BIBRA*, Vol.16, No.2, pp. 151-158, ISSN 0887-2333

Yoshimura, K. & Nezu, E. (1991). Dynamic changes in the rate of amylase release induced by various secretagogues examined in isolated rat parotid cells by using column perifusion. *The Japanese Journal of Physiology*, Vol.41, No.3, pp. 443-459, ISSN 0021-521X

Yu, H.; Klonowski-Stumpe, H. & Lüthen, R. (2002). Glutathione might exert an important function in caerulein-stimulated amylase release in isolated rat pancreatic acini. *Pancreas*, Vol.24, No.1, pp. 53-62, ISSN 0885-3177

Section 5

Antioxidants

10

Probiotics and Oxidative Stress

Tiiu Kullisaar, Epp Songisepp and Mihkel Zilmer
University of Tartu, Bio-competence Centre of Healthy Dairy products LCC, Estonia

1. Introduction

A large number of reports about the health benefits of probiotics could be found in the PubMed database. Very little information is available about probiotics possessing physiologically relevant antioxidative properties. Quite scarce is information on the influence of probiotics on human body oxidative stress status and a limited number of clinical trials have been conducted on the effect of antioxidative lactic acid bacteria on human oxidative stress-driven cardiovascular disease-related aspects. In this chapter possibilities of antioxidative probiotics to influence on oxidative stress status in human body are discussed.

2. Short survey of probiotics

The potential life-lengthening properties of lactic acid bacteria (LAB) were proposed by Metchnikoff already at the beginning of the 20th century. The term "probiotic" is an etymological hybrid derived from Greek and Latin meaning "for life" (Hamilton-Miller et al., 2003). Today probiotics are defined as live microorganisms which, when consumed in appropriate amounts, confer a health benefit on the host (FAO/WHO, 2002). Genera most commonly used as probiotics are *Lactobacillus* and *Bifidobacterium*, but other LAB such as lactococci, streptococci, enterococci as well as propionibacteria, bacilli (e.g. *Bacillus subtilis*) and yeasts (e.g. *Saccharomyces boulardii*) are applied. However, probiotics are usually LAB. Introducing a new probiotic into the market involves a step-wise process in order to obtain a functional and safe product (Saarela et al., 2000; Vankerckhoven et al., 2008). Exact requirements are set for probiotic bacteria. Centuries-long use of LAB in the food industry has proven their safety. Nevertheless, it is compulsory to test the safety of each new potential probiotic. The recommendations include an absence of hemolytic activity and the transferable antibiotic resistance of the selected strain, while safety should be proven in various animal models (FAO/WHO, 2002; Vesterlund et al., 2007; Kõll et al., 2010). There is a necessity for pilot clinical trials on healthy volunteers to exclude the adverse effects of probiotic administration on gut health, biochemical and cellular indices of blood reflecting the proper functions of human organs (Reid, 2005; Rijkers et al., 2010). Probiotics must be able to resist stomach acid, bile and the effects of digestive enzymes. Thus, potential probiotic candidates will be selected mostly from human normal microflora. The ability to survive in the GI tract, adhere to intestinal epithelium and maintain its metabolic activity is directly related to the manifestation of probiotic properties in the human body. Probiotic properties are strictly strain-specific. Even the related microbial species may have very different clinical effects. Thus, one cannot arbitrarily attribute the properties of one probiotic strain to another, even within the same species (Vaughan et al., 2005).

Probiotic effects have a dosage threshold. The minimum effective dose, which affects the intestinal environment and provides beneficial effects on human health, is considered to be 10^6-10^9 live microbial cells per day. The minimum dose depends on the particular strain and the type of foodstuffs (Reid, 2005, Williams, 2010, Champagne et al., 2011). Probiotics have been demonstrated to be effective in a variety of conditions including the relaxation of intestinal discomfort (bloating and pain), the alleviation of chronic intestinal inflammatory diseases, the prevention and treatment of pathogen-induced diarrhea, lowering lactose intolerance and food allergies, the lowering of cholesterol levels, the prevention of urogenital infections and the reduction of atopic diseases (Andersson et al. 2001; Chapman et al., 2011). The important area of human physiology that is relevant to functional food science according to the ILSI and FUFOSE (the European Commission Concerted Action on Functional Food Science in Europe) is, among others, the modulation of the defence against high-grade oxidative stress. The latter is one of the principal players in the pathogenesis of CVD and other diseases. Thousands of reports reflecting the abovementioned different health benefits of probiotics could be found in the databases. However, scarce information is available regarding probiotics possessing physiologically relevant antioxidative properties and a limited number of clinical trials on the effect of antioxidative LAB on human CVD-related aspects have been conducted.

3. Short survey of oxidative stress

A net of pro-oxidants and the potency of an antioxidant defence system normally balanced in the body. Principal pro-oxidants are reactive species (including free radicals) divided into reactive oxygen species (ROS) and reactive nitrogen species (RNS) and they mediate the main effects of other pro-oxidative factors (Sies, 1991; Halliwell & Gutterridge, 1999). In the organisms the crucial ROS are superoxide radical, hydroxyl radical, lipid peroxyl radical and non-radical hydrogen peroxide (the latter is produced from superoxide by superoxide dismutase) and the principal RNS are nitric oxide and non-radical peroxynitrite. The pathological efficiency of the hydroxyl radical is the most potent and it is rapidly generated via the Fenton cycle where free iron (a very potent pro-oxidant) reacts with hydrogen peroxide (Halliwell & Gutterridge, 1999). Most of the mentioned reactive species (RS) come from endogenous sources as by-products of normal essential metabolic processes, while exogenous sources involve exposure to cigarette smoke, environmental pollutants, radiation, drugs, bacterial infections, excess of food iron, dysbalanced intestinal microflora, etc. Several diseases are associated with the toxic effect of the transition metals (iron, copper, cadmium). Thus, abnormal formation of the RS can occur *in vivo* and that leads to the damage of lipids, proteins, nucleic acids and carbohydrates of cells and tissues. An excessive production of RS causes an imbalance in the pro-oxidants/antioxidants system. Any imbalance in favour of the pro-oxidants potentially leading to damage was termed 'oxidative stress' (Sies, 1991). Recently an additional adapted conception of oxidative stress (OxS) was advanced as "a disruption of redox signalling and control"(Jones, 2006), emphasizing an impact of the redox ratio as good tools for the quantification of OxS. It is remarkable that the glutathione redox ratio has a crucial impact concerning this conception. A large body of evidence confirms that high-grade OxS is one of the crucial players in the pathogenesis of disorders/diseases (cf below). To maintain the physiological grade of OxS needed for a number's biofunctions like intracellular messaging, growth, cellular differention, phagocytosis, immune response, etc the human body has an integrated antioxidative defence system (IADS, Table. 1). Several antioxidative components for this human IADS are derived from foodstuffs and provided by GI microbiota. Interestingly, it became more and more apparent that the IADS of the host and GI microbiota are tightly linked and some specific strains with physiologically effective antioxidative activity may have a great impact on the management of the OxS level in the gut lumen, inside mucosa cells and even in the host blood, to support the functionality of the IADS of the human body. Thus, experiments to find out strains with physiologically relevant antioxidative properties/effects as well as trials (including special clinical trials) using capsules of such strains or foodstuffs enriched with antioxidative strains are needed. Unfortunately, scientific data on probiotic LAB with physiologically relevant antioxidative properties is very limited and the data of experiments/trials about both intestinal antioxidative protection/influence and systemic antioxidative protection/influence (effects of OxS-related indices) are scarce.

Oxidative stressors (pro-oxidants)	Integrated antioxidative defence system
Ischemia/reperfusion Smoking, Inflammation, xenobiotics PUFA megadoses Iron or copper excess Radiation, Exhaustive exercises Prolonged severe emotional stress	Vitamin E, C, Q, A Enzymes as antioxidants (SOD, GPx, CAT, HO1) Other antioxidants (GSH, plasma albumin, uric acid, Bilirubin, carotenoids, etc)

Table 1. A net effect of oxidative stressors and the potency of the integrated antioxidant defence system (IADS) of the body are normally balanced. An imbalance leads to potentially harmful oxidative stress. PUFA, polyunsaturated fatty acids; SOD, superoxide dismutase; GPx, glutathione peroxidase; CAT, catalase; HO1, haem oxygenase1; GSH, reduced glutathione.

However, as a certain progress has been made during recent years and we will give a summarized overview about probiotics and OxS.

4. Short survey of oxidative stress-related pathological states (CVD, metabolic syndrome, allergy, atopic dermatitis, radiation induced problems in the intestinal tract)

A large body of evidence exists that high-grade OxS has one of the crucial roles in the pathogenesis of disorders/diseases of the vascular system (atherosclerosis, myocardial infarction, stroke, peripheral artery disease), the nervous system (Alzheimer's disease, Parkinson's disease), the liver (cirrhosis, ethanol damage), the skin (dermatoses), the pancreas (diabetes mellitus), metabolic syndrome, obesity, premature ageing, the eyes (age-related macular degeneration, retinopathy), development of some tumors and the GI (inflammatory bowel disease, coeliac disease, etc), etc (Halliwell & Gutterridge, 1999; Stocker & Keaney, 2004; Kals et al., 2006; Stojiljkovic et al., 2007; Krzystek-Korpacka et al., 2008; Tsukahara, 2007; Suzuki et al., 2007; Vincent et al., 2007; Castellani et al., 2008). It has recently reviewed that harmful GI consequences of radiation therapy have OxS-related background (Spyropoulos et al., 2011). Firstly, radiolysis of water molecules causes rapid production of ROS, secondly, an increase in oxygen radical production in the vascular wall has shown already 2h after irradiation with a more intense OxS observed at 6h, this second burst being produced mainly by infiltrating inflammatory cells (Molla & Panes, 2007).

Prolonged excessive ROS/RNS production can trigger chemical chain reactions with all major biomolecules such as DNA, proteins, and membrane lipids. DNA is affected with a variety of lesions like oxidized bases, stand brakes, as well as DNA–DNA and DNA–protein cross-links (Barker et al., 2005). Oxidatively damaged proteins are characterized by formation of carbonyl groups (Stadtman, 1992). Hydroxyl radicals depolymerize hyaluronic acid, degrade collagen, inactivate enzymes and transport proteins via sulfhydryl oxidation. RNS may induce nitration of protein tyrosine residues. Lipid peroxidation is the oxidative degradation of membrane lipids and oxidation that can cause severe impairment of membrane function through changes in membrane permeability and fluidity, its protein oxidation, ultimately leading to cell lysis (Halliwell & Gutterridge, 1999). Lipid peroxidation also damages blood lipoproteins. Therefore, prolonged high-grade OxS causes damages in biomolecules, cells, tissue and organ functionality. Reactive species-damage can be evaluated via markers for oxidized proteins (i.e. nitrated tyrosine, protein carbonyls); oxidized nucleic acid bases (8-oxo-2-deoxyguanosine), oxidized carbohydrates (glycated products) and oxidized lipids (F2-isoprostanes, oxidized low-density lipoproteins (oxLDL), etc). Additional approach for investigations of OxS is an assay of the capacity of IADS (i.e. assay of total antioxidative status or response (TAS, TAR), etc). All these markers are informative but they are not still ultimately accepted as new risk markers yet. However, recently pathogenetic relevance of isoprostanes and oxLDL has been accepted (Statements of European Food Safety Authority). A large number of articles shows that oxLDL level is associated with development of cardiovascular diseases (CVD). Thus, to describe both process and status of atherosclerosis common risk markers like low-density lipoprotein or LDL-cholesterol, HDL-cholesterol, fasting triglycerides (TG), plasma homocysteine as well as by new additional OxS- and inflammation-related indices (oxLDL, 8-isoprostanes, highly sensitive C-reactive protein) should be used. All these markers are also diet-related markers (Mensink et al., 2003). It is reviewed that OxS indices (oxLDL, urine 8-isoprostanes, etc) together with the increased inflammatory markers (white blood cells (WBCs), highly sensitive C-reactive protein have been shown to be characteristic of patients with atherosclerotic lesions of the vascular system (Stocker, Keaney, 2004). Consequently, probiotics with physiologically relevant multivalent antioxidative properties/effects expressed via a positive influence both on a GI and systemic OxS level may have impact concerning the pathogenesis of different disorders/diseases, particularly CVD.

5. Properties of probiotics necessary to have an influence on oxidative stress status

5.1 Role of probiotics in intestinal antioxidative protection (possible action mechanisms)

The most documented effects of LAB in humans are the stimulation of the immune system, the prevention and the reduction of the intensity and duration of diarrhea, and reduction of lactose intolerance (Wolvers et al., 2010). LAB also have some other beneficial effects such as vitamin synthesis, improvement of mineral and nutrient absorption, degradation of antinutritional factors, and/or modulation of GI physiology and reduction of pain perception. Special probiotic strains may induce the expression of receptors on epithelial cells that locally control the transmission of nociceptive information to the GI nervous system (Rousseaux et al., 2007). Beneficial bacteria have enzymatic equipment which enables them to break down a wide variety of food constituents that cannot be metabolized by the host such as galactooligosaccharides, inulin, resistant starches, and antinutritional factors such as tannins or phytates responsible for the chelation of minerals including iron, zinc, magnesium and calcium (Gilman & Cashman, 2006; Songre-Quattara et al., 2008; Cecconi et al., 2009). They can also modify the host gut physiology by increasing the production of growth factors (Alberto et al., 2007). LAB may thus be of benefit to health and help protect against diseases, like CVD, diabetes, metabolic syndrome, etc. As far as OxS is at least one of the components of initiation and/or the development of the mentioned diseases thus any kind of agent which can prevent the development of harmful OxS has a principal impact. Probiotics involve LAB or bifidobacteria of human origin. They can during the consumption period adhere to the epithelial cells of GI modulating the human physiological status via the gut associated immune system and/or directly due to the expression of receptors of GI and/or systematically. LAB beneficial effects are strain-specific. *In vitro* and cellular models, the probiotic properties of lactobacilli have been limited to few parameters such as the ability to survive low (pH 2-3) and bile salts, to produce pathogen inhibitory compounds (including hydrogen peroxide), to compete with energy availability or adhesion sites, and to enhance immune response (Ryan et., 2008; Todorov et al., 2008; Pfeiler & Klaenhammer, 2009). Along with the probiotics themselves, there are metabiotics i.e, the metabolic by-products of probiotics. Metabiotics are beneficial in promoting a healthy GI by creating an environment most favorable to probiotics, by nourishing the enterocytes, reinforcing mucosal barrier function, by maintaining or supporting epithelial integrity or signaling the immune system to limit inflammatory responses both in the gut and through influencing T-cells throughout the body. The principal metabiotics are short-chain fatty acids but also other substances like polyamines (putrescine, spermidine, spermine) have an impact (Larqué et al., 2007). It has been demonstrated, that NO produced by LAB protects mucosa for damages and excessive permeability (Payne et al., 1993; Korhonen et al., 2001).

Since 1993 when Kaizu and co-workers discovered antioxidative activity of LAB, a few of them have had effects in clinical human trials (Kaizu et al., 1993). One of them is antioxidative-antiatherogenic and antimicrobial probiotic *Lactobacillus fermentum* ME-3 (LfME-3). Tartu University has patented this strain in Estonia, Russia, USA and Europe. LfME-3 (DSM 14241) is of human origin (Sepp et al., 1997) and has proven its safety as a probiotic exhibiting both antimicrobial and antioxidative benefits under different *in vitro* and *in vivo* conditions (Kullisaar et al., 2002, 2003; Truusalu et al., 2004; Songisepp et al., 2005; Järvenpää et al., 2007). What makes this strain such a powerful multivalent antioxidant? It is confirmed that *in vitro* the superoxide anion scavenging efficiency of LfME-3 was more than 80-100 times stronger as compared with trolox or ascorbic acid (Ahotupa, personal communication). LfME-3 expresses Mn-superoxide dismutase (MnSOD) activity, can effectively eliminate hydroxyl and peroxyl radicals, and has the complete glutathione system (reduced glutathione, glutathione peroxidase, GPx, glutathione reductase, GRed) necessary for glutathione recycling, transporting and synthesis (Kullisaar et al., 2002, 2010). Mn-SOD is very important in the control of LP. Manganese and Mn-SOD activity of LAB (not possessing catalase) is important for their survival in the oxidative milieu (milk, host) created by the production of hydrogen peroxide (Sanders et al., 1995). It has been shown that some LAB (*L.gasseri*) engineered to produce SOD reduce the inflammation in the case of colitis in interleukin-10-deficient mice (Carroll et al., 2007).

Glutathione (*L*-gamma-Glu-*L*-Cys-Gly or GSH) is a major cellular non-enzymatic antioxidant. It eliminates lipid- and hydroperoxides, hydroxyl radical and peroxynitrite mainly via cooperation with Se-dependent glutathione peroxidase (Zilmer et al., 2005). The GI surface is an important host organism-environment boundary and the interactions of gut microbes inside the intestinal lumen and mucosal cells are important for the host. An impaired environment such as the imbalance of GI microbiota, but also the increase of LP and decrease of the reduced GSH both at the GI surface and in the GI cells, are the mighty modulators causing different unhealthy outcomes in the host. In this process the involvement of the glutathione system is crucial as GSH, besides its role as a crucial antioxidant, is the principal redox controller for a number of processes in cells. Glutathione-related data has impact for LfME-3 regarding at several aspects (cf. 5.2). Thus, confirmation of the presence of all glutathione system components in a specific concrete LAB gives very valuable information as it shows that a specific LAB strain has especially high oxygen and ROS tolerance under different stress conditions. An essential physiological trait for probiotics is tolerance to stress in the GI as well as during the production of functional foods (Ross et al., 2005). Beside that GSH has essential role in maintaining mucosal integrity. Studies have shown diminished GSH levels in inflammatory diseases of intestine and GSH supplementation has beneficial effect (Coskun et al., 2010).

Evidently some probiotics are able to promote an elevation of the level of beneficial bacteria in the GI. In experiments and clinical trials, the administration of the LfME-3 strain has led to the improvement of the GI microbial ecology. More than a 10-fold increase of total lactobacilli counts in comparison with the individually different initial count was registered in the collected faecal samples (Mikelsaar & Zilmer, 2009). The metabolites secreted by LfME-3 into the GI tract could be used as a substrate by other lactobacilli. Adding LfM-3 as a probiotic into a dairy product (yoghurt, cheese, milk) also suppressed the putative contaminants of food (*Salmonella spp., Shigella spp.,*). The secreted substantial amount of hydrogen peroxide and the production of NO by LfME-3 are the main antimicrobial agents (Mikelsaar & Zilmer, 2009. Animal studies have confirmed that the increase in total LAB counts as much as the specific LfME-3 strain antioxidative action in the GI eradicated live salmonellas and prevented the formation of typhoid nodules in experimental *Salmonella Typhimurium* infections, resembling typhoid fever in humans (Truusalu et al., 2004, 2008). It was the first time that the antibiotic therapy of an invasive infection like enteric fever was shown to be more effective if administered together with a probiotic.

5.2 Role of probiotics for systemic antioxidative defence (possible action mechanisms)

Such information is limited. However, some specific multifunctional probiotics may have an influence on systemic (blood) antioxidative defence and the OxS status of host. Thus, to characterize the role of high-grade OxS in the pathogenesis of CVD, we will give an overview about the possible action mechanisms of probiotics on OxS-related indices of CVD.

On the basis of simplified general understandings it can be speculated that there are several factors that may have an impact on OxS. This is only one of the examples. It can be speculated that the suppression of *Helicobacter pylori* infection by some LAB (Wang et al., 2004; Cruchet et al., 2003; Linsalata et al., 2004) may have a certain effect on the host OxS-related indices in blood. However, such approaches are actually only speculations. Why? An analysis of scientific literature allows one to conclude that for a real effect on the systemic OxS-related indices of a host, a specific probiotic strain should have multifunctional bioquality: a) to have positive effects on GI total lactobacilli counts; b) to be able to suppress putative contaminants of food; c) to have biovaluable different antioxidative properties; d) to have a positive effect on OxS-related CVD markers, like TG, oxLDL, etc. In section 5.1. it was explained that the probiotic LfME-3 carries first three types (a,b,c) of properties. Thus, these multifunctional properties of LfME-3 may protect the host against both food-derived infections and help in the prevention of the oxidative damage of food. For example, the antioxidative protection provided by the LfME-3 strain for the prevention of the oxidative spoilage of semi-soft cheeses was found out (Järvenpaa et al., 2007). Thus, points a, b and c have an impact on the role of probiotics for systemic antioxidant defence. However, it is crucial also to have data (according to point d) about the specific influence of probiotics on OxS-related CVD markers. Since LfME-3 has been carefully investigated, concerning the latter we will use gathered information as a model to discuss possible mechanisms on how probiotics may have an influence on the OxS-driven CVD risk markers of a host.

We repeatedly showed that administering a food products to humans comprising strain LfME-3 enhances the systemic antioxidative activity of sera (increases total antioxidative activity, TAA, or total antioxidative status, TAS), enhances the lag phase of LDL (increases oxi-resistance of LDL particles, i.e. suppresses production of atherogenic oxLDL) and decreases level of oxidized glutathione (pro-oxidant), oxLDL and BCD-LDL of sera (Kullisaar et al., 2003, 2011; Mikelsaar et al., 2007). Clinical trials showed that the strain LfME-3 alleviates inflammation and OxS-associated shifts in gut, skin and blood (Kullisaar et al., 2003, 2008; Kaur et al.2008). This realizes via complicated cross-talk between probiotic and host body cells via the integrated influence of several factors of strain LfME-3 like having complete glutathione system, the expression of antioxidative enzymes, the production of CLA and NO by strain LfME-3, etc (Mikelsaar & Zilmer, 2009; Kullisaar et al., 2010, 2011). This strain survives in different fermentation processes of milk due to its good tolerance to low temperature, acid and salt (Songisepp et al., 2004; Songisepp, 2005) and is able for temporal colonization of the GI tract of the consumer. All this is very important as the GI surface is a crucial host organism-environment boundary and the interactions of GI microbes inside the intestinal lumen and mucosal cells have impact for the metabolic activity both microbes and host cells. An impaired environment (the imbalance of GI microbiota, the increase of LP, decrease of the GSH) both at the intestinal surface and in the intestinal cells, are substantial modulators causing unhealthy outcomes in the host. In addition, data that these cellular modulators of the intestinal mucosal status can be repaired by applying of strain LfME-3 was confirmed by using a mouse model of experimental *S. Typhimurium* infection (Truusalu et al., 2004, 2008). Concerning this process the involvement of the glutathione system is crucial as GSH, in addition its role as a crucial cellular antioxidant, is the principal redox controller for a number of cellular processes. Glutathione-related information has impact for LfM-3 regarding next information: a) a recent adapted conceptions of OxS is advanced as "a disruption of redox signalling and control" (Jones, 2006) or "steady-state ROS" (Lushchak,

2011) that emphasize an impact of GSH and its redox ratio for the quantification of OxS and the signalling role of GSH, described previously (Karelson et al., 2002; Zilmer et al., 2005); and b) there exists the possibility for the effective participation of LfM-3 in both enzymatic and non-enzymatic glutathione-driven protection as this strain carries all components needed for functionality of complete glutathione system (Kullisaar et al, 2010). It is interesting to add that recently it has been shown that just *L. fermentum* as a species significantly counteracted the depletion of colonic GSH content induced by some inflammatory processes (Peran et al., 2007) that also supported our understandings. There exists also a correlation between the glutathione redox ratio and DNA oxidative damages (de la Asuncion et al., 1996). Thus, consumption of multivalent probiotic LfME-3, which produces glutathione and has complete glutathione redox cycle enzymes (GPx and GRed), may contribute to the reduction of lipid hydroperoxides in the GI tract and in hepatocytes and prevent them from entering the circulation (Kullisaar et al., 2010). This may lead to an improvement of systemic picture in the host organism.

Data showed that the improvement of the intestinal extra- and intracellular environment yielded beneficial changes of some general (systemic) biochemical indices of the host organism. The administration of LfME-3 to healthy volunteers and atopic adults results in a reduction of LP and a counterbalance of the glutathione system both in blood and in skin. In addition, in several trials LfME-3 has beneficial effect on the blood LDL fraction: the prolongation of its resistance to oxidation, the lowering of the content of oxLDL (a potent inflammatory and atherogenic factor) and BDC-LDL and the enhancement of the TAS of sera (Kullisaar et al., 2003, 2011; Songisepp et al., 2005; Mikelsaar et al., 2008). In trial on elderly persons the lower content of oxLDL was significantly predicted by the higher count of live lactobacilli in the GI tract. Evidently, both the number special antioxidative characteristics of strain LfME-3 and the increase in lactobacilli counts induced by administration of LfME-3 are responsible for such effect on host lipoprotein circulation/metabolism. As we mentioned before, the status of OxS and blood lipoproteins are both related to the pathogenesis of different diseases, including inflammation-related diseases and CVD. Dzau et al (2006) presented in *Circulation* the pathophysiological continuum showing that traditional CVD risk factors all promote OxS and endothelial dysfunction as the first steps in a cascade of pathological events. Elevated OxS leads to the overproduction of oxLDL and the latter has accepted as one of the new systemic markers of the development of CVD (Bonaterra et al., 2010). The higher levels of circulating oxLDL are strongly (much more than LDL-cholesterol) associated with an increased incidence of metabolic syndrome already in people who are currently young and healthy according to a large population-based study (Holvoet et al., 2008). Next, oxLDL is an important determinant of structural changes of the arteries already in asymptomatic persons (Kals et al., 2006; Kampus et al., 2007). An increased production of atherogenic and inflammatory oxLDL within the vessel wall suppresses immunity-related cells, including regulatory T cells (George, 2008) exerting antiatherogenic and antiallergic effects.

The influence of strain LfM-3 on host systemic OxS markers has been showed also via the decline of the values of isoprostanes and 8-OHdG in urine (Kullisaar et al., 2003, 2008; Songisepp et al., 2005). These indices are very informative for systemic OxS burden (Halliwell & Gutteridge, 1999). Evidently the systemic antioxidative effect of strain LfME-3 begins from the alleviation of the OxS- and inflammation-related abnormalities in the GI cells that lead to the assembling of particles of chylomicrons, LDL and HDL with a higher bioquality (with lower levels of harmful oxidation products) and higher concentrations of antioxidant factors/enzymes. The higher bioquality of assembled lipoprotein particles leads to improvement of their metabolism/circulation in the host body. This is one of the explanations why strain LfME-3 exerted the prolonged resistance of the blood lipoprotein fraction to oxidation, lowered the level of oxLDL and enhanced the TAC of sera in both healthy and problematic consumers (Kullisaar et al., 2003, 2006, 2008, 2011; Songisepp et al., 2005). Recently it was showed that administration of strain LfME-3 alleviated the postprandial elevation of TG levels in the blood, and improves HDL bioquality (elevates of paraoxonase level in HDL particles) (Kullisaar et al., 2006; 2008; 2011). The antioxidant activity of HDL can be expressed via several mechanisms (Bruckert & Hansel B, 2009). Paraoxonase (PON), an antioxidant enzyme associated with HDL, hydrolyzes oxidized phospholipids and inhibits the LDL oxidation that is an important step in atherogenesis. In animals, the addition of oxidized lipids to the circulation reduces PON activity, and diets rich in oxidized fat accelerate the development of aterogenesis (Sutherland et al., 1999). Removal and inactivation of lipid peroxides accumulating during LDL oxidation may be the central mechanism accounting for HDL antioxidative properties and when HDL particles have poor bioquality (low antioxidant properties and anti-atherosclerotic potency), they may have even inflammatory effect (Navab et al., 2006). The increase in PON activity after LfME-3 consumption shows that protection of LDL particles against oxidative modification by ROS is improved. PON inhibits atherogenesis by hydrolyzing lipid hydroperoxides and cholesterol ester hydroperoxides, reducing peroxides to the hydroxides, and hydrolyzing homocycteine thiolactone which prevents protein homocycteinylation (Beltowski et al., 2003; Durrington et al., 2005). Therefore, an elevation of PON activity should decrease the level of oxLDL. Antioxidant action of HDL is noted as one of the principal mechanisms mediating its cardioprotective effect (Hansel et al., 2006). It should be noted that HDL-associated antioxidant activity information is also supported both by data of anti-inflammatory effects of strain LfM-3 on the liver (Truusalu, et al., 2008) and by a hepato-protective role for PON against inflammation and liver disease mediated by OxS (Marsillach et al., 2009). Next, it is accepted that postprandial abnormal events are crucial concerning the development of CVD (Lopez-Miranda et al., 2006). Recently a postprandial decrease of three different OxS-related parameters (oxLDL, BCD-LDL, Beta2-GPI-OxLDL) was established (Kullisaar et al., 2011). Thus, the foodstuffs enriched with LfME-3 substantially improves postprandial indices both of lipid/lipoproteins and OxS (Kullisaar et al., 2006; 2008; 2011). The beneficial influence of such enriched food on the postprandial lipid metabolism and OxS is important as many links between OxS and metabolic syndrome occur during the postprandial period. These include an excessive and prolonged elevation of blood TG levels, impairment of the endothelial function, an intestinal overproduction of chylomicrons, a redundant load for insulin production, the elevation of levels of atherogenic oxLDL and possible disturbances in the antioxidative activity of HDL (Bae et al., 2001; Jackson et al., 2007; Perez-Martinez et al., 2009; Hopps et al.,2010). To summarize, a positive modulation of the postprandial situation, including postprandial OxS, is an important target for dietary preventive actions concerning cardiovascular diseases.

6. Possibilities of the oxidative stress-targeted administration of probiotics

6.1 Functional food and capsules

Functional foods are foods or dietary components (incl. probiotics) that may provide a health benefit beyond basic nutrition. Probiotic products may be conventional foods consumed for nutritional purposes, but also for the probiotic effect or "medical foods" - the primary purpose is that food formulation is a delivery vehicle for probiotics or metabiotics (beneficial by-products of probiotics). Probiotics are also available as dietary supplements in capsule, powder or liquid extract form. In functional food products no more than two probiotic strains are used in combination as a rule. Probiotic dietary supplements can consist of one single strain or mixed cultures of two or even more strains. There is some evidence that multi-strain probiotic mixtures could be more effective than single strains, including strains that are components of the mixtures themselves (Chapman et al., 2011).

Many functional foods can be found in a form of synbiotics. Synbiotics have been defined as mixtures of probiotics and prebiotics (dietary fiber) (Schrezenmeir & de Vrese, 2001; Saulner et al., 2007). One of the main benefits of synbiotics is the increased persistence of probiotics in the GI tract. Probiotic dietary supplements (capsules, powders and chewing tablets) often additionally contain amino acids, vitamins and/or prebiotics. Probiotic functional foods could be fermented or non-fermented foods. Traditionally dairy products are the carriers of probiotics. A large variety of probiotic dairy products with particular functional properties are available on the market worldwide. Fermented dairy products, especially yoghurts and yoghurt-like products are most widely used. There is a technological reason for using dairy products as probiotic carriers: dairy products have been optimized for the survival of starter cultures (mostly LAB) and are relatively easily adapted to grant the survival of probiotic strains as well. Besides, dairy products have other advantages over other formulations. Dairy foods are refrigerated. Probiotic bacteria in cultured dairy products benefit, as they remain the most stable in a refrigerated storage condition.

Cheese is used as a probiotic vehicle to a lesser extent than fermented milk products (Songisepp et al., 2004; Ross et al., 2005, Ibrahim et al., 2010). Cheese (especially cheddar) may offer certain advantages over other probiotic products such as yogurt or milk. The cheese is a protective environment for the microbes, as the anaerobic conditions, relatively high fat content and buffering capacity of the cheese matrix helps to protect the probiotic strain in the product. The longer cheese is aged for, the higher density of probiotic microbes and metabiotics it will contain. Although the sensitivity of probiotics to physical and chemical stress, heat and acidity makes the product development challenging for other type of food products, probiotics in addition to dairy have been applied in nontraditional foods such as chocolate, energy bars, juices, smoothies, vegetables, breakfast cereals and even meat products like salami etc (Saarela et al., 2000, Siro et al., 2008).

The physiological state of the bacteria in a functional product is an important factor for the survival of the probiotic strain in the product, but most important is the manifestation of functional/health promoting properties in the human body after ingestion. There is a crucial difference between functional food and dietary supplements concerning the physiological state of the probiotic culture. Microbes are often freeze dried by the process of lyophilization before being manufactured as a dietary supplement (free-flowing powders, capsules, tablets). The dryness keeps the probiotic in a quiescent state during storage, while in food products the bacteria are in a vegetative state with a potentially active metabolism. Besides, dried probiotic cultures may have undergone several stressful processes during their production that damage the cells and may affect their viability (Champagne et al., 2011). Milk as a delivery vehicle has a dual effect on the probiotic additive: the buffering capacity of milk protects the viability of the strain against the stomach's acidic conditions. In addition to the protective effect, which affects the survival of the ingested probiotic, milk contains lactose, minerals, vitamins and bioactive peptides, which enhance the metabolic activity of the ingested probiotic strain in the GI tract.

6.2 Special clinical trial with lactobacillus strains concerning oxidative stress

Probiotics have been advocated for the prevention and treatment of a wide range of diseases, and there is a growing evidence for their efficacy in some clinical scenarios. Probiotics are now widely used in many countries by consumers and in clinical practice. Given the increasingly widespread use of probiotics, a thorough understanding of their benefits is imperative. The properties of different probiotic species vary and can be strain-specific. Therefore, the effects of one probiotic strain should not be generalized to others without confirmation in separate studies. The proposed health benefits of probiotics have undergone increasingly rigorous scientific evaluation in recent years, and there is now strong evidence for their use in treating and preventing some human diseases.

A meta-analysis of randomized controlled trials (RCTs) has shown that many probiotics are effective in preventing antibiotic-associated diarrhoea (McFarland, 2006; Ruszczynski et al., 2008). A separate meta-analysis of RCTs has shown a variety of probiotics to be effective in the treatment of infective diarrhea in both adults and children (Allen et al., 2011) acute waterly diarrhoea (Dutta et al., 2011), *C. difficile* diarrhoea (Plummer et al., 2004), ulcerative colitis and necrotizing enterocolitis (Sari et al., 2011). There is also support from RCTs for the efficacy of a probiotic mix in patients with inflammatory bowel disease (Kajander et al., 2007; Hovyeda et al., 2009). Nevertheless the evidence to date suggests that the major clinical effects of probiotics are seen in prevention GI disorders, probiotic therapy has also been explored in non-GI diseases, including the treatment of atopic eczema in children and adults (Kalliomaki et al., 2001, 2007; Kaur et al., 2008). A specialprobiotic, LfME-3, offers a good potential also in cardiovascular health management. LfME-3 an antioxidative-antiatherogenic and antimicrobial probiotic decreases OxS level in human body. The foodstuffs enriched with this probiotic decrease the level of oxidized LDL, increases the level of HDL, modulates postprandial lipid profile and OxS, and decreases the level of 8-isoprostanes in urine (the markers of systemic OxS) and body overall OxS-load, indicating an atherogenic potential (Kullisaar et al., 2002, 2003, 2011; Songisepp et al., 2005; Mikelsaar, Zilmer, 2009, Table 2).

Intense physical activity increases oxygen consumption and inflammation induced by tissue damage and the probiotic consumption decreased the OxS level (Martarelli et al., 2011). The emerging evidence of a role for GI microbiota on central nervous system functions suggests that the oral intake of probiotics may have beneficial consequences on mood and psychological distress by the competitive exclusion of deleterious GI pathogens, decreases in proinflammatory cytokines and communication with the CNS, leading to changes in neurotransmitter level or function (Logan; Katzman, 2005; Messaoudi et al., 2011). Probiotics are widely used to promote host health. Despite the huge amount of *in vitro* and *in vivo* studies (including cell culture, animal and human studies) we still lack data on the exact mechanisms involved. Our recent results by using MALDI-TOF spectrometry proteomic analysis confirmed that the concentration of glutathione in the blood of the probiotic LfME-3 users increases substantially; that is in good correlation with earlier results. Thus, new proteomic and metabolomic data about LAB and the relation between the colonic microbiota and host status could give new information regarding the mechanism of probiotic beneficial effects, including the effects on the OxS status of a host organism.

It has been demonstrated that functional food products with special *Lactobacillus sp.* strains have the potential to lower blood pressure (Naruszewicz et al., 2002). We demonstrated that semi-hard Edam-type cheese comprising the strain *L. plantarum* TENSIA (DSM 21380, property of Bio-Competence Centre of Healthy Dairy Products LLC) helps to maintain normal systolic and diastolic blood pressure in healthy adults and elderly subjects, thus supporting the functions of the cardiovascular system (Songisepp et al. 2009). Lately we have found that a 3-week consumption of 50g of probiotic cheese comprising *L. plantarum* TENSIA (daily dose 10^{10} of probiotic viable cells per serving) decreased both diastolic (diapason of change: -3.6±7.1 (median -2.3; p=0.01) and systolic (diapason of change: -4.4±8.2 (median -4.0, p=0.01) blood pressure in adult subjects with high normal blood pressure (130-139.5 mmHg).

Marker	oxLDL	HDL-Chol	BCD-LDL	Glutathione redox ratio (GSSG/GSH)	Total anti-oxidative activity	8-iso-prostanes
Number of participants	169	63	106	54	130	63
Decrease* or increase** of level compared to baseline	*16% p<0.03	**7% p<0.003	*17% p<0.02	*33% p<0.03	**20% p<0.005	*26% p<0.03

Table 2. Effects of foodstuffs enriched with probiotic LfME-3 on both oxidative stress-related indices and HDL-cholesterol level of human body. oxLDL, oxidized low-density lipoprotein; BDC-LDL, baseline conjugated diene in LDL; GSSG, oxidized glutathione; GSH, reduced glutathione; HDL-Chol, high-density lipoprotein cholesterol

In the elderly, the consumption of the same amount of probiotic cheese in a somewhat lower daily dose of probiotic viable cells per serving (10^8) decreased both diastolic (diapason of change -4.0 ± 5.2mmHg, median -5mm Hg; p= 0.004) and systolic (diapason of change: -5.9±13.4, median -12 mmHg; p=0.038).

It is repeatedly declared that new approaches in global CVD risk reduction are needed (Elliott, 2008). It is stated that for the prevention of CVD risk the anti-inflammatory agents and antioxidants are considered as a possible "third great wave"' (Bhatt, 2008). Evidently, the prevention complexes of several diseases could become more successful by including probiotics with a special multivalent (including antioxidative properties) biopotency.

7. References

Alberto, MR., Arena, ME., & Manca de Nadra, M.C. (2007). Putrescine production from agmatine by Lactobacillus hilgardii: effect of phenolic compounds. *Food Control,* 18, 898–903.

Allen, SJ., Martinez, EG., Gregorio, GV., & Dans, LF. (2011). Probiotics for treating acute infectious diarrhoea. *Sao Paulo Med J*, 129, 185.

Andersson,H., Asp, NG., Bruce, A., Roos, S., Wadström, T., & Wold, AE. (2001). Health effects of probiotics and prebiotics: a literature review on human studies. *Scand J Nutr*, 45, 58–75.

Bae, JH., Bassenge, E., Kim, KB., Kim, YN., Kim, KS., Lee HJ., Moon, KC., Lee, MS., Park, KY., & Schwemmer, M. (2001). Postprandial hypertriglyceridemia impairs endothelial function by enhanced oxidative stress. *Atherosclerosis*, 155, 517-523.

Barker, S., Weinfeld, M., Zheng, J., Li L., & Murray. D. (2005). Identificationof mammalian proteins cross-linked to DNA by ionizing radiation. *J Biol Chem*, 280, 33826–33838.

Beltowski, J., Wojcicka, G., & Jamroz, A. (2003), Leptin decreases plasma paraoxonase 1 (PON1) activity and induces oxidative stress: the possible novel mechanism for proatherogenic effect of chronic hyperleptinemia. *Atherosclerosis*, 2003, 170, 21-29.

Bhatt, DL. (2008). Anti-inflammatory agents and antioxidants as a possible "third great wave" in cardiovascular secondary prevention. *Am J Cardiol*, 101 (10A),4D-13D.

Bonaterra, GA., Zügel, S., & Kinscherf, R. (2010). Novel systemic cardiovascular disease biomarkers. *Curr Mol Med*, 10, 180-205.

Bruckert, E., & Hansel, B. (2009). HDL-c is a powerful lipid predictor of cardiovascular diseases, *Int J Clin Pract*, 61, 1905-1913.

Carroll, IM., Andrus, JM., Bruno-Barcena, JM., Klaenhammer, TR., Hassan, HM., & Threadgill, DS. (2007). Anti-inflammatory properties of Lactobacillus gasseri expressing manganese superoxide dismutase using the interleukin 10-deficient mouse model of colitis. *Am J Physiol Gastrointest Liver Physiol*, 293, G729 –G738.

Castellani, RJ., Lee, HG., Zhu, X., Perry, G., & Smith, MA. (2008). Alzheimer disease pathology as a host response. *J Neuropathol Exp Neurol*, 67,523-531.

Cecconi, D., Cristofoletti, M., Milli, A., Antonioli, P., Rinalducci, S., Zolla, L., & Zapparoli, G. (2009). Effect of tannic acid on *Lactobacillus plantarum* wine strain during starvation: a proteomic study. Electrophoresis, 30, 957–965.

Champagne, CP., Ross, RP., Saarela, M., Hansen, KF., & Charalampopoulos, D. (2011). Recommendations for the viability assessment of probiotics as concentrated cultures and in food matrices. *Int J Food Microbiol*, 149, 185–193.

Chapman, CM., Gibson, GR., & Rowland, I. (2011). Health benefits of probiotics: are mixtures more effective than single strains? *Eur J Nutr*, 50,1-17.

Coskun, S., Aslim, B., & Yuksekgad, ZN. (2010). Effect of two strains of *Lactobacillus delbruckii* subsp. *bulgaricus* on nitric oxide generation and antioxidant status of rat intestine. *Medicinal Chem Res*, 19,1082-1091.

Cruchet, S., Obregon, MC., Salazar, G., Diaz, E., & Gotteland, M. (2003). Effect of the ingestion of a dietary product containing *Lactobacillus johnsonii La1* on *Helicobacter pylori* colonization in children. *Nutrition*, 19, 716-721.

de la Asuncion, JG., Millan, A., Pla, R., Bruseghini, L., Esteras, A., Pallardo, FV., Sastre, J., & Vina, J. (1996). Mitochondrial glutathione oxidation correlates with age-associated oxidative damage to mitochondrial DNA. *FASEB J*, 10, 333-338.

Durrington, P.N., Mackness, B., & Mackness, M.I. (2005). Paraoxonase and atherosclerosis, Arterioscler Thromb Vasc Biol, 21, 473-480.

Dutta, P., Mitra, U., Dutta, S., Rajendran, K., Saha, TK., & Chatterjee, MK. (2011). Randomised controlled clinical trial of *Lactobacillus sporogenes* (*Bacillus coagulans*), used as probiotic in clinical practice, on acute watery diarrhoea in children. *Trop Med Int Health*,16, 555-561.

Dzau, VJ., Antman, EM., Black, HR., Hayes, DL., Manson, JE., Plutzky, J., Popma, JJ., & Stevenson, W. (2006). The cardiovascular disease continuum validated: clinical evidence of improved patient outcomes: part I: Pathophysiology and clinical trial evidence (risk factors through stable coronary artery disease). *Circulation*, 114, 2850-2870.

Elliott, WJ. (2008). Renovascular hypertension: an update. *J Clin Hypertens*, 10, 522-533.

FAO/WHO. (2002). Guidelines for the evaluation of probiotics in food. Food and Agriculture Organization of the UN and WHO Working Group Report: *http://www.who.-int/foodsafety/fs_management/en/probiotic_guidelines.pdf.*

George, J. (2008). Mechanisms of disease: the evolving role of regulatory T cells in atherosclerosis. *Nat Clin Pract Cardiovasc Med*, 5, 531-540.

Gilman, J., & Cashman, KD. (2006). The effect of probiotic bacteria on transepithelial calcium transport and calcium uptake in human intestinal-like Caco-2 cells. *Curr Issues Intest Microbiol*, 7, 1–5.

Halliwell, B., & Gutteridge, JMC. (1999). *Free radicals in biology and medicine*. New York: Oxford University Press.

Hamilton-Miller, JMT., Gibson, GR., & Bruck W. (2003). Some insights into the derivation and early uses of the word 'probiotic'. *Br J Nutr*, 90, 845.

Hansel, B., Giral, P., Nobecourt, E., Clantepie, S., Bruckert, E., Chapman, M.J., & Kontush, A. (2006). Metabolic syndrome is associated with elevated oxidative stress and dysfunctional dense high-density lipoprotein particles displaying impaired antioxidative activity. *J Clin Endocrin Met*, 89, 4963-4971.

Holvoet, P., Lee, DH., Steffes, M., Gross, M., & Jacobs, DR. (2008). Association between circulating oxidized low-density lipoprotein and incidence of the metabolic syndrome. *JAMA*, 299, 2287-2293.

Hopps, E., Noto, D., Caimi, G., & Averna, M.R. (2010). A novel component of the metabolic syndrome: the oxidative stress, *Nutr Metab Cardiovasc Dis*, 20, 72-77.

Hoveyda, N., Heneghan, C., Mahtani, KR., Perera, R., Roberts, N., & Glasziou, P (2009). A systematic review and meta-analysis: probiotics in the treatment of irritable bowel syndrome. *BMC Gastroenterol*, 9, 1-15.

Ibrahim, F., Ruvio, S., Granlund, L., Salminen, S., Viitanen, M., & Ouwehand, A. (2010). Probiotics and immunosenescence: cheese as a carrier. *FEMS Immun Med Microbiol*,59, 53–59.

Jackson, KG., Armah, CK., & Minihane, AM. (2007). Meal fatty acids and postprandial vascular reactivity, *Biochem Soc Trans*, 35, 451-453.

Jarvenpää, S., Tahvonen, RL, Ouwehand, AC., Sandell, M., Jarvenpää, E., & Salminen, S. (2007). A probiotic, *Lactobacillus fermentum ME-3*, has antioxidative capacity in soft cheese spreads with different fats. *J Dairy Sci*, 90, 3171-3177.

Jones, DP. (2006). Redefining oxidative stress. *Antioxid Redox Signal*, 8, 1865-1879.

Kaizu, M., Sasaki, M., Nakajima, H., & Suzuki, Y. (1993). Effect of antioxidative lactic acid bacteria on rats fed a diet deficient in vitamin E. *J Dairy Sci*, 76, 2493-2499.

Kajander, K., Myllyluoma, E., Rajilić-Stojanović, M., Kyrönpalo, S., Rasmussen, M., Järvenpää, S., Zoetendal, EG, de Vos, WM., Vapaatalo, H., & Korpela R. (2008). Clinical trial: multispecies probiotic supplementation alleviates the symptoms of irritable bowel syndrome and stabilizes intestinal microbiota. *Aliment Pharmacol Ther*, 27, 48-57.

Kalliomaki, M., Salminen, S., Arvilommi, H., Kero, P., Koskinen, P., & Isolauri E. (2001). Probiotics in primary prevention of atopic disease: a randomised placebo-controlled trial. *Lancet*, 357, 1076-1079.

Kalliomaki, M., Salminen, S., Poussa, T., & Isolauri, E. (2007). Probiotics during the first 7 years of life: a cumulative risk reduction of eczema in a randomized, placebo-controlled trial. *J Allergy Clin Immunol*, 119, 1019-1021.

Kals, J., Kampus, P., Kals, M., Zilmer, K., Kullisaar, T., Teesalu, R., Pulges, A., & Zilmer, M. (2006). Impact of oxidative stress on arterial elasticity in patients with atherosclerosis. *Am J Hypertens*, 19, 902-908.

Kampus, P., Kals, J., Ristimäe, T., Muda, P., Ulst, K., Zilmer, K., Salonen, RM., Toumainen, TP., Teesalu, R., & Zilmer, M. (2007). Augmentation index and carotid intima-media thickness are differently related to age, C-reactive protein and oxidized low-density lipoprotein. *J Hypertens*, 25, 819-825.

Karelson, E., Mahlapuu, R., Zilmer, M., Soomets, U., Bogdanovic, N., & Langel, Ü. (2002). Possible signaling by glutathione and its novel analogue through potent stimulation of fontocortical G proteins in normal aging and in Alzheimer's disease. *Ann N Y Acad Sci*, 973, 537-540.

Kaur, S., Kullisaar, T., Mikelsaar, M., Eisen, M., Rehema, A., Vihalemm, T., Zilmer, K., & Zilmer, M. (2008). Successful management of mild atopic dermatitis in adults with probiotics and emollients. *Cent Eur J Med*, 3, 215-220.

Kõll, P., Mändar, R., Smidt, I., Hütt, P., Truusalu, K., Mikelsaar, RH., Shchepetova, J., Krogh-Andersen, K., Marcotte, H., Hammarström, L., & Mikelsaar, M. (2010). Screening and Evaluation of Human Intestinal Lactobacilli for the Development of Novel Gastrointestinal Probiotics. *Curr Microbiol*, 61, 560-556.

Korhonen, K., Reijonen, TM., Remes, K., Malmström, K., Klaukka, T., & Korppi, M. (2001). Reasons for and costs of hospitalization for paediatric asthma: A prospective 1-year follow-up in a population-based setting. *Pediatr Allergy Immunol*, 12, 331-338.

Krzystek-Korpacka, M., Neubauer, K., Berdowska, I., Boehm, D., Zielinski, B., Petryszyn, P., Terlecki, G., Paradowski, L., & Gamian, A. (2008). Enhanced formation of advanced oxidation protein products in IBD. *Inflamm Bowel Dis*, 14, 794-802.

Kullisaar, T., Songisepp, E., Mikelsaar, M., Zilmer, K., Vihalemm, T., & Zilmer, M. (2003). Antioxidative probiotic fermented goats' milk decreases oxidative stress-mediated atherogenicity in human subjects, *Br J Nutr*, 90, 449-456.

Kullisaar, T., Shepetova, J., Zilmer, K., Songisepp, E., Rehema, A., Mikelsaar, M., & Zilmer, M. (2011). An antioxidant probiotic redudes postprandial lipeemia and oxidative stress. *CEJB*, 6, 32-40.

Kullisaar, T., Mikelsaar, M., Kaur, S., Songisepp, E., Zilmer, K., Hütt, P., Lukmann, A., Maaroos, J., Kals, J., & Zilmer, M. (2008). Probiotics, oxidative stress and diseases. Curic´ D, editor. , ed. *Proceedings of the 2008 Joint Central European Congress: 4th Central European Congress on Food and 6th Croatian Congress of Food Technologists, Biotechnologists and Nutritionists*, 15-17 May, Cavtat, Croatia, 1, 367-373, ISBN 978-953-6207-87-9.

Kullisaar, T., Songisepp, E., Aunapuu, M., Kilk, K., Arend, A., Mikelsaar, M., Rehema, A., & Zilmer, M. (2010). Complete glutathione system in probiotic *L. fermentum* ME-3. *Appl Biochem Microbiol,* 46, 527-531

Kullisaar, T., Zilmer, K., Vihalemm, T., & Zilmer, M. (2006). Consumption of antioxidative probiotic probiotic L. fermentum ME-3 and its impact on post-prandial lipid profile in human volunteers. *Third Central European Congress on Food,* May, Sofia, Bulgaria, p. 208.

Kullisaar, T., Zilmer, M., Mikelsaar, M., Vihalemm, T., Annuk, H., Kairane, C., & Kilk, A. (2002). Two antioxidative lactobacilli strains as promising probiotics. *Int J Food Microbiol,* 72, 215-224.

Larqué M., Sabater-Molina, S., & Zamora, E. (2007). Biological significance of dietary polyamines. Nutrition, 23, 87-95.

Linsalata, M., Russo, F., Berloco, P., Caruso, ML., Matteo, GD., Cifone, MG., Simone, CD., Ierardi, E., & Di Leo, A. (2004). The influence of *Lactobacillus brevis* on ornithine decarboxylase activity and polyamine profiles in *Helicobacter pylori*-infected gastric mucosa. *Helicobacter,* 9,/65-172..

Logan, AC., & Katzman, M. (2005). Major depressive-disorder: probiotics may be an adjuvant therapy. *Med Hypotheses,* 64, 533-538.

Lopez-Miranda, J., Perez-Martinez, P., Marin, C., Moreno, JA., Gomez P., & Perez-Jimenez, F. (2006). Postprandial lipoprotein metabolism, genes and risk of cardiovascular disease. *Curr Opin Lipidol,* 17, 132-138.

Lushchak, VI. (2011). Adaptive response to oxidative stress: Bacteria, fungi, plants and animals. *Comp Biochem Physiol C Toxicol Pharmacol,*153,175-190.

Marsillach, J., Camps, J., Ferre´, N., Beltran, R., Rull, A., Mackness, B., Mackness, M., & Joven, J. (2009). Paraoxonase-1 is related to inflammation, fibrosis and PPAR delta in experimental liver disease. *BMC Gastroenterol,* 9, 3-10.

Martarelli, D., Verdenelli, MC., Scuri, S., Cocchioni, M., Silvi, S., Cecchini, C., & Pompei, P. (2011). Effect of a probiotic intake on oxidant and antioxidant parameters in plasma of athletes during intense exercise training. *Curr Microbiol,* 62, 1689-1696.

McFarland, LV. (2006). Meta-analysis of probiotics for the prevention of antibiotic associated diarrhea and the treatment of *Clostridium difficile* disease. *Am J Gastroenterol,* 101,812-822.

Mensink, RP., Aro, A., Den Hond, E., German, JB., Griffin, BA., ten Meer, HU., Mutanen, M., Pannemans, D., & Stahl, W. (2003). PASSCLAIM - Diet-related cardiovascular disease. Eur *J Nutr,* 42, 16-27.

Messaoudi, M., Lalonde, R., Violle, N., Javelot, H., Desor, D., Nejdi, A., Bisson, JF., Rougeot, C., Pichelin, M., Cazaubiel, M., & Cazaubiel, JM. (2011). Assessment of psychotropic-like properties of a probiotic formulation (*Lactobacillus helveticus R0052* and *Bifidobacterium longum R0175*) in rats and human subjects. *Br J Nutr,* 105, 755-764.

Mikelsaar, M., & Zilmer, M. (2009). *Lactobacillus fermentum ME-3* an antimicrobial and antioxidative probiotic. *Microb. Ecol. Health Dis,* 21, 1-27.

Mikelsaar, M., Hütt, P., Kullisaar, T., Zilmer, K., & Zilmer, M. (2008). Double benefit claims for antimicrobial and antioxidative probiotic. *MEHD,* 20, 184-188.

Mikelsaar, M., Zilmer, M., Kullisaar, T., Annuk, H., & Songisepp, E. (2007). Strain of microorganism *Lactobacillus fermentum ME-3* as novel antimicrobial and antioxidative probiotic. *USA Patent* 7, 244, 424 B2. Approved July 17, 2007.

Naruszewicz, M., Johansson, ML., Zapolska-Downar, D., & Bukowska, H. (2002). Effect of *Lactobacillus plantarum* 299v on cardiovascular disease risk factors in smokers. *Am J Clin Nutr*, 76, 1249–1255.

Navab, M., Anantharamaiah, GM., Reddy, ST., Van Lenten, BJ., Ansell, BJ., & Fogelman AM. (2006). Mechanisms of disease: proatherogenic HDL - an evolving field. *Nat Clin Pract Endocrin Met*, 2, 504-511.

Payne, D., & Kubes, P. (1993). Nitric oxide donors reduce the rise in reperfusion-induced intestinal mucosal permeability. *Am J Physiol*, 265, G189–G195.

Peran, L., Sierra, S., Comalada, M., Lara-Villoslada, F., Bailon, E., Nieto, A., Concha, A., Olivares, M., Zarzuelo, A., Xaus, J., & Gálvez, J. (2007). A comparative study of the preventative effects exerted by two probiotics, *Lactobacillus reuteri* and *Lactobacillus fermentum*, in the trinitrobenzenesulfonic acid model of rat colitis. *Br J Nutr*, 97, 96-103.

Perez-Martinez, P., Ordovas, J.M., Garcia-Rios, A., Delgado-Lista, J., Delgado-Casado, N., Cruz-Teno, C., Camargo, A., Yubero-Serrano, EM., Rodriguez, F., Perez-Jimenez, F., & Lopez-Miranda, J. (2011). Consumption of diets with different type of fat influences triacylglycerols-rich lipoprotein particle number and size during the postprandial state. *Nutr Metab Cardiovasc Dis*, 21, 39-45.

Pfeiler, EA., & Klaenhammer, TR. (2009). Role of transporter proteins in bile tolerance of *Lactobacillus acidophilus*. *Appl Environ Microbiol*, 75, 6013–6016.

Plummer, S., Weaver, MA., Harris, JC., Dee, P., & Hunter, J. (2004). Clostridium difficile pilot study: effects of probiotic supplementation on the incidence of *C. difficile* diarrhoea. *Int Microbiol*, 7, 59-62.

Reid, G. (2005). The Importance of Guidelines in the Development and Application of Probiotics. *Curr Pharm Des*, 11, 11-16.

Rijkers, GT., Bengmark, S., Enck, P., Haller, D., Herz, U., Kalliomaki, M., Kudo, S., Lenoir-Wijnkoop, I., Mercenier, A., Myllyluoma, E., Rabot, S., Rafter, J., Szajewska, H., Watzl, B., Wells, J., Wolvers, D., & Antoine, JM. (2010). Guidance for Substantiating the Evidence for Beneficial Effects of Probiotics: Current Status and Recommendations for Future Research. *J Nutr*, 140, 671S-676S.

Ross, R., Fitzgerald, G., Collins, JK., Sullivan, GC., & Stanton, C. (2005). Process of manufacture of probiotic cheese. US Patent 6872411 Accessed 20.07.2011, *http://www.freepatents-online. com/6872411.html*.

Ross, R., Desmond, C., Fitzgerald, G., & Stanton, C. (2005). Overcoming the technological hurdles in the development of probiotic foods. *J Appl Microbiol*, 98, 1410-1417.

Rousseaux, C., Thuru .X., Gelot, A., Barnich, N., Neut, C., Dubuquoy, L., Dubuquoy, C., Merour, E., Geboes, K., Chamaillard, M., Ouwehand, A., Leyer, G., Carcano, D., Colombel, JF., Ardid, D., & Desreumaux, P. (2007). *Lactobacillus acidophilus* modulates intestinal pain and induces opioid and cannabinoid receptors. *Nat Med*, 13, 35-37.

Ruszczyński, M., Radzikowski, A., & Szajewska, H. (2008). Clinical trial: effectiveness of *Lactobacillus rhamnosus* (strains E/N, Oxy and Pen) in the prevention of antibiotic-associated diarrhoea in children. *Aliment Pharmacol Ther*, 28, 154-161.

Ryan, KA., Jayaraman, T., Daly, P., Canchaya, C., Curran, S., Fang, F., Quigley, E.M., & O'Toole, P.W. (2008). Isolation of lactobacilli with probiotic properties from the human stomach. *Lett Appl Microbiol*, 47, 269–274.

Saarela, M., Morgensen, G., Fondén, R., Mättö, J., & Mattila-Sandholm, T. (2000). Probiotic bacteria: safety, functional and technological properties. *J Biotechnol*, 84, 197-215.

Sanders, JW., Leenhouts, KJ., Haandrikman, AJ., Venema, G. & Kok, J. (1995). Stress response in *Lactococcus lactis*: cloning, expression analysis, and mutation of the lactococcal superoxide dismutase gene. *J Bacteriol*, 177, 5254–5260.

Sari, FN., Dizdar, EA., Oguz, S., Erdeve, O., Uras, N., & Dilmen, U. (2011). Oral probiotics: *Lactobacillus sporogenes* for prevention of necrotizing enterocolitis in very low-birth weight infants: a randomized, controlled trial. *Eur J Clin Nutr*, 65, 434-439.

Saulnier, DMA., Hütt, P., Mikelsaar, M., Bosscher, D., Gibson, G., & Kolida, S. (2007). Effects of a symbiotic on biomarkers of oxidative stress and fecal microbiota in healthy adults: results of a cross-over double-blind placebo controlled trial. *Proc Nutr Soc*, 66,101A.

Schrezenmeir, J., & de Vrese, M. (2001). Probiotics, prebiotics, and synbiotics-approaching a definition. *Am J Clin Nutr*, 73,361S–364S

Sepp, E., Julge, K., Vasar, M., Naaber, P., Björksten, B., & Mikelsaar, M. (1997). Intestinal microflora of Estonian and Swedish infants. *Acta Pediatr*, 86, 956-961.

Sies, H. (1991). Oxidative stress: oxidants and antioxidants. London, Academic Press.

Siró, I., Kápolna, E., Kápolna, B., & Lugasi, A. (2008). Functional food. Product development, marketing and consumer acceptance. *Appetite*, 51, 456–467.

Songisepp, E., Kullisaar, T., Hütt, P., Elias, P., Brilene, T., Zilmer, M. & Mikelsaar, M. (2004) A New probiotic cheese with antioxidative and antimicrobial activity. *J Dairy Sci*, 87, 2017-2023.

Songisepp, E. (2005) Evaluation of technological and functional properties of the new probiotic *Lactobacillus fermentum ME-3*. Diss Med Univ Tartuensis.

Songisepp, E., Kals, J., Kullisaar, T., Mändar, R., Hütt, P., Zilmer, M., & Mikelsaar, M. (2005). Evaluation of the functional efficacy of an antioxidative probiotic in healthy volunteers. *Nutr J*, 4, 22.

Songisepp, E., Mikelsaar, M., Rätsep, M., Zilmer, M., Hütt, P., Utt, M., Zilmer, K., Üksti, J., & Kõljalg, S. (2009). Isolated microorganism strain *Lactobacillus plantarum Tensia* DSM 21380 as antimicrobial and antihypertensive probiotic, food product and composition comprising said microorganism and use of said microorganism for preparation of antihypertensive medicine and method for suppressing pathogens and non-starter lactobacilli in food product", International patent application WO2009138091. Published: 19.11.2009.

Songré-Ouattara, LT., Mouquet-Rivier, C., Icard-Vernière, C., Humblot, C., Diawara, B., & Guyot, JP. (2008). Enzyme activities of lactic acid bacteria from a pearl millet fermented gruel (ben-saalga) of functional interest in nutrition. *Int J Food Microbiol*, 128, 395–400.

Spyropoulos, BG., Misiakos, EP., Fotiadis, C., &Stoidis CN. (2011). Antioxidant properties of probiotics and their protective effects in the pathogenesis of radiation-induced enteritis and colitis. *Dig Dis Sci*, 56, 285–294.

Stadtman, ER. (1992). Protein oxidation and aging. *Science*, 257,1220–1224.

Stocker, R., & Keaney, JF. (2004). Role of oxidative modifications in atherosclerosis. *Physiol Rev*, 84,1381_1478.

Stojiljković, V., Todorović, A., Radlović, N., Pejić, S., Mladenović, M., Kasapović, J., & Pajović, SB. (2007). Antioxidant enzymes, glutathione and lipid peroxidation in peripheral blood of children affected by coeliac disease. *Ann Clin Biochem*, 44, 537-543.

Sutherland, WH., Walker, RJ., de Jong, SA.,van Rij, AM., Phillips, V., & Walker H.L. (1999). Reduced postprandial serum paraoxonase activity after a meal rich in used cooking fat. *Arterioscler Thromb Vasc Biol*, 19, 1340-1347.

Suzuki, M., Kamei, M., Itabe, H., Yoneda, K., Bando, H., Kume, N., Tano, Y. (2007). Oxidized phospholipids in the macula increase with age and in eyes with age-related macular degeneration. *Mol Vis*, 13, 772-778.

Todorov, SD., Botes, M., Guigas, C., Schillinger, U., Wiid, I., Wachsman, MB., Holzapfel, WH., & Dicks, LM. (2008). Boza, a natural source of probiotic lactic acid bacteria. *J Appl Microbiol*, 104, 465–477.

Truusalu, K., Mikelsaar, RH., Naaber, P., Karki, T., Kullisaar, T., Rehema, A., Zilmer, M., & Mikelsaar, M. (2008). Eradication of Salmonella Typhimurium infection in a murine model of typhoid fever with the combination of probiotic Lactobacillus fermentum ME-3 and ofloxacin. *BMC Microbiol*, 8,132.

Truusalu, K., Naaber, P., Kullisaar, T., Tamm, H., Mikelsaar, RH., Zilmer, K., Zilmer, M., & Mikelsaar. M. (2004). The influence of antibacterial and antioxidative probiotic lactobacilli on gut mucosa in mouse model of Salmonella infection. *MEHD*,16, 180-187.

Tsukahara, H. (2007). Biomarkers for oxidative stress: clinical application in pediatric medicine. *Curr Med Chem*, 14, 339-351.

Vankerckhoven, V., Huys, G., Vancanneyt M, Vaela C, Klared I, Romonde M-B, Entenzaf J M., Moreillonf P, Windg RD, Knolg, J., Wiertzh, E., & Poti, *B.* (2008). Biosafety assessment of probiotics used for human consumption: recommendations from the EU-PROSAFE project Trends Food Sci Technol. 19:102-114.

Vaughan, EE., Heilig, HG., Ben-Amor, K., & de Vos, WM. (2005). Diversity, vitality and activities of intestinal lactic acid bacteria and bifidobacteria assessed by molecular approaches. *FEMS Microbiol Rev*, 29, 477-490.

Vesterlund, S., Vankerckhoven, V., Saxelin, M., Goossens, H., Salminen, S., & Ouwehand, AC. (2007). Safety assessment of Lactobacillus strains: presence of putative risk factors in faecal, blood and probiotic isolates. *Int J Food Microbiol*, 116, 325–331.

Vincent, HK., Innes, KE., & Vincent, KR. (2007). Oxidative stress and potential interventions to reduce oxidative stress in overweight and obesity. *Diabetes Obes Metab*, 9, 813-839.

Wang, KY., Li, SN., Liu, CS., Perng, DS., Su, YC., Wu, DC., Jan, CM., Lai, CH., Wang, TN., & Wang, WM. (2004). Effects of ingesting Lactobacillus- and Bifidobacterium containing yogurt in subjects with colonized *Helicobacter pylori*. *Am J Clin Nutr*, 80, 737-741.

Williams, NT. (2010). Probiotics. *Am J Health Syst Pharm*, 67, 449-458.

Wolvers, D., Antoine, JM., Myllyluoma, E., Schrezenmeir, J., Szajewska, H., & Rijkers, G.T. (2010). Guidance for substantiating the evidence for beneficial effects of probiotics: prevention and management of infections by probiotics. *J Nutr*, 140, 698S–712S.

Zilmer, M., Soomets, U., Rehema, A., & Langel, Ü. (2005). The glutathione system as an attractice therapeutic target. *Drug Design Reviews-Online,* 2, 121-127.

11

Diabetes, Oxidative Stress and Tea

B. Alipoor[1*], A. Homayouni Rad[2] and E. Vaghef Mehrabany[3]
[1]*Department of Nutrition, Faculty of Health and Nutrition, Tabriz University of Medical Sciences, Tabriz,*
[2]*Department of Food Science and Technology, Faculty of Health and Nutrition, Tabriz University of Medical Sciences, Tabriz,*
[3]*Department of Nutrition, Faculty of Health and Nutrition, Tabriz University of Medical Sciences, Tabriz, Iran*

1. Introduction

Diabetes mellitus is the most common serious metabolic disorder in the world. Diabetes is characterized by a hyperglycemia that results from an absolute or relative insulin deficiency and is associated with long term complications affecting the eyes, kidneys, heart and nerves (Baydas *et al.*, 2003). Oxidative stress is defined as imbalance between the generation of reactive oxygen species and antioxidant defense capacity of the body that is closely associated with aging and a number of diseases including cancer, cardiovascular diseases, diabetes and diabetic complications (Atalay *et al.*, 2002). Irregular cellular metabolism in diabetes leads to production of free oxygen radicals and imbalanced antioxidant capacity (oxidative stress) of the body (Vincent *et al.*, 2004).

Recent studies have shown that both types of diabetes can increase oxidative stress in blood and treatment with antioxidants such as vitamin E and flavonoids may be used for decreasing of oxidative stress and diabetic complications (Baydas *et al.*, 2003; Vincent *et al.*, 2004). There is good evidence that tea flavonoids intake have a role in protection against degenerative diseases and long-term intake of tea flavonoids can prevent obesity in high fat diet. Also it has positive effects against glucose metabolism disorders and diabetes-induced fat disorders that lead to lowering the risk of diabetes complications (Crespy *et al.*, 2004). Flavonoids have antioxidant properties, and tea is one of the main sources of flavonoids. Tea (from the plant Camellia Sinensis) is the most popular beverage next to water, consumed by over two-thirds of the world's population. About three billion kilograms of tea are produced and consumed yearly (Yang, 2000; Gupta *et al.*, 2002; Crespy *et al.*, 2004). Regular intake of tea is associated with an improved antioxidant status in vivo conditions that may contribute to the lowering risk of certain types of cancer, coronary heart disease, atherosclerosis, stroke, reduced mutagenicity and inflammation, protection against neurodegenerative diseases and increasing insulin sensitivity (Luximon-Ramma *et al.*, 2005; Alipoor et al., 2011).

* Corresponding Author

2. Diabetes

Diabetes resembles fasting, especially regarding the responses of the liver, muscle cells, and adipose tissues. With low serum ratios of insulin to glucagon and high levels of fatty acids, the liver produces glucose, whereas other tissues use fatty acids and ketones instead of glucose. Muscle cells and adipose tissue respond by using ketones and fatty acids. Although these resemblances between fasting and diabetes are striking, pathologically low serum insulin levels disrupt the efficiency seen during fasting. With low insulin levels, key glycolytic enzyme activities decrease. Glucose use decreases to levels far below those seen during fasting. Concurrently, hepatic gluconeogenic enzyme activities and gluconeogenic rates increase.Bombarded with free fatty acids, the liver increases gluconeogenesis, secreting large amounts of very low density lipoproteins (VLDLs) and accumulating fatty acids in droplet form. A long-term toxic effect of diabetes is the accumulation of 25% more lipid than normal. In the diabetic state, the liver oxidizes these fatty acids and produces acetone, acetoacetate, and β-hydroxybutyrate. Muscle cells and adipose tissue also show major metabolic changes in diabetes. Muscle glycogen almost disappears, and muscle protein is broken down to support gluconeogenesis. Cardiac and skeletal muscles meet their energy needs from ketones and fatty acids. Fat cells actively release fatty acids under the lipolytic stimuli of glucagon, catecholamines, and insulin deficiency (Shils *et al.*, 2006).

2.1 Historical overview

Diabetes mellitus is a chronic disease that has affected mankind throughout the world. The records of the ancient civilizations of Egypt, India, Japan, Greece, and Rome describe the symptoms of the disease and usually include recommendations for treatment. The wasting away of flesh, copious urination, and the sweet taste of the urine were frequently noted by the ancient medical writers. Aretaeus of Cappadocia, who lived between A.D. 30 and 90, not only named the disease diabetes, which means "to run through or to siphon" but also recommended, "The food is to be milk and with in the cereals, starch, autumn fruits and sweet wines". The term mellitus, which means honey like, was added by a London physician, Willis, in 1675 (Robinson, 1972).

2.2 Epidemiology and etiology

Diabetes mellitus has reached epidemic proportions worldwide. There is an apparent epidemic of diabetes which is strongly related to lifestyle and economic change (WHO). Over the next decade the projected number will exceed 200 million. Most will have type2 diabetes, and all are at risk of the development of complications. Diabetes mellitus is a heterogeneous group of diseases that develops dangerously and characterized by a state of chronic hyperglycemia, resulting from a diversity of etiologies, environmental and genetic. Diabetes mellitus is increasing due to population growth, aging, consequences of industrialization and urbanization, preference of high fat containing fast foods, sedentary life and obesity. Given the enormous public health and economic burden posed by the global epidemic of type 2 diabetes mellitus (T2DM), intervention in the pre-diabetes stage of disease to prevent progression to T2DM and its vascular complications seems the most sensible approach. Prudent lifestyle changes have been shown to significantly reduce the risk of progression in individuals with impaired fasting glucose (IFG) and impaired glucose tolerance (IGT). Although lifestyle modifications are difficult to maintain, there is evidence that intensive intervention results in continued preventive benefit after the stopping of structured counseling (Bharati *et al.*, 2011).

The prevalence of diabetes for all age-groups worldwide was estimated to be 2.8% in 2000 and 4.4% in 2030. The total number of people with diabetes is projected to rise from 171 million in 2000 to 366 million in 2030. Quantifying the prevalence of diabetes and the number of people affected by diabetes, now and in the future, is important to allow rational planning and allocation of resources (Wild *et al.*, 2011).

2.3 Classification and diagnosis

Diabetes is a heterogeneous disorder both genetically and clinically and is hyperglycemia, attributable to either insulin insufficiency or insulin resistance. The traditional classification separates out hyperglycemic conditions into these groups: insulin-dependant diabetes mellitus (IDDM or type 1), non-insulin-dependent diabetes mellitus (NIDDM or type 2), other specific types of diabetes and gestational diabetes mellitus (GDM). Type 1 diabetes accounts for approximately 5% of diabetes and is manifested by insulin deficiency caused by destruction of the pancreatic β cells. Type 2 diabetes accounts for about 90% of diabetes and is characterized by two primary defects: insulin resistance (diminished tissue sensitivity to insulin) and impaired β-cell function (delayed or inadequate insulin release). Other causes account for the remaining 5% of diabetes. Classic symptoms such as polydipsia, polyuria, and rapid weight loss associated with gross and unequivocal elevation of blood glucose (≥200 mg/dl) make the diagnosis of Diabetes mellitus. A fasting plasma glucose level greater than or equal to 126 mg/dl on two occasions is diagnostic of diabetes (Shils *et al.*, 2006).

2.4 Diabetic complication

Associated to insulin-dependent diabetes (type 1), makes the disease one of the worst by considering the human suffering and the socio-economic trouble. In developed countries the number of diabetic patients is increasing all the time and both inability and mortality values are staggering. There is a dedication of studies aiming first to block or slow down the onset of type 1 diabetes, secondly to identify the numerous environmental and genetic factors causing type 2 diabetes and thirdly to suggest possible ways for the prevention or the postponement of crippling complications. The initial problem of diabetes is the hyperglycemia due to the inability of several control systems to maintain a normal glycemic plasma level. A first question is: can diabetic complications be prevented or delayed by normalizing hyperglycemia? This can be achieved at least in part if a meticulous control of glycemia is kept with an appropriate diet, oral anti-diabetic drugs, or insulin administration associated with daily exercise and a correct lifestyle. However, owing to genetic factors and in spite of a serious control, complications are found even in patients with a transitory and slight hyperglycemia. Circulatory abnormalities are the common denominator and they are present under the form of micro- and macro-vascular disease. Throughout the years the following complications may develop with different intensity and localization:

1. Diabetic retinopathy is a leading cause of blindness in about 85% of patients aged 20-75 years.
2. Diabetic nephropathy occurs in 20-40% of patients and when the GFR is <15 ml/min, the end stage renal disease (ESRD) is a leading cause of disability and premature death.
3. Diabetic foot disease normally caused by several factors such as peripheral vascular disease (PVD), altered biomechanics, possibly polyneuropathy and infected foot ulcers.
4. Neuropathy involving both the somatic and autonomic nervous system with neuromuscular dysfunction and muscular wasting is another major cause of morbidity.
5. Accelerated atherosclerosis frequently manifests itself with myocardial infarction, stroke and limb vascular occlusion complicated with necrotic ulcers.
6. Lipodistrophy, seemingly due to ineffective leptin activity or/and fatty acids dysmetabolism, represents another aspect of the metabolic syndrome (Bocci *et al.*, 2011).

Early detection and appropriate management of diabetes is essential to reduce major morbidity and mortality, however these strategies are not implemented in many countries of the world. In the diabetes centre in Isfahan, I.R. Iran, the rate of complications among approximately 4000 type 2 diabetes patients have been recorded as: ischemic heart disease 34%, hypertension 50%, congestive heart failure 12%, retinopathy 44%, cataract 5%, bacteriuria 27%, nephropathy 19%, neuropathy 27%,depression 60%, diabetic foot 2.5%, hypercholesterolemia 37%, and hypertriglyceridemia 37%. Among 296 cases of non-traumatic amputations, 38% were diabetes-related; 27% of stroke cases (cerebrovascular accident), 15% of patients with acute myocardial infarction and 15% of dialysis patients were also diabetics (Azizi *et al.*, 2003).

3. Oxidative stress

Oxidative stress happens in a cellular system when the production of free radical moieties exceeds the antioxidant capacity of that system. If cellular antioxidants do not remove free radicals, radicals attack and damage proteins, lipids and nucleic acids. The oxidized or nitrosylated products of free radical attack have decreased biological activity, leading to loss of energy metabolism, cell signaling, transport, and other major functions. These altered products also are objected for proteosome degradation, further decreasing cellular function. Accumulation of such injury ultimately leads a cell to die through necrotic or apoptotic mechanisms (Vincent *et al.*, 2004).

3.1 Historical overview

The presence of free radicals in biological materials was discovered less than 50 years ago. Thereafter, Denham Harman hypothesized that oxygen radicals may be formed as by-products of enzymatic reactions in vivo. In 1956, he described free radicals as a Pandora's Box of evils that may account for gross cellular damage, mutagenesis, cancer, and last but not least, the degenerative process of biological aging. The science of free radicals in living organisms entered a second time after McCord and Fridovich discovered the enzyme superoxide dismutase (SOD) and, finally convinced most colleagues that free radicals are important in biology. Numerous researchers were now inspired to investigate oxidative damage inflicted by radicals upon DNA, proteins, lipids, and other components of the cell. A third period began with the first reports describing advantageous biological effects of free radicals. Mittal and Murard provided suggestive evidence that the superoxide anion, through its derivative, the hydroxyl radical, stimulates the activation of guanylate cyclase and formation of the "second messenger" cGMP. Similar effects were reported for the superoxide derivative hydrogen peroxide. It was discovered that nitric oxide (NO) has independently role as a regulatory molecule in the control of smooth muscle relaxation and in the inhibition of platelet adhesion (Droge, 2002).

Also it is found that in activated T-cells the superoxide anion or low micromolar concentrations of hydrogen peroxide increase the production of the T-cell growth factor, interleukin-2 which is an immunologically important T-cell protein. Studies have shown that hydrogen peroxide induces the expression of the heme oxygenase (HO-1) gene and hydrogen peroxide has induction effects on various genes in bacteria, as well as activation of the transcription factor nuclear factor κB (NF-κB) in mammalian cells. At the beginning of the 21st century, there is a large amounts of evidence showing that living organisms have not only adapted to an unfriendly coexistence with free radicals but have, in fact, developed mechanisms for the advantageous use of free radicals. Important physiological functions that involve free radicals or their derivatives include the following: regulation of vascular tone, sensing of oxygen tension and regulation of functions that are controlled by oxygen concentration, enhancement of signal transduction from various membrane receptors including the antigen receptor of lymphocytes, and oxidative stress responses that ensure the maintenance of redox homeostasis (Droge, 2002).

The field of redox regulation is also receiving growing attention from clinical colleagues in view of the role that oxidative stress has been found to play in numerous disease conditions. These pathological conditions demonstrate the biological relevance of redox regulation. The delicate balance between the advantageous and detrimental effects of free radicals is clearly an important aspect of life. The science of biological "redox regulation" is a rapidly growing field of research that has impact on diverse disciplines including physiology, cell biology, and clinical medicine (Droge, 2002).

3.2 Biomarkers of oxidative stress

Measuring biomarkers of oxidative stress is an essential step toward better understanding the pathogenesis and developing treatments for diabetic. There are several approaches that may be adopted, including measurements of the depletion of antioxidant reserves, changes in the activities of antioxidant enzymes, free radical production, and presence of protein, lipid, and DNA free radical adducts. For the purposes of clinical assessment, measurements of end products of free radical attack may be the most reliable determination of the occurrence of oxidative stress because enzyme activities and cellular antioxidants are likely to display transient changes. The enzymes responsible for detoxifying free radicals or regenerating antioxidant molecules can provide an indication of the stress level experienced in a cell or tissue. These enzymes are usually measured by *in vitro* activity assays, although changes in transcription can also provide evidence of cell stress. In long-term diabetes, catalase, GSH reductase, GSH peroxidase, and SOD decrease in complication-prone tissue (Vincent *et al.*, 2004).

3.3 Oxidative stress and disease

There is a growing awareness that oxidative stress plays a role in various clinical conditions. Malignant diseases, diabetes, atherosclerosis, chronic inflammation, human immunodeficiency virus (HIV) infection, ischemia reperfusion injury, and sleep apnea are important examples. These diseases fall into two major categories. In the first category, diabetes mellitus and cancer show commonly a pro-oxidative shift in the systemic thiol/disulfide redox state and impaired glucose clearance, suggesting that skeletal muscle mitochondria may be the major site of elevated reactive oxygen species (ROS) production. These conditions may be referred to as "mitochondrial oxidative stress." Without therapeutic intervention these conditions lead to massive skeletal muscle wasting, reminiscent of aging-related wasting. The second category may be referred to as "inflammatory oxidative conditions" because it is typically associated with an excessive stimulation of NAD(P)H oxidase activity by cytokines or other agents. In this case increased ROS levels or changes in intracellular glutathione levels are often associated with pathological changes indicative of a dysregulation of signal cascades and/or gene expression, exemplified by altered expression of cell adhesion molecules (Droge, 2002).

3.4 Oxidative stress and diabetes

Increased oxidative stress is widely accepted as a major role player in both development and progression of diabetes (Maritim *et al*, 2002). Figure 1 summarizes the relationship between oxidative stress, development of diabetes and resulting complications.

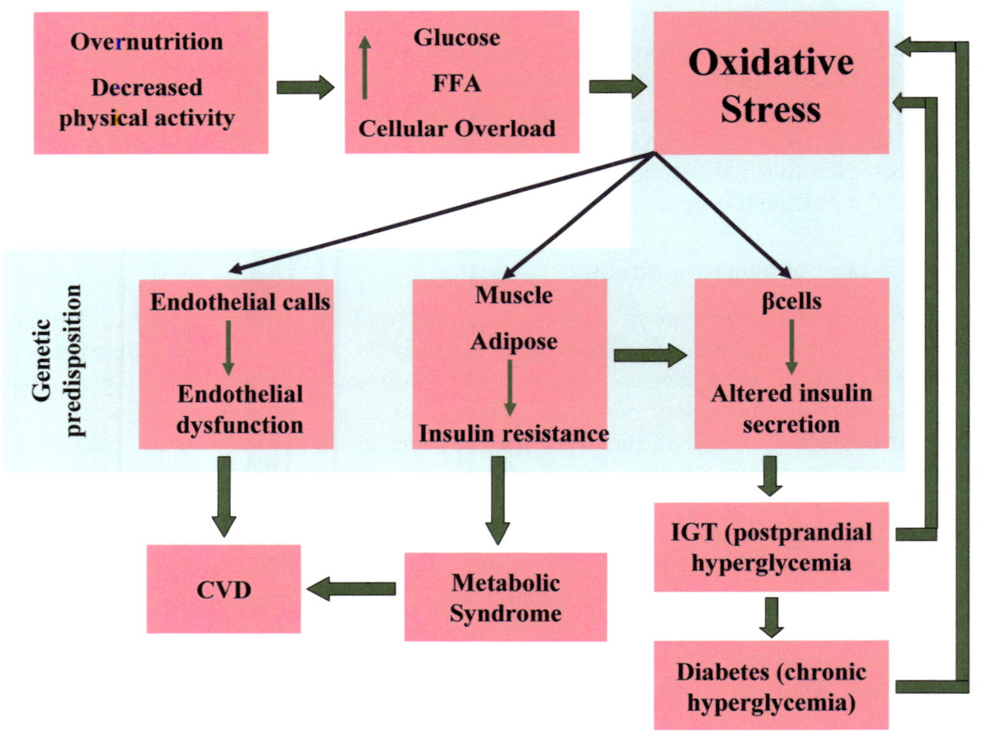

Fig. 1.Overnutrition leads to oxidative stress which in turn results in diabetes and its complications (FFA: Free Fatty Acids, IGT: Impaired Glucose Tolerance, CVD: Cardiovascular Disease)

To better understand the role of tea antioxidants in either preventing diabetes or reducing its complications, one must first know the mechanisms through which oxidative stress contributes to the development of this chronic disease and disorders following it. Some evidence on how tea antioxidants, in particular, can prevent diabetes development and its progression will be presented next.

As mentioned previously, type1 diabetes mellitus which is less prevalent than type 2 is a genetic autoimmune disorder affecting the islet cells leading to insulin deficiency and thus hyperglycemia. Type 2 diabetes however is a multi-factorial disease. Insulin resistance most often precedes the onset of this type by many years and can be caused by acquired factors. Elevations in glucose and free fatty acids have been shown to induce oxidative stress which in turn can play a key role in causing insulin resistance and β-cell dysfunction (Evans *et al.*, 2002).

One most favored hypothesis on how hyperglycemia and elevated free fatty acids (FFA) can lead to oxidative stress is that as energy intake exceeds energy expenditure, generation of excess mitochondrial NADH (mNADH) and the level of reactive oxygen species (ROS) is increased due to the greater activity of citric acid cycle, induced by the abundance of substrates. Reducing ROS formation or increasing its removal is the way through which cells can protect themselves. The mechanism of preventing excessive mNADH generation may be inhibiting insulin-stimulated nutrient uptake and preventing the entrance of pyruvate and fatty acids into the mitochondria. Either of glucose or FFA enters the citric acid cycle after being converted to acetyl-CoA which then combines with oxaloacetate to form citrate. Greater availability of substrates will result in greater production of mNADH which is beyond the capability of oxidative phosphorylation to dissipate it all, leading to increased mitochondrial proton gradient. Thus single electrons are transferred to molecular oxygen forming free radicals, especially superoxide anion.

One way the cells can reduce free radical generation is inhibition of FFA oxidation. This will increase intracellular FFA which in turn leads to reduced GLUT4 translocation to the plasma membrane resulting in resistance to insulin-stimulated glucose uptake in muscle and adipose tissue. *In vitro* studies have shown that antioxidants may have role in reducing insulin resistance (Ceriello *et al.*, 2004).

Chronic exposure to abnormally high levels of glucose and FFA leads to toxic effect on β-cells of pancreas. Furthermore as aforementioned, hyperglycemia and high levels of FFA leads to increased oxidative stress. ß-cells are particularly susceptible to the damages inflicted by oxidative stress; since they are low in free radical quenching enzymes such as catalase, glutathione peroxidase and superoxide dismutase. ß-cells are responsible for the sensing glucose and secreting appropriate amount of insulin in response to glucose boots. This process is pretty complex, but the critical significance of mitochondrial glucose metabolism in linking stimulus to secretion is well established. This is one reason that oxidative stress can blunt insulin secretion due to its ability to damage mitochondria (Evans *et al.*, 2003; Robertson *et al.*, 2004).

Chronic oxidative stress can also affect insulin gene expression. At least two critical proteins that activate the insulin promoter are involved in defects in insulin gene expression. One is PDX-1 and the other is RIPE-3b1 activator recently identified as MafA. Glucose toxicity and lipotoxicity both of which lead to increased oxidative stress have been shown to leave deleterious effects on islet cells (Robertson, 2004).

Apoptosis is one other way through which oxidative stress can cause beta-cell dysfunction. There is some evidence that NF-κB is in part responsible for the induction of apoptosis in β cells; NF-κB production is stimulated by oxidative stress (Evans *et al.*, 2002; Robertson, 2004).

Uncoupling proteins (UCP) are carriers expressed in the mitochondrial inner membrane that uncouple oxygen consumption by the respiratory chain from ATP synthesis and can play a significant role in diabetes. These proteins can control ROS production in the mitochondria. UCP2 and UCP3 are expressed in adipose tissue and skeletal muscles, the tissues important for thermogenesis and substrate oxidation. Elevated expression of UCP2 has been shown to exert negative regulation of β-cell insulin secretion and contribute to the impairment of β-cell function. UCP3 level has been reported to be decreased in diabetic patients and is assumed to facilitate fatty acid oxidation and minimize ROS production (Maiese *et al.*, 2007).

As mentioned previously, hyperglycemia increases peroxide generation in mitochondria which then through many different routes results in endothelial dysfunction and other diabetic complications in the end. Figure 2 illustrates how oxidative stress induced by hyperglycemia leads to the downstream events.

Superoxide overproduction decreases eNOS activity, but increases iNOS expression through NF-κB and protein kinase C (PKC); the final effect is greater NO generation and strong oxidant peroxynitrite which in turn produces in iNOS and eNOS, an uncoupled state resulting in the production of superoxide rather than NO, and damages DNA. DNA damage is necessary for the activation of the nuclear enzyme poly (ADP-ribose) polymerase (PARP). This reduces the intracellular concentration of NAD^+ which it uses as a substrate. The rate of glycolysis, electron transport and ATP formation reduces as a result of decreased NAD^+ and an ADP-ribosylation of the GAPDH (glyceraldehydes 3-phosphate dehydrogenase) occurs. This process results in acute endothelial dysfunction in diabetic blood vessels, which contributes to the development of diabetic complications. NF-κB activation also induces a proinflammatory conditions and overexpression of the adhesion molecules overexpression. All these alterations end in the diabetic complications, and cardiovascular disorders.

NF-κB, one major intracellular target of hyperglycemia and oxidative stress which can be activated by a number of stimuli including hyperglycemia, elevated FFA, ROS, TNF-α, IL-1β, and other proinflammatory cytokines, AGE (advanced glycation end product)-binding to RAGE (receptor for AGE), DNA damage, viral infection and UV irradiation, regulates the expression of a large number of genes, including growth factors (vascular endothelial growth factor (VEGF), proinflammatory cytokines like TNF-α and IL-1β, RAGE, adhesion molecules like vascular cell adhesion molecule-1, and many others).

VEGF has been identified as a primary initiator of proliferative diabetic retinopathy and as a potential mediator of nonproliferative retinopathy. It is also involved in the development of nephropathy and neuropathy. Thus VEGF seems to play an important role in the etiology of several complications of diabetes (Ishii *et al.*, 2001; Evans *et al.*, 2002; Esposito *et al.*, 2002; Ceriello, 2003; Ceriello, 2006; Negrean *et al.*, 2007)

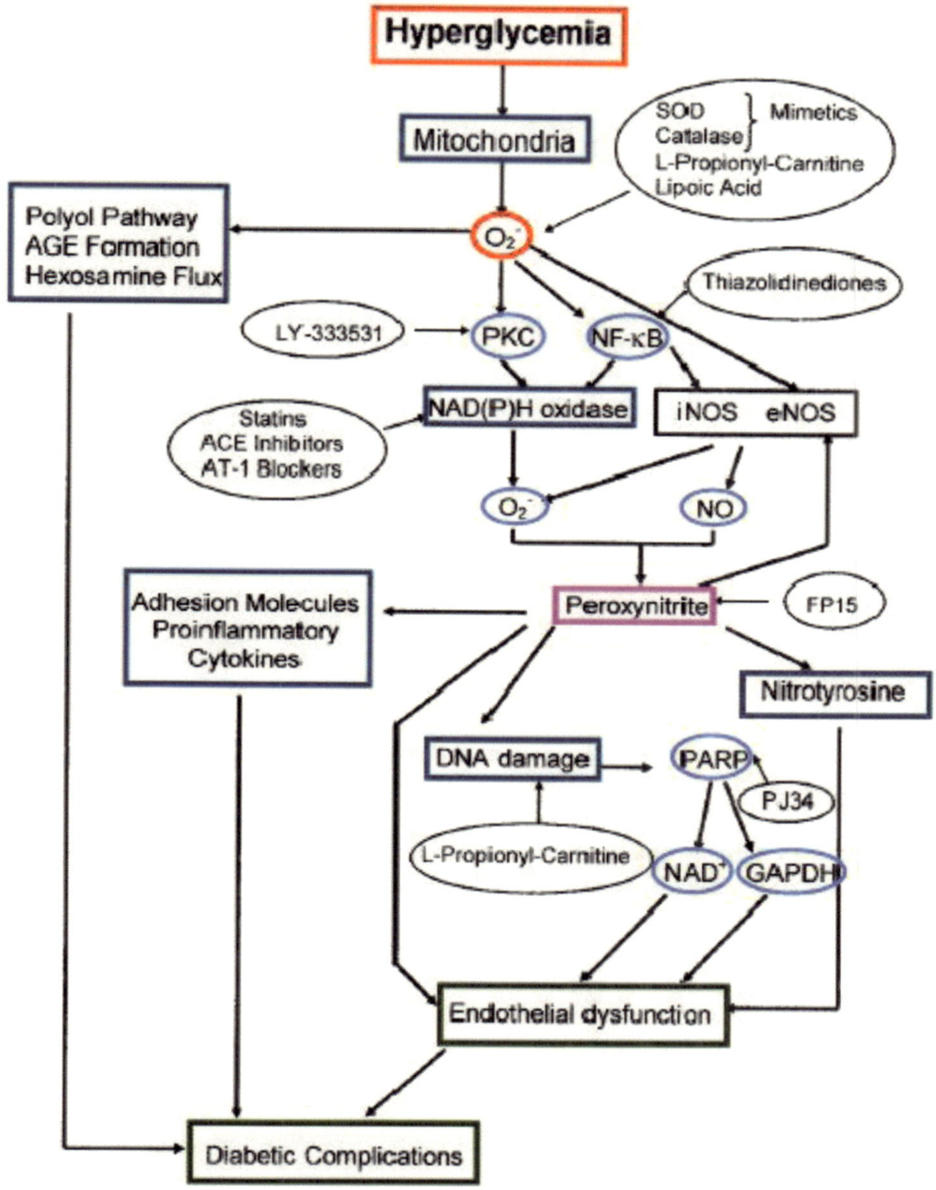

Fig. 2. Hyperglycemia results in endothelial dysfunction and diabetic complications through causing oxidative stress

3.5 Oxidative stress and antioxidants

Several free radical species are normally produced in the body to perform specific functions. O_2^{-}., H_2O_2 and NO are three free radical reactive oxygen species (ROS) that are essential for normal physiology, but are also believed to accelerate the process of aging and to mediate cellular degeneration in disease states. These agents together produce highly active singlet oxygen, hydroxyl radicals, and peroxynitrite that can attack proteins, lipids, and DNA. Antioxidants are defined as any compound that can donate at least one hydrogen atom to a free radical, resulting in the termination of radical chain reactions. An alternative type of antioxidant is defined by its ability to prevent the initiation of a free radical chain reaction rather than to terminate them. This latter type of antioxidant is usually dependent upon the ability to bind metal ions and includes ceruloplasmin, transferrin and albumin. Cells must maintain the levels of antioxidants, often defined as antioxidant potential, through dietary uptake or *de novo* synthesis. Excess production of free radicals can reduce the intracellular antioxidants, resulting in oxidative stress. In brief, acute hyperglycemic episodes such as an oral glucose tolerance test or a meal can decrease the antioxidant capacity of plasma in both normal and diabetic subjects and increase oxidative stress in diabetic patients. As a type 2 diabetic patient ages, increased basal levels of free radical production and decreased antioxidants are even further intensified by elevated plasma glucose. Analysis of individual vitamin and enzyme components of the antioxidant system in man reveals significant changes in diabetes. The levels of vitamins A and E and catalase activity are decreased in both type 1 and 2 patients compared with controls. Whereas GSH-metabolizing enzymes are decreased in type 1 but not type 2 patients, SOD activity is lower in type 2 but not type 1 (Vincent *et al.*, 2004).

4. Tea

The scientific name given to tea, in the first volume of the book "Species Plantarum" by Carl Linnaeus, was "Thea Sinensis"; but in the second volume of the very book, the tea tree is addressed as "Camelia". Later in 1762, Linnaeus assuming black and green tea to be obtained from two different shrubs, chose the names "Thea bohea" and "Thea vividis" for black and green tea respectively. Now it is revealed that it is "Thea bohea" from which, both black and green tea are attained. Also the scientists have merged the two genuses "Camelia" and "Thea". Today the international scientific expression for tea is "*Camelia Sinensis (L) O.kuntze*", Camelia and Sinensis indicating the genus and the variety respectively, (L) regarding Linneaus, the first botanist to give tea a scientific name and O.kuntze being the one who combined the names used for black and green tea. Camelia Sinensis is an evergreen plant which can grow into a tree of up to 30 meters if left undisturbed; but cultivated plants usually have a height around 50-70 centimeters (Hara, 2001; Moxham, 2009).

4.1 Historical background

It may always remain in mist, when tea first stepped into man's life. General consensus attributes the birth of the tea bush to the area we now call eastern China. But the discovery of a tea bush deep in Assam, India with leaves much larger than the Chinese one, caused controversy, as far as it concerns the birthplace of Camelia Sinensis. Today it is assumed that the tea bush was first found in the southwestern China, centered in the Yunnan district (Hara, 2001). Tea was first carried westwards during 5th century by Turkish traders (Alkan *et al.*, 2009).

The question as to when the man first consumed tea is unanswered as well. According to Chinese mythology, it was the emperor Shen Nung who discovered tea for the first time in 2737 B.C.; but this is not in consistence with the first credible documentary reference on tea which was made in 59 B.C. (Hara, 2001; Gupta *et al.*, 2002).

It is probable that our forbears used tea in response to their instinctive seek for a material to calm them; because tea is rich in an alkaloid called caffeine which acts as an opioid in the nervous system, relaxing the consumer.

On tropical and subtropical climates and regions on which precipitation is coordinate according to months and where summers and winters are lukewarm, tea production is realized. Sour and humid land structure is crucial to growing tea as well (Alkan *et al.*, 2009). Based on the data generated by the food and agriculture organization (FAO) of the united nations as of January 2010, China was the leading country in tea production in 2006, 2007 and 2008, followed by India, Kenya, Sri Lanka, Turkey, Vietnam and Indonesia. Other main tea producing countries are Japan, Argentina, Iran, Bangladesh, Malawi and Uganda Figure (3). The global tea production growth rate in 2006 extended more than 3% to reach an estimated 3.6 million tons, China, Viet Nam and India being the main counties to have contributed to this rise. It is predicted that world black tea production rate decreases in the current century, due to slowing down of production growth in Africa. India followed by Kenya and Sri Lanka are projected to be the main contributors to black tea production by 2017 which is estimated to reach 3.1 million tons (http://en.wikipedia.org/wiki/Tea; Hicks, 2009).

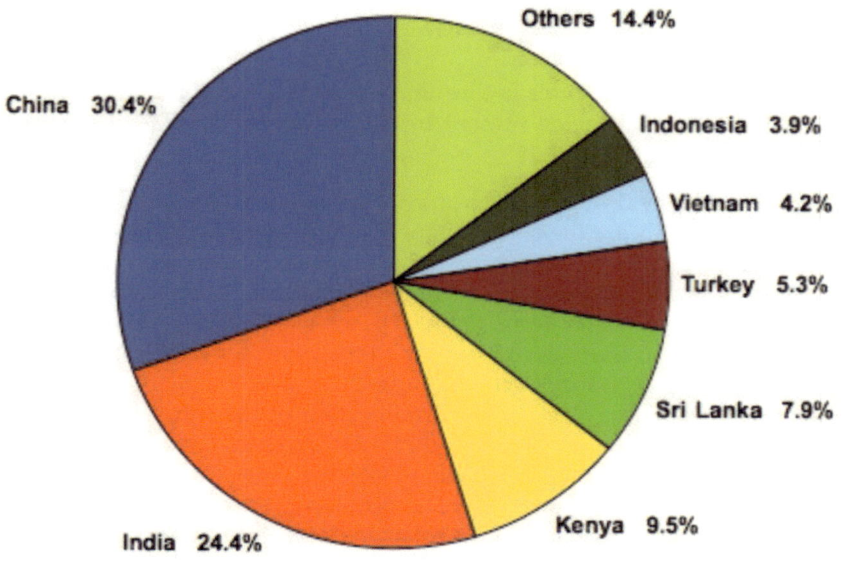

Fig. 3. The tea producing regions in 2007

4.2 Chemical compounds in tea

In contrast to the history of tea drinking which is ancient, the chemical components of tea have quite recently been investigated. Teas acquired from different regions may have different chemical components in different amounts. The agents found in tea are classified as primary or subordinate (Table 1):

Primary components	subordinate components
Alkaloids (xanthines)	Mineral acids
Polyphenols	Organic acids
Vitamins	Proteins
Enzymes	Pectin
Volatile oils (essence)	Lignans
	Amino acids

Table 1. The classification of primary and subordinate tea agents

The quality of a tea is related to its content of alkaloids (caffeine), flavonoids (catechins), phenolic acids (gallic acid, coumaric acid, caffeic acid and chlorogenic acid) and volatile oils (essences) (Table 1) (Bendini *et al.*, 1998; Wang *et al.*, 2000).

Alkaloids: In 1827, caffeine which is present in a few other plants was discovered in tea. By then it was given the name "Theine" which was dropped as its structure was proven to be exactly the same as that of caffeine, in 1820. The mean content of caffeine in tea ranges between 1.9 and 4.5 and is negatively correlated with the age of the leaves (Wanger *et al.*, 1996; Hara, 2001).

Polyphenols: Theanine and flavonoids (catechins in particular) are the main polyphenols found in tea constituting 30% of its agents.

Theanine: A unique substance in tea is theanine which is a kind of amino acid comprising more than half of the amino acids present in tea. It has an "umami" or sweet taste and constitutes 2% of tea (Hara, 2001).

Flavonoids: Flavanols and their derivatives including flavan-3-ols (catechins and epicatechins) and flavonols are the chief flavonoids in tea. Under mild oxidation, flavan 3-4 diol derivatives of flavonoids are converted to catechins and its isomers. Green tea is a great source of catechins and thus exerts antioxidant properties. These catechins change into oligomeric quinones under the fermentation process of black tea which reduces its antioxidant capacity by 2-6 times in comparison to the green tea (Hara, 2001). Each gram of green tea contains 123.8- 206.3 milligrams of catechins which is 10-30 percent of the dry weight of the green leaves. In black tea, 79.3 milligrams of catechins is found in one gram (Wang *et al.*, 2000; Bronner *et al.*, 1998; Keys, 1976). All catechins have 2 asymmetric carbons, thus there are four isomers of them: catechin (C)(+), catechin gallate (CG)(-), gallocatechin (GC) and gallocatechin gallate (GCG) (-). The number of hydroxyl group on the B ring differs for the derivatives of catechins. Like catechins, epicatechins are the monomers of the condensed thanines, are derived from flavan 3-4 diols and have two asymmetric carbons in their structure resulting in four isomers. These isomers include: epicatechin (EC)(-), epigallocatechin (EGC)(-), Epicatechin gallate (ECG)(-) and epigallocatechin gallate (EGCG)(-). Catechins and epicatechins are the major polyphenols found particularly in green tea (Figure 4).

Epigallocatechin gallate (EGCG), being the greatest in amount in tea compared to the other catechins, makes up to 50% of its catechins. EGCG is more abundant in green tea and its quantity is negatively correlated with the age of the leaves (Hara, 2001; Leung *et al.*, 2001).

Vitamins and minerals: Vitamin C was discovered in 1924 in fresh tea leaves. Tea is a great source of fluoride too (Hara, 2001). Other vitamins and minerals may be present in tea at little amounts as well.

Enzymes: The enzymes in tea which catalyze the oxidation processes are called Thease. During fermentation in which tea pectins are demetylated, polyphenolic compounds are decomposed which as a result of the quinone appearance, turn into some colorful agents including theaflavin and thearubigin, both of which are plentiful in black tea (Figure 4) (Hara, 2001; Leung *et al.*, 2001; Cadenas, 2002).

Volatile oils: More than 600 volatile agents have been established in tea, most of which have a yellow color and a characteristic scent. Linalool is the main essence in tea, other of lesser importance ones being dihydroactinide iolido paravinile phenol, hexenol, hexenal, aldehydes, phenyl ethyl alcohols, phenols and geraniols (Hara, 2001).

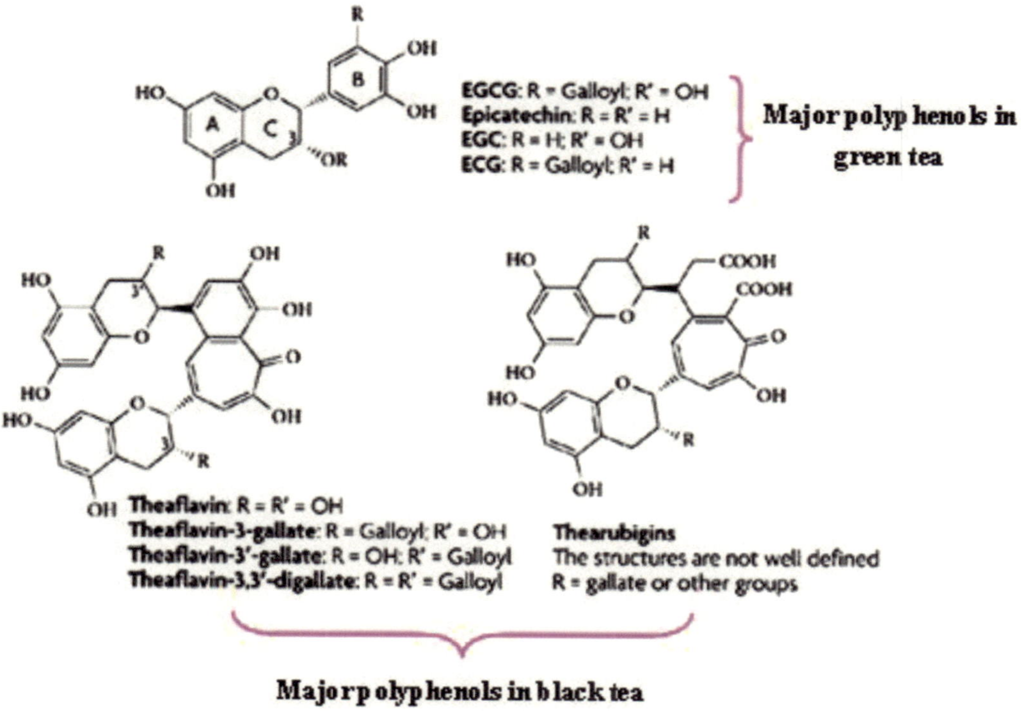

Fig. 4. Major polyphenols in green and black tea

Based upon the preparation method, the degree to which it is fermented and the steps it goes under during the production, different types of tea consumed all over the world are classified into at least six categories (Figure 5). The less processed the tea, the greater the polyphenols content will be, which the extent of oxidation accounts for (Santana-Rios *et al.*, 2001).

1. White tea: White tea is manufactured only from the buds or first leaves of *C.sinensis*. It is the least processed type of tea and is simply steamed and dried without a prior withering stage; therefore the concentrations of EGCG and also methylxanthines (like caffeine) are enriched in white tea compared with green and black tea.
2. Yellow tea: It usually implies a special tea processed in a similar way to green tea; but the drying process takes place at a slower rate. The damp tea leaves are allowed to sit and yellow. Its taste resembles that of green and white teas.
3. Green tea: To manufacture green tea, first the fresh leaves are steamed, then primary drying-rolling, rolling, secondary drying-rolling, final drying-rolling and at last drying are performed. No fermentation takes place in this type of tea.
4. Oolong tea: Fresh leaves undergo solar withering at the first step, indoor withering and rolling, pan firing, rolling, mass breaking and drying are the steps to be taken, to produce oolong tea. In this kind of tea, partial fermentation occurs after the rolling.
5. Black tea: The manufacturing process for black tea includes withering of fresh leaves, rolling, fermenting and drying. Thorough fermentation is done in black tea.
6. Pu-erh: Pu-erh is applied to old tea with extreme fermentation in it (Hara, 2001; Santana-Rios *et al.*, 2001; Kuo *et al.*, 2005; Lin *et al.*, 2006; Sohle *et al.* 2009; Wang *et al.*, 2008; http://en.wikipedia.org/wiki/Tea; http://www.tea-of-chinese.com).

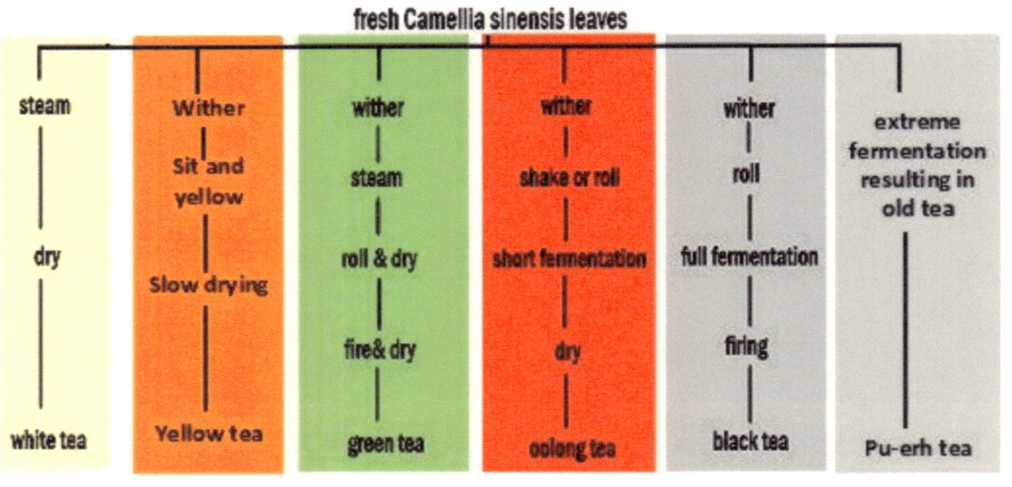

Fig. 5. The production of different types of tea

4.3 Tea and diseases

Tea being a great source of phytoestrogens and fluoride, both of which play a major role in bone health, is reported to prevent osteoporosis and the lower prevalence of the very disease in Japanese postmenopausal women in comparison with American and European ones, is attributed to greater amounts of tea consumed by Japanese (Adlercreutz *et al.*, 1991; Adlercreutz *et al.*, 1992; Johnell *et al.*, 1995; Kanis *et al.*, 1999).

Tea due to its content of polyphenols has been found to be effective in preventing many types of cancer including liver, small intestine and lung. Polyphenols increase the catalytic activity of enzymes involved in glutathione and quinone synthesis and remove the free radicals of hydrogen peroxide and superoxide anions. Tea consumption also inhibits metastasis of human lymphoid leukemia cells through stimulation of apoptosis and hindrance of platelet aggregation (Bronner *et al.*, 1998; Dulluge *et al.*, 1998; Integrative Medicine, 2000; Springhouse, 2001).

Studies have shown that tea is beneficial in delaying cardiovascular disorders. Some mechanisms described are: inhibiting the progression of atherosclerosis and thrombosis, preventing hypertension by either exerting effects similar to those of beta-blockers or stimulating diuresis, decreasing postprandial blood cholesterol and triglycerides, inhibition of LDL oxidation and improvement of endothelial function. Also, decreasing the activity of lipoxygenase enzymes and stimulating central nervous system, tea can improve heart muscle function, circulation in coronary vessels and respiration (Yamamoto, 1997; Robbers *et al.*, 1999; Leung *et al.*, 2001; Alipoor et al., 2008). Hypertension is another disorder which can be corrected by tea and its polyphenols. This has been attributed to its role in regulating renin-angiotensin system (RAS) and improving endothelial function (Adlercreutz, 1991).

Since tea and its polyphenols have been observed to reduce digestion and absorption of fats and carbohydrates, and due to their role in controlling food intake, increasing energy expenditure, modifying the activity of liver, muscle, gastrointestinal tract and fat cells, weight loss and prevention of diabetes mellitus could be one advantage of drinking appropriate amounts of tea (Watanabe *et al.*, 1998; Kuo *et al.*, 2005; Ynng *et al.*, 2006). How tea can play a major role in prevention and treatment of many complications of diabetes mellitus will be presented more precisely in the next section.

Other disorders which tea can play a role in prevention or treatment of, includes inflammation, migraine, nausea, diarrhea, maldigestion, sore throat, depression, prostatitis, hemochromatosis, neurodegenerative diseases like Parkinson and Alzheimer, cataract, dental carries and some viral and bacterial infections including influenza, polio, herpes simplex and AIDS (Duke, 1985; Robertson *et al.*, 1991; Hertog *et al.*, 1993; Cummings *et al.*, 1995; Tavani *et al.*, 1996; Van Het Hof *et al.*, 1997; Integrative Medicine, 2000; Mills *et al.*, 2000; McKay *et al.*, 2002; Wright, 2005; Kao *et al.*, 2006; Sasso *et al.*, 2006; Alipoor et al., 2011).

4.4 Tea antioxidants and oxidative stress

There is a considerable amount of evidence indicating the benefits of tea consumption to prevent diabetes and reducing its resulting complications. The less processed the tea, the more its antioxidant content; which may explain why most studies have been conducted using green tea as the supplement. Recently, white tea which is not fermented either, has been studied for its impact on diabetes too. Some ways through which tea and its bioactive compounds affect diabetes are not related to the antioxidant properties of tea, thus not within the scope of this chapter; and won't be discussed here.

Many studies have shown that different types of tea are potentially effective in reducing oxidative stress and related diseases. Attempts have been made to manufacture products containing tea bioactive compounds for prevention and treatment of mentioned diseases. In order to design such product, the effective compounds of tea and the safe dose of them must be first identified. For instance, EGCG has been shown to act as a prooxidant when administered in high doses and lead to apoptosis. Furthermore compounds other than catechins may exert the desired effects as well (Liao *et al.*, 2001; Mandel *et al.*, 2004; Kao *et al.*, 2006). To determine the very compounds acting as antioxidants in black tea, Alipoor et al., (2009) performed a study in which diabetic rats were supplemented total extract of black tea and its fractions. Total extract and fractions were attained by hydromethanol method and solid phase extraction using Sep-Pak respectively. Results of this study showed that injection of total extract and 20% fraction of black tea decreased malondialdehyde (MDA) and increased total antioxidant, Super oxide Dismotase (SOD), Glutathione Peroxides (GPX) and Glutathione in diabetic rats. To find out the major substances in the 20% fraction, Analytical HPLC, Preparative HPLC (High Performance Liquid Chromatography) and NMR (Nuclear Magnetic Resonance) (CNMR and HNMR) were employed. Caffeine, Epicatechin Gallate, Quercetin and Kampferol were the main compounds capable of combating oxiadative stress, to be determined in 20% fraction of tea (Figure 6) (Alipoor *et al.*, 2010).

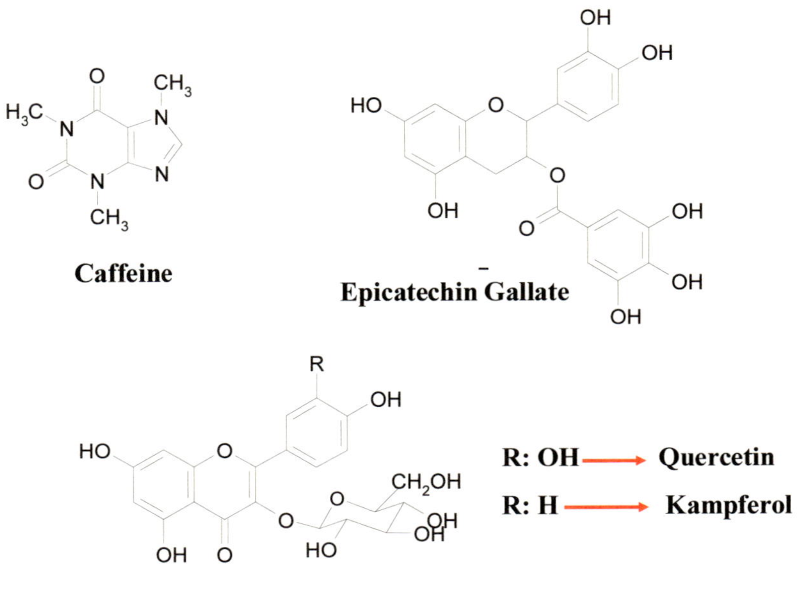

Fig. 6. Major antioxidants in 20% fraction of black tea

Caffeine is a strong antioxidant and its activity being equal to that of glutathione and exceeding that of vitamin C (Devasagayam *et al.*, 1996; Kamat *et al.*, 2000; Nikolic *et al.*, 2003). The free radical scavenging capacity of flavonoids is due to the 3', 4' dihydroxyl and 3' hydroxy in the β ring (Amic *et al.*, 2003). The 20% fraction of black tea has been shown to be more effective than the other fractions which may be explained by the high concentration of the aforementioned compounds in it and absence of polyphenol antagonists in the very extract prepared (Alipoor *et al.*, 2009).

Tea polyphenols have been found to induce expression of phase II enzymes and endogenous antioxidants that defend cells from oxidative stress. The promoter regions of the phase II genes contain specific DNA sequences, termed the antioxidant response elements (AREs) or the electrophile response elements (EREs) that are required for induction by chemopreventive compounds, oxidative stress or electrophiles. In an attempt to find the transcription factors that bind to ARE, NF-E2-related factor 2 (Nrf2) was identified (Zhang, 2006). Nrf2 binds to Kelch-like ECH-associated protein 1 (Keap1) under nonstressed conditions. Keap1 in complex with cullin3, Roc1 and E2 proteins provides ubiquitination followed by proteasomal degradation. When oxidative stress occurs, oxidation of Keap1 leads to inability to bind Nrf2 protein by forming intramolecular disulfide bonds. Then Nrf2 migrates into the nucleus and binds a protein of Maf family (like sMaf) and CBP/p. This complex is formed on ARE promoter region of certain genes leading to transcription activation. Phosphorylation of by protein kinases which may be activated by oxidants is one way to provide Nrf2 migration in nucleus (Lushchak, 2011).

4.4.1 Evidence from animal studies

Tea polyphenols have been demonstrated to improve lipid profile in diabetic and nondiabetic models. In this section, we follow the antioxidant properties of tea. Improved glucose tolerance and increased plasma insulin concentrations by tea, have been found in some studies which may be, in part, explained by the effect of tea antioxidants on insulin resistance and β-cell function. EGCG has also been shown to suppress cytokine-induced β-cell damage; this may also contribute to glucose lowering effect of tea (Gomes *et al.*, 1995; Han *et al.*, 1999; Kao *et al.*, 2000; Sabu *et al.*, 2002; Wu *et al.*, 2004; Tsuneki *et al.*, 2004; Babu *et al.*, 2006; Wolfram *et al.*, 2006; Igarashi *et al.*, 2007; Badawoud *et al.*, 2007; Ostad Rahimi *et al.*, 2007; Potenza *et al.*, 2007).

Lipid peroxidation is an indicator of oxidative stress and plays major role in development of some complications of diabetes. Animal studies have shown that green tea administration can reduce lipid peroxidation in diabetic animals (Yamaguchi *et al.*, 1991; Tijburg *et al.*, 1997; Vinson *et al.*, 1998; Miura *et al.*, 2001; Guleria *et al.*, 2002; Kasaoka *et al.*, 2002; Nakagawa *et al.*, 2002; Liuji *et al.*, 2002; Sabu *et al.*, 2002; Skrzydlewska *et al.*, 2002; Babu *et al.*, 2006). Black tea has been reported to be an efficient reducer of peroxidation of lipoproteins as well (Tijburg *et al.*, 1997; Vinson *et al.*, 1998; Sur-Altiner *et al.*, 2000; Yokozawa *et al.*, 2002; Vinson *et al.*, 2005; Alipoor et al., 2008). Some studies have investigated the effects of purified tea polyphenols and drawn similar results (Quine *et al.*, 2005; Yamabe *et al.*, 2006).

MDA is another important indicator of oxidative stress that is usually measured in diabetics. Based on the results of studies, tea seems to affect this factor too and reduce its plasma concentration (Durate *et al.*, 2001; Skrzydlewska *et al.*, 2002; Chander *et al.*, 2003; Sürmen-Gür *et al.*, 2006; Alipoor *et al.*, 2009).

The activity of antioxidant enzymes such as catalase, superoxide dismutase, glutathione reductase and glutathione peroxidase has been shown to increase by supplementation of tea or its polyphenols as well. Actually some enzymes showed greater activity after the very supplementation in one study but not the other which seems to be due to different doses of supplementation, design and duration of the study (Khan *et al.*, 1992; Lin *et al.*, 1998; Durate *et al.*, 2001; Sabu *et al.*, 2002; Skrzydlewska *et al.*, 2002; Chander *et al.*, 2003; Kuo *et al.*, 2005; Babu *et al.*, 2006; Alipoor *et al.*, 2009).

Glutathione is another parameter which has been measured in some studies and seems to increase in diabetics receiving tea intervention (Sohn *et al.*, 1994; Lin *et al.*, 1998; Durate *et al.*, 2001; Sabu *et al.*, 2002; Skrzydlewska *et al.*, 2002; Babu *et al.*, 2006; Alipoor *et al.*, 2009).

4.4.2 Evidence from human studies

Human studies are not as conclusive as animal ones. There is some evidence that in countries with higher tea consumption like Japan, diabetes is less prevalent (Iso *et al.*, 2006). Some studies have shown a negative correlation between tea consumption and heart disorders and its consequent death (Stensvold *et al.*, 1992; Imai *et al.*, 1995; Duffy *et al.*, 2001; Geleijnse *et al.*, 2002; Hodgson *et al.*, 2002; Hakim *et al.*, 2003) which can be, in large part, attributed to the effects of tea on endothelial function through reducing oxidative stress; but there have been studies in which no relation was observed (Brown *et al.*, 1993; Sesso *et al.*, 2003)

Tea consumption decreased lipid peroxidation in some clinical trials (Klaunig *et al.*, 1999), but was not as effective in the others (Van Het Hof *et al.*, 1999; Hodgson *et al.*, 2000; Rumpler *et al.*, 2001; Hodgson *et al.*, 2002).

Results for malondialdehyde were inconsistent as well: some investigations indicating a negative relation between tea and MDA (Freese *et al.*, 1999; Hirano-Ohmori *et al.*, 2005; Nagao *et al.*, 2005), others showing no significant relationship (Rumpler *et al.*, 2001; Davis *et al.*, 2003).

The activity of antioxidant enzymes or oxidative status of the serum were improved by tea intervention in some studies (Serafini *et al.*, 1996; Nakagawa *et al.*, 1999; Leenen *et al.*, 2000; Sung *et al.*, 2000; Young *et al.*, 2002), but remained unchanged in the others (Van Het Hof *et al.*, 1997; Princen *et al.*, 1998; Freese *et al.*, 1999; Miura *et al.*, 2000; Davis *et al.*, 2003; Henning *et al.*, 2004; Davis *et al.*, 2005).

As reviewed above there is some evidence that tea and its fractions can act against development of diabetes and its complications but some studies have shown insignificant results. More detailed and precise clinical trials are essential to better understanding of tea's role in diabetes through its capacity to reduce oxidative stress.

Although animal studies provide great deal of evidence on usefulness of tea and its polyphenols against oxidative stress and its consequences in diabetes, human studies are not conclusive and limited research has not generally revealed significant decreases in biomarkers of *in vivo* oxidative damage. Far wider genetic variations in the response of humans to oxidative stress in comparison with animals may be one important factor obscuring small changes in biomarkers induced by tea and its polyphenols. Another reason may be that, though the dose of tea and its effective compounds used in animal and human studies do not differ much. Much higher doses relative to body weight is used in animal studies (Frei *et al.*, 2003).

5. Conclusion and future trends

Oxidative stress has been showed to play an important role in initiation and progression of diabetes and its accompanying complications. Thus to prevent the very consequences of oxidative stress, it seems logical to take the necessary steps to reduce it. Antioxidants have been reported to be effective in fulfilling this goal. Tea is a great source of a group of antioxidants so called flavonoids. Animal studies have been done to detect which compounds in tea are responsible for its effects on oxidative stress but human studies are lacking.

Animal studies have strongly supported the idea of tea being an efficient suppressor of oxidative stress in diabetic animals but human studies have faced inconsistency which may be rooted in factors like the design and time course of the study, the dose supplemented, the oxidative status of the subjects at baseline, the type of the tea studied, the stage of the disease, confounding factors not considered in some studies. It is recommended that well designed controlled clinical trials be done taking into account all the factors affecting the oxidative status of the patients and using sensitive and specific indicators of oxidative stress.

6. References

Adlercreutz H, Hamalaine E, Gorbach S, Goldin B. (1992). Dietary phyto-eostrogens and the menopause in Japan. Lancet, 339, 1233.

Adlercreutz H, Hojo H, Higashi A, Fotsis T, Hamalainen E, Hasegawa T, Okada H. (1991). Urinary excretion of lignans and isoflavonoid phytoestrogens in Japanese men and women consuming a traditional Japanese diet, Am J clin Nutr 54: 1093-10100.

Alipoor B, Alipoor Aghiri S, Ostad Rahimi A, Delazar A, Mesghary M. (2011). Effects of total extract and its different hydromethanol fractions of Irainian black tea on inflammatory factors and glycosilated hemoglubin (HbA1c) in STZ-induced diabetic rats, Urmia Med J 21(5): 2710-2737.

Alipoor B, Alipoor S, Delazar A, Ostadrahimi A,Bamdad Mogadam S. (2010). Recognition of the most components in the most effective hydromethanol fraction of Iranian black tea on lipid profile, oxidative stress and inflammatory factors in type I Diabetic rats, Pharmaceutical Sciences 16(2): 90-98.

Alipoor B, Delazar A, Ostad Rahimi A, Alipoor Aghiri S, Mesghary M. (2009). The effect of total extract and hydromethanol fractions of Iranian black tea on oxidative stress of type1 diabetic rats, Iranian J Diabe & Lipid 9(2): 130-136.

Alipoor B, Ostad Rahimi A, Delazar A, Meskary M, Osnaashary S, Vatankhah A.M, Alipoor Aghiri S, Safaian A. (2008).The Effects of Total Extract and its Different Hydromethanole Fractions of Iranian Black Orthodox Tea on Blood Lipid Profile in Type I Diabetic Rats . Med J Tabriz University of Medical Sciences 30(3):83-87.

Alkan Işil, Koprulu O, Alkan B. (2009). Latest advances in world tea production and trade, Turkey's aspect, World Journal of Agricultural Sciences 5(3): 345-349.

Amic D, Davidovic-Amic D, Beslo D, Trinajstic N. (2003). Structure-radical scavenging activity relationships of flavonoids, CROATIC CHEMICA ACTA CCACAA 76(1): 55-61.

Atalay M, Laaksonen D.E. (2002). Diabetes, oxidative stress and physical exercise, Journal of Sports Science and Medicine 1: 1-14.

Azizi F, Gouya M.M, Vazirian P, Dolatshahi P, Habibian S. (2003). Screening for type 2 diabetes in the Iranian national programme: a preliminary report, Eastern Mediterranean Health Journal 9: Nos 5/6.

Babu P.V.A, Sabitha K.E, Shyamaladevi C.S. (2006). Green tea impedesdyslipidemia, lipid peroxidation, protein glycation and ameliorates Ca2-ATPase and Na/K-ATPase activity in the heart of streptozotocin-diabetic rats, Chem Biol Interact 162: 157-164.

Babu P.V.A, Sabitha K.E, Shyamaladevi C.S. (2006). Therapeutic effect of green tea extract on advanced glycation and cross-linking of collagen in the aorta of streptozotocin diabetic rats, Clinical and Experimental pharmacology and physiology 33: 351-357.

Babu P.V.A, Sabitha K.E, Shyamaladevi C.S. (2006). Therapeutic effect of green tea extract on oxidative stress in aorta and heart of streptozotocin diabetic rats, Chemico – Biological Interactions 162: 114-120.

Badawoud M.H, Al-Saggaf S.M, Hagrasi M.M. (2007). Effects of green tea on the oxidative stress and glucose level of diabetic rats, Med. Sci 14(3): 3-11.

Baydas G, Canatan H, Turkoglu A. (2002). Comparative analysis of the protective effects of melatonin and vitamin E on streptozocin-induced diabetes mellitus, J. Pineal Res 32: 225-230.

Baydas G, Nedzvetskiib V.S, Nerushc P.A, Kirichenkob S.V, Yoldasd T. (2003). Altered expression of NCAM in hippocampus and cortex may underlie memory and learning deficits in rats with streptozotocin-induced diabetes mellitus, Life Sciences 73: 1907-1916.

Bendini A, Toschi T.G, Vanzini M. (1998). Green tea catechins: Analysis by RP-HPLC, Department of food science, sect 40126(7):1-8.

Bharati D.R, Pal R, Rekha R, Yamuna T.V. (2011). Evaluation of the burden of type 2 diabetes mellitus in population of Puducherry, South India, Diabetes & Metabolic Syndrome, Clinical Research & Reviews 5: 12-16.

Bocci V, Zanardi I, huijberts M.S.P, Travagli V. (2011). Review: Diabetes and chronic oxidative stress. A perspective based on the possible usefulness of ozone therapy, Diabetes & Metabolic Syndrome, Clinical Research & Reviews 5: 45-49.

Bronner W.E, Beecher G. R. (1998). Method for determining the content of catechins in tea infusions by HPLC, J of chromatograohy 805: 137-142.

Brown C.A, Bolton-Smith C, Woodward M, Tunstall-Pedoe H. (1993). Coffe and tea consumption and the prevalence of coronary heart disease in men and women, Results from the Scottish heart study 47: 171-175.

Cadenas E, Packer L. (2002). Handbook of antioxidants. 2nd ed., marcel dekker, INC, 371-399.

Ceriello A. (2003). New insight on oxidative stress and diabetic complications may lead to a "causal" antioxidant therapy, Diabetes Care 26: 1589-1596.

Ceriello A. (2006). Oxidative stress and diabetes-associated complications, Endocr Pract 12: S60-S62.

Ceriello A, Motz E. (2004). Is oxidative stress the pathogenic mechanism underlying insulin resistance, diabetes, and cardiovascular disease? The common soil hypothesis revisited, Arterioscler Thromb Vasc Biol 24: 816-823.

Chander V, Singh D, Chopra K. (2003). Catechin, anatural antioxidant protects against rhabdomyolysis - induced myoglobinuric acute renal failure, Pharmacological Research 48: 503-509.

Crespy V, Williamson G. (2004). A Review of the Health Effects of Green Tea Catechins in In Vivo Animal Models, J. Nutr 134: 3431S-3440S.

Cummings S.R, Nevitt M.C, Browner W.S, Stone K, Fox K.M, Ensrud K, Cauley J, Black D, vogt T.M. (1995). Risk factors for hip fracture in white women, N Engl J Med 332: 767-73.

Davis M.J, Judd J.T, Baer D.J, Clevidence B.A, Paul D.R, Edwards A.J, Wiseman S.A, Muesing R.A, Chen S.C. (2003). Black Tea Consumption Reduces Total and LDL Cholesterol in Mildly Hypercholesterolemic Adults, J. Nutr 133(10): 3298S-3302S.

Devasagayam T.P.A, Kamat J.P, Mohan H, Kesavan P.C. (1996). Caffeine as an antioxidant: inhibition of lipid peroxidation induced by reactive oxygen species, Biochimica et Biophysica Acta (BBA) - Biomembranes 1282(1): 63-70.

Droge W. (2002). Free Radicals in the Physiological Control of Cell Function, Physiol Rev 82: 47-95.

Duffy S.J, Vitta J.A, Holbrook M, Swerdloff P.L, Keaney J.F. (2001). Effect of Acute and chronic tea consumption on platelet aggregation in patients with coronary artery disease, Arteriosclerosis, Thrombosis, and vascular Biology 21: 1084-1089.

Dulluge J.J, Nelson B.C, Thomas J.B, Sander L.C. (1998). Selection of column and gradient elution system for the separation of catechins in green tea using HPLC, J of chromatography 793: 265-279.

Duke J. A. (1985). Handbook of medicinal herbs. CRC, Bocaraton, 93-94.

Durate J, Galisteo M, Ocete M.A, Perez F, Zarzuelo A. (2001). Effects of chronic quercetin treatment on hepatic oxidative status of spontaneously hypertensive rats, Mol Cell Biochem 221(1-2): 155-60.

Esposito K, Nappo F, Marfella R, Giugliano G, Giugliano F, Ciotola M, Quagliaro L, Ceriello A, Giugliano D. (2002). Inflammatory cytokine concentrations are acutely increased by hyperglycemia in humans, role of oxidative stress, Circulation 106: 2067-2072.

Evans J.L, Goldfine I.D, Maddux B.A, Grodsky G.M. (2003). Are oxidative stress-activated signaling pathways mediators of insulin resistance and β-cell dysfunction?, Diabetes 52: 1-8.

Evans J.L, Goldfine I.D, Maddux B.A, Grodsky G.M. (2002). Oxidative stress and stress-activated signaling pathways: A unifying hypothesis of type 2 diabetes, Endocrine Reviews 23(5): 599-622.

Freese R, Basu S, Hietanen E, Nair J, Nakachi K, Bartsch H, Mutanen M. (1999). Green tea extract decreases plasma malondialde hyde concentration but does not affect other indicators of oxidative stress, nitric oxide production, or hemostatic factors during a high- lenoleic acid diet in healthy females, Eur J Nutr 38: 149-157.

Frei B, Higdon J.V. (2003). Antioxidant activity of tea polyphenols in vivo: evidence from animal studies, J Nutr 133: 3275S-3284S.

Geleijnse J.M, Launer L.J, Kuip D.A, Hofman A, Witteman j.C.M. (2002). Inverse association of tea and flavonoid intakes with incident myocardial in farction: the Rotterdam study, Am. J. Clin. Nutr 75: 880-886.

Gomes J.R, Vedasiromoni J.R, Das M, Sharma R.M, Ganguly D.K. (1995). Journal of Ethnopharmacology 45(3): 223-226.

Guleria R.S, Jain A, Tiwari V, Misra M.K. (2002). Protective effect of green tea extract against the erythrocytic exidative stress injury during mycobacterium tuberculosis infection in mice, Mol. Cell. Biochem 236(1-2): 173-81.

Gupta S, Saha B, Giri A.K. (2002). Comparative antimutagenic and anticlastogenic effects of green tea and black tea: a review, Mutation Research 512: 37-65.

Hakim I.A, Alsaif M.A, Alduwaihy M, Al-Rubeaan K, Al-Nuaim A.R, Al-Attas O. (2003). Tea consumption and the prevalence of coronary heart disease in Saudi adults: results from a Saudi national study, Prev. Med 36(1): 64-70.

Han L.K, Takaku T, Kimura Y, Okuda H. (1999). Anti – obesity of oolong tea, Int J obes Relat Metab Disord 23(1): 98-105.

Hara Y. (2001). Green tea, Health benefits and applications, Marcell Dekker Inc, New York, 1-11, 27-43, 53-83, 97-119.

Henning S.M, Niu Y, Lee N.H, Thames G.D, Minutti R.R, Wang H, Go V.L, Heber D. (2004). Bioavailability and antioxidant activity of tea flavonols after consumption of green tea, or a green tea extract supplementation, Am J Clin Nutr 80(6): 1558-1564.

Hertog M.G, Feskens E.J, Kromhout D, Hertog M.G.L, Hollman P.C.H, Hertog M.G.L, Katan M.B. (1993). Dietary antioxidant flavonoids and risk of coronary heart disease: the zutphen Elderly study, Lancet 342: 1007-1011.

Hicks A. (2009). Current status and future development of global tea production and tea products, Au J.T 12(4): 251-264.

Hirano-Ohmori R, Takahashi R, Momiyama Y, Taniguchi H, Yonemura A, Tamai S, Umegaki K, Nakamura H, Kondo K, Ohsuzu F. (2005). Green Tea Consumption and Serum Malondialdehyde-Modified LDL Concentrations in Healthy Subjects, J Am Coll Nutr 24(5): 342-346.

Hodgson J.M, Croft K.D, Mori T.A, Burke V, Beilin L.J, Puddey I.B. (2002). Regular ingestion of tea does not inhibit in vivo lipid peroxidation in humans, J Nutr 132: 55-58.

Hodgson J.M, Puddey I.B, Burke V, Watts G.F, Beilin L.J. (2002). Regular ingestion of black tea improves brachial artery vasodilator function, Clin Sci 102(2): 195-201.

http://en.wikipedia.org/wiki/Tea, Accessed on 14 August 2011.

http://www.tea-of-chinese.com, Accessed on 12 August 2011.

Igarashi K, Honma K, Yoshinari O, Nanjo f, Hara Y. (2007). Effects of dietary catechins on glucose tolerance, blood pressure and oxidative status in goto-kakizaki rats, J Nutr Sci Vitaminol 53: 496-500.

Imai K, Nakachi k. (1995). Cross sectional study of effects of drinking green tea on cardiovascular and liver diseases, BMJ 310: 693-696.

Integrative Medicine. (2000). Professional guide to condition, Herbs & supplements, S.L, Integrative Medicine Communications, Newton, 318-319.

Ishii N, Patel K.P, Lane P.H, Taylor T, Bian K, Murad F, Pollock J.S, Carmines P.K. (2001). Nitric oxide synthesis and oxidative stress in the renal cortex of rats with diabetes mellitus, J Am Soc Nephrol 12: 1630-1639.

Iso H, Date C, Wakai K, Fukui M, Tamakoshi A, JACC study group. (2006). The relationship between green tea and total caffeine intake and risk for self-reported type2 diabetes among Japanese adults, Annals of Internal Medicine 144: 554-562.

Johnell O, Gullberg B, Kanis JA, Allander E, Elffors L, Dequeker J, Dilsen G, Gennari C, Lopes Vaz A, Lyritis G, Mazzuoli G, Miravet L, Passeri M, Cano P.R, Rapado A, Ribot C. (1995). Risk factors for hip fracturei European women. The MEDOS study. Mediterranean osteoporosis study, J Bone Miner Res 10: 1802-1815.

Kammat J.P, Boloor N.K, Devasagayam T.P, Jayanshree B, Kesavan P.C. (2000). Differential modification by caffeine of oxygen-dependant and independent effects of gamma-irradiation on rat liver mitochondria, Int J Radiat Biol 76(9): 1281-1288.

Kanis J, Johnell O, Gullberg B, Allander E, Elffors L, Ranstam J, Dequeker J, Dilsen G, Gennari C, Vaz AL, Lyritis G, Mazzuoli G, Miravet L, Passeri M, Perez Cano R, Rapado A, Ribot C. (1999). Risk factors for hip fracture in men from southern Europe the MEDOS study, Osteoporos Int 9: 45-54.

Kao Y.H, Chang H.H, Lee M.J, Chen C.L. (2006). Tea, Obesity, and diabetes, Mol. Nutr Food Res 50: 188-210.

Kao Y.H, Hiipakka R.A, Liao S. (2000). Modulation of endocrine systems and food intake by green tea epigallocatechin gallate, Endocrinology 141(3): 980-987.

Kasaoka S, Hase K, Morita T, Kiriyama S. (2002). Green tea flavonoids inhibit the LDL oxidation in osteogenic disordered rats fed a marginal ascorbic acid in diet, Journal of Nutritional Biochemistry 13: 96-102.

Keys J. D. (1976). Chinese Herbs: Their botany chemistry and pharmcodynamics. Tuttle company, Rut land, 189-190.

Khan S.G, Katiyar S.K, Agarwal R, Mukhtar H. (1992). Enhansment of antioxidant and phase II enzymes by oral feeding of green tea polyphenols in drinking water to SKH-1 hairless mice. Cancer Research 52: 4050-4052.

Klaunig J.E, Xu Y, Han C, kamendulis L.M, Chen J, Heiser C. (1999). The effect of tea consumption on oxidative stress in smokers and nonsmokers, Experimental Biology and Medicine 220: 249-254.

Kuo k.L, Weng M.S, Chiang C.T, Tsai Y.J, Lin-Shiau S.Y, Lin J.K. (2005). Comparative Studies on the hypolipidemic and growth suppressive effects of Oolong, Black, Pu-erh, and green tea leaves in rats, J.Agric.Food hem 53: 480-489.

Leenen R, Roodenburg A.J, Tijburg L.B, Wiseman S.A. (2000). A single dose of tea with or without milk increases plasma antioxidant activity in humans, Eur. J. Clin. Nutr 54(1): 87-92.

Leung L.K, Su Y, Chen R, Zhang Z, Huang Y, Chen Z.Y. (2001). Theaflavins in black tea and catechins green tea are equally effective antioxidants, The J of nutrition 131(9): 2248-51.

Liao S, Kao Y.H, Hiipakka R.A. (2001). Green tea: biochemical and biological basis for health benefits, Vitam Horm 62: 1-94.

Lin Y, Cheng C, Lin P. (1998). Hypolipidemic effect of green tea leaves through induction of antioxidant and phase II enzymes including superoxide dismutase, catalase, and Glutathione S-transferase in rats, Journal of Agricultural and food chemistry 46(5): 1893-1899.

Lin J.K, Sy L.S. (2006). Mechanisms of hypolipidemic and anti-obesity effects of tea and tea polyphenols, Molecular Nutrition and Food Research, 50(2): 211-217.

Liuji C, Xianqiang Y, Hongli J, Baolu Z. (2002). Tea catechins protect against lead-induced cytotoxicity, lipid peroxidation, and membrane fluidity in HepG2 Cells, Toxicological sciences 69: 149-156.

Lushchak V.I. (2011). Adaptive response to oxidative stress: Bacteria, fungi, plants and animals, Comparative Biochemistry and Physiology 153: 175-190.

Luximon-Ramma A, Bahorun T, Crozier A, Zbarsky V, Datla K.P, Dexter D.T, Aruoma O.I. (2005). Characterization of the antioxidant functions of flavonoids and proanthocyanidins in Mauritian black teas, Food Research International 38: 357-367.

Maiese K, Morhan S.D, Chong Z.Ż. (2007). Oxidative stress biology and cell injury during type 1 and type 2 diabetes mellitus, Curr Neurovasc Res 4(1): 63-71.

Mandel S, Youdim M.B. (2004). Catechin polyphenols: neurodegeneration and neuroprotection in neurodegenerative diseases Free Radic Biol Med 37(3): 304-317.

Maritim A.C, Sanders R.A, Watkins III J.B. (2003). Diabetes, oxidative stress and antioxidants: A review, J Biochem Molecular Toxicology 17(1): 24-34.

McKay D.L, Blumberg J.B. (2002). The role of tea in human health, An update, J Am Colle Nutr 21(1): 1-13.

Mills S, Bone K. (2000). Principles and practice of phytotherapy: (Modern Herbal Medicine). Churchill Livingstone, Eding burgh 70: 34-37.

Miura Y, Chiba T, Miura S, Tomita I, Umegaki K, Ikeda M, Tomita T. (2000). Green tea polyphenols (flavan-3-ols) prevent oxidative modification of low density lipoproteins: An ex vivo study in humans, J. Nutr. Biochem 11: 216-222.

Miura Y, Chiba T, Tomita I, Koizumi H, Miura S, Umegaki K, Hara Y, Ikeda M. (2001). Tea catechins prevent the development of atherosclerosis in apoprotein E deficient mice, J. Nutr 131: 27-32.

Moxham R. (2009). A brief history of tea, Running press, Michigan.

Nagao T, Komine Y, Soga S, Meguro M, Hase T, Tanaka Y, Tokimitsu I. (2005). Ingestion of a tea rich in catechins leads to a reduction in body fat and malondialdehyde-modified LDL in men, Am J Clin Nutr 81(1): 122-129.

Nakagawa K, Ninomiya M, Okubo T, Aoi N, Juneja L.R, Kim M, Yamanaka K, Miyazawa T. (1999). Tea catechin supplementation increases antioxidant capacity and prevents phospholipids hydroperoxidation in plasma of humans, J. Agric. Food.Chem 47(10): 3967-3973.

Nakagawa T, Yokozawa T. (2002). Direct scavenging of nitric oxide and superoxide by green tea, Food and chemical Toxicology 40: 1745-1750.

Negrean M, Stirban A, Stratmann B, Gawlowski T, Horstmann T, Gotting C, Kleesiek K, Mueller-Roesel M, Koschinsky T, Uribarri J, Vlassara H, Tschoepe D. (2007). Effects of low- and high-advanced glycation endproduct meals on macro- and microvascular endothelial function and oxidative stress in patients with type 2 diabetes mellitus, Am J Clin Nutr 85: 1236-1243.

Nikolic J, Bjelakovic G, Stojanovic I. (2003). Effect of caffeine on metabolism of L-arginine in the brain, Molecular and Cellular biochemistry 244: 125-128.

Ostad Rahimi A, Delazar A, Alipoor B, Meskary M, Osnaashary S, Vatankhah A.M, Alipoor Azhiri S, Safaian A. (2007). The effect of total extract and fractions of Iran Ortodox tea on body weight, food intake and blood glucose of type1 diabetic rats, Pharmacological sciences 2: 23-29.

Potenza M.A, Marasciulo F.L, Tarquinio M, Tiravanti E, Colantuono G, Federici A, Kim J, Quon M.J, Montagnani M. (2006). EGCG, a green tea polyphenol improves endothelial function and insulin sensitivity, reduces blood pressure and protects against myocardial I/R injury in SHR, Am J Physiol Endocrinol Metab 292: E1378-E1387.

Princen H.M.G, Duyvenvoorde W.V, Buytenhek R, Blonk C, Tijburg L.B.M, Langius J.A.E, Meinders A.E, Pijl H. (1998). No effects of consumption of green tea and black tea on plasma lipid and antioxidant levels and LDL oxidation in smokers, Arteriosclerosis, Thrombosis and Vascular Biology 18: 833-841.

Quine S.D, Raghu P.S. (2005). Effects of (-)-epicatechin, a flavonoid on lipid peroxidation and antioxidants in streptozotocin-induced diabetic liver, kidney and heart, Pharmacol Rep 57: 610-615.

Robertson R.P. (2004). Chronic oxidative stress as a central mechanism for glucose toxicity in pancreatic islet beta cells in diabetes, The Journal of Biological Chemistry 279(41): 42351-42354.

Robertson J, Donner A.P, Trevithick J.R. (1991). A possible role for vitamins C and E in cataract prevention. Am J Clin Nutr 53: 346S-351S.

Robertson R.P, Harmon J, Tran P.O.T, Poitout V. (2004). β-cell glucose toxicity, lipotoxicity and chronic oxidative stress in type 2 diabetes, Diabetes 53: S119-S124.

Robinson C.H, Proudfit F.T. (1972). Normal and Therapeutic Nutrition. Macmillan Publishing, New York, 508.

Rumpler W, Seale J, Clevidence B, Judd J, Wiley E, Yamamoto S, Komatsu T, Sawaki T, Ishikura Y, Hosoda K. (2001). Oolong tea increases metabolic rate and fat oxidation in man, J. Nutr 131: 2848-2852.

Sabu M.C, Smitha K, Kuttan R. (2002). Anti-diabetic activity of green tea polyphenols and their role in reducing oxidative stress in experimental diabetes, J Ethnopharmacol 83: 109-116.

Sasso F.C, nicola L.D, Carbonara O, Nasti R, Minutolo R, Salvatore T, Conte G, Torella R. (2006). Cardio Vascular risk factors and disease management in type 2 diabetic patients with diabetic nephropathy, Diabetes care 29: 498-503.

Serafini M, Ghiselli A, Ferro-Luzzi A. (1996). In vivo antioxidant effect of green and black tea in man, Eur J clin Nutr 50: 28-32.

Sesso H.D, Gaziano J.M, Liu S, Buring J. (2003). Flavonoid intake and the risk of cardiovascular disease in women, Am. J. Clin. Nutr 77: 1400-1408.

Sesso H.D, Paffenbarger RS.JR, Oguma Y, Lee I.M. (2003). Lack of association between tea and cardiovascular disease in college alumni, Inter. J. Epidemiol 32: 527-533.

Shils M.E, Shike M, Ross A.C, Caballero B, Cousins R.C. (2006). Modern Nutrition in Health and Disease. 10th ed. Lippincott Williams & Wilkins, Philadelphia, 1044-1048.

Skrzydlewska E, Ostrowska J, Farbiszewski R, Michalak K. (2002). Protective effect of green tea agains lipid peroxidation in the rat liver, blood serum and the brain, Phytomedicine 9(3): 232-238.

Sohn O.S, Surace A, Fiala E.S, Richie J.P, Colosimo S. (1994). Effects of green and black tea on hepatic xenobiotic metabolizing systems in the male F344 rat, Xenobiotica 24(2): 119-127.

Springhouse. (2001). Nursing Herbal Medicine Handbook, Springhouse Pub co, 216-217.

Stensvold I, Tvrdal A, Solvoll k, Foss O.P. (1992). Tea consumption. Relationship to cholesterol, blood pressure, and coronary and total mortality, Prev Med 21(4): 546-553.

Sung H, Nah J, Chun S, Park H, Yang S.E, Min W.K. (2000). In vivo antioxidant effect of green tea, Eur. J. Clin. Nutr 54(7): 527-529.

Sur-Altiner D, Yenice B. (2000). Effect of black tea on lipid peroxidation in carbon tetrachloride treated male rats, Drug Metabol. Drug Interact 16: 123-128.

Sürmen-Gür E, Gülten T, Serdar Z, Colakoğullari M. (2006). Chronic black tea administration protects plasma proteins, plasma, liver and kidney lipids against oxidation, Med Sci Monit 12(3): 102-105.

Tavani A, Negri E, Vecchia C.L. (1996). Food and nutrient intake and risk of cataract, Ann Epidemiol 6: 41-46.

Tijburg L.B.M, Wiseman S.A, Meijer G.W, Weststrate J.A. (1997). Effects of green tea, black tea and dietary lipophilic antioxidants on LDL oxidizability and atherosclerosis in hypercholesterolaemic rabbits, Atherosclerosis 135(1): 37-47.

Tsuneki H, Ishizuka M, Terasawa M, Wu J.B, Sasaoka T, Kimura I. (2004). Effect of green tea on blood glucose levels and serum proteomic patterns in diabetic mice and on glucose metabolism in healthy humans, BMC pharmacology 4: 18-28.

Tyler V, Robbers J. (1999). Tyler's herbs of choice: The therapeutic use of phytomedicinals, 2nd ed., CRC Press, New York, 248-251.

Van Het Hof K, Boer H.S, Wiseman S.A, Lien N, Westrate J.A, Tijburg L.B. (1997). Consumption of green or black tea doesnot increase resistance of low- density lipoprotein to oxidation in humans, Am J Clin Nutr 66: 1125-1132.

Van Het Hof K.H, Wiseman S.A, Yang C.S, Tijburg L.B. (1999). Plasma and lipoprotein levels of tea catechins following repeated tea consumption, Proc Soc Exp Biol Med 220(4): 203-209.

Vincent A.M, Russell J.W, Low P, EVA L. Feldman E.L. (2003). Oxidative Stress in the Pathogenesis of Diabetic Neuropathy, Endocrine reviews 25(4): 612-628.

Vinson J.A, Dabbagh Y.A. (1998). Effect of green and black tea supplementation on lipids, lipid oxidation and fibrinogen in the hamster: mechanisms for the epidemiological benefits of tea drinking, FEBS Letters 433: 44-46.

Vinson J.A, Zhang J. (2005). Black and green teas equally inhibit diabetic cataracts in a streptozotocin-induced rat model of diabetes, J Agric Food Chem 53: 3710-3713.

Wang H, Provan G.J. (2000). Tea Flavonoids: Their functions, utilisation and Analysis, Trendsin Food science and Technology 11: 152-160.

Wanger H, Bladt S. (1996). Plant Drug Analysis: A thin layer chromatography atlas. 2nd Ed. Springer, verlag . berlin , 13, 207.

Watanabe J, Kavabata J, Niki R. (1998). Isolation and identification of acetyl-CoA carboxylase inhibitorsfrom green tea, Biosci Biotechol Biochem 62: 532-534.

Wild S, Roglic G, Green A, Sicree R, King H. (2004). Global Prevalence of Diabetes Estimates for the year 2000 and projections for 2030, Diabetes Care 27: 1047-1053.

Wolfram S, Raederstorff D, Preller M, Wang Y, Teixeira S.R, Riegger C, Weber P. (2006). Epigallocatechin gallate supplementation alleviates diabetes in rodents, J. Nutr 136: 2512-2518.

Wright L.P. (2005). Biochemical analysis for identification of quality in black tea (camellia sinensis). University of Pretoria etd-wright, LP. South Africa.

Wu L, Juan C.C, Ho L.T, Hsu Y.P, Hwang L.S. (2004). Effect of green tea supplementation on insulin sensitivity in Sprague-Dawley rats, J. Agric. Food. Chem 52(3): 643-648.

Yamabe N, Yokozawa T, Oya T, Kim M. (2006). Therapeutic potential of (-)-epigallocatechin 3-O-gallate on renal damage in diabetic nephropathy model rats, J Pharmacol Exp Ther 319: 228-236.

Yamaguchi Y, Hayashi M, Yamazoe H, Kunitomo M. (1991). Preventive effects of green tea extract on lipid abnormalitities in serum, liver and aorta of mice fed an atherogenic diet, Nippon Yajurigaku zasshi 97(6): 329-337.

Yamamoto T. (1997). Chemistry and applications of green tea. 1st ed., green tea technology Book from C.H.I.P.S, 1-2.

Yang C.S, Landau J.M. (2000). Effects of Tea Consumption on Nutrition and Health, J. Nutr 130: 2409-2412.

Yokozawa T, Nakagawa T, Kitani K. (2002). Antioxidant activity of green tea polyphenol in cholesterol fed rats, J. Agric. Food. Chem 50(12): 3549-3552.

Young J.F, Dragstedt L.O, Haraldsdottir J, Danesh Van B, Kall M.A, Loft S, Nilsson L, Nielsen S.E, Mayer B, Skibsted L.H, Huynh-Ba T, Hermetter A, Sandström B. (2002). Green tea extract only affects markers of oxidative status postprandially: lasting antioxidant effect of flavonoid-free diet, Br. J. Nutr 87(4): 343-55.

Zhang D.D. (2006). Mechanistic studies of the NRF2-KEAP1 signaling pathway, Drug Metabolism Reviews 38: 769-789.

12

Flavonoid Treatment for Mustard Agents' Toxicity

Rajagopalan Vijayaraghavan and Anshoo Gautam
Defence Research and Development Establishment, Gwalior India

1. Introduction

The weapons of mass destruction, chemical, biological and nuclear warfare are the most brutal created by the humans. They kill and incapacitate not only the armed forces but also the innocent public, without any mercy. The Chemical Weapons Convention prohibits the production, storage and use of toxic chemicals during warfare. In fact, the use of "Any chemical which through its chemical action on life processes can cause death, temporary incapacitation or permanent harm to humans and animals" as a method of warfare is discouraged by the Chemical Weapons Convention and many of such toxic chemicals are listed in its three Schedules for verification purpose (OPCW). The chemical warfare agents are extremely toxic chemicals. They act in very small quantities and very rapidly, and death may occur in minutes, like the nerve agents and the blood agents (Somani, 1992). Some of them like the blistering agents, though may not cause immediate lethality, but are highly incapacitating (Dacre & Goldman, 1996). The nerve agents are organophosphorous compounds that include tabun, sarin, soman and Vx. They inhibit acetylcholinesterase enzyme resulting in the accumulation of acetylcholine leading to muscarinic and nicotinic receptor stimulation (Bajgar, 2004). The blood agents include the cyanides. They inhibit cytochrome oxidase enzyme leading to cellular hypoxia (Way, 1984). Though the nerve agents and the blood agents are immediately lethal, specific antidotes are available for use in the field as First Aid Kit (Vijayaraghavan et al, 2011). For nerve agent poisoning the recommended antidotes are atropine sulphate and pralidoxime chloride that are administered by autoinjectors (Friedl, 1989; Vijayaraghavan et al, 2007). For cyanide poisoning the recommended antidotes are amyl nitrite inhalation, and sodium nitrite and sodium thiosulphate injection (Chen & Rose, 1952; Bhattacharya & Vijayaraghavan, 2002).

The blistering agents are the sulphur mustard (SM) and the nitrogen mustards (NM). They cause severe toxicity with delayed clinical symptom. In the biological system they undergo an intramolecular cyclisation and produce highly reactive electrophiles that have strong affinity for a variety of macromolecules. They are extremely toxic to rapidly dividing cells, resulting in multiorgan failure (Papirmeister et al, 1991). Unlike the nerve agents and the blood agents no specific treatment is available for the mustard agents. A wide variety of molecules are being evaluated as antidote for mustard agent toxicity. Antidote against mustard agents require few major characteristics: (a) molecules should be strong neucleophiles because mustard agents are highly reactive electrophiles, (b) molecules may be effective only prophylactically and may be difficult or less effective after exposure due to the cascade of events following mustard agent exposure, (c) molecules should have delayed clearance and low protein binding as the chain reactions following mustard agent poisoning is long-lasting and fast, and (d) as free drug concentration in the blood is required for a long time, molecule should be highly safe and non toxic even at higher doses. At present decontamination by physical adsorption and by chemical degradation is the main methodology adopted for reducing the toxicity of the mustard agents.

The effects of SM are more prominent on the eyes, skin and lung. Toxic effects of the mustard agents (SM and NM) also occur in bone marrow, the central nervous system and the gastrointestinal tract. Oxidative stress is one of the many causes of the mustard agents' toxicity and antidotes directed against them have proven to be beneficial. Antioxidants can enhance survival time, protect liver and lung from oxidative damage and also reduce accumulation of purine metabolites in SM toxicity (Kumar et al, 2001; Vijayaraghavan et al, 2009). Among the various types of nucleophiles, the flavonoids appear to be very promising. They are polyphenolic compounds present in several plants, inhibit lipid peroxidation and also act as a free radical scavengers. The flavonoids are in current choice of drug for various disease conditions like cardiovascular disease, diabetic nephropathy, hypertension, colorectal cancer and aging, as an adjunct to other drugs (Narayana et al, 2001).

2. Blister inducing chemical warfare agents

The blistering inducing chemical warfare agents are also known as vesicants and are of three types namely the mustards, arsenicals and phosgene oxime. The mustard agents belong to two distinct classes, the sulphur mustard (halogenated thioether) and the nitrogen mustards (halogenated tertiary amines). Chemically, Sulphur mustard is bis (2-chloroethyl) sulphide and generally known as mustard gas (Figure 1). Sulphur mustard was first synthesised in 1822 and has the highest military significance since its first use in 1915 in World War I. It is one of the frequently used chemical warfare agents and the latest use was in the Iran-Iraq war (Kehe & Szinicz, 2005). It is considered as the king of the chemical warfare agents. As a chemical warfare agent, three nitrogen mustards are important (HN-1, HN-2 and HN-3). The nitrogen mustards were synthesised in the 1930s but were not produced in large amounts for warfare. Among the nitrogen mustards HN-3, which is chemically, tris (2-chloroethyl) amine is more toxic than others. HN-2, also known as mechlorethamine is used as a cancer chemotherapeutic agent and has remained the standard drug for many years. Lewisite was synthesised in 1918 for military purpose and no report is available of its use in battlefield.

In the pure form SM is colorless and odorless liquid. With impurities it has a characteristic smell similar to mustard or garlic. It has low volatility. It is soluble in organic solvents readily but very less soluble in water (0.8 $g.L^{-1}$). In pure state the nitrogen mustards are also colorless liquids. They are less volatile, less soluble and more resistant to oxidising agents than SM, but are less stable on storage. Pure lewisite is also colorless liquid with metallic odor and its solubility in water is similar to SM. Lewisite is relatively unstable. Thus, nitrogen mustards and lewisite lack the basic requirement, i.e. storage stability of a chemical warfare agent (Ganesan et al, 2010).

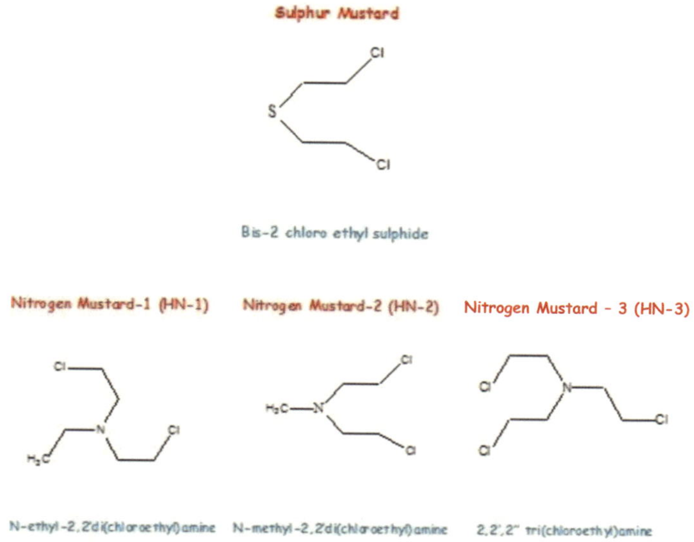

Fig. 1. Structure of sulphur mustard and nitrogen mustards.

The mustard agents are radiomimetic and are extremely toxic to dividing cells. They are highly lipophilic and readily penetrate the skin, and even cotton fabric, latex and rubber. Sulphur mustard cyclises inside the body to a highly reactive sulphonium ion. The most important target for mustard agent is DNA. Mustard agents alkylate the purine bases of DNA and damage them (Papirmeister et al, 1991). They irreversibly alkylate DNA, RNA and protein, causing cell death. At the cellular level mustard agents cause cytostasis and mutation. Death will occur if 4 to 5 grams of SM falls on the bare skin. HN-3 is as toxic as SM. The respiratory lethal dose of SM is estimated to be 1500 $mg.min.m^{-3}$.

3. Toxic effects of sulphur mustard

In the form of vapour or aerosol, SM attacks the eyes, lungs and skin. If the concentration and the duration of exposure are larger than systemic effects will occur. In small laboratory animals SM is highly lethal (Vijayaraghavan et al, 2005). The characteristic of SM poisoning is the delayed appearance of toxic effects. The victim knows about the exposure only after a lapse of 3 to 6 hours, though the damage begins within 1 to 2 minutes after contact. The eye is more vulnerable and sensitive to SM vapour. Exposure to a concentration of 0.001 $mg.L^{-1}$ for one hour can cause conjunctivitis. This concentration is not easily recognisable by odour. The effect of mustard gas on the eye can be classified as mild, moderate and severe. If the exposure is mild there will be itching, lacrimation and a sensation of a dust particle in the eye. This is followed by a burning sensation and photophobia. There will be hyperemia of the conjunctivae. Edema of the lids may also be present. In moderate exposure there will be complete closure of eyes, because of spasm and swelling of the lids leading to blurring of the vision. Edema of the conjunctiva, mild iritis and edema of the cornea will be present. Blepharospasm and edema of the lid will be so severe that the patient cannot open the eyes. If liquid SM directly falls in the eye, the cornea and iris may be affected very severely. In this type the latent period will be very short. Severe edema of the lids, and marked hyperemia and edema of the conjunctiva will be present. The epithelium and stroma of the cornea will be damaged, and edema will develop later. Iritis and mucoserous discharge will also be present. In severe cases, blindness may occur (Safarinejad, 2001). The development of SM induced ocular lesions in rabbits is similar to the lesions described in human casualties (Kadar *et al.*, 2001). The lesions produced by HN-3 on the eye will be more severe and will appear in a shorter time than SM.

The effects of SM on the skin resemble those of burn injuries. It depends upon the weather condition and the degree of exposure. In a warm humid climate, the effect will be more severe. Similarly, the lesions will be more severe in damp and warm parts of the body like finger folds, groin and axilla. Normally, the symptoms appear after a latent period of 6 - 12 hours. Sulphur mustard is a potent cutaneous vesicant. Sulphur mustard penetrates rapidly through the skin, causing prolonged injuries and leading to severe incapacitation (Kadar et al, 2000). The proliferating basal cells in the skin are metabolically very active and are more sensitive to SM (Ray et al., 2000). Sulphur mustard produces blisters with a severe inflammatory reaction in skin of exposed individuals (Casbohm et al, 2004). The first symptom will be continuous itching at the site of contact. Then erythema will appear gradually followed by vesication. This is due to the necrosis of cells in the lower layers of the epidermis, and exudation of tissue fluids into the spaces so formed. In animal models frank blisters are not seen, but microblisters appear. The blister may rupture in due course with the possibility of infection. The healing will take place with a small scar, except in very severe or infected burns. Sulphur mustard burns usually are followed by a persistent brown pigmentation except at the site of actual vesication, where there may be a temporary depigmentation. The effect of SM on the respiratory tract also depends upon the degree of exposure. If the exposure is mild, swelling and erythema will be present, in the nose, larynx and trachea. This will be followed by sloughing and fibrous exudation. The laryngeal edema and necrosis may lead to respiratory obstruction. If the exposure is severe, bronchioles and alveoli may be damaged. In more severe cases the pulmonary parenchyma shows congestion, mild patchy edema, emphysema and focal atelectasis. There is a danger of bacterial infection of the lungs which will result in bronchopneumonia. The latter may be responsible for the death in humans following SM exposure. SM vapours induces sensory irritation in mice during exposure. The respiratory frequency decreased on subsequent days of exposure depending upon the exposure concentration, and the breathing pattern was characteristic of bronchial obstruction (Vijayaraghavan, 1997). Ingestion of food or water contaminated with liquid SM, may produce nausea, vomiting, pain and diarrhoea. Even exposure to the skin alone can cause malaise, anuria, vomiting and cardiac abnormalities. When the amount approaches lethal dose, injury to the haematopoietic tissues may occur. SM in small animals will cause systemic toxicity distal to the site of exposure. Probably, SM is the only chemical showing more toxicity through percutaneous route compared to subcutaneous route and oral route in animal models (Vijayaraghavan et al, 2005).

4. Treatment of sulphur mustard toxicity

The treatment for SM lesions is similar to that of burn injuries. There is no specific antidote for SM intoxication. A variety of compounds have been evaluated as antidotes using *in vitro* and *in vivo* models but none of them are recommended as standard therapy (Vijayaraghavan et al, 2009). In the case of eye contamination the eye should be washed with water. If the eye lids are sticky then sterile petroleum jelly can be applied. Photophobia and blepharospasm can be relieved by instilling one drop of atropine sulphate solution (1 %) 3 - 4 times a day. Topically applied steroid treatment and anti-inflammatory drugs are potential therapies, and this can be supplemented with an antibacterial agent like ciprofloxacin eye drops.

Sulphur mustard is a lipophillic compound and rapidly penetrates the skin. Hence, upon contact the skin should be decontaminated immediately and completely. Decontamination, five minutes after SM contact may not be beneficial as sufficient quantity would have been absorbed by the skin (Vijayaraghavan et al, 2002). A number of proprietary formulations are available that can efficiently decontaminate SM. The personal decontamination kit (PDK) is a multipurpose physical and chemical decontamination of toxic chemicals and microbes (Chilcott et al, 2001; Vijayaraghavan et al, 2011). It consists of PDK-1 and PDK-2 that are Fuller's earth, PDK(CC2) suspension, which is chemically N,N'-dichloro-bis(2,4,6) trichlorophenyl urea, a chemical decontaminant and RDP wipe which is surfactant soaked napkin for removal of solid particles. The sequence of decontamination is PDK-1 (if contact is less) or PDK-2 (if contact is more), followed by PDK(CC2) and then RDP wipe (Figure 2). Fuller's earth will physically remove the chemical and biological particles due to its adsorptive effect and PDK(CC2) will oxidise them. The surfactant soaked napkin will wipe off all the liquid and solid matter. In humans the treatment of skin lesions are mostly symptomatic. Several studies showed that topical treatment with iodine or povidone-iodine ointment significantly reduced the skin lesions induced by SM and that the ointment should be applied immediately after SM exposure (Brodsky *et al*, 2006; Wormser *et al*, 1997). The combination of anti-inflammatory agents and iodine increased the counter-irritating activity. Povidone-iodine preparation combined with an anti-inflammatory agent is better for the skin lesions induced by SM at relatively long intervals between exposure and treatment (Wormser *et al.*, 2004). Pruritus, is a common problem among SM exposed veterans. A number of treatments like antihistamines, local anesthetics, and corticosteroids are prescribed in order to control pruritus in Iranian patients (Shohrati *et al.*, 2007). Once the blister breaks, topical antibiotics like framycetin or mupirocin skin ointment can be applied.

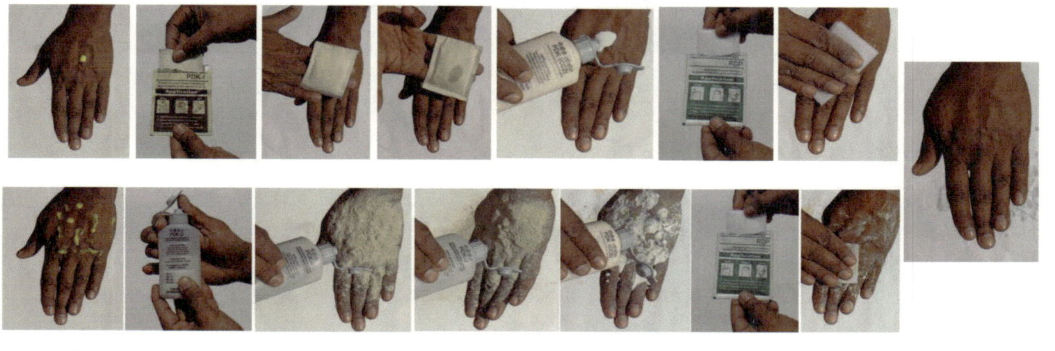

Fig. 2. Decontamination of chemicals with fuller's earth (PDK-1 or PDK-2) followed by PDK (CC-2) and RDP wipe.

Pharyngitis can be relieved by taking saline gargle. Intratracheal injection of SM in guinea pigs induced airway epithelial damage and corticosteroids like betamethasone showed significant effect on airway epithelium (Calvet *et al.*, 1996). Chronic bronchitis is the most frequent late respiratory disease among Iranians exposed to SM during the Iran-Iraq war. Oral and intravenous corticosteroid therapy was investigated in improving the lung function in SM induced chronic bronchitis patients (Ghanei *et al.*, 2005). Doxycycline may be a promising therapeutic agent for SM (Guignabert *et al.*, 2005). For the lung lesions and systemic toxicity of SM, paracetamol-ibuprofen combination for controlling pain and fever, beclomethasone inhaler as an anti-inflammatory corticosteroid, codeine phosphate or sulphate as a cough suppressant, doxycycline as an antibacterial agent, and N-acetyl cysteine, as a mucolytic and glutathione sparing drug are recommended (Lindsay *et al.*, 2008, Bobb et al, 2005; Ghanei et al, 2007, Vijayaraghavan et al, 2011). If haemoconcentration and shock are present either whole blood transfusion or transfusion of plasma expanders should be considered.

5. Classification of flavonoids

The flavonoids are secondary metabolites of plants and once they were considered to be waste products of plant metabolism. Flavonoids are the derivatives of benzopyrone and are widespread in photosynthesising plants. More than 4000 different flavonoids are known. The flavonoids are polyphenolic compounds with the basic skeleton of a phenyl benzopyrone (Flavone) ring. Various substitutions, mainly phenolic OH groups, take place in 3, 5, 7, 3′ and 4′ positions of the flavone nucleus. Flavonoids are generally found in nature as glycosides with sugar moiety attached to the 3 or 7 position. Depending upon the position of OH groups and of the pyrone nucleus, flavonoids are classified into various categories, viz., isoflavones, flavonol, flavone, flavonones, flavans and chalcones (Bohm, 1998). The isoflavones have their phenyl group (B ring) attached to the 3 position of the benzopyrone nucleus. Isoflavones have been shown to possess potential estrogenic actions. Flavonols have a OH group attached to the 3 position of the ring. Quercetin, gossypin and rutin are important members of this group. Flavanones result from the saturation of the double bond in the 2-3 position of the benzopyrone nucleus. Reduction of the carbonyl group of the pyrone ring and subsequent saturation of this ring gives flavans. Catechin and epicatechin are typical examples of this group. At higher pH pyrone ring in 1-2 position of flavonone opens resulting in chalcones. Hesperidin methyl chalcone is the best example of this class.

6. Flavonoids in the treatment of diseases

The use of plants in many forms has been recorded in the history. Knowledge of the usefulness of certain plants for medicinal purposes has been passed from generation to generation by word of mouth. Flavonoids were identified in search for physiologically active natural products. They have been shown to possess remarkable physiological activity in mammalian system. The mechanism of action is not known in most cases, but with the growing sophistication of analytical tools particularly X-ray analysis of flavonoid enzyme interaction, continues to reveal more details. It is difficult to predict any underlying structural feature of flavonoids which lead to specific biological function. No structure function relationship has emerged that why certain flavonoids can function as cytotoxic in nature while others function as cytoprotective agent.

Reports suggest that flavonoids are known to have hepatoprotective effect from ancient time onwards. Wagner (1986) described studies of plants as a source of liver protecting drugs from Indian folk medicine, viz., *Butea monosperma, Eclipta alba* and *Wedelia calandulae.* Hydnocarpin has been shown to be active against several human tumors and Ehrlich ascites tumor in mouse, and also exhibit anti-inflammatory and hypolipidemic activity. Studies on human immune response and inflammatory reactions showed the involvement of nitric oxide as a major participant. It has now been demonstrated that natural and synthetic flavonoids have significant inhibitory effect on the production of nitric oxide (Lee & Kim, 2010). There are reports that kaempferol and quercetin can efficiently suppress oxygen induced cytotoxicity. Quercetin derivative, rutin has been found to be effective in reducing toxic effects of iron overloading in experimental animals (Gao et al, 2006). The flavonoids are thought to form complex with iron atoms thus preventing them from catalysing the conversion of superoxide ions to harmful hydroxyl radicals. Flavonoids can exert their antioxidant activity by various mechanisms, including scavenging free radicals, which initiate lipid peroxidation, by binding metal ions, and by inhibiting enzymes responsible for free radical generation (Haleagrahara, et al, 2011).

Among the various actions of flavonoids, the anti inflammatory action has been extensively studied by several workers (Gabor, 1986; Parmar & Ghosh, 1978). Flavonoids were found to exert a beneficial effect in rheumatoid arthritis and also in gingival inflammatory conditions (Carvel & Halperin, 1961). Flavonoids are effective in acute as well as chronic inflammatory conditions (Lee et al, 2004; Martinez et al, 1997; Agarwal, 1982). Flavonone glycosides like naringin and hesperidins when administered intraperitoneally showed good response in both acute and chronic inflammatory models (Perriera et al, 2007). 5,7-di hydroxy 7-methoxy flavone and Wogonin (5,7-dihydroxy 8-methoxy flavone) were reported to have moderate inhibitory activity of prostaglandins by inhibiting COX-2 enzyme (Daott et al, 2003). Several flavonoids were investigated on various *in vitro* assays for lipoxygenase inhibitory activity. A significant anti-inflammatory and anti-arthritic activity of silymarin, a mixture of flavano lignans, was reported in animal model of inflammation by inhibition of 5- Lipoxygenase (Gupta et al, 2000).

Morin and fisetin were shown to inhibit the oxidative modification of low density lipoproteins. Flavonoids could improve the capillary resistance in scorbutic animals more effectively than pure vitamin C (Rusznyak & Szent Gyorgi, 1936). Further studies conducted in rats revealed an unequivocal effect of flavonoids in maintaining capillary integrity in these species (Benko et al, 1970). Hesperidine methyl chalcone increased the capillary resistance of small intestine, large intestine and kidney of guinea pigs kept on a scorbutic diet (Gabor et al, 1968). Reports suggest that flavonoids also have vasodilatory activity. Hesperidine and catechin could develop collateral circulation after left coronary occlusion (Brkic & Laszt, 1972). Perflavone has been suggested to be useful in the treatment of angina, atherosclerosis and myocardial infarction (Wagner, 1977). The cataract observed in diabetic and galactosemic conditions can be treated by flavonoids as they inhibit the enzyme aldose reductase in the lens (Varma et al, 1977; Parmar and Ghosh, 1979). Gastric anti-ulcer effect of various flavonoids has been extensively studied by Parmar (1977). The antihistaminic effect, histidine decarboxylase inhibition and mast cell stabilising effect of flavonoids may play an important role in the anti ulcer and anti secretory property of these compounds (Reimann et al, 1977; Fewtrell & Gomperts, 1977; Ramaswamy et al, 1979).

Cisplatin, a widely used anticancer drug causes undesirable side effects such as nephrotoxicity and hepatotoxicity. The protective effect of silymarin against cisplatin induced hepatotoxicity was evaluated in rats. Cisplatin caused an increase in serum alanine aminotransferase and aspartate aminotransferase, elevation of malondialdehyde and nitric oxide in liver tissues as well as decrease in reduced glutathione and the activities of antioxidant enzymes, including superoxide dismutase and glutathione peroxidase in liver tissues. Silymarin significantly reduced the hepatotoxicity induced by cisplatin (Mansour et al, 2006). Doxorubicin is a widely used anthracycline anticancer agent. Doxorubicin causes cardiotoxicity which is the major limitation of its clinical use. 7-monohydroxyethyl-rutoside protected the negative inotropic action of doxorubicin *in vitro* in electrically paced mouse left atrium model and *in vivo* in the ST-interval lengthening in the ECG. This protection did not affect the antitumor effect (Bast et al, 2007). Natural antioxidants like catechin are now known to detoxify toxic metabolites of xenobiotics. The modulatory and protective effect of catechin on tamoxifen, an anticancer drug induced disruption of glutathione metabolism and antioxidant enzyme was evaluated. Tamoxifen treatment resulted in a significant increase in the lipid peroxidation, hydrogen peroxide generation and protein carbonyl content in liver and kidney, and catechin treatment significantly protected them. The decrease in reduced glutathione content was also corrected by catechin. Catechin pre-treatment showed restoration in the level of cytochrome P450 content and in the activities of glutathione metabolizing enzymes, viz., glutathione-S-transferase, glutathione reductase and glutathione peroxidase, and other antioxidant enzymes such as, glucose-6-phosphate dehydrogenase, catalase and superoxide dismutase in both liver and kidney. There is good evidence that catechin supplementation can reduce tamoxifen induced toxicity as prophylactic and post treatment (Parvez et al, 2006).

Interestingly, several flavonoids and flavonoid derivatives have also been found to possess anti-mitotic behaviour which suggests possible use as anti-cancer drugs. These flavonoids exhibited inhibition of the conversion of tubulin into microtubules (Panda & Srivastav, 2007). The isoflavone genistein and orobol have been shown to induce mammalian topoisomerase II dependent DNA cleavage with activities comparable to those of known antitumor agents (Bandele & Osheroff, 2008). Quercetin and fisetin had activities similar to that of drug adriamycin.

Flavonoids have been found to be active against a variety of viruses. Among the animal viruses are polio virus, herpes simplex, and pseudorabies virus (Ogra et al, 1985). One of the standard approaches for determining the potential medicinal use of flavonoids is to screen them for activity against a range of bacteria and pathogenic fungi (Sasaki et al, 2011). The range of organisms against which certain flavonoids are active is quite impressive and in many cases not trivial. Anti fungal activity of flavonoids has also been demonstrated clearly (Parvez et al, 2009). Many flavonoids both natural and synthetic have been tested as potential medicinal agents against human disease including malaria (Montip et al, 2011) and HIV (Zhao et al, 2010).

Though, the plants are source of abundant flavonoids their usefulness could not be established. The activity of the flavonoids will depend to the large extent on their pharmacokinetics. It is a widespread belief that some plant preparations have almost magical properties. It is almost certain that many do contain potentially useful compounds but, the issues of effective dose, duration of effect, transport to site of action, life time of the effective agent and other factors must be considered before the claim can be said to have any pharmacological validity.

7. Flavonoids as promising cytoprotectants for sulphur mustard and nitrogen mustard induced oxidative stress

Sulphur mustard is a bifunctional alkylating agent and readily reacts with a variety of macromolecules including nucleic acids, proteins and lipids, as well as small molecular weight metabolites such as glutathione. By alkylating subcellular components, SM disrupts metabolism and can lead to oxidative stress and multi-organ failure. Increased formation of reactive oxygen species, the presence of lipid peroxidation products and oxidised proteins, and increase in antioxidant enzymes such as superoxide dismutase, catalase, and glutathione-S-transferase are the effects of SM and other mustard agents. Antioxidants will be very useful as prophylactic as well as post treatment for SM and other mustard agent toxicity (Laskin et al, 2010). Antioxidants could enhance survival time, protect liver and lung from oxidative damage and reduce accumulation of purine metabolites in blood following SM intoxication. Oxidative stress and damage to macromolecules in the human body is considered as one of the major consequences of SM toxicity. Intradermal administration of SM in rats resulted in increased thiobarbituric-acid-reactive substances, plasma protein carbonyls and ferric-reducing antioxidant power. Subcutaneous administration of melatonin protected these effects (Pohanka et al, 2011a). Flavonoids inhibit lipid peroxidation and also act as free radical scavengers. Sulphur mustard is a potent cytotoxic compound and many studies suggest that flavonoids are strong cytoprotectants (Figure 3).

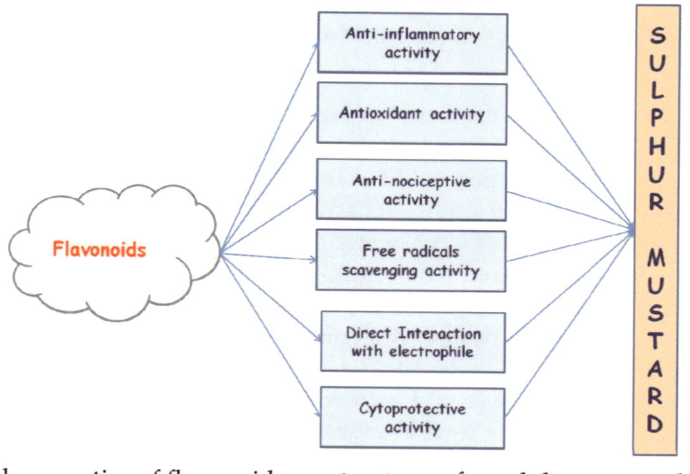

Fig. 3. Beneficial properties of flavonoids as a treatment for sulphur mustard toxicity.

When rats were exposed to SM at a dose of 20 or 80 mg/kg, depletion of reduced glutathione was observed. Epigallocatechin pretreatment did not show any appreciable protection, probably due to the use of high dose of SM (Pohanka et al, 2011b). The effect of dermal application of high dose of SM (155 mg/kg) on hepatic lipid peroxidation and the protective effect of flavonoids was investigated in mice. Sulphur mustard depleted reduced glutathione in blood and liver. Thiobarbituric acid reactive substances (TBARS) levels in the liver showed an increase indicating lipid peroxidation. Administration of vitamin E or flavonoids, gossypin or hydroxyethyl rutoside after dermal application of SM did not alter depletion of reduced glutathione but did reduce the malondialdehyde level significantly. The survival time of mice was increased by gossypin and hydroxyethyl rutoside to a greater extent than by vitamin E or sodium thiosulphate (Vijayaraghavan et al, 1991).

Protective effect of intraperitoneal administration of various antioxidants, trolox (water soluble analogue of vitamin E), quercetin and glutathione, was studied against SM by percutaneous administration and inhalation in mice. Survival time increased significantly following trolox and quercetin treatments when SM was exposed either through inhalation or by percutaneous route. The protection was better than intraperitoneal administration of glutathione. Significant decrease in reduced glutathione and increase in the level of malondialdehyde indicated oxidative damage to liver and lung tissues following SM inhalation and percutaneous exposure. Trolox and quercetin protected the liver and lung tissues from oxidative damage caused by SM exposure through inhalation and percutaneous routes (Kumar et al, 2001). Quercetin was administered intraperitoneally to mice along with SM and oxidative stress parameters were evaluated after 7 days. Sulphur mustard decreased the body weight significantly and quercetin protected the mice significantly, in a dose dependent manner. The protection was better only when quercetin was administered as pretreatment or simultaneous treatment. The decrease in reduced and oxidised glutathione levels, and the increase in malondialdehyde level following dermal application of SM was protected by quercetin, when it was administered as pre-treatment or simultaneous treatment. The histological lesions induced by sulphur mustard on liver, spleen and skin were also significantly protected by quercetin as a pretreatment or as simultaneous treatment. This clearly proved that percutaneous administration of SM induces oxidative stress and quercetin can protect it only as a prophylactic agent (Gautam et al, 2007).

Gossypin is a water soluble and naturally occurring bioflavonoid. The protection of varying doses of gossypin administered intraperitoneally was studied, prior to, simultaneous and 2 hr after percutaneous administration of SM in mice. The protection against systemic toxicity of SM was better when gossypin was given with lipophylic solvents (polyethylene glycol-300 or dimethyl sulphoxide) than with water. Good protection (8.0 folds) was observed when gossypin was administered (200 mg/kg in PEG-300; i.p.) at 30 min prior or simultaneous to SM exposure, but no protection was observed when gossypin was administered 2 hr post to SM exposure. A significant decrease in total antioxidant status of plasma, liver glutathione level (reduced and oxidised) and the activities of glutathione peroxidase, glutathione reductase and superoxide dismutase were observed after SM administration. Sulphur mustard treated mouse liver also showed necrosis. A significant protection was observed in these variables when gossypin was administered as pre-treatment or simultaneous treatment, and not as post treatment (Gautam and Vijayaraghavan 2007).

Hippophae rhamnoides (Linn) also known as sea-buckthorn is a high altitude (Indian Continent) and plain land (eg. Ukrain) growing shrub yielding small berries. The fruit is rich in vitamin C, antioxidants and flavonoids. Extracts of *H. rhamnoides* leaf and fruit, and *H. rhamnoides* flavone from fruit were evaluated against percutaneously administered SM in mice. Significant protection was observed with oral administration of ethanolic extract of *H.*

rhamnoides leaf and *H. rhamnoides* flavone. Reduced glutathione and oxidised glutalthione levels were decreased, and malondialdehyde was elevated after percutaneous administration of SM. Oral administration of ethanolic extract of *H. rhamnoides* and *H. rhamnoides* flavone significantly protected theses variables (Vijayaraghavan et al, 2006). Comparative evaluation of bioflavonoids viz., quercetin, gossypin and *H. rhamnoides* flavone, and tocopherol acetate (vitamin E) were carried out against the systemic toxicity of SM in mice (Figure 4). Quercetin, gossypin, *H. rhamnoides* flavone and tocopherol acetate were administered intraperitoneally prior to percutaneous administration of SM, and protection against the SM lethality and biochemical parameters were evaluated. The protection against the lethality of SM was very good with the flavonoids in the order gossypin showed 6.7 folds, *H. rhamnoides* flavone showed 5.6 folds and quercetin showed 4.7 folds protection. Tocopherol acetate did not give any protection (0.7 fold). Percutaneous administration of SM showed decrease in reduced and oxidised glutathione levels, and an increase in malondialdehyde level. Antioxidant enzymes like glutathione peroxidase, glutathione reductase and superoxide dismutase were significantly decreased. The total antioxidant status was also significantly decreased. Quercetin, gossypin and *H. rhamnoides* flavone significantly protected the reduced and oxidised glutathione level, and malondialdehyde level. Tocopherol acetate failed to offer any protection (Table 1). This study supported that SM induces oxidative stress and flavonoids are promising cytoprotectants against the toxic effects of SM better than vitamin E, but they all are effective only as a pretreatment (Vijayaraghavan et al., 2008).

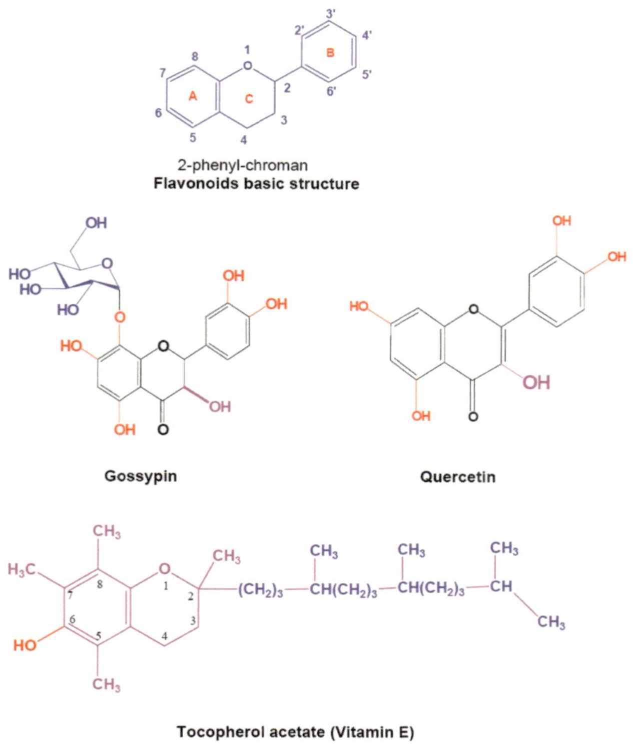

Fig. 4. Chemical structure of flavonoids and tocopherol acetate.

Groups	Dose (mg/kg)	Route	PI
Flavonoids:			
Quercetin	200	i.p.	4.7
Gossypin	200	i.p.	6.7
H. rhamnoides -flavone	200	i.p.	5.6
Herbal Extract:			
Aloe vera Leaf Extract	1000	p.o.	1.9
Ginkgo Biloba Extract	1000	p.o.	4.0
H. rhamnoides Leaf Extract	1000	p.o.	2.4
H. rhamnoides Fruit Extract	1000	p.o.	1.7
Other antioxidants:			
Tocopherol Acetate	200	i.p.	0.7
N-acetyl cysteine	125	i.p.	2.8
Melatonin	125	i.p.	5.6

Table 1. Protective efficacy of flavonoids and herbal extracts against sulphur mustard induced toxicity compared with other antioxidants in mice. LD50 of sulphur mustard by percutaneous route in mice is 8.1 mg/kg.

Nitrogen mustards are also alkylating agents and damage cellular nuclear DNA after penetrating the tissue. Rats were exposed to nitrogen mustard and treated with proanthocyanidin. A segment of the cortical tissue was prepared and evaluated by electron microscopy. Degeneration of the cortical neural cell nuclei with edema and axonal degeneration in the subcortical neural tissue were observed. Proanthocyanidin treatment showed less edema and degeneration (Tekiner et al, 2009). Hesperidin was investigated in mouse bone marrow cells against the genotoxicity induced by cyclophosphamide. Mice were orally pretreated with hesperidin and injected intraperitoneally with cyclophosphamide. Hesperidine could significantly protect the toxicity induced by cyclophosphamide on occurrence of the micronucleated polychromatic erythrocytes. Hesperidin due to its antioxidant effect reduced the oxidative stress and genotoxicity induced by cyclophosphamide in mouse bone marrow cells (Ahmadi et al, 2008). The preventive effect of hawthorn (*Crataegus microphylla*) fruit extract which is a rich source of flavonoids was investigated in mouse bone marrow cells against genotoxicity induced by cyclophosphamide. Hawthorn contains high amounts of phenolic compounds, chlorogenic acid, epicatechin and hyperoside. Mice were given hawthorn extract orally and cyclophosphamide intraperitoneally. Hawthorn extract could significantly protect the toxicity induced by cyclophosphamide on the occurrence of micronucleated polychromatic erythrocytes (Hosseinimehr et al, 2008).

8. Conclusion

Flavonoids are promising as antidotes, due to their plethora of beneficial effects that can mitigate the varied symptoms of sulphur mustard and nitrogen mustard. As a chemical they are very safe and the only limitation is its poor absorption by oral route. If their structure is modified to improve the bioavailability through oral route, and also with a stronger nucleophilic property by simplifying the structure and adding electron donating functional groups, they can be recommended as treatment and also as supportive therapy for a variety of diseases. They also will be highly useful for scavenging variety of toxic chemicals.

9. Acknowledgement

Authors are grateful to Dr. S. J. S. Flora, Head, Division of Pharmacology and Toxicology, Defence Research and Development Establishment, for keen interest and valuable suggestions.

10. References

Agarwal, O.P., 1982, The antiinflammatory action of nepritin. Agents and actions., 12, 298-332.

Ahmadi, A., Hosseinimehr, S.J., Naghshvar, F., Hajir, E., & Ghahremani, M., 2008, Chemoprotective effects of hesperidin against genotoxicity induced by cyclophosphamide in mice bone marrow cells. Arch Pharm Res., 31, 794 - 797.

Bajgar, J., 2004, Organophosphates nerve agent poisoning: Mechanism of action, diagnosis, prophylaxis and treatment, Adv. Clin, Chem., 38, 151 - 216.

Bandele, O.J., & Osheroff, N., 2008, The efficacy of topoisomerase II-targeted anticancer agents reflects the persistence of drug-induced cleavage complexes in cells. Biochemistry, 11, 47, 11900-11908.

Bast. A., Kaiserova, H., Den Hartog, G.J., Haenen, G.R., & van der Vijgh, W.J., 2007, Protectors against doxorubicin-induced cardiotoxicity: flavonoids. Cell Biol. Toxicol., 23, 39 - 47.

Benko, S., Gabor, M., Varkonyi, T., Antal, A., & Foldi, M., 1970, Brain oedema and sub pleural haemorrhage in experimental P-avitaminosis. Physiol Chem and Phy., 2, 110-116.

Bhattacharya, R., & Vijayaraghavan, R. 2002, Promising role of alpha-ketoglutarate in protecting against the lethal effects of cyanide, Hum. Exptl. Toxicol., 21, 297 - 303.

Bobb, A.J., Arfsten, D.P., & Jederberg, W.W., 2005, N-acetyl-L-Cysteine as prophylaxis against sulfur mustard. Mil. Med., 170, 52 - 56.

Bohm, B.A., 1998, Introduction to flavonoids, Harwood Acad. Pub., The Netherlands.

Brodsky, B., Trivedi, S., Peddada, S., Flagler, N., Wormser, U., & Nyska, A., 2006. Early effects of iodine on DNA synthesis in sulfur mustard-induced skin lesions. Arch. Toxicol., 80, 212 - 216.

Calvet, J.H., Coste, A., Levame, M., Harf, A., Macquin-Mavier, I., & Escudier, E., 1996, Airway epithelial damage induced by sulfur mustard in guinea pigs, effects of glucocorticoids. *Hum. Exp. Toxicol..* 15, 964 - 971.

Carvel, R.I., & Halperin, V.C., 1961, Therapeutic effect of water soluble bioflavonoids in gingival inflammatory conditions. Oral Surg Oral Med and Oral Path. 14, 847-855.

Casbohm, S.L., Rogers, J.V., Stonerock, M.K., Martin, J.L., Ricketts-Kaminsky, K.M., Babin, M.C., Casillas, R.P., & Sabourin, C.L., 2004, Localization of substance P gene expression for evaluating protective countermeasures against sulfur mustard. *Toxicology,* 204, 229 - 239.

Chen, K.K., & Rose, C.L. 1952, Nitrite and thiosulphate therapy in cyanide poisoning, JAMA, 149, 113 – 119.

Chilcott, R.P., Jenner, J., Hotchkiss, S.A., & Rice, P., 2001, *In vitro* skin absorption and decontamination of sulphur mustard: comparison of human and pig-ear skin. J. Appl. Toxicol., 21, 279 - 283.

Dacre, J.C., & Goldman, M. 1996, Toxicology and pharmacology of the chemical warfare agent sulphur mustard, Pharmacol. Rev., 48, 289 – 326.

Daott, C., Kim, J., Kim, H.P., Kim, S., & Park H., 2003, Synthesis and PGE2 inhibitory activity of 5, 7 – dihydroxy flavones and their O-methylated flavones analogs. Arch Pharm Res., 6, 245-250.

Friedl, K.E., Hannan, C.J. Jr., Schadler, P.W., & Jacob, W.H. 1989, Atropine absorption after intramuscular administration with 2-pralidoxime chloride by two automatic injector devices. J. Pharm. Sci., 78, 728 - 731.

Fewtrell, C.M.S. & Gomperts, B.D., 1977, Effect of flavone inhibitors of transport ATpases on histamine secretion from mast cells. Nature, 265, 635-636.

Gabor, M., 1986, Anti-inflammatory and anti-allergic properties of flavonoids. Prog Clin Biol Res., 213, 471-480.

Ganesan, K., Raza, S.K., & Vijayaraghavan, R., 2010, Chemical warfare agents, J. Pharm. Bioallied. Sci., 2, 166-178.

Gautam, A., & Vijayaraghavan, R., 2007, The prophylactic effect of gossypin against percutaneously administered sulfur mustard, Biomed. Environ. Sci., 20, 250 – 259.

Gautam, A., Vijayaraghavan, R., Pant, S.C., Kumar, O., Singh, S., & Satish Kumar, H.T., 2007, Protective effect of quercetin against sulphur mustard induced oxidative stress in mice, Def. Sci. J., 57, 707 - 720.

Ghanei, M., Khalili, A.R., Arab, M.J., Mojtahedzadeh, M., Aslani, J., Lessan-Pezeshki, M., Panahi, Y., & Alaeddini, F., 2005, Diagnostic and therapeutic value of short-term corticosteroid therapy in exacerbation of mustard gas-induced chronic bronchitis. *Basic Clin. Pharmacol. Toxicol.,* 97, 302 - 305.

Ghanei, M., Shohrati, M., Harandi, A.A., Eshraghi, M., Aslani, J., Alaeddini F., & Manzoori H., 2007, Inhaled corticosteroids and long-acting beta 2-agonists in treatment of patients with chronic bronchiolitis following exposure to sulfur mustard. Inhal. Toxicol., 19, 889 - 894.

Guignabert, C., Taysse, L., Calvet, J.H., Planus, E., Delamanche, S., Galiacy, S., & d'Ortho, M.P., 2005, Effect of doxycycline on sulfur mustard-induced respiratory lesions in guinea pigs. Am. J. Physiol. Lung. Cell. Mol. Physiol., 289, L67 - 74.

Gupta, O.P., Singh, S., Bani, S., Sharma, N., Malhotra, S., Gupta, B.D., Banerjee, S.K., & Handa, S.S., 2000, Antiinflammatory and anti arthritic activities of silymarin acting through inhibition of 5-lipoxygenase. Phytomed., 1, 21-24.

Gutteridge, C.E., Thota, D.S., Curtis, S.M., Kozar, M.P., Li, Q., Xie, L., Zhang, J., & Melendez, V., 2011, Asher CO, Luong TT, Gerena L, Nichols DA,Montip G. In vitro biotransformation, in vivo efficacy and pharmacokinetics of antimalarial chalcones. Pharmacol., 87, 96-104.

Haleagrahara, N., Siew, C.J., Mitra, N.K., & Kumari, M., 2011, Neuroprotective effect of bioflavonoid quercetin in 6-hydroxydopamine-induced oxidative stress biomarkers in the rat striatum. Neurosci Lett. 2011, 500(2), 139-143.

Hosseinimehr, S.J., Azadbakht, M., & Abadi, A.J., 2008, Protective effect of hawthorn extract against genotoxicity induced by cyclophosphamide in mouse bone marrow cells. Environ. Toxicol. Pharmacol., 25, 51 - 56.

Hossion, A.M., Zamami, Y., Kandahary, R.K., Tsuchiya, T., Ogawa, W., Iwado, A., & Sasaki, K., 2011, Quercetin diacylglycoside analogues showing dual inhibition of DNA gyrase and topoisomerase IV as novel antibacterial agents. J Med Chem., 54, 3686-3703.

Hussain, T., Siddiqui, H.L., Zia-ur-Rehman, M., Masoom, Yasinzai, M., & Parvez, M., 2009, Anti-oxidant, anti-fungal and anti-leishmanial activities of novel 3-[4-(1H-imidazol-1-yl) phenyl]prop-2-en-1-ones. Eur J Med Chem., 44, 4654-4660.

Jiang, F., Chen, W., Yi, K., Wu, Z., Si, Y., Han, W., & Zhao, Y., 2010, The evaluation of catechins that contain a galloyl moiety as potential HIV-1 integrase inhibitors. Clin Immunol. 137, 347-356.

Kadar, T., Fishbeine, E., Meshulam, Y., Sahar, R., Chapman, S., Liani, H., Barness, I., & Amir, A., 2000, Treatment of skin injuries induced by sulfur mustard with calmodulin antagonists, using the pig model. J. Appl. Toxicol. 20, S133 - 136.

Kadar, T., Turetz, J., Fishbine, E., Sahar, R., Chapman, S., & Amir, A. (2001). Characterization of acute and delayed ocular lesions induced by sulfur mustard in rabbits. Curr. Eye Res. 22, 42-53.

Kaul, T.N., Middleton, E. Jr., & Ogra, P.L., 1985, Antiviral effect of flavonoids on human viruses. J Med Virol., 15, 71-79.

Kehe, K., & Szinicz, L. 2005, Medical aspects of sulphur mustard poisoning. Toxicology, 214, 198 - 209.

Kumar, O., Sugendran, K., & Vijayaraghavan, R., 2001, Protective effect of various antioxidants on the toxicity of sulphur mustard administered to mice by inhalation or percutaneous routes, Chemico-Biol. Int., 34, 1 – 12.

Laskin, J.D., Black, A.T., Jan, Y.H., Sinko, P.J., Heindel, N.D., Sunil, V., Heck, D.E., & Laskin, D.L., 2010, Oxidants and antioxidants in sulfur mustard-induced injury. Ann. N. Y. Acad. Sci., 1203, 92 - 100.

Lee, H.G., Kim, H., Oh, W.K., Yuk, A., Choe, Y.K., Ahn, J.S., Kim, S.H., Dinarello, C.A., Kim, K., & Yoon, D.Y., 2004, Tetramethoxy hydroxy flavone PYF downregulates inflammatory mediators via the inhibition of Nuclear factor kappa B, Ann N Y Acad Sci, 1030, 555-568.

Lee, J.H., & Kim, G.H., 2010, Evaluation of antioxidant and inhibitory activities for different subclasses flavonoids on enzymes for rheumatoid arthritis. J Food Sci., 75, 7

Lindsay, C. D., Gentilhomme, E., & Mathieu, J. D., 2008, The use of doxycycline as a protectant against sulphur mustard in HaCaT cells. *J. Appl. Toxicol., 28,* 665 - 673.

Mansour, H.H., Hafez, H.F., & Fahmy, N.M., 2006, Silymarin modulates Cisplatin-induced oxidative stress and hepatotoxicity in rats. J. Biochem. Mol. Biol., 39, 656 - 661.

Martinez, J., Silvan, A.M., Abad, M.J., Bermejo, P., Villar, A., & Sollhuber, M., 1997, Isolation of two flavonoids from Tanacetum microphyllum as PMA-induced ear edema inhibitors. J.Nat Prod. 60, 142-147.

Narayana, K.R., Reddy, S.M., Chaluvadi, M.R., & Krishna, D.R., 2001, Bioflavonoids classification, pharmacological, biochemical effects and therapeutic potential, Indian J. Pharmacol., 33, 2 – 16.

OPCW, Convention on the prohibition of the development, production, stockpiling and use of chemical weapons and destruction, Technical Secretariat of The Organisation for the Prohibition of Chemical Weapons, Johan de Wittalaan, The Hague, The Netherlands.

Papirmeister, B., Feister, A.J., Robinson, S.I., & Ford, R.D., 1991, Mustard gas. Toxic mechanisms and pharmacological implications. Boca Raton, FL: CRC Press.

Parmar, N.S., & Ghosh, M.N., 1978, Antiinflammatory activity of gossypin, a bioflavonoid isolated from Hibiscus Vitifolius Linn. Ind J pharmacol, 10, 277-293.

Parmar, N.S., & Ghosh, M.N., 1979, Effect of gossypin a bioflavonoid on the formation of galactose induced cataracts in rats. Exp Eye Res., 29, 229- 232.

Pohanka, M., Sobotka, J., Jilkova, M., & Stetina, R., 2011a, Oxidative stress after sulfur mustard intoxication and its reduction by melatonin: efficacy of antioxidant therapy during serious intoxication. Drug Chem. Toxicol., 34, 85 - 91.

Pohanka, M., Sobotka, J., & Stetina, R., 2011b, Sulfur mustard induced oxidative stress and its alteration by epigallocatechin gallate. Toxicol. Lett., 201, 105 - 109.

Parvez, S., Tabassum, H., Rehman, H., Banerjee, B.D., Athar, M., & Raisuddin, S., 2006, Catechin prevents tamoxifen-induced oxidative stress and biochemical perturbations in mice. Toxicology, 225, 109 - 118.

Periera, M.S.R., Andrades, E.D.N., Paulino, N., Sawaya, F.H.C.A., & Eberlin, N.M., 2007, Synthesis and characterization of a metal complex containing Naringin and Cu, and its antioxidant, antimicrobial,anti-inflammatory and tumour cell cytotoxicity. Molecules, 12, 1352-1366.

Ramaswamy, S., Padmanabha, Pillai, N., & Parmar, N.S., 1979, The antihistaminic activity of (+) cyanidanol-3 on isolated guinea pig ileum. Indian J pharmacol, 11, 135-138.

Ray, R., Benton, B.J., Anderson, D.R., Byers, S.L., & Petrali, J.P., 2000, Intervention of sulfur mustard toxicity by downregulation of cell proliferation and metabolic rates. *J. Appl. Toxicol.* 20, S87 - 91.

Riemann, H.J., Lorenz, W., Fischer, M., Frolich, R., & Meyer, H.J., 1977, Histamine and acute haemorrhagic lesions in rat gastric mucosa; Prevention of stress ulcer formation by catechin, an inhibitor of specific histidine decarboxylase in vitro. Agents and Actions, 7, 69-72.

Rusznyak, I., & Gyorgi,. S.A., 1936, Vitamin P: Flavonol as vitamins. Nature, 27, 138-142.

Safarinejad, M. R., Moosavi, S. A., & Montazeri, B. (2001). Ocular injuries caused by mustard gas: diagnosis, treatment, and medical defense. Mil. Med. 166, 67-70.

Shohrati, M., Davoudi, S. M., Keshavarz, S., Sadr, B., & Tajik, A., 2007, Cetirizine, doxepine, and hydroxyzine in the treatment of pruritus due to sulfur mustard: a randomized clinical trial. *Cutan. Ocul. Toxicol.*, 26, 249 - 255.

Somani, S.M., 1992. Chemical Warfare Agents. Academic Press Inc, London.

Srivastava, P., & Panda, D., 2007, Rotenone inhibits mammalian cell proliferation by inhibiting microtubule assembly through tubulin binding. FEBS J., 274, 4788-4801.

Tekiner, A., Yucel, O., Sargin, A.K., Genc, O., Can, B., Karayilanoglu, T., Karakaya, J., & Bayar, M.A., 2009, The effect of nitrogen mustard on the rat brain and the therapeutic value of proanthocyanidin. Turk. Neurosurg., 19, 360 - 366.

Varma, S.D., Mizuno, A., & Kinoshita, J.H., 1977, Diabetic cataracts and flavonoids. Science, 195, 205-206.

Vijayaraghavan, R., 1997, Modifications of breathing pattern induced by inhaled sulphur mustard in mice. *Arch. Toxicol.*, 71, 157 - 164.

Vijayaraghavan, R., Anshoo Gautam, Om Kumar, Pant, S.C., Manoj Sharma, Seema Singh, Satish Kumar, H.T., Anand Kumar Singh, Manisha Nivsarkar, Kaushik, M.P., Sawhney, R.C., Chaurasia, O.P., & Prasad, G.B.K.S., 2006, The protective effect of ethanolic and water extracts of Hippophae rhamnoides against the toxic effects of mustard gas Protective effect of ethanolic and water extracts of Sea Buckthorn (Hippophae rhamnoides L.) against the toxic effects of mustard gas, Indian J Exptl. Biol., 44, 821 - 831. 2006.

Vijayaraghavan, R., Bhaskar, A.S.B., Anshoo Gautam, Gopalan, N., Anil Kumar Singh, Beer Singh, & Flora, S.J.S., 2011, A convenient first aid kit for chemical and biological agents and radiation exposure, J. Environ. Biol., In press.

Vijayaraghavan, R., Gautam, A., & Sharma, M. 2009, Medical countermeasures and other therapeutic strategies for sulphur mustard toxicity, Chapter 60, In: Handbook of Toxicology of Chemical Warfare Agents (Gupta RC, Ed), Academic Press, New York, 897 - 918.

Vijayaraghavan, R., Gautam, A., Sharma, M., Satish, H.T., Pant, S.C., & Ganesan, K., 2008, Comparative evaluation of some flavonoids and tocopherol acetate against the systemic toxicity of sulfur mustard, Indian J. Pharmacol., 40, 114 - 120.

Vijayaraghavan, R., Kulkarni, A.S., Pant, S.C., Kumar, P., Lakshmana Rao, P.V., Gupta, N., Gautam, A., & Ganesan, K., 2005, Differential toxicity of sulfur mustard administered through percutaneous, subcutaneous and oral routes, Toxicol. Appl. Pharmacol., 202, 180 - 188.

Vijayaraghavan, R., Kumar, P., Dubey, D.K., Ram Singh, Sachan, A.S., Deo Kumar, Sugendran, K., Om Kumar, Singh, M., Pant, S.C., & Bhattacharya, R., 2002. Acute toxicity studies of CC2: an effective chemical decontaminant of sulphur mustard in hydrophilic formulation. *Ind. J. Pharmacol.*, 34, 321 - 331.

Vijayaraghavan, R., Neeti Jain, Anshoo Gautam, Manoj Sharma, Seema Singh, Deo Kumar, Ram Singh, Pravin Kumar, Gupta, A.K., & Suman Jain, 2007, Evaluation of the antidotal efficacy of atropine sulphate and pralidoxime chloride given by autoinjectors against nerve agent (sarin) toxicity, J. Med. C.B.R. Def. 5, e1 - 12.

Vijayaraghavan, R., Sugendran, K., Pant, S.C., Husain, K., & Malhotra, R.C., 1991, Dermal intoxication of mice with bis (2 chloroethyl) sulphide and the protective effect of flavonoids, Toxicology, 69, 35 – 42.

Way, J.L., 1984, Cyanide intoxication and its mechanism of antagonism. Annu. Rev. Pharmacol. Toxicol., 24, 451 – 481.

Wagner, H., Geyer, B., Fiebig, M., Kiso, Y., & Hikino, H., 1986, Isobutrin and butrin, the antihepatotoxic principles of Butea monosperma flowers. Planta Med., 2, 77-79.

Wormser, U., Brodsky, B., Green, B.S., Arad-Yellin, R., & Nyska, A., 1997, Protective effect of povidone-iodine ointment against skin lesions induced by sulphur and nitrogen mustards and by non-mustard vesicants. *Arch. Toxicol.,* 71, 165 - 170.

Wormser, U., Sintov, A., Brodsky, B., Casillas, R. P., & Nyska, A., 2004, Protective effect of topical iodine containing anti-inflammatory drugs against sulfur mustard-induced skin lesions. *Arch. Toxicol.,* 78, 156 - 166.

Zhang, Y., Li, H., Zhao, Y., & Gao, Z., 2006, Dietary supplementation of baicalin and quercetin attenuates iron overload induced mouse liver injury. Eur J Pharmacol., 27, 535, 263-9.

13

The Effects of Propolis in Animals Exposed Oxidative Stress

Pinar Tatli Seven[1], Seval Yilmaz[2],
Ismail Seven[3] and Gulizar Tuna Kelestemur[4]
[1]*University of Firat, Faculty of Veterinary Medicine,*
Department of Animal Nutrition and Nutritional Diseases, Elazig,
[2]*University of Firat, Faculty of Veterinary Medicine,*
Department of Biochemistry, Elazig,
[3]*University of Firat, Vocation School of Sivrice,*
Department of Beekeping, Elazig,
[4]*University of Firat, Faculty of Fisheries,*
Department of Aquaculture, Elazig,
Turkey

1. Introduction

This chapter expresses the effects of propolis on oxidative stress in animals. The term "stress" was first coined by the endocrinologist Hans Selye (1936) more than 70 years ago to define the physiological adaptive responses of the organism to emotional or physical threats (stressors), whether real or perceived (Selye, 1936). Factors causing stress include physiological factors, such as climate, environment, nutrition, and diseases, and physical conditions, such as cage density and transport. Under stress, rapid and temporary changes occur in the body initially; with continuous stress, these are followed by permanent and irreversible changes (Tatli Seven, 2008). Stress responses are characterized as primary, secondary and tertiary. The primary stress response is a neuroendocrine response leading to corticosteroid and catecholamine release. The secondary stress response includes changes in plasma and tissue ion and metabolite levels induced by neuroendocrine hormones. The changes in disease resistance, growth, condition factor, and behaviors at a whole organism level are tertiary responses (Wedemeyer et al., 1990). Finally, a decline in yield and resistance to diseases may occur. Animals under stress become ill more easily, and excess medicine may be necessary to maintain health. As a result, drug residues increase in animal products and threaten public health directly. Stock health and welfare management are key factors in animal health and food safety. For this reason, stress conditions in animals need to be examined carefully (Tatli Seven, 2008). Reactive oxygen species (ROS) are chemically reactive molecules containing oxygen. During times of environmental stress (e.g. ultraviolet or heat exposure, environmental pollutant), ROS levels can increase dramatically.This may result in significant damage to cell structures.This cumulates into a situation known as oxidative stress.

This chapter was written to demonstrate the importance of propolis that have effects antioxidant, antibacterial, antitumor, anti-inflammatory etc. in animals under oxidative stress.

2. Oxidative stress

Oxidative Stress is a general term used to describe the effect of oxidation in which an abnormal level of ROS, such as the free radicals (e.g. hydroxyl, nitric acid, superoxide) or the non-radicals (e.g. hydrogen peroxide (H_2O_2), lipid peroxide) lead of damage (called oxidative damage) to specific molecules with consequential injury to cells or tissue. Oxidative stress is caused by an imbalance between the production of reactive oxygen and a biological system's ability to readily detoxify the reactive intermediates or easily repair the resulting damage (Bulger & Helton, 1998). Oxidative stress occurs when the generation of ROS in a system exceeds the system's ability to neutralize and eliminate them. The imbalance can result from a lack of antioxidant capacity caused by disturbance in production, distribution, or by an over-abundance of ROS from an environmental or behavioral stressor. This damage can affect a specific molecule or the entire organism. If not regulated properly, the excess ROS can damage a cell's lipids, protein or DNA, inhibiting normal function. Because of this, oxidative stress has been implicated in a growing list of diseases as well as in the aging process (Sies, 1985).

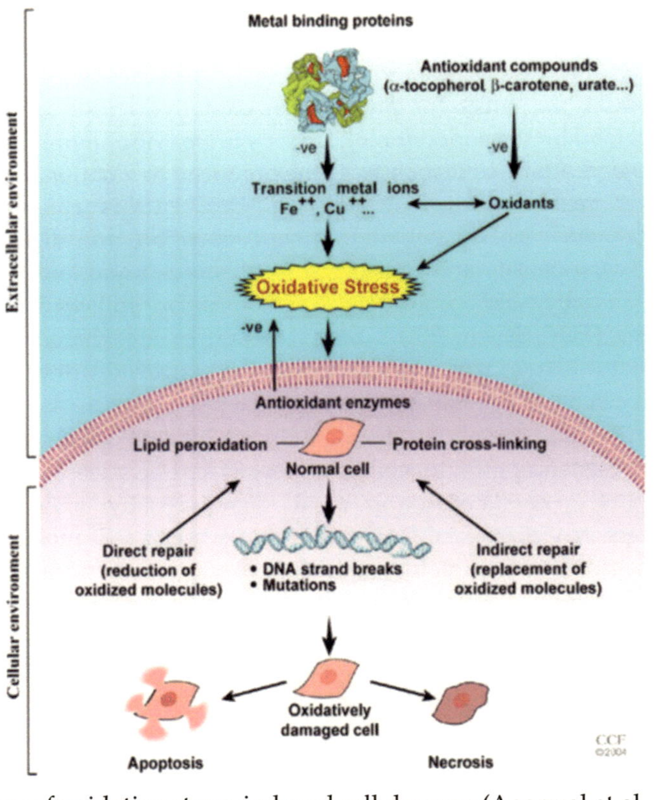

Fig. 1. Mechanisms of oxidative stress-induced cell damage (Agarwal et al., 2005).

ROS can impair lipids, proteins, carbohydrates and nucleotides, which are important parts of cellular constituents, including membranes, enzymes and DNA. Radical damage can be significant because it can generally proceed as a chain reaction (Chen and Pan, 1996; Wejil et al., 1997). These radicals can damage cell membranes inducing lipid peroxidation of polyunsaturated fatty acids in the cell membranes and other complexes (Fang et al., 2002; Stephan et al., 1995) Malondialdehyde (MDA) is one of the the final products of lipid peroxidation. The concentration of MDA is the direct evidence of toxic processes caused by free radicals (Talas & Gulhan, 2009; Tatli Seven et al., 2009). Damaged lipids lead to rigid cell membranes; oxidized cholesterol often leads to hardening of the arteries and poorly repaired DNA chains lead to cell mutation (future generation of cells) as implicated in cancer and aging. Scientific research has established that the root cause of more than seventy chronic degenerative diseases is due to oxidative stress development, i.e. cell damage caused by free radicals (Davies, 1995).

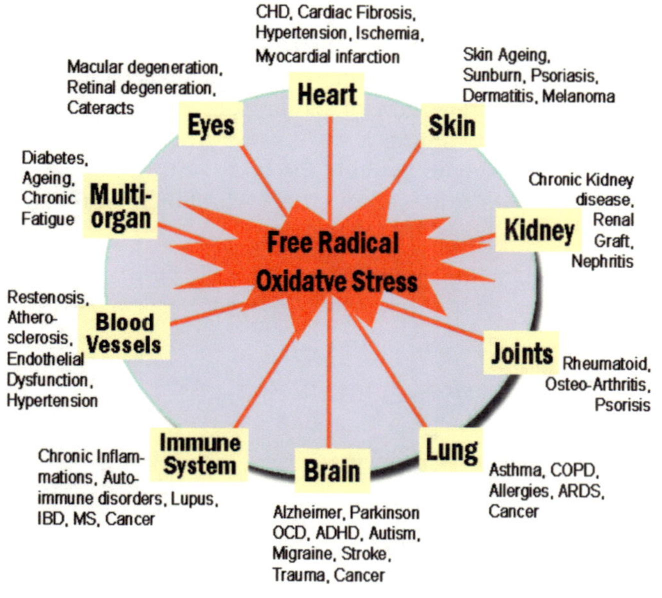

Fig. 2. Diseases related to oxidative stress (http://www.enzoprofessional.com/default).

The intensity of oxidative stress is determined by the balance between the rate at which oxidative damage is induced (input) and the rate at which it is efficiently repaired and removed (output). The balance provides certain steady-state ROS level (Lushchak, 2011). The rate at which damage caused is determined by how fast ROS are generated and then eliminated by endogenous defense agents called antioxidants. The rate at which damage is removed depends on the efficiency of repair enzymes (Sies, 1985). Detoxification of ROS is one of the prerequisites of aerobic life, and hence an elaborate antioxidant system has evolved (Sies, 1991). Antioxidants are agents that scavenge ROS, prevent their formation, or repair the damage they cause (Halliwell, 1991). Antioxidants are effective because they are capable to donate their own electrons to free radicals. When a free radical gains the electron from an antioxidant, it no longer attack the cell and the chain reaction of oxidation is broken. After donating an electron, an antioxidant becomes a free radical by definition. Antioxidants in this state are not harmful because they have the ability to accommodate the change in electrons without becoming reactive. The body has an elaborate antioxidant defense system. Antioxidants are produced within the body and can also be received from food such as fruits, vegetables, seeds, nuts, meats, and oil. There are two lines of antioxidant defense within the cell. The first line, found in the fat-soluble cellular membrane consists of vitamin E, beta-carotene, and coenzyme Q. Of these, vitamin E is considered the most potent chain breaking antioxidant within the membrane of the cell (Yilmaz et al., 2006). Inside the cell, water soluble antioxidant scavengers are present. These include vitamin C, glutathione (GSH), glutathione peroxidase (GPx), superoxide dismutase (SOD), and catalase (CAT) (Atessahin et al., 2005).

This complex system consists of GSH, ancillary enzymes (such as glutathione reductase (GR), glutathione S-transferase (GST), glucose-6-phosphate dehydrogenase (G6PD)), metal-binding proteins (transferrin, ceruloplasmin, and albumin), flavonoids, and urate (Sies, 1991; Halliwell, 1994). In addition to those antioxidants herbs, such as bilberry, blueberry, turmeric (curcumin), grape seed or pine bark extracts, licopen, propolis and ginkgo can also provide powerful antioxidant protection for the body. Thus, there is a delicate balance between the generation and elimination of oxidant agents, which may be beneficial or deleterious to the organism. Consuming a wide variety of antioxidant enzymes, vitamins and minerals may be the best way to provide the body with the most complete protection against free radical damage (Sies & Masumoto, 1997; Yaralioglu Gurgoze et al., 2005; Yilmaz et al., 2006).

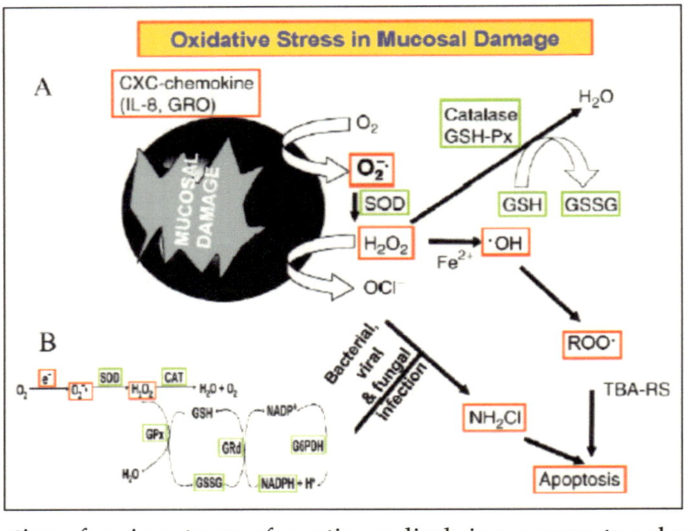

Fig. 3. The formation of various types of reactive radicals in response to enhanced steady-state ROS levels related to mucosal damage and antioxidative enzymes involved in neutralization of free radicals (Czesnikiewicz-Guzik et al., 2007).

In order to protect against lipid peroxidation and oxidative damage, all living organisms have evolved an interdependent antioxidant system that includes enzymatic and non-enzymatic components in the liver and erythrocytes. The major antioxidant enzymes are SOD, CAT, and GPx. GSH, melatonin, ceruloplasmin (Cp), and albumin are non-enzymatic antioxidants. Antioxidant enzymes play a vital role in protecting cellular damage from the harmful effects of ROS. In addition, the stimulation in lipid peroxidation decreases with the addition of some antioxidant matters (Seven et al., 2010).

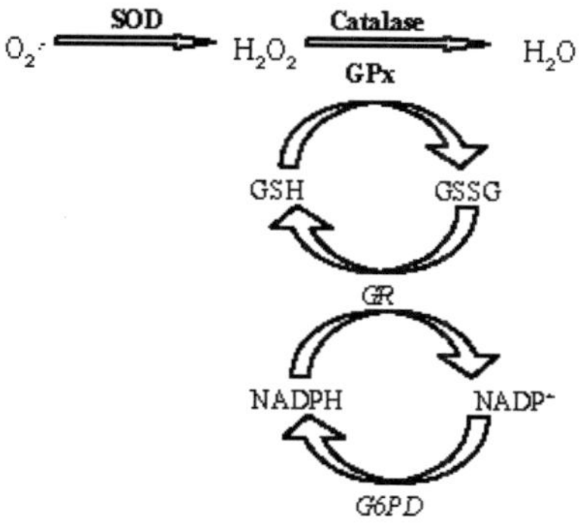

Fig. 4. Detoxification of ROS by the primary antioxidant enzymes (bold) (Kahlos, 1999). GR and G6PD are ancillary enzymes of the antioxidant system.

3. Antioxidant supplements in animal diets

Antioxidant dietary supplement greatly helps in boosting the immune system and thus aids in preventing the onset of degenerative diseases. Vitamin C antioxidant dietary supplement is perhaps the most famous form of antioxidant available. Also known as ascorbic acid, bottles or pills with this antioxidant dietary supplement can be found in any pharmacy or health food store. Another popular form of antioxidant dietary supplement is vitamin E. This antioxidant dietary supplement works best when taken with vitamin C as it seems that both vitamins have synergistic effect when taken in combination. Besides vitamins, antioxidant dietary supplements may be in the form of botanicals. Antioxidant sources are rich of the flavonoid derivatives (polyphenols). Antioxidant polyphenols are chemical compounds that are naturally found in plants. Their function is to hunt down free radicals and neutralize them. In so doing, they not only prevent free radicals from causing damage but repair any damages (www.webcontent.com/articles/52/1/Antioxidant-Dietary-Supplement/Page1.html). Propolis is recently a most important dietary supplement as antioxidant compound (Tatli Seven, 2008; Tatli Seven et al., 2008; Tatli Seven et al., 2009; Seven et al., 2011).

3.1 Propolis and its properties

Propolis (bee glue) is an adhesive, dark yellow to brown colored balsam that smells like resin. It is collected from the buds, leaves and similar parts of trees and other plants like pine, oak, eucalyptus, poplar, chestnut, and so on by bees and mixed with their wax (Seven et al., 2010). Propolis is not a new discovery. It has been used for folk medicine and foods since ancient times in many parts of the world. The use of propolis goes back to ancient times, at least to 300 BC, and it has been used as a medicine in local and popular medicine in many parts of the world, both internally and externally. Egyptians, Greeks and Romans reported the use of propolis for its general healing qualities and for the cure of some lesions of the skin. Propolis has always been reputed as an anti-inflammatory agent and to heal sores and ulcers. Ancient Egyptians used it to embalm their dead, and more recently it was used during the Boer War for healing wounds and tissue regeneration (Ghisalberti, 1979). However, its use continues today in remedies and personal products, and the list of preparations and uses is endless. It is still one of the most frequently used remedies in the Balkan States (Bankova, 2005), and it has only been in the last decades that scientists have investigated its constituents and biological properties. Propolis is a complex resinous material that honey bees (*Apis mellifera*) produce from the exudates of various plants. Beeswax is derived from the buds and bark of certain trees and other plants. The substance populus, a favour ingredient, has been confused with propolis. Etymologically, the Greek word propolis means *pro*, for or in defense, and *polis*, the city, that is "defense of the hive". Bees use it to seal holes in their honeycombs, smooth out internal walls as well as to cover carcasses of intruders who died inside the hive in order to avoid their decomposition (Sforcin, 2007).

Fig. 5. Collecting propolis from plant of honeybee

Fig. 6. The propolis collecting trap

Fig. 7. Raw propolis

Compounds	TB	TBA	TA	TT
Aromatic alcohols				
Benzyl alcohol	0.38	0.57	0.19	0.89
Phenyl ethanol	0.66	0.59	0.88	0.83
2-methoxy-4-vinylphenol	–	1.74	–	0.24
2-napthalenemethanol	2.18	1.45	0.87	0.30
5-azulenementhanol	0.80	0.04	–	–
1-naphtlenemethanol	1.20	0.50	–	1.09
Bisabolol-alpha	–	0.20	0.53	0.33
2-phenanthrenol	–	0.41	–	–
Aromatic acids				
Benzoic acid	0.96	1.20	0.53	4.30
Benzenepropanoic acid	–	–	0.04	–
4-pentenoic acid, 5-phenyl	2.40	–	–	0.03
Ferulic acid	–	0.60	–	0.12
Caffeic acid	1.20	0.44	0.05	0.61
2-propenoic Acid,3-phenyl	2.23	0.81	1.06	1.53
2-propenoic acid, 3-(4-methoxyphenyl)	1.21	0.39	0.32	0.16
1-phenanthrenecarboxylic acid	0.30	0.21	0.18	0.41
Aromatic aldehydes				
Benzaldehyde	–	–	0.04	–
Cinnamic acid and its esters				
Cinnamyl cinnamate	5.28	1.32	0.23	0.86
Benzyl cinamate	0.14	0.45	0.12	0.37
Benzyl benzoate	0.32	0.13	0.05	0.02
Cinnamic acids	–	–	–	–
1-3-hydroxy-4-methoxycinnamic acid	0.80	0.80	0.08	0.85
Fatty acids				
Lauric acid	–	0.07	–	–
Myristic acid	–	0.04	–	0.03
Palmitic acid	0.22	0.42	0.20	0.21
Oleic acid	–	1.10	–	0.47
Stearic acid	–	1.26	1.78	0.16
Linoleic acid	0.26	0.37	0.67	0.35
Linear hydrocarbons and their acids				
Cyclohexadecane	0.18	0.75	0.10	2.10
Hexadecane	–	–	–	–
Nonadecane	0.40	0.18	–	–
Octadecane	–	–	0.11	0.20
Octadecanoic acid	0.41	0.41	–	–
Flavanone				
Isalpinin	6.17	5.76	4.97	5.04
Pinocembrin	13.61	14.76	7.01	16.26
Pinostropin	13.06	11.45	4.46	2.26
Naringenin	6.20	1.40	0.90	6.20
40,5-dihydroxy-7-methoxyflavanone	1.79	–	0.84	0.69
Chrysin	1.45	2.29	3.11	9.86
3,40,7-trimethoxy flavanone	–	0.31	0.12	0.51
Hexadecanol	–	0.11	–	–
Flavonones				
Pinobanksin and its derivatives	4.3	11.5	8.3	7.6
Quercetin and its derivatives	5.1	6.2	9.1	1.1
Galangine and its derivatives	0.9	3.1	3.4	1.6
Apigenin and its derivatives	0.2	3.2	3.8	2.6

The yields of dry propolis extracts were; 44.80% (w/v) for Bartın (TBA), 36.63% (w/v) for Trabzon (TT), 31.58% (w/v) for Bursa (TB) and 20.51% (w/v) for Ankara (TA) using 96% ethanol as solvent.

Table 1. Chemical compositions of ethanol extract of Turkish propolis samples (% of total ion current) (Uzel et al., 2005)

Propolis has antioxidative, cytostatic, antimutagenic and immunomodulatory properties. These properties of propolis are based on its rich, flavonoid, phenolic acid and terpenoid contents (Seven et al., 2010). Propolis antioxidant, antibacterial and antifungal properties, combined with the fact that several of its constituents are present in food and/or food additives, and are recognized as Generally recognized as Safe (Burdock, 1998), make it an attractive candidate as a natural preservative in new food applications. This meets the demand for natural antioxidants and antimicrobials, fuelled by the increasing consumer awareness for natural, minimally processed foods with traditional preservatives absent or at very low concentrations (Han & Park, 1995; Tosi et al., 2007). Most recent studies have shown that natural preventive compounds have gained popularity day by day as some of the widely used synthetic pharmaceuticals and therapeutics might have some undesirable effects. One can think that certain natural food ingredients would be better and safer than synthetic ones. Many of these compounds, such as plant phenolics, often exhibit antioxidant activities; therefore the addition of these compounds into food products may be helpful to the health of consumers and also to the stabilization of food products.

Due to the presence of some of these effective compounds such as flavonoids (flavones and flavanones), phenolic acids and their esters in propolis and propolis extract, if the positive physiological properties and the non-toxicity of the propolis sample are proven it could be used as a mild antioxidant and preservative (Talas & Gulhan, 2009).

Characteristics	Mean value
Balsam, %	57
Phenolics, %	28
Flavones and flavonols, %	8
Flavanones and dihydroflavonols, %	6
MIC, $\mu g.mL^{-1}$ [1]	210

[1] MIC (Minimum Inhibitory Concentration)

Table 2. Characteristics of poplar propolis samples, based on 114 samples (Popova et al., 2007)

After administration to mice or to humans, propolis does not appear to have side effects (Sforcin, 2007). Although few in number, some events of propolis allergy and contact dermatitis have been informed (Callejo et al., 2001), differently from the common allergy to honey, which contains allergens obtained flowers. Ethanol and water extracts of propolis possess antiallergic action, restraining histamine release in rat peritoneal mast cells (Miyataka et al., 1998). However, in higher concentrations (300 µg/ml), propolis directly activated mast cells, promoting inflammatory mediator release, what could be linked to allergic processes in propolis-sensitive individuals (Orsi et al., 2005). In a study (Kashkooli et al., 2011), to determine the possible toxicity and side effects of propolis, fishes (Rainbow Trout) were fed on diets containing 0, 0.5, 1.5, 4.5 and 9 g propolis/kg diet for 8 weeks. Their results showed that all dosages induced no significant alterations in growth parameters and the levels of total protein, albumin, globulin, low-density lipoprotein cholesterol, high-density lipoprotein cholesterol, triglycerides and activities of glutamic pyruvic transaminase (ALT), glutamic oxaloacetic transaminase (AST), alkaline phosphatase (ALP) and lactate dehydrogenase (LDH), when compared to the control group. On the basis of their findings, propolis is a non-toxic substance for Rainbow Trout and its long-term administration might not have any side effects. Recently, the presence of radioactive particles in propolis samples was investigated, since these particles may be concentrated in the soil, contaminating the plants, insects and its products, and, consequently, humans as well. Cesium (Cs_{137}) was not found in the samples, and only natural radioactive particles such as potassium (K_{40}) and beryllium (Be_7) were found. These data suggested that propolis may be studied as an environmental pollution indicator in order to understand the soil-plant-bee-propolis chain (Orsi et al., 2006).

Antioxidative effect of propolis extracts has been reported in different methods including iodometric method, thiobarbituric acid (TBA) method and free radical scavenging ability with reduction of radical diphenylpicrylhydrazyl (DPPH) (Mohammadzadeh et al., 2007), but Mohammadzadeh et al. (2007) reported that the ferric reducing ability of plasma (FRAP) assay the reagents are inexpensive and simple to prepare, results are fast and reproducible and the equipment required is of a type commonly found in biochemical laboratories. FRAP assay is based on ferric to ferrous ion reduction at low pH. In this method the ferric reducing ability of antioxidant compound is measured. At low pH, ferric-tripyridyltriazine (Fe^{+3}-TPTZ) complex is reduced to the ferrous (Fe^{+2}) blue colour complex with an absorption maximum at 593 nm. Test conditions favour reduction of the complex and, thereby, colour development, provided that an antioxidant is present (Mohammadzadeh et al., 2007).

3.2 The effects of propolis supplementation in animals

3.2.1 Antioxidant effects of propolis

Propolis contains about 300 constituents (Türkez et al., 2010). Latterly, propolis has gained popularity in connection with oxidative stress (Tatli Seven, 2008; Tatli Seven & Seven, 2008; Tatli Seven et al., 2008; Tatli Seven et al., 2009) and used widely in healthy drinks and foods to recuperate health and prevent diseases such as inflammation, heart disease, diabetes and even cancer (Burdock, 1998; Banskota et al., 2000). Because of such broad spectrum of biological properties and their different uses, there is a renewed interest in its biological activities. Several investigations on propolis in Eastern Europe and South America have showed that flavonoids concentrated in propolis are powerful antioxidants which are capable to scavenge free radicals (Basnet et al., 1997; Banskota et al., 2000).

Flavonoids of propolis are one of the most important compounds. Flavonoids are thought to be responsible for many of its biological and pharmacological activities including anticancer, anti-inflammatory, antimicrobial and antioxidant effects. Active free radicals, together with other factors are responsible for cellular aging and many conditions such as cardiovascular diseases, cancer, diabetes, arthritis, heat stress, environmental pollution (Tatli Seven et al., 2008; Seven et al., 2010; Sforcin & Bankova, 2011). The antioxidant serves as a defensive factor against free radicals in the body. Enzymes such as SOD, CAT and GPx are the main system that opposes oxidation. If production free radicals overwhelm the capacity of enzymatic system, the second line of defense (vitamins) may come to action (Tatli Seven, 2008; Tatli Seven et al., 2009). Such as antioxidants vitamins C and E extinguish free radicals and become oxidized and non-active (Halliwell, 1994). Flavonoids and various phenolics in propolis have been appeared to be capable of scavenging free radicals and thereby defending lipids and other compounds such as vitamin C from being oxidized or destroyed during oxidative damage (Tatli Seven et al., 2009). Besides, flavonoids inhibit lipid peroxidation, platelet aggregation, capillary permeability and fragility, and the activity of enzyme systems including cyclooxygenase (COX) and lipoxygenase (Havsteen, 2002). Cardio protective effects have also been reported for flavonoids (Celle et al., 2004). Chopra et al. (1995) reported that doxorubicin-induced cardiomyopathy in rats followed by treatment with propolis induced a significant reduction of creatine phosphokinase, aspartate aminotransferase, blood and tissue GSH levels and TBA-reactive substances. It was also observed a decreased degeneration of cardiac fibers in propolis-treated rats, suggesting that this effect could be due to flavonoids present in propolis composition (Pinchuk & Lichtenberg, 2002).

Chyrisin is one of the propolis compounds which is has hepatoprotective and antioxidant activities in rats (Sathiavelu et al., 2009). Benzoic acid derivate exhibits antioxidant effects using inhibition assays of luminol luminescence, 2,2-diphenyl-1-picrylhydrazyl, and lipoperoxidation. Particularly caffeic acid, caffeoylquinic acid and cinnamic acid are effective O_2^{-} scavenging activity (Christov et al., 2006; Nakajima et al., 2007).

Propolis is effective in neurotoxicity. Kwon et al. (2004) examined the effects of the antioxidant propolis on seizures induced by kainic acid (KA) in rats. They found that KA induced increases in the levels of MDA and protein carbonyl, and decreases in the ratio of GSH/GSSG. In addition, the researchers determined that these changes in oxidative stresses markers at neuronal degenerations were significantly attenuated by pretreatment with propolis, and that the neuroprotective effects of propolis appeared to be counteracted by adenosine receptor antagonists. Their results suggest that the protective effect of propolis against KA-induced neurotoxic oxidative damage is, at least in part, via adenosine A_1 receptor modulation. Thus, they postulate that propolis significantly blocks seizure-induced neuronal loss by attenuating the impairment of GSH metabolism via, in part, adenosine A_1 receptor modulation. The novel antioxidant/anticonvulsant effect could be an important contribution to extending the neuroprotective potential of propolis.

A recent study (Mannaa et al., 2011), it was found that oral administration of propolis in epileptic animals which received the anticonvulsant drug valproate, resulted in significant improvements in the neurotransmitter levels in both hippocampus and in serum. The results obtained in the mentioned study, regarding the lipid peroxide (LPO) level and total antioxidant capacity (TAC) in the hippocampus homogenate of epileptic rats showed a significant increase and a significant decrease in both parameters. This may be explained by the fact that generalized epilepsy is a chronic disorder characterized by recurrent seizures which can increase the content of ROS and superoxide generation in the brain. Free radical generation can induce seizure activity by direct inactivation of glutamine synthase thereby permitting an abnormal construct of excitatory neurotransmitter glutamic acid (Oliver et al., 1990). The onset of oxygen-induced convulsions in animals correlated with a decrease in the cerebral content of neurotransmitter gamma-aminobutyric acid (the main inhibitory neurotransmitter) because of the inhibition of enzyme glutamate decarboxylase by oxygen free radicals. Thus, it appears that free radicals may be responsible for the development of convulsions. The mentioned study showed that propolis improved the effect of valproate on LPO level towards the normal values (Mannaa et al., 2011).

Brain oxidative injury, resulting from excessive generation of free radicals, likely contribute to the initiation and progression of epilepsy after brain injury. Therefore, antioxidant therapies aimed attenuation of oxidative stress have received considerable attention in the treatment of epilepsy. The researchers demonstrated that propolis possessed neuroprotective effects both against neurotoxicity in cell cultures and against ischemic neuronal damage (Shimazawa et al., 2005). The neuroprotective effects of propolis may be related to its constituents, such as 3,4-di-*O*-caffeoylqunic acid, 3,5-di-*O*-caffeoylqunicacid and/or *p*-coumaric acid (Inokuchi et al., 2006).

Propolis effects were analyzed on macrophages of BALB/c mice submitted to immobilization stress as well as on the histopathological analysis of the thymus, bone marrow, spleen and adrenal glands. Stressed mice showed higher H_2O_2 generation by peritoneal macrophages, and propolis treatment (200 mg/kg) potentiated H_2O_2 generation and inhibited nitric oxide production by these cells. Histopathological analysis of stressed mice showed no alterations in the thymus, bone marrow and adrenal glands, but an increase in germinal centers in the spleen was seen. Propolis treatment counteracted the alterations found in the spleen of stressed mice (Missima & Sforcin, 2008).

The chemical content of Turkish propolis was investigated with the focus on protective effect against alcohol-induced oxidative stress (Kolankaya et al., 2002). The authors declared that the ethanolic propolis extract of propolis, at dose of 200 mg/kg body weight/day, was given, by gavage, to male rats for 15 days. It was found that HDL level decreased and LDL level increased in the alcohol group, while HDL level increased and LDL level decreased in the alcohol+propolis group. There were decreases in cholesterol and triglyceride levels in the alcohol+propolis group. Also, there were decreases in activities of serum ALP and AST, but increases in LDH activity in the propolis treated group compared to the alcohol group. No toxic effects of Turkish propolis were found, while it caused an increase in HDL level and a decrease in LDL level. They suggest that these effects are protective against degenerative diseases and against alcohol-induced oxidative stress via free radicals (Kolankaya et al., 2002).

Heat stress is an important stressor resulting in the reduced welfare of birds. Heat stress increased lipid peroxidation as a consequence of increased free radical generation. The rise of lipid peroxidation increases the MDA level in blood and tissues (Tatli Seven et al., 2009). Tatli Seven et al. (2009) were found that heat stress-induced oxidative stress was indicated by increased plasma, liver and muscle MDA levels. Dietary propolis and vitamin C supplementation significantly decreased plasma, liver and muscle MDA levels. It may be considered that dietary vitamin C and high dose of propolis attenuated lipid peroxidation. Besides, 3 g/kg dietary propolis was found to be more effective than dietary vitamin C, on especially liver and muscle MDA levels. Likewise, Okonenko et al. (1988) reported that propolis had more pronounced antioxidant action compared to that of vitamin E that has a similar activity to vitamin C. Living organisms are able to adapt to oxidative stress by inducing the synthesis of antioxidant enzymes and damage removal/repair enzymes (Tatli Seven et al., 2009).

Antioxidant enzyme activities such as SOD and CAT under stimulation of lipid peroxidation may sometimes decrease (Wohaieb & Godin, 1987; Ozkaya et al., 2002) or increase (Huang et al., 1999; Aliciguzel et al., 2003). The increase of antioxidant enzyme activities such as SOD, CAT and GSH may be considered as a protective mechanism against heat-induced free radical production and lipid peroxidation (Tatli Seven et al., 2009).

Exposure of broilers to heat stress resulted in a significant increase in SOD and CAT (Altan et al., 2003). Moreover, significant differences between enzymes were obtained in antioxidant enzyme responses to heat treatment. A similar response has been reported in many human diseases, in which MDA concentrations increased concomitantly with an increase in antioxidant enzyme activities. McArdle and Jackson (2000) have also demonstrated a significant increase in free radical production together with an increase in the expression of antioxidant enzymes during a period of non-damaging exercise in muscle. These increases in antioxidant enzyme activities have been considered as a protective response against oxidative stress (Altan et al., 2003). A previous study (Okutan et al., 2005), investigated the effects of caffeic acid phenethyl ester (CAPE) which is a component of propolis on lipid peroxidation and antioxidant enzymes in diabetic rat heart. They found that in untreated diabetic group, the SOD activities and CAT levels have significantly decreased, while GSH-Px activity was increased in the CAPE-treated diabetic rats compared to those observed in untreated diabetic rats. The GSH-Px activities in blood, liver and kidneys of heat stressed birds were significantly reduced, while SOD, CAT and GSH activities were increased in blood and some tissues. This may be explained by GSH-Px inhibition at increased free radical levels in tissues (Nakazawa et al., 1996). It can be concluded that in broilers heat stress induced oxidative stress in blood and tissues. Dietary propolis decreased lipid peroxidation and regulated antioxidant enzymes activities in the broilers exposed to heat stress. The protective role of propolis might be related to its antioxidant effect. Researchers suggest that propolis and especially propolis in dose supplemented 3 mg/kg diet might be considered to prevent oxidative stress in the broilers exposed to heat stress (Tatli Seven et al., 2009).

Heavy metal pollution is provoked cardio toxicity, nephrotoxicity and neurotoxicity and they are show pro-oxidative effects. They demonstrate adverse effects, such as the production of ROS, disruption of tissue oxidant/antioxidant balance, and alteration of lipid metabolism (Seven et al., 2010; Türkez et al., 2010). For example, aluminum induced changes in biochemical parameters, stimulated lipid peroxidation and decreased the activities of the antioxidant enzymes in plasma and different tissues of male rabbits and rats. Also, aluminum chloride caused deterioration in sperm quality, enhancement of free radical levels and alterations in antioxidant enzymes in both *in vivo* and *in vitro*. The mechanisms of Al-induced toxicity may be attributed to the potentiation of Fe^{+2} oxidationto Fe^{+3} to cause oxidative damage (Xie & Yokel, 1996). Türkez et al. (2010) found effectiveness of propolis (50 mg propolis/kg of body weight (BW)) in modulating the $AlCl_3$ (34 mg $AlCl_3$/kg BW) was genotoxic and hepatotoxic in liver of rats. $AlCl_3$ significantly increased the amount of micronucleated hepatocytes ALP, activities of transaminases (AST and ALT) and LDH. Furthermore, severe pathological damages such as sinusoidal dilatation, congestion of central vein, lipid accumulation and lymphocyte infiltration were found in liver. On the contrary, the researchers mentioned that treatment with propolis alone did not cause any adverse effect on above parameters. The physiological effects of propolis in hepatocytes are not clear; a hypothesis is that Al-induced genetic damage can be prevented by inductive effects of propolis on antioxidant capacity. Because the toxic effects of Al appear to be intervened, at least in part, by free radicals (Abubakar et al., 2003). As known, genetic damages mainly develops related with oxidative stress. Propolis can be proposed to prevent Al toxicity as a nutritional supplement or a functional food component (Türkez et al., 2010).

Seven et al. (2010) investigated the effects of propolis in broilers exposed to lead-induced oxidative stress. The authors found that the addition of lead increased the plasma MDA level. The authors found that the addition of propolis significantly decreased blood SOD activity and the CAT activity of heart tissue. The researchers suggested that propolis (1 g/kg) supplementation in broiler diets might overcome the adverse effects of oxidative stress induced dietary lead. Antioxidant effects of propolis are based on flavonoids and CAPE. It was reported that CAPE decreased MDA levels by blocking ROS production as an antioxidant (Seven et al., 2010).

Propolis supports liver metabolism under oxidative stress. Seven et al. (2010) observed that dietary propolis significantly decreased the blood triglyceride levels in the lead supplemented group. According to this finding, the decrease in triglyceride level of propolis might indicate that the addition of propolis relieved the adverse effects on triglyceride level of oxidative stress. Seven et al. (2010) suggested that using 1 g/kg of propolis supplementation in maize soybean meal type broiler diets may attenuate the adverse effects of oxidative stress on the antioxidant defense system.

Heat stress stimulates lipid peroxidation as a consequence of increased free radical generation. The increase in lipid peroxidation decreases antioxidant levels such as vitamin C and vitamin E in tissues (Tatli Seven et al., 2008). Performance of animals in heat stress is decreased (Tatli Seven, 2008; Tatli Seven et al., 2008; Seven et al., 2011). Antioxidants such as vitamin C, vitamin E and propolis are used poultry diet because of their anti-stress effects and because their levels is reduced during heat stress (Tatli Seven, 2008; Tatli Seven et al., 2008). Propolis prevented negative effects caused by heat stress on performance, digestibility and egg qualities (Tatli Seven, 2008). The authors reported that supplementation with propolis (5 g/kg diet) was the most efficient treatment, and increased feed intake and improved hen day egg, egg weight and digestibility (of dry matter, organic matter, crude protein (CP) and ether extract) in laying hens. Tatli Seven (2008) explained that the positive results appeared due to palatable and antioxidant properties of propolis. Especially, effects on performance and digestibility of propolis dietary supplementation may appear more powerful under stress. Moreover, it was declared that propolis supplementation increased egg shell thickness and egg shell weight in heat stressed laying hens. It was due to improved calcium digestibility and absorption resulting from the acid derivates such as benzoik, 4-hydroxy-benzoic, etc., which are found in propolis (Haro et al., 2000; Tatli Seven, 2008).

3.2.2 Antimicrobial, anti-inflammatory and antitumor effects of propolis

In addition to antioxidant properties, propolis demonstrate other beneficial effects. Especially its antibacterial, anti-inflammatory and antitumor effects are very important for human and animal health, and animal production.

Itavo et al. (2011) indicated that propolis is an alternative to the use of dietary antibiotics. Propolis has bacteriostatic activity against gram-positive and some gram-negative bacteria. The action of propolis is likely related to changes in the bioenergetic status, which inhibits bacterial motility. This is similar to the action of ionophores. The chemical composition of propolis is complex and variable because it is intrinsically related to the floristic and ecological composition of the environment visited by the bees. The combination of these factors affects the pharmacological properties of propolis, which is in fact classified into different types such as brown, green and red propolis.

Tatli Seven & Seven (2008) reported that propolis stimulated immune system and decreased mortality rate by improving immunity in broilers. They remarked that propolis supplementation in poultry diets as alternative to antibiotics may be recommended in broilers in heat stress conditions.

Epidemiological studies provide evidences that propolis protect humans against cancer, and also results from animal experiments and *in vitro* microorganism manipulations showed its efficiency to reduce pathogenic bacteria. Moreover, propolis showed substantial protection against cariogenic bacteria and oral pathogens suggesting its valuable clinical use. Kouidhi et al. (2010) declared that Tunisian propolis presented potential activity against oral *Streptococci* causing dental diseases as well as different kind of cancer cell proliferations. Those excellent activities could be due to specific bioactive compounds of the Tunisian propolis.

Also, propolis is effective on the rumen environment in ruminants. Brodiscou et al. (2000) determined that propolis affected fermentation and methanogenesis of continuous microbial culture in ruminants; it decreased protozoa population and raised propionate levels by 10.3%. However, some negative propolis effects could be related to high flavonoid levels (14.9 g/kg propolis) reported (Itavo et al., 2011). Lambs fed by the diet supplemented with propolis spent more time resting and less time ruminating. Researchers (Itavo et al., 2011) explained that this may have had a negative on digestibility because this diet provided the highest food intake with the worst feed conversion. The higher flavonoid and phenol content of their diet was probably toxic to the ruminal environment of lambs, which accelerated food passage through the gastrointestinal tract, causing lower nutrient absorption and increased dry matter intake, corroborating Van Soest (1994). Rumination rate was therefore higher in lambs fed propolis because of their increased neutral detergent fiber intake and reduced rumination time. However, improvement in rumination rate itself was not enough to raise the productivity of these lambs, which had the lowest weight gain. Rumination quality is essential to optimize food use, but this was apparently compromised by the high flavonoid and total phenolic content of propolis, resulting in reduced lamb performance (Van Soest, 1994).

Propolis also affects bone metabolism (Amany et al., 2008). In the study, it was used orally fish liver oil and propolis as protective natural products against the effect of the antiepileptic drug valproate on immunological markers of bone formation in rats. Propolis increased the bone formation markers and decreases the bone resorption ones. it also increased the osteoprotegrin and decreased tumor necrosis factor-alpha-α (TNF-α), and NF kappa-B ligand which inhibited the osteoclastogenesis. The researchers recommended the use of propolis as a prophylactic treatment for epileptic patients using valproate against the side effect of valproate on bone.

Propolis is effective in inflammation-related diseases such as rheumatoid arthritis (RA). The main characteristic of RA is the ongoing damage in arthrosis of cartilage and bone, and at the same time with a disturbance of immune function. In the context of RA, there exist neutrophils, activation macrophages, lymphocytes and other elements associated with the abduction, activation and releasing of cytokine, which is perhaps one of the mechanisms of RA development (Hu et al., 2005). In the case of RA, the concentration of cytokine derived from T cells was generally low, whereas that of mononuclear macrophage levels were significantly higher (Shuyun, 1996). Propolis inhibited the increase of inflammatory medium and decreased the activation and inducing effects of cytokines, which indicated that both extracts exhibited the same anti-inflammatory effects (Hu et al., 2005).

Propolis was used in a tumor event (Nada & Ivan, 2003). The tumor was a transplantable mammary carcinoma of mouse. Metastases in the lung were induced by injection of $2x10^5$ viable tumor cells i.v. and propolis was given intraperitoneally at doses of 50 or 150 mg/kg before or after tumor cell inoculation. Researchers (Nada & Ivan, 2003) demonstrated that therapies reduced the number of metastases in the lung and tumor growth was suppressed significantly by propolis. They commented that it is likely antimetastatic activity of the propolis is mainly mediated by immunomodulatory activity. Flavonoids in propolis stimulated macrophages to produce lymphocyte activating factor, a factor relevant for control of immune cell cooperation.

The observed anti-inflammatory effects of propolis have been attributed to its flavonoid, phenolic acid and caffeic acid contents. Flavonoids were reported to inhibit the activity of enzymes involved in the conversion of membrane polyunsaturated fatty acids such as phospholipase A_2, COX and lipoxygenase, to inhibit the release of lysosomal enzymes from rabbit polymorphonuclear leucocytes and scavenge free radicals. Aqueous extracts of propolis were formed to have inhibitory effects on dihydrofolate reductase similar to the well-known non-steroidal anti-inflammatory drugs. CAPE, which is an active component of propolis extract, was found to inhibit 5-lipoxygenase in micro molar concentrations, and to block the production of ROS in neutrophils and xanthine/xanthine oxidase system. It was also believed to contribute to the anti-inflammatory activity of propolis by being both a lipoxygenase–cyclooxygenase inhibitor and an antioxidant (Onlen et al., 2007).

Sy et al. (2006) investigated the activities of propolis using an OVA-induced asthma animal model. Mice were immunized and sensitized by exposure to ovalbumin (OVA) antigen and administered with low (65 mg/kg body weight) and high-dose (325 mg/kg body weight) propolis water extracts by tube feeding. The serum OVA-specific IgE titer and cytokine profiles in cultured splenocytes and bronchoalveolar lavage fluids (BALF) were analyzed. The number of eosinophils in BALF was counted. Here we demonstrate that propolis extracts can suppress the serum levels of OVA-specific IgE and IgG1, and airway hyper responsiveness in OVA-sensitized mice. Results suggest that propolis extracts may be a potential novel therapeutic agent for asthma. CAPE, an anti-inflammatory component of propolis, is known to be an inhibitor of nuclear factor-kappa B and significantly reduces the levels of pro-inflammation cytokines (TNF-α and Interleukin 1, beta) in rats (Fitzpatrick et al., 2001).

CAPE is essential for the anti-inflammatory activity of propolis. Because CAPE derivatives are more lipophilic, thus easily facilitate their entry into cells (Michaluart et al., 1999). It has been reported that (Michaluart et al., 1999) CAPE, possesses anti-inflammatory activity by inhibiting the release of arachidonic acid from cell membrane, suppressing the enzyme activities of cyclooxygenase-1 (COX-1) and cyclooxygenase-2 (COX-2), and inhibiting the activation of COX-2 gene expression. Propolis with CAPE and CAPE produce a significant inhibition of both exudate volume and leukocytes migration induced by carrageenin injection into the pleural cavity (Borrelli et al., 2002). Furthermore, the observed anti-inflammatory activity of propolis appears to be due to the presence of flavonoids, especially galangin. Indeed, this flavonoid has shown to inhibit COX and lipooxygenase activity, decrease prostaglandin E-2 release and the expression of the inducible isoform of COX- 2 (Gabor & Razga, 1991; Raso et al., 2001).

4. Conclusion

As a result, propolis began to attract the attention of scientists extremely (Table 3). It is used in researches related to important diseases such as cancer and diabetes. Besides, propolis is used as diet supplement in poultry researches. Many animal researches's results showed that propolis might be relieve the negative effects of oxidative stress. But, new researches should be made related to propolis.

Table 3. The Effects of Propolis in Animals

Form used of Propolis	Animal	Oxidative Stress	Given shape	Dose of propolis	Activity	Significantly		Authors
Commercial preparation	Mice	Induced in Ehrlich ascitis carcinoma cell line	Orally	160 mg/kg BW	Antitumoral and antioxidant characteristics of propolis the reduction of viability and the number of tumor cells. Induction of apoptosis.	MDA (-), GST (+), GSH (+), Total protein (-), DNA (-) and RNA (-) concentrations, Tumor volume (-), Viability of tumor cells (-)	P<0.05	El-khawaga et al., 2003.
Commercial preparation	Rat	Pilocarpin Epileptic rats	Orally	50mg/kg BW + volproate (400 mg/ kg BW)	Neuroprotective, antioxidant characteristics of propolis	Hippocampus serotonin (-), Serum dopamin (-) and serotonin (-)	P<0.05	Manna et al., 2011.
						Hippocampus MDA (-) and TAC (+)	P<0.05	
EEP	Fish (Rainbo wtrout)	No stress	Dietary	0.5, 1.5, 4.5, 9.0 g/kg diet	Non toxic effect of propolis	BWG, SGR, ALT, AST, ALP, albumin, globulin, HDL, LDL	P>0.05	Kashkooli et al., 2011.
Alchol extraction	Lambs	No stress	Dietary	199 g/kg	Bacteriostatic effects on Rumen medium	DMI (-), DWG (-), NDFI (-), FCR (+)	P<0.01	Itavo et al., 2011.
						Ruminating, Resting	P>0.05	
EEP	Laying Hens	Heat stress	Dietary	3g/kg	antibacterial antioxidant, antiseptic, palatable characteristics of propolis	EW (+), FCR (+), digesibility of CP (+), EST (+) and ESW (+)	P<0.05	Seven et al., 2011.
					Rises values of EST and ESW increasing digestibilty of Ca and P due to acid derivates such as benzoik, 4-hydroxy-benzoik which are found in propolis			
EEP	Laying Hens	Heat stress	Dietary	5 g/kg	antioxidant, immuno stimulatory, palatable characteristics of propolis	FI (+), BW (+), FCR (+), HEP (+), EW (+), Digestibility (+), EST and ESW (+)	P<0.05	Tatli Seven, 2008.
					antioxidant, immuno stimulatory			
					Rises values of EST and ESW increasing digestibilty of Ca and P due to acid derivates such as benzoik, 4-hydroxy-benzoik which are found in propolis			
EEP	Broiler	Heat stress	Dietary	0,5 g/kg 1 g/kg 3 g/kg	Antioxidant effects on lipit peroxidation of flavonoids and CAPE in propolis	Plasma MDA (-) and SOD (-); serum AST (-); kidneys (-) and heart (-) CAT; kidneys (-) and heart (-) GSH; blood (+), liver (+) and kidneys (+) GSHPx	P<0.05	Tatli Seven et al., 2009.
						Serum ALT; liver (-), muscle (-) MDA; blood (-) and liver (-) CAT; muscle (-) GSH	P<0.01	
EEP	Broiler	Heat stress	Dietary	3 g/kg	antioxidant and palatable characteristics of propolis	BWG (+) and FI (+), BWG (+), Digesibility of CP (+), EE (+), Tryglyceride (-)	P<0.05	Tatli Seven et al., 2008.
EEP	Broiler	Heat stress	Dietary	5 g/kg	antibacterial, antioxidant, palatable, growth promoter characteristics of propolis			Tatli Seven and Seven, 2008.
EEP	Broiler	Lead-Induced	Dietary	1 g/kg	Antioxidant effects of flavonoids and CAPE in propolis	Plasma MDA (-), SOD (-) and heart CAT (-)	P<0.01	Seven et al., 2010.

BWG: Body Weight Gain; **DMI:** Dry Matter Intake; **DWG:** Daily Weight Gain; **EEP:** Ether Extract of Propolis; **EST:** Egg Shell Thickness; **ESW:** Egg Shell Weight; **FCR:** Feed Convertion Ratio; **FI:** Feed Intake; **HEP:** Hen-day Egg Production; **NDFI:** Neutral Detergent Fiber Intake; **SGR:** Specific Growth Rate; **GST:** Glutathione S-Transferase; **TAC:** Total Antioxidant Capacity; **BW:** Body Weight; **CP:** Crude Protein; **CAPE:** CAT; **(+):** Significatly Increased; **(-):** Significatly Decreased.

Table 3. The Effects of Propolis in Animals

5. References

Abubakar, M.G.; Taylor, A. & Ferns, G.A. (2003). Aluminum administration is associated with enhanced hepatic oxidant stress that may be offset by dietary vitamin E in the rat. *Int J Exp Pathol*, 84, 49–54.

Agarwal, A.; Gupta, S. & Sharma R.K. (2005). Role of oxidative stress in female reproduction. *Reproductive Biology and Endocrinology*, 3, 28–47.

Aliciguzel, Y.; Ozen, I.; Aslan, M. & Karayalcin, U. (2003). Activities of xanthine oxidoreductase and antioxidant enzymes in different tissues of diabetic rats. *J Lab Clin Med*, 142, 172–177.

Altan, O.; Pabuccuoglu, A.; Altan, A.; Konyalioglu, S. & Bayraktar, H. (2003). Effect of heat stress on oxidative stress, lipid peroxidation and some stress parameters in broilers. *Br Poult Sci*, 44. 545–550.

Amany, S.E.E.; Karima A.I. E. & Sibaii H. (2008). Fish liver oil and propolis as protective natural products against the effect of the anti-epileptic drug valproate on immunological markers of bone formation in rats. *Epilepsy Research*, 80, 47–56.

Atessahin, A.; Yilmaz, S.; Karahan, I.; Ceribasi, A.O. & Karaoglu, A. (2005). Effects of lycopene against cisplatin-induced nephrotoxicity and oxidative stress in rats. *Toxicology*, 212, 116–123.

Basnet, P.; Matsuno, T. & Neidlein, R. (1997). Potent free radical scavenging activity of propolis isolated from Brazilian propolis. *Z Naturforsch C*, 52, 828–833.

Bankova, V., 2005. Chemical diversity of propolis and the problem of standardization. *Journal of Ethnopharmacology*, 100, 114–117.

Banskota, A.H.; Tezuka, Y.; Adnyana, I.K.; Midorikawa, K.; Matsushige, K.; Message, D.; Huertas, A.A.G. & Kadota, S. (2000). Cytotoxic, hepatoprotective and free radical scavenging effects of propolis from Brazil, Peru, the Netherlands and China. *Journal of Ethnopharmacology*, 72, 239–246.

Borrelli, F.; Maffia, P.; Pinto, L.; Ianaro, A.; Russo, A.; Capasso, F. & Ialenti, A. (2002). Phytochemical compounds involved in the anti-inflammatory effect of propolis extract. *Fitoterapia*, 73, 53–63.

Brodiscou, L.P.; Papona, Y. & Brodiscou, A.F. (2000). Effects of dry plant extracts on fermentation and methanogenesis in continuous culture of rumen microbes. *Anim Feed Sci Technol*, 87, 263–277.

Bulger, E.M. & Helton, W.S. (1998). Nutrient antioxidants in gastrointestinal diseases. *Clinical Nutrition*, 27, 403–419.

Burdock, G.A. (1998). Review of the biological properties and toxicity of bee propolis (Propolis). *Food and Chemical Toxicology*, 36, 347–363.

Callejo, A.; Armentia, A.; Lombardero, M. & Asensio, T. (2001). Propolis, a new bee-related allergen. *Allergy*, 56, 579.

Celle, T.; Heeringa, P.; Strzelecka, A.E.; Bast, A.; Smits, J.F. & Janssen, B.J. (2004). Sustained protective effects of 7-monohydroxyethylrutoside in an *in vivo* model of cardiac ischemia-reperfusion. *European Journal of Pharmacology*, 494, 205–212.

Chen, C.N. & Pan, S.M. (1996). Assay of superoxide dismutase activity by combining electrophoresis and densitometry. *Bot Bull Acad Sin*, 37, 107-111.

Chopra, S.; Pillai, K.K.; Husain, S.Z. & Giri, D.K. (1995). Propolis protects against doxorubicin-induced myocardiopathy in rats. *Experimental and Molecular Pathology*, 62, 190–198.

Christov, R.; Trusheva, B.; Popova, M.; Bankova, V. & Bertrand, M. (2006). Chemical composition of propolis from Canada, its antiradical activity and plant origin. *Nat Prod Res,* 20, 531-536.

Czesnikiewicz-Guzik, M.; Konturek, S.J.; Loster, B.; Wisniewska, G. & Majewski, S. (2007). Melatonin and its role in oxidative stress related diseases of oral cavity. *J Physiol Pharmacol,* 58, 5-19.

Davies, K.J. (1995). Oxidative stress: The paradox of aerobic life. *Biochem Soc Symp,* 61, 1-31.

El-khawaga, O.A., Salem, T.A., Elshal, M.F. (2003). Protective role of Egyptian propolis against tumor in mice. *Clin Chim Acta.,* 338(1-2):11-6.

Fang, Y.Z.; Yang, S. & Wu, G. (2002). Free radicals, antioxidants and nutrition. *Nutrition,* 18, 872–879.

Fitzpatrick, L.R.; Wang, J. & Le, T. (2001). Caffeic acid phenethyl ester, an inhibitor of nuclear factor-κB, attenuates bacterial peptidoglycan polysaccharide-induced colitis in rats. *J Pharmacol Exp Ther,* 299, 915–20.

Gabor, M. & Razga, Z. (1991). Effect of benzopyrone derivatives on simultaneously induced croton oil ear oedema and carrageenin paw oedema in rats. *Acta Physiol Hung,* 77, 197–207.

Ghisalberti, E.L. (1979). Propolis a Review. *Bee World,* 60, 59–84.

Halliwell, B. (1991). Reactive oxygen species in living systems: Source, biochemistry, and role in human disease. *Am J Med,* 91, 14–22.

Halliwell, B. (1994). Free radicals and antioxidants, and human disease: Curiosity, cause, or consequence. *Lancet,* 344, 721–724.

Han, S.K. & Park, H.K. (1995). A study on the preservation of meat products by natural propolis: Effect of EEP on protein change of meat products. *Korean Journal of Animal Science,* 37, 551–557.

Haro, A.; López-Aliaga, I.; Lisbona, F.; Barrionuevo, M.; Alférez, M.J.M. & Campos, M.S. (2000). Beneficial effect of pollen and/or propolis on the metabolism of iron, calcium, phosphorus, and magnesium in rats with nutritional ferropenic anemia. *J Agric Food Chem,* 48, 5715–5722.

Havsteen, B. (2002). The biochemistry and medical significance of the flavonoids. *Pharmacology & Therapeutic,* 96, 67–202.

Hu, F.; Hepburn, H.R.; Li, Y.; Chen, M.; Radloff, S.E. & Daya, S. (2005). Effects of ethanol and water extracts of propolis (bee glue) on acute inflammatory animal models. *Journal of Ethnopharmacology,* 100, 276–283.

Huang, W.C.; Juang, S.W.; Liu, I.M.; Chi, T.C. & Cheng, J.T. (1999). Changes of superoxide dismutase gene expression and activity in the brain of streptozotocin-induced diabetic rats. *Neurosci Lett,* 5, 25-28.

Inokuchi, Y.; Shimazawa, M.; Nakajima, Y.; Suemori, S.; Mishima, S. & Hara, H. (2006). Brazilian green propolis protects against retinal damage *in vitro* and *in vivo. Evid Based Complement Alternat Med,* 3, 71–77.

Itavo, C.C.B.F.; Morais, M.G.; Costa, C.; Ítavo, L.C.V.; Franco, G.L.; Da Silva, J.A. & Reis, F.A. (2011). Addition of propolis or monensin in the diet: Behavior and productivity of lambs in feedlot. *Animal Feed Science and Technology,* 165, 161–166.

Kahlos, K. (1999). The expression and possible role of manganese superoxide dismutase in malignant pleural mesothelioma. Department of Internal Medicine, University of Oulu, FIN-90401 Oulu, Finland.

Kashkooli, O.B.; Dorcheh, E.E.; Mahboobi-Soofiani, N. & Samie, A. (2011). Long-term effects of propolis on serum biochemical parameters of rainbow trout (Oncorhynchus mykiss). *Ecotoxicology and Environmental Safety*, 74, 315–318.

Kolankaya, D.; Selmanoğlu, G.; Sorkun, K. & Salih, B. (2002). Protective effects of Turkish propolis on alcohol-induced serum lipid changes and liver injury in male rats. *Food Chemistry*, 78, 213–217.

Kouidhi, B.; Zmantar, T. & Bakhrouf, A. (2010). Anti-cariogenic and anti-biofilms activity of Tunisian propolis extract and its potential protective effect against cancer cells proliferation. *Anaerobe*, 16, 566-571.

Kwon, Y.S.; Park, D.H.; Shin, E.J.; Kwon, M.S.; Ko, K.H.; Kim, W.K.; Jhoo, J.H.; Jhoo, W.K.; Wie, M.B.; Jung, B.D. & Kim, H.C. (2004). Antioxidant propolis attenuates kainate-induced neurotoxicity via adenosine A_1 receptor modulation in the rat. *Neuroscience Letters*, 355, 231–235.

Lushchak VI. (2011). Adaptive response to oxidative stress: Bacteria, fungi, plants and animals. *Comp Biochem Physiol C Toxicol Pharmacol*. 153,175-190.Mannaa, F.; El-Shamy, K.A.; El-Shaikh, K.A. & El-Kassaby, M. (2011). Efficacy of fish liver oil and propolis as neuroprotective agents in pilocarpine epileptic rats treated with valproate. *Pathophysiology*, 18, 287-294.

Raso, G.M.; Meli, R.; Di Carlo, G.; Pacilio, M. & Di Carlo, R. (2001). Inhibition of inducible nitric oxide synthase and cyclooxygenase-2 expression by flavonoids in macrophage J774A.1. *Life sciences*, 68, 921-931.

McArdle, A. & Jackson, M.J. (2000). Exercise, oxidative stress and ageing. *J Anat*, 197, 539-541.

Michaluart, P.; Masferrer, J.L.; Carothers, A.M.; Subbaramaiah, K.; Zweifel, B.S.; Koboldt, C.; Mestre, J.R.; Grunberger, D.; Sacks, P.G.; Tanabe, T. & Dannenberg, A.J. (1999). Inhibitory effects of caffeic acid phenethyl ester on the activity and expression of cyclooxygenase-2 in human oral epithelial cells and in a rat model of inflammation. *Cancer Res*, 59, 2347-2352.

Missima, F. & Sforcin, J.M. (2008). Green Brazilian propolis action on macrophages and lymphoid organs of chronically stressed mice. *Evidence-Based Complementary and Alternative Medicine*, 5, 71-75.

Miyataka, H.; Nishiki, M.; Matsumoto, H.; Fujimoto, T.; Matsuka, M.; Isobe, A. & Satoh, T. (1998). Evaluation of propolis (II): Effects of Brazilian and Chinese propolis on histamine release from rat peritoneal mast cells induced by compound 48/80 and concanavalin A. *Biological & Pharmaceutical Bulletin*, 21, 723–729.

Mohammadzadeh, S.; Sharriatpanahi, M.; Hamedi, M.; Amanzadeh, Y.; Ebrahimi, S.E.S. & Ostad, S.N. (2007). Antioxidant power of Iranian propolis extract. *Food Chemistry*, 103, 729–733.

Nakazawa, H.; Genka, C. & Fujishima, M. (1996). Pathological aspect of active oxygens/free radicals. *Jpn J Physiol*, 46, 15-32.

Nakajima, Y.; Shimazawa, M.; Mishima, S. & Hara, H. (2007). Water extract of propolis and its main constituents, caffeoylquinic acid derivatives, exert neuroprotective effects via antioxidant actions. *Life Sci*, 80, 370-377.

Okonenko, L.B.; Aidarkhanov, B.B.; Rakhmetova, A.A.; Zhakisheva, S.S.H. & Iksymbaeva, Z.H.S. (1988). Vitamin E and propolis as antioxidants after excessive administration of polyunsaturated fatty acids. *Vopr Pitan*, 4, 68-70.

Okutan, H.; Ozcelik, N.; Yilmaz, H.R. & Uz, E. (2005). Effects of caffeic acid phenethyl ester on lipid peroxidation and antioxidant enzymes in diabetic rat heart. *Clin Biochem,* 38, 191-196.

Oliver, C.N.; Starke-Reed, P.E.; Stadtman, E.R.; Liu, G.J.; Carney, J.M. & Floyd, R.A. (1990). Oxidative damage to brain proteins, loss of glutamine synthetase activity and production of free radicals during ischemia/reperfusion induced injury to gerbil brain. *Proc Natl Acad Sci USA,* 87, 5144–5147.

Onlen, Y.; Tamer, C.; Oksuz, H.; Duranc, N.; Altug, M.E. & Yakan, S. (2007). Comparative trial of different anti-bacterial combinations with propolis and ciprofloxacin on Pseudomonas keratitis in rabbits. *Microbiological Research,* 162, 62–68.

Orsi, R.O.; Sforcin, J.M.; Funari, S.R.C. & Gomes, J.C. (2005). Effect of propolis extract on guinea pig lung mast cell. *The Journal of Venomous Animals and Toxins,* 11, 76–83.

Orsi, R.O.; Funari, S.R.C.; Barbattini, R.; Giovani, C.; Frilli, F.; Sforcin, J.M. & Bankova, V. (2006). Radionuclides in honeybee propolis (*Apis mellifera* L.). *Bulletin of Environmental Contamination and Toxicology,* 76, 637–640.

Nada, O. & Ivan, B. (2003). Immunomodulation by water-soluble derivative of propolis: A factor of antitumor reactivity. *Journal of Ethnopharmacology,* 84, 265–273.

Ozkaya, Y.G.; Agar, A.; Yargicoglu, P.; Hacıoglu, G.; Bılmen-Sarıkcıoglu, S.; Ozen, I. & Alıcıguzel, Y. (2002). The effect of exercise on brain antioxidant status of diabetic rats. *Diabetes Metab,* 28, 377-384.

Pinchuk, I. & Lichtenberg, D. (2002). The mechanism of action of antioxidants against lipoprotein peroxidation, evaluation based on kinetic experiments. *Progress in Lipid Research,* 41, 279–314.

Popova, M.P.; Bankova, V.S.; Bogdanov, S.; Tsvetkova, I.; Naydenski, C.; Marcazzan, G.L. & Sabatini, A.G. (2007). Chemical characteristics of poplar type propolis of different geographic origin. *Apidologie,* 38, 306–311.

Sathiavelu, J.; Senapathy, G.J.; Devaraj, R. & Namasivayam, N. (2009). Hepatoprotective effect of chrysin on prooxidant-antioxidant status during ethanol-induced toxicity in female albino rats. *J Pharm Pharmacol,* 61, 809-817.

Selye, H. (1936). A syndrome produced by diverse nocuous agents. *Nature,*138, 32-34.

Seven, İ.; Aksu, T. & Tatli Seven, P. (2010). The effects of propolis on biochemical parameters and activity of antioxidant enzymes in broilers exposed to lead-induced oxidative stress. *AJAS,* 23, 1482-1489.

Seven, İ.; Tatli Seven, P. & Silici, S. (2011). Effects of dietary Turkish propolis as alternative to antibiotic on growth and laying performances, nutrient digestibility and egg quality in laying hens under heat stress. *Revue Med Vet,* 162, 186-191.

Sforcin, J.M. (2007). Propolis and the immune system: A review. *Journal of Ethnopharmacology,* 113, 1–14.

Sforcin, J.M. & Bankova, V. (2011). Propolis: Is there a potential for the development of new drugs? (Review). *Journal of Ethnopharmacology,* 133, 253–260.

Shimazawa, M.; Chikamatsu, S.; Morimoto, N.; Mishima, S.; Nagai, H. & Hara, H. (2005). Neuroprotection by Brazilian green propolis against *in vitro* and *in vivo* ischemic neuronal damage. *Evid Based Complem Alternat Med,* 2, 201–207.

Shuyun, X. (1996). Ten years study on the pharmacology of anti-inflammatory reactions and immunity. *Bulletin of Chinese Pharmacology,* 12, 1–6.

Sies, H. (1985). Oxidative stress: introductory remarks. ed. SIES, H., In: Oxidative Stress, Academic Press, London, pp. 1-8.

Sies, H. (1991). Oxidative stress: introduction. ed. SIES, H., In: Oxidative Stress: Oxidants and Antioxidants, Academic Press, London, pp. XV-XXİİ.

Sies, H. & Masumoto, H. (1997). Ebselen as a glutathione peroxidase mimic and as a reactant with peroxynitrite. *Advances in Pharmacology*, 38, 229-246.

Stephan, G.; Guillaume, J. & Lamour, F. (1995). Lipid peroxidation in turbot (Scophthalmus Maximus) tissue: Effect of dietary vitamin E and dietary n-6 or n-3 polyunsaturated fatty acid. *Aquaculture*, 130, 251–268.

Sy, L.B.; Wu, Y.L.; Chiang, B.L.; Wang, Y.H. & Wu, W.M. (2006). Propolis extracts exhibit an immunoregulatory activity in an OVA-sensitized airway inflammatory animal model. *International Immunopharmacology*, 6, 1053–1060.

Talas, Z.S. & Gulhan, M.F. (2009). Effects of various propolis concentrations on biochemical and hematological parameters of rainbow trout (*Oncorhynchus mykiss*). *Ecotoxicology Environmental Saffet*, 72, 1994–1998.

Tatli Seven, P. (2008). The effects of dietary Turkish propolis and vitamin C on performance, digestibility, egg production and egg quality in laying hens under different environmental temperatures. *AJAS*, 21, 1164-1170.

Tatli Seven, P. & Seven, İ. (2008). Effect of dietary Turkish propolis as alternative to antibiotic on performance and digestibility in broilers exposed to heat stress. *J Appl Anim Res*, 34, 193-196.

Tatli Seven, P.; Seven, İ.; Yılmaz, M. & Şimşek, Ü.G. (2008). The effects of Turkish propolis on growth and carcass characteristics in broilers under heat stress. *Animal Feed Science and Technology*, 146, 137-148.

Tatli Seven, P.; Yılmaz, S.; Seven, İ.; Çerçi, İ.H.; Azman, M.A. & Yılmaz, M. (2009) The effect of propolis on selected blood indicators and antioxidant enzyme activities in broilers under heat stress. *Acta Vet Brno*, 78, 75-83.

Tosi, E.A.; Ré, E.; Ortega, M.E. & Cazzoli, A.F. (2007). Food preservative based on propolis: Bacteriostatic activity of propolis polyphenols and flavonoids upon Escherichia coli. *Food Chemistry*, 104, 1025–1029.

Türkez, H.; Yousef, M.I. & Geyikoglu, F. (2010). Propolis prevents aluminum-induced genetic and hepatic damages in rat liver. *Food and Chemical Toxicology*, 48, 2741–2746.

Uzel, A.; Sorkun, K.;Özçağ, Ö.; Çoğulu, D.; Gençay, Ö. & Salih, B. (2005). Chemical compositions and antimicrobial activities of four different Anatolian propolis samples. *Microbiological Research*, 160, 189-195.

Van Soest, P.J. 1994. Nutritional Ecology of the Ruminant, Comstock Publishing Associates, 2nd ed. Cornell University Press, Ithaca, New York, USA.

Yaralioglu Gurgoze, S.; Cetin, H.; Cen, O.; Yilmaz, S. & Atli, M.O. (2005). Changes in malondialdehyde concentrations and glutathione peroxidase activity in purebred Arabian mares with endometritis. *The Veterinary Journal*, 170, 135-137.

Yilmaz, S.; Beytut, E.; Erisir, M.; Ozan, S. & Aksakal, M. (2006). Effects of additional vitamin E and selenium supply on G6PDH activity in rats treated with high doses of glucocorticoid. *Neuroscience Letters*, 30, 85-89.

Wedemeyer, G.A.; Barton, B.A. & McLeay, D.J. (1990). Stress and acclimation. In: Methods for fish biology, eds. C.B. Schreck and P.B. Moyle, American Fisheries Society, Bethesda, Maryland, USA, 451–489.

Wejil, N.I.; Cleton, F.J. & Osanto, S. (1997). Free radicals and antioxidants in chemotherapy induced toxicity. *Cancer Treat,* 23, 209-240.

Wohaieb, S.A. & Godin, D.V. (1987). Alterations in free radical tissue-defense mechanisms in streptozocin-induced diabetes in rat: effect of insulin treatment. *Diabetes,* 36,1014-1018.

Xie, C.X. & Yokel, R.A. (1996). Aluminum facilitation of iron mediated lipid peroxidation is dependent on substrate, pH and aluminum and iron concentrations. *Arch Biochem Biophys,* 327, 222–226.

http://www.webcontent.com/articles/52/1/Antioxidant-Dietary-Supplement/Page1.html (Date of access: September, 08, 2011).

14

Antioxidants in Thai Herb, Vegetable and Fruit Inhibit Hemolysis and Heinz Body Formation in Human Erythrocytes

Warin Sangkitikomol
Chulalongkorn University, Center for Excellence in Omics-Nano Medical Technology Development Project, Department of Clinical Chemistry, Faculty of Allied Health Sciences, Bangkok, Thailand

1. Introduction

Over centuries, Thai traditional herbal beverage and herbal tea have played an important role not only in quenching the thirst but also in providing therapy for common ailments. Recently, ready-to-drink beverages; tea, herbal-tea and fruit juice, are become very popular for Thai people, especially the teenagers. Because of the media presents results of promoting health and preventing many diseases by intake fruit, vegetable and herb which contained high level of antioxidant activities. These plants are rich of antioxidants those are capable to terminate free radical reactions and prevent oxidative damage. Vegetables and fruits are important sources of antioxidant in diet. Scientists have been searching for more than 2 decades to identify the specific ingredients in fruit vegetable and herb that account for their many health promotion benefits. That search points to plant metabolites, many of which are antioxidants, phytochemicals, mixtures of vitamins and fiber content. Thousands experimental studies have examined the role of specifics flavonoids or phenolics in disease prevention(Sen et al., 2010; Virgili & Marino, 2008). Different phytochemicals in herbal products are safer than synthetic medicine and beneficial in the treatment of diseases caused by free radicals. Multiple biological effects of them have been described, among them; antioxidants, cellular signals, cardioprotective effects, antibiotics, antiinflammation, antiallergic, anticoagulation, antineoplastic, anti-mutagenesis, anti-carcinogenesis (Ames et al.,1993; Lin and Liang, 2000; Ziegler, 1991; Packer and Colman,1999; Halliwell and Gutteridge, 2001; Nakamura, 1997; Cook & Samma, 1996; Chen and Yen, 1997; Lin & Liang, 2000).

The major fraction of those plants is nutraceuticals, flavonoids, for neutralizing stress induced free radicals. Flavonoids are effective antioxidants and may protect against several chronic diseases. Flavonoids are divided into six different classes (flavanols, flavanones, flavones, isoflavones, flavonols and anthocyanins) depended on different molecular structure. The flavonol quercetin and the flavone apigenin are found in many fruits and vegetables such as onions, apples, broccoli, and berries. Catechins are main flavonoids in green tea. Cyanidin and anthocyanidins are the pigment of black rice, berries and grapes. Genistein is an isoflavone found in legumes. Hesperitin, Naringenin and Eriodictyol are flavanones in citrus fruits. The group of flavanol catechins is derived from green tea, red grapes, red wine and chocolate (Beecher, 2003). The antioxidant activities of phenolics are mainly due to their redox properties which allow them to act as reducing agents, hydrogen donors, and singlet oxygen quenchers. In addition they have a metal chelation potential (Halliwell and Gutteridge, 2001). Flavonoids may suppress LDL oxidation and inflammatory progression in the artery wall (Hertog et al, 1995; Peluso, 2006). It was found that catechins promote many biological functions, including prevention of cardiovascular diseases (Langley-Evans, 2001) and cancer (Kohlmeier et al, 1997; Steinmetz & Potter, 1991). Polyphenolic phytochemicals such as epigallocatechin from tea, the flavonoids quercetin and genistein from onions and soya, curcumin in curry spice and resveratrol from red grapes are diet constituents with high efficacy in preclinical carcinogenesis of colorectum, breast and prostate (Thomasset et al., 2006; Lambert et al., 2005; Surh, 2003). Low cerebrovascular disease was associated with high intake of the flavonol kaempferol and of the flavonones naringenin and hesperitin and there was a trend of reduction in type 2 diabetes was associated with higher quercetin and myricetin intake (Knekt et al., 2002). The other reported a strong inverse association between the sum of quercetin, myricetin, luteolin, and apigenin intaked and stroke (Keli et al.,1996) and low risk of lung cancer at high flavonoids intake (Knekt et al.,1997). Black rice anthocyanins reduced oxidative damage (Sangkitikomol et al., 2010a), anti-inflammation (Wang et al., 2007), enhanced LDL-receptor (Sangkitikomol et al., 2010b) and promoted cardiovascular health status (Ling et al., 2002; Wang et al., 2007). Aging is the major risk factor for neurodegenerative diseases such as Parkinson's diseases and Alzheimer's. Polyphenolic compounds could affect on cells not only due to their antioxidant activities but also due to their modulation of different pathways including signaling cascades, antiapoptotic processes and the synthesis of the amyloid β peptide (Ramassamy, 2006).

CytochromeP450 isoenzymes (CYP450s) are major enzymes in phase I of biotransformation system which involved in the metabolism of various endogenous chemicals such as; fatty acids, steroids, hormones, bile acids, eicosanoids and exogenous chemicals such as; xenobiotics, carcinogens, mutagens and environmental pollutants. A major function of CYP450s is catalyzed the reactions then conversion these nonpolar compounds into polar metabolites which can be conjugated by phase II enzymes. The oxygen activation by CYP450s' catalytic function generate the reactive oxygen species (ROS) (Guengerich, 2008). Flavonoids may be beneficial for health protection by reducing oxidative damage and minimizing toxicity by regulating mRNA CYP450s expression for suitable production or inhibition of CYP450s isoenzymes, thereby maintaining xenobiotic biotransformation balance. Flavonoids from St. John's wort can selectively inhibit CYP1 enzymes may be useful as chemoprotective agents in prostate cancer (Chaudhary & Willett, 2006). Inhibition of PAH-induced carcinogenesis using cancer chemoprevention; methoxylated flavones and stilbene resveratrol, could effectively inhibit the benzo[*a*]pyrene-DNA binding and CYP 1A induction which were the early step in molecular levels of cencer prevention (Tsuji & Walle, 2007). Emerging evidence indicates that transcriptional activation of the antioxidant response element (ARE) plays a crucial role in modulating oxidative stress and providing cytoprotection against prooxidant stimuli (Nguyen et al., 2003). Several chemopreventive agents, such as curcumin, caffeic acid phenethyl ester, rectinoic acid, (–)-epigallocatechin-3-gallate and (–)-epicatechin-3-gallate from tea, directed to protect DNA and other important cellular molecules by inducing the synthesis of phase II detoxifying genes and antioxidant genes via the Nrf2–ARE signaling pathway. Thereby enhancing those genes transcription and stimulating carcinogen detoxification/inactivation. The Nrf2 (nuclear transcription factor erythroid 2p45 (NF-E2)-related factor 2), is a regulator of genes encoding antioxidants and phase II enzymes such as glutathione S-transferase, NAD(P)H:quinine oxidoreductase 1, UDP-glucuronosyl-transferase, γ-glutamate cysteine ligase, and hemeoxygenase-1. The Nrf2 is known to mediate detoxification and/or to exert antioxidant functions thereby protecting cells from genotoxic damage (Lee & Surh, 2005; Zhang & Gordon, 2004; Hayes & McMahon, 2001). NF-κB regulated genes such as inducible nitric oxide synthase (iNOS) and cyclooxgenase-2 (COX-2) which are the inflammatory mediators and may promote cancinogenesis (Greten & Karin, 2004; Surh, 2003). Nrf2 encodes for antioxidant and general cytoprotection genes, while NF-κB regulates the expression of proinflammatory genes. A variety of antiinflammatory or anticarcinogenic phytochemicals suppress NF-κB signalling and activate the Nrf2-ARE pathway (Bellezza et al., 2010)

There is no evidence that different types of tea, herbal tea, fruit and vegetable currently available in the markets of Thailand have any significant antioxidant contents. The aim of this study was to carry out a survey of relative levels of total antioxidant activities by TEAC assay (Re et al., 1999) with reference to their total phenolic contents (Singleton et al., 1999). In order to find the natural sources of antioxidants, thereafter some of them were selected to study the inhibition effect on hemolysis (Sangkitikomol et al.,2010a) as a marker of lipid peroxidation, induced by 2, 2′-azobis (2-amidinopropane) hydrochloride and the inhibition effect on Heinz body formation, as a marker of protein oxidation, induced by N-acethylphenylhydrazine (Sangkitikomol et al., 2001; 2010a) by using normal fresh human red blood cells.

2. Measuring antioxidant activities of tea, herbal tea, fruit and vegetable

2.1 Chemicals

ABTS [2,2′-azinobis-(3-ethylbenzothiazoline-6-sulphonic acid)] diammonium salt, TPTZ (2,4,6-tripyridyl-s-triazine), gallic acid, Trolox(6-hydroxy-2,5,7,8-tetramethylchroman-2-carboxylic acid), AAPH[2,2'-azobis(2 amidinopropane) dihydrochloride], Folin Ciocalteu's phenol reagent and APHZ (N-acethylphenylhydrazine) were purchased from Sigma-Aldrich, St. Louis, MO,USA. All other basic reagents were of analytical grade.

2.2 Preparation of plant extracts

Different types of Thai plants were purchased from Thai-herb shops, health food shops, and local markets. These samples were crushed into small pieces and some were ready-made, they were 152 samples of fruit, vegetable and herb, 33 brands of tea. A hundred milligram of dry material was extracted with 2 x 10 ml of solvent using ultrasonic bath for 5 minutes sonication, centrifuged at 3000 rpm for 10 minutes and the combined extracts were kept in deep freezer at -80°C until used. Using 2 kinds of solvent for extraction, one is 80 % aqueous methanol and the other is deionized reverse osmosis water. The plant water-extracts were used to determine lipid and protein oxidation in human red blood cells, and the other were used to analyze the antioxidant activity. Fourteen samples of ready-to-drink beverages were purchased from supper markets and directly used for analysis.

2.3 Preparation of fresh whole blood

Human blood samples were obtained from The National Blood Centre, Thai Red Cross Society, Bangkok, Thailand. Fresh blood was collected in heparinised tubes and centrifuged at 252 × g for 3 min. Plasma was carefully removed by aspiration in order to obtain a hematocrit of approximately 50% (packed red blood cells; RBCs) for hemolysis test and Heinz body formation. The blood was stored at 4°C and used within 3 h.

2.4 Determination of antioxidants in plant extracts

The total antioxidant determination and total phenolic contents were performed on the Shimadzu spectrophotometer model UV160A (Tokyo, Japan). All determinations were carried out at least three times of the standards and samples. And the correlation analysis had been done by using program Microsolf Excel 2007.

2.4.1 Trolox Equivalent Antioxidant Capacity (TEAC) Assay

Total antioxidant activity is measured by TEAC assay of Re et al. (1999). It is a screening of lipophilic and hydrophilic antioxidants activity by decolorization assay. The radical monocation of 2,2'-azinobis-(3-ethylbenzothiazoline-6-sulfonic acid) (ABTS•) is generated by oxidation of ABTS with potassium persulfate. Then the radical ABTS• is reduced in the presence of hydrogen-donating antioxidants from the plant extracts. The percentage inhibition of absorbance at 734 nm is calculated and plotted as a function of concentration of antioxidants and of Trolox for the standards. The results were expressed as Trolox equivalent in mM per kilogram of dry weight (TE mM/kg.dw).

2.4.2 Folin cioculteau phenol assay (Folin assay)

Modified Folin Cioculteau Phenol assay (Singleton et al., 1999) is used to determine reducing properties of phenolics contents. Briefly, 500 µl of samples or standards was mixed with 500 µl of 10% Folin reagent, let stand for 20 min, added 10 mM Na_2CO_3 350 µl, let stand for 20 min for the solution turned blue color and then measured the absorbance at 750 nm by using gallic acid as the standards. The results were expressed as Gallic acid equivalent (GE) mM/kg.dw.

2.4.3 Determination of AAPH-Induced oxidative damage of erythrocyte membrane

The erythrocyte susceptibility to oxidative destruction was evaluated *in vitro* by subjecting the cells to oxidative stress. Since a peroxyl radical initiator, AAPH, has been proved to cause an oxidation of cell membrane proteins and lipids resulting in lysis of red blood cells (RBCs), oxidative hemolysis induced by AAPH (Sangkitikomol et al.,2010a) was used as a tool in the present study. Briefly, positive control was added 0.100 ml of fresh whole blood into pre-incubated medium [1.00 ml of phosphate buffer saline (PBS) pH 7.4], then incubated at 37 °C for 5 min. One ml of PBS solution with 200 mM AAPH was added to the whole blood suspension. The reaction mixture was shaken gently at 37 °C under aerobic condition for three and a half hours. Reaction mixture (0.100 ml) was withdrawn to 1.5 ml of ice cold PBS at 120, 150, 180 and 210 minutes after added AAPH and centrifuged at 1006 × g for 10 min. The extent of hemolysis was determined by measuring the absorbance of hemolysate at 540 nm. Negative control was done the same as positive control without AAPH presented. And the plant extracts was done the same as positive control excepted only using the extract 0.5 ml mixed with 0.5 ml of double strength of PBS in stead of 1.00 ml pre-incubated medium. Similarly, 0.100 ml of reaction mixture of negative control was treated with 1.5 ml of distilled water to yield complete hemolysis. Percentage hemolysis was calculated according to the equation. % hemolysis = absorbance of the sample aliquot divided by absorbance of the complete hemolysis x 100. Data was represented as the time required to achieve 50 % hemolysis (T_{50} min).

2.4.4 Determination of APHZ-Induced Heinz body formation in cytosol of erythrocytes

APHZ was used as a free radical initiator inside RBCs to oxidize proteins mostly hemoglobin based on the modification of protocol described previously (Sangkitikomol et al. 2001). Using Heinz body formation was a marker of protein oxidation for testing the antioxidants properties of plant extracts. Briefly, positive control was added 0.1 ml of fresh whole blood into 2.0 ml of reaction medium [contained 1.0 g/L of APHZ, 2.0 g/L of glucose in phosphate buffer (1.3 parts of 1/15 M of KH_2PO_4 and 8.7 parts of 1/15 M of Na_2HPO_4, pH 7.4)] and incubated for 2 hours at 37°C under aerobic condition. Heinz bodies are precipitates of oxidized or denatured hemoglobin that adhere strongly to the red blood cell membrane. Negative control was done without APHZ added. And the plant extracts were done the same as positive control excepted only using the plant extract 1.0 ml mixed with 1.0 ml of double strength of reaction medium. Staining Heinz bodies in RBC with 3 drops of crystal violet solution (10 g/L in 0.73 % of normal saline) and 3 drops of RBC from reaction medium for 5 minutes at room temperature. Made blood smear on glass slides and counted RBC with Heinz body inside/1,000 of RBC using light microscope. The results were reported in % inhibition of Heinz body formation.

2.4.5 Results

Antioxidant activities of 152 herbs and 33 brands of tea using TEAC assay varied considerably with the types of plants and the types of tea.The range of antioxidant activities and total phenolic contents were several hundred-fold. The plants which are very good sources of antioxidants are *Quercus infectoria, Areca catechu Linn., Terminalia spp., Phyllanthus emblica Linn., Punica granutum Linn., Eugenia caryophyllus* (Table 1) and tea (*Camellia sinensis*) (Table 2). Correlation analysis (R) between the antioxidant activity measurements of plant-extracts and tea-extracts with the measured total phenol concentration were calculated to be 0.988 and 0.902, respectively (Figure 1-2). Antioxidant activities varied considerably with the types of tea (Table 2). In the TEAC assay tea-extracts with the highest to the lowest activity were green tea, oolong tea and black tea, respectively. The range of antioxidant activity was 15.6-fold for the TEAC analysis and 6.6-fold for Folin assay. Fourteen samples of ready-to-drink beverages were found small amount of polyphenolics and antioxidant contents (Table 3). Selected 30 samples with high level of total antioxidant activities, within the range 141-9490 mM TE/kg.dw., were used to study the inhibition effect of lipid peroxidation and protein oxidation. The results showed that 13 out of 30 samples showed inhibition effect on Heinz body formation and the other had no effect may be caused by antioxidant contents in the extracts was not enough to inhibit Heinz body formation inside red blood cells. All samples showed prolong T_{50} min. of hemolysis assay, and most of them had very strong inhibition effect which showed the T_{50} min. were longer than 3 and a half hours. Time required to get T_{50} min. of control positive was between 120-150 min. (Table 4).

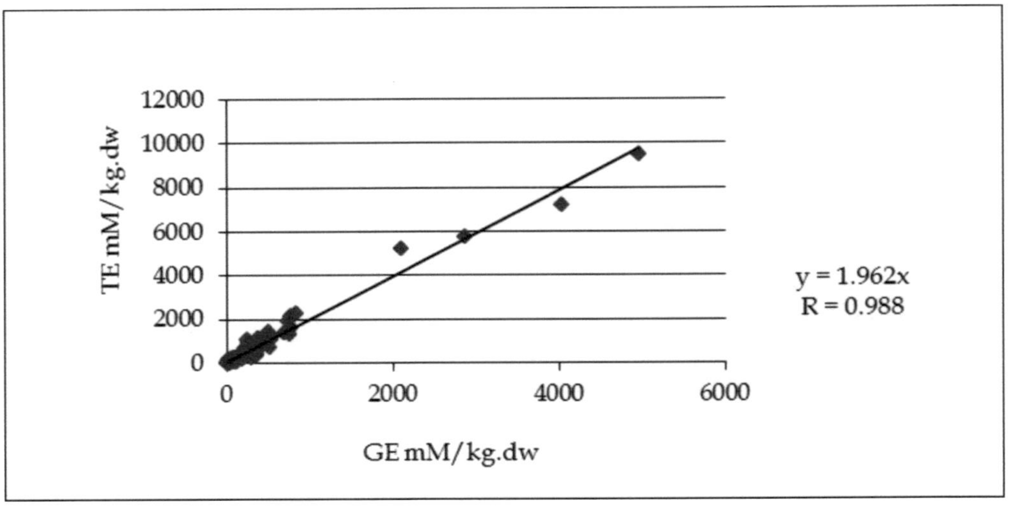

Fig. 1. Correlation between polyphenolics using Folin assay with total antioxidant activities using Trolox equivalent antioxidant capacity in 152 plant-extracts. Both assays are expressed in dry weight.

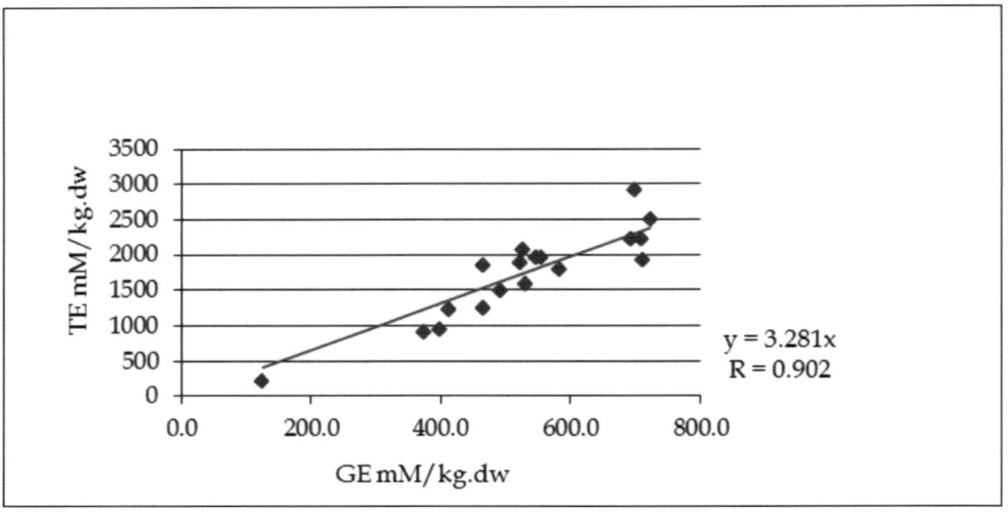

Fig. 2. Correlation between 33 polyphenolics using Folin assay with total antioxidant activities using Trolox equivalent antioxidant capacity in 33 tea-extracts. Both assays are expressed in dry weight.

No	Scientific Name of plants	Folin Assay GE mM/kg.dw.	TEAC Assay TE mM/kg.dw.
1	*Abroma augusta Linn.*	63 ± 3	40 ± 3
2	*Acacia catechu Willd.*	4038 ± 106	7183 ± 204
3	*Acacia rugata Merr.*	94 ± 10	52 ± 1
4	*Adhatoda vasica (L.) Nees*	47 ± 3	56 ± 10
5	*Aiiium ascalonicum Linn.*	221 ± 30	357 ± 20
6	*Albizia myriophylla Benth.*	61 ± 8	144± 7
7	*Allium tuberosum Roxb.*	36 ± 4	80.8 ± 3.5
8	*Amomum cadamomum*	5 ± 1	12.0 ± 1.0
9	*Anethum graveolens Linn.*	47 ± 4	55 ± 4
10	*Angelina sinensis*	25 ± 2	46 ± 4
11	*Areca catechu Linn.*	2876 ± 68	5771 ± 46
12	*Artimisia pallens Wall. Ex Bess.*	35 ± 5	44 ± 5
13	*Artocapus heterophyllus Lamk.*	116 ± 10	29 ± 4
14	*Artocarpus lakoocha Roxb.*	744 ± 32	1923 ± 201
15	*Atractylodes lyrata Sieb.et Zucc.*	19 ± 3	31 ± 6
16	*Azodirachta indica A Juss.* (stem)	237 ± 10	467± 39
17	*Azodirachta indica A Juss.* (flower)	53 ± 2	99 ± 8
18	*Azodirachta indica A Juss.* (leaf)	188 ± 10	322 ± 35
19	*Baliospermum montanum Willd.*	38 ± 2	64 ± 6
20	*Baliospermum montanum Willd.*	157 ± 4	212 ± 15
21	*Bambusa spp.*	42 ± 3	74± 10
22	*Boesenbergia pandurate Holtt.*	107 ± 6	221 ± 15
23	*Bougainvillea spectabilis Willd.*	157 ± 10	218 ± 12
24	*Brassica camprestris L ssp.*	59 ± 2	144 ± 11
25	*Bridelia siamensis Craib.*	98 ± 8	129 ± 10
26	*Caesalpinia sappan Linn.*	520 ± 15	1292 ± 102
27	*Cantharanthus roseus* (flower)	95 ± 10	134 ± 12
28	*Cantharanthus roseus* (leaf)	107 ± 7	151 ± 23
29	*Carica papaya Linn.* (leaf)	47 ± 2	90 ± 3
30	*Carum carvi Linn.*	53 ± 5	66 ± 10
31	*Casicum frutecens Linn.* (fruit)	58 + 3	89 ± 4
32	*Casicum frutecens Linn.* (seed)	40 ± 5	15 ± 3
33	*Cassia sophera Linn.*	59 ± 7	88 ± 7
34	*Cinnamomum cassia*	47 ± 4	38 ± 6
35	*Cinnamomum zeylanicum Linn.*	219 ± 22	426 ± 31
36	*Citrus hystrix DC.* (leaf)	56 ± 10	88 ± 6
37	*Citrus hystrix DC.* (peel)	90 ± 9	254 ± 29
38	*Citrus medica Linn. Var.*	53 ± 6	102 ± 10
39	*Cladogynos orientalis Zipp. Ex Span*	27 ± 3	32 ± 5
40	*Combretum extensum*	61 ± 7	158 ± 15
41	*Combretum quadrangulare Kurz.*	164 ± 10	399 ± 29
42	*Conioselinum univitatum*	49 ± 8	73 ± 10
43	*Connarus ferugineus*	417 ± 31	767 ± 60
44	*Cuminumc cyminum Linn.*	60 ± 5	94 ± 10
45	*Curcuma longa Linn.*	352 ± 20	395 ± 41

No	Scientific Name of plants	Folin Assay GE mM/kg.dw.	TEAC Assay TE mM/kg.dw.
46	*Cymbopogon citratus stapf.*	50 ± 4	64 ± 5
47	*Cyperus rotundus Linn.*	22 ± 1	189 ± 15
48	*Diospyros decandra Lour.*	33 ± 4	66 ± 9
49	*Diosyrros mollis Griff*	38 ± 2	49 ± 4
50	*Dracaena loureiri Gagnep.*	169 ± 9	292 ± 27
51	*Emita sonchifolia DC.*	32 ± 3	49 ± 7
52	*Erythrina suberosa Roxb.*	105 ± 9	165 ± 20
53	*Eugenia caryophyllus*	687 ± 15	1400 ± 302
54	*Eurycoma longifolia Jack*	23.4 ±1.6	34.5 ± 4.1
55	*Garcinia mangostana* (peel)	243 ± 12	1095 ± 206
56	*Glycyrrhiza glabra Linn.*	295 ± 20	242 ± 30
57	*Hamisonia perforata (Lour.) Merr.*	405 ± 31	887 ± 41
58	*Heracleumsiamnicum Craib.*	42 ± 4	55 ± 5
59	*Hibiscus abelmoschus Linn.*	22 ± 2	30 ± 2
60	*Illicium verum Hook*	100 ± 9	171 ± 16
61	*Impormoca reptans Poir.*	377 ± 25	751 ± 32
62	*Ixora cibdela craib*	84 ± 7	134 ± 10
63	*Jatropha gossypifolia Linn.*	41± 3	69 ± 6
64	*Jusminum sambac Ait.*	49 ± 3	71 ± 9
65	*Leucaena leucocephala de Wit*	143 ± 10	316 ± 25
66	*Levisticum officinale Koch*	52 ± 3	64 ± 5
67	*Lycium chinensis*	62 ± 9	66 ± 8
68	*Mammea siamensis*	133 ± 15	291 ± 25
69	*Mangifera indica Linn.* (leaf)	539 ± 32	1118 ± 90
70	*Mesua ferrea Linn.*	222 ± 38	357 ± 30
71	*Millingtonia hortensis Linn.*	100 ± 15	133 ± 12
72	*Mimusops elengi Linn.*	52 ± 9	114 ± 11
73	*Molindia citrifolia Linn.* (leaf)	41 ± 7	79 ± 9
74	*Momordica charantia Linn.*	28 ± 3	29 ± 3
75	*Murdandia lorifomis (Hassk.)*	36 ± 3	56 ± 4
76	*Musa sapientum Linn.* (leaf)	28 ± 2	38 ± 4
77	*Myristica fragrans Houtt.*	77 ± 6	247 ± 15
78	*Myristica fragrans Houtt.*	8 ± 1	21 ± 4
79	*Nelumbo Mucifera Gaertn* (flower)	500 ± 3	1014 ± 101
80	*Nelumbo Mucifera Gaertn* (leaf)	308 ± 24	714 ± 35
81	*Nicotiana tabacum*	97 ± 12	106 ± 14
82	*Nigella sativa Linn.*	83 ± 5	118 ± 11.
83	*Ocimum basilicum*	56 ± 4	76 ± 8
84	*Ocimum sanctum Linn.* (flower)	95 ± 7	130 ± 10
85	*Ocimum sanctum Linn.* (leaf)	53 ± 4	64 ± 6
86	*Pandanus odoratissimus Linn.*	29 ± 3	36 ± 4
87	*Panicum repens Linn.*	21 ± 4	21 ± 2
88	*Petroselium crispum (Mill.)*	107 ± 10	200 ± 14
89	*Pimpinella anisum Linn.*	59 ± 5	107 ± 10
90	*Pinus mercukusii Jungh & de Vriese*	65 ± 5	83 ± 7

No	Scientific Name of plants	Folin Assay GE mM/kg.dw.	TEAC Assay TE mM/kg.dw.
91	*Piper aurantiacum Mig*	75 ± 10	108 ± 12
92	*Piper betle Linn.*	360 ± 15	375 ± 20
93	*Piper chaba Hunt.*	31 ± 3	37 ± 4
94	*Piper nigrum Linn*(Black pepper)	24 ± 4	34 ± 4
95	*Piper sarmentosum Roxb.*	48 ± 4	71 ± 7
96	*Pithecellobium tenue Craib*	507 ± 32	1433 ± 102
97	*Plantago ovata Forskal P.*	17 ± 2	22 ± 2
98	*Plumbago indica Linn.*	66 ± 6	100 ± 11
99	*Plumeria alba Linn.*	21 ± 3	35 ± 4
100	*Polygonium odoratum Lour.*	168 ± 12	380± 20
101	*Psidium guajava*	376 ± 29	1121 ± 100
102	*Punica granutum Linn.*	778 ± 60	2157 ± 180
103	*Quercus infectoria*	4962 ± 202	9490 ± 390
104	*Rosa domescena Mill.*	428 ± 24	1035 ± 95
105	*Sidarhombifolia Linn.*	97 ± 18	113 ± 11
106	*Smilax corbulalia Kunta*	752 ± 80	1308 ± 99
107	*Smilax micro-china T. Koyama*	162 ± 11	239 ± 16
108	*Spilanthes acmella Linn.*	44 ± 3	71 ± 8
109	*Spilanthes acmella Linn.*	39 ± 5	73 ± 9
110	*Stemona toberosa Lour.*	14 ± 2	26 ± 2
111	*Streblus asper Lour.*	16 ± 1	20 ± 2
112	*Strychnos lucida R. Br.*	81 ± 3	91 ± 8
113	*Syzygium gratum*	227 ± 27	718 ± 66
114	*Tamaridus indica Linn.*	57 ± 6	104 ± 11
115	*Tamarindus indica Linn.*	255 ± 20	932 ± 46
116	*Terminalia chebula Retz.*	772 ± 65	1640 ± 112
117	*Terminalia spp.*	2100 ± 100	5216 ± 150
118	*Tiliacora triandra Diels*	83 ± 8	86 ± 10
119	*Tinospora cordifolia Miers*	65± 8	77 ± 6
120	*Urceola minutiflora Pierre*	172 ± 10	283 ± 18
121	*Xylinbaria minutiflora Pierre*	539 ± 50	1050 ± 97
122	*Zingiber cassumunar Roxb.*	97 ± 11	113 ± 11
123	*Ziziphus mauritiana*	74 ± 7	29 ± 3
124	*Andrographis paniculata Nee*	140 ± 9	164 ± 10
125	*Bulbostylis barbata Clarke*	116 ± 11	119 ± 10
126	*Carcinia atroviridis Griff.*	22 ± 2	34 ± 3
127	*Carthamus tinctorius Linn*	142 ± 10	186 ± 10
128	*Cassia alata Linn.*	124 ± 10	208 ± 14
129	*Cassia angustifilia vohl.*	84 ± 9	139 ± 11
130	*Cassia tora Linn.*	111 ± 10	203 ± 18
131	*Centella asiatica Urban*	78 ± 6	136 ± 10
132	*Chrysanthenum indicum*	96 ± 8	114 ± 10
133	*Clitorea ternatea Linn.*	86 ± 6	156 ± 10
134	*Cymbopogon citratus stapf.*	50 ± 4	64 ± 6
135	*Ganoderma lucidum* (sample 1)	21 ± 2	29 ± 2

No	Scientific Name of plants	Folin Assay GE mM/kg.dw.	TEAC Assay TE mM/kg.dw.
136	*Gonoderma lucidum* (sample 2)	19 ± 2	57 ± 3
137	*Ginkgo biloba*	216 ± 11	652 ± 40
138	*Mimosa pudica Linn.*	123 ± 10	193 ± 19
139	*Morus alba Linn.* (sample 1)	56 ± 4	126 ± 9
140	*Morus alba Linn.*(sample 2)	119 ± 10	139 ± 10
141	*Nelumbo Mucifera Gaertn* (pollen)	143 ± 12	290 ± 18
142	*Oroxylum indicum Vent.*	511 ± 20	710 ± 32
143	*Pandanus amaryllifolius Roxb*	89 ± 9	141 ± 11
144	*Phyllanthus emblica Linn.*	841 ± 40	2288 ± 91
145	*Phyllanthus urinaria Linn.*	219 ± 29	330 ± 25
146	*Rhinacanthus nasutus Kurz Sn.*	16 ± 1	41 ± 3
147	*Schefflera leucantha* (sample 1)	38 ± 4	88 ± 9
148	*Schefflera leucantha*(sample 2)	57 ± 4	82 ± 6
149	*Thunbergia iaurifolia Linn.*	189 ± 12	190 ± 11
150	*Tinospara erispa Miers*	20 ± 2	31 ± 4
151	*Zingiber officenale Roscoe*	194 ± 11	220 ± 19
152	*Zingiber officinale Roscoe* (young)	37 ± 3	60 ± 4

All determinations were carried out at least three times .

*Polyphenolics are expressed as the mean ± SD in mM of catechin equivalents per kilogram of dry weights.

**Total antioxidant activities are expressed as the mean ± SD in mM of Trolox equivalents per kilogram of dry weights.

Table 1. The polyphenolic contents and total antioxidant activities of 152 plant-extracts using Folin assay and TEAC assay, respectively.

Type of Tea	Folin Assay GE mM/kg.dw	TEAC Assay TE mM/kg.dw.
Green tea (brand 1)	818 ± 66	3307 ± 112
Green tea (brand 2)	807 ± 42	2740 ± 96
Green tea (brand 3)	752 ± 52	2541 ± 69
Green tea (brand 4)	697 ± 37	2452 ± 100
Green tea (brand 5)	633 ± 26	1848 ± 74
Green tea (brand 6)	636 ± 46	2252 ± 56
Green tea (brand 7)	622 ± 41	2369 ± 78
Green tea (brand 8)	698 + 43	2836 ± 82
Green tea (brand 9)	506± 40	1704 ± 40
Green tea (brand 10)	563 ± 52	1901 ± 70
Green tea (brand11)	466 ±12	1800 ± 36
Green tea (brand 12)	444 ±16	1650 ± 45
Green tea (brand 13)	438 ± 28	1236 ± 56
Green tea (brand 14)	422± 25	1459 ± 55
Oolong (brand 1)	722 ± 43	2500 ± 85
Oolong (brand 2)	711 ± 44	1926 ± 38
Oolong (brand 3)	709 ± 28	2214 ± 83
Oolong (brand 4)	698 ± 17	2906 ± 97

Type of Tea	Folin Assay GE mM/kg.dw	TEAC Assay TE mM/kg.dw.
Oolong (brand 5)	693 ± 36	2222 ± 80
Oolong (brand 6)	600 ± 22	1994 ± 71
Oolong (brand 7)	583 ± 22	1793 ± 38
Oolong (brand 8)	526 ± 29	2062 ± 71
Oolong (brand 9)	555 ± 26	1963 ± 24
Oolong (brand 10)	547± 35	1963 ± 67
Oolong (brand 11)	523 ±18	1878 ± 41
Oolong (brand 12)	464 ± 30	1840 ± 31
Oolong (brand 13)	412+ 20	1222 ± 38
Oolong (brand 14)	464 ± 19	1237 ± 43
Black tea (brand 1)	530 ± 21	1578 ± 38
Black tea (brand 2)	491 ± 20	1489 ± 34
Black tea (brand 3)	398 ± 30	942 ± 18
Black tea (brand4)	374 ± 21	904 ± 26
Black tea (brand 5)	125 ± 34	212 ± 6

All determinations were carried out at least three times .

*Polyphenolics are expressed as the mean ± SD in mM of catechin equivalents per kilogram dry weight.

**Total antioxidant activities are expressed as the mean ± SD in mM of Trolox equivalents per kilogram dry weight.

Table 2. The polyphenolic contents and total antioxidant activities of 33 tea-extracts using Folin assay and TEAC assay, respectively.

Kind of beverages	Folin Assay* GE μM/L	TEAC Assay** TE μM/L
1. Green tea(sample1)	7,800 ± 500	11,400 ± 1,000
2. Green tea(sample2)	4,200 ± 200	9,000 ± 600
3. Green tea(sample3)	5,400 ± 300	8,900 ± 700
4. Green tea(sample4)	3,500 ± 100	5,000 ± 300
5. Grape juice (sample1)	8,400 ± 300	12,200 ± 400
6. Grape juice (sample2)	3,500 ± 100	4,600 ± 200
7. Guava juice	4,200 ± 200	4,200 ± 100
8. Carrot mixed fruit juice	3,400 ± 100	4,200 ± 300
9. Passion fruit tea	2,800 ± 100	3,600 ± 400
10. Pine apple juice(sample1)	700 ± 200	2,700 ± 200
11. Pine apple juice(sample2)	1,000 ± 100	2,400 ± 100
12. Orange juice	3,000 ± 100	2,400 ± 200
13. Apple juice	1,900 ± 100	2,000 ± 100
14. Litchi juice	600 ± 100	900 ± 100

All determinations were carried out at least three times .

*Polyphenolics are expressed as the mean ± SD in mM of catechin equivalents per liter of beverage.

**Total antioxidant activities are expressed as the mean ± SD in mM of Trolox equivalents per liter of beverage.

Table 3. The polyphenolic contents and total antioxidant activities of 14 beverages using Folin assay and TEAC assay, respectively.

Name of plants	TEAC Assay TE mM/kg.dw	50%Hemolysis* T_{50} min.	Heinz body** % inhibition
1. *Quercus infectoria*	9,490	>210	100
2. *Areca catechu Linn.*	5,771	>210	100
3. *Terminalia spp.*	5,216	>210	100
4. *Phyllanthus emblica Linn.*	2,288	>210	100
5. *Punica granutum Linn.*	2,157	>210	100
6. *Camellia sinensis (green tea)*	2,906	>210	100
7. *Eugenia caryophyllus*	1,400	>210	100
8. *Mangifera indica Linn*	1,118	>210	100
9. *Camellia sinensis(Oolong tea)*	2,204	>210	100
10. *Rosa domescena Mill.*	1,035	>210	100
11. *Tagetes erecta Linn.*	1,129	>210	100
12. *Psidium guajava*	1,121	>210	100
13. *Syzygium gratum*	718	>210	100
14. *Impormoca reptans Poir.*	751	>210	0
15. *Nelumbo Mucifera Gaertn*	1,014	>210	0
16. *Smilax corbulalia Kunth*	1,308	>210	0
17. *Nelumbo Mucifera Gaertn*	714	>210	0
18. *Tamarindus indica Linn.*	932	180	0
19. *Garcinia mangostana*	1,095	>210	0
20. *Cinnamomum cassia*	686	>210	0
21. *Mesua ferrea Linn.*	357	>210	0
22. *Ginkgo biloba (leaf)*	652	>210	0
23. *Aiiium ascalonicum Linn.*	357	>210	0
24. *Azodirachta indica A Juss.*	467	>210	0
25. *Phyllanthus urinaria Linn.*	330	>210	0
26. *Polygonium odoratum Lour.*	380	>210	0
27. *Leucaena leucocephala*	316	>210	0
28. *Erythrina suberosa Roxb.*	165	>210	0
29. *Citrus hystrix DC.*	254	>210	0
30. *Pandanus amaryllifolius Roxb*	141	>210	0
31. Positive control		120-150	0
32. Negative control		>210	100

*T50 min. = the time required to achieve 50 % hemolysis.

** Blood smear on glass slides and counted RBC with Heinz body inside/1,000 of RBC. The results were reported in % inhibition of Heinz body formation.

Table 4.Total antioxidant activities of 30 plant-extracts with inhibition effect on hemolysis, $T_{50\ min.}$, and inhibition effect on Heinz body formation.

3. Conclusion

For our knowledge this current work is the first report of phenolic contents and total antioxidant activities of Thai herb. The value of antioxidant activities of plant polyphenols showed varied widely with the various types of herb. The range of antioxidant activities and total phenolic contents were several hundred-fold for herbal extracts and 15.6-fold using TEAC analysis and 6.5-fold using Folin assay for 33 brands of tea. Nevertheless there was a very good relationship between the results from the TEAC assay which can measure the antioxidants activities and polyphenols in herbal extracts ($r = 0.988$) and in tea extracts ($r = 0.902$). It is suggested that the process of oxidation to make black tea and oolong tea may cause decreasing their antioxidant activities, and the correlation between the antioxidant activities and total phenolic contents of tea extracts was lower than that of herbal extracts.

Tea is one of the most popular and widely consumed as daily beverage in the world. At present, tea has become an important agricultural product that more than 40 countries in the world, especially Thailand, grow tea trees within Asian countries producing 90 % of the world output. All tea trees have their origin directly or indirectly in China and come from the same plant, *Camellia sinensis*. The composition of tea varies with the age of the leaf (plunking position), climate, horticultural practices and the process of storage (Lin et al., 1996, 1998). The different types of tea result from variation in processing of the leaves after they are harvested. The difference in processing results in different types of polyphenolics profiles between oxidation and no oxidation tea. There are different methods in manufacturing tea that give green tea (no oxidation) contains catechins. (Lin et al, 1998) black tea (fully oxidation tea) contains the polymeric compounds, thearubigins and theaflavins, and oolong tea (semi-oxidation tea) contains a mixture of the monomeric polyphenols and higher molecular weight theaflavins (Graham, 1992). Several epidemiological studies have shown beneficial effects of tea in cancer, cardiovascular, and neurological diseases (Zaveri, 2006).

Selected 30 types of herb with various polyphenol levels to study antioxidant activities by the modified methods which were developed by the author. For the test of AAPH induced hemolysis, most studies used isolated erythrocyte suspension for the simplicity of the system and the data interpretation. However fresh whole blood was employed as the *ex vivo* sample for this study to ensure the erythrocytes were tested in the least modified state. To study the effect of herbal extracts against oxidative damage to lipids and proteins in human erythrocytes, free radical initiators, AAPH and APHZ, were employed for inducing oxidative stress. AAPH was source of free radical formation capable of inducing oxidation of lipid and protein structurally located on erythrocyte membranes (Chaudhuri et al., 2007). The results showed that erythrocytes were more resistant to AAPH-induced oxidation when herbal extracts were added, most of them had very strong inhibition effect. It was showed that the T_{50} mins were longer than 3 and a half hours. Plant polyphenols are membrane-active antioxidant agents (Saija et al., 1995) and are the cell metabolism regulators by modulating the fluidity of lipid bilayer, since they have been demonstrated to control cell signal pathways by targeting receptors on the cell surface or by intercalating the lipid bilayer of membranes (Tarahovsky et al., 2008). Polyphenols interact with the surface of bilayer through hydrogen bonding have been shown to reduce the accessibility of oxidants, thus protecting the structure and function of membranes (Oteiza et al., 2005).

APHZ was a source of free radical formation inside cytosol of erythrocytes leading to induction of proteins mostly hemoglobin and lipid bilayer of membranes oxidation (Sangkitikomol et al., 2001). Phenylhydrazine in the presence of oxidase or peroxidase , it reacts with oxyhemoglobin to form phenylhydrazine radicals. Phenylhydrazine slowly oxidizes in aqueous solution to form O_2 •- and H_2O_2 and the end products of the reaction are benzene and N_2. Phenylhydrazine radical is the most damaging agent that can denature hemoglobin molecules. The oxidation of reactive sulfhydryl (S-H) groups creates disulfide bonds that may change the conformation of globin chains, resulting in precipitation of the hemoglobin molecules called Heinz bodies, then follows by membrane lipids oxidation and causing hemolysis (Winterbourn, 1985). Bioavailability differs greatly from one polyphenol to the other; therefore the most abundant polyphenols in plants are not leading to the highest concentrations of active metabolites in target cells. Gallic acid and isoflavones are the most absorbed polyphenols, followed by catechins, flavanones, and quercetin glucosides. The least absorbed polyphenols are proanthocyanidins, galloylated tea catechins, and anthocyanins (Manach et al., 2005). The results showed that some herbal extracts could inhibit Heinz body formation inside erythrocytes, but the other could not. This finding suggested that the process of proteins and lipids oxidation was still taking place as a result of insufficiency of absorbed polyphenols inside erythrocytes to inhibit oxidation.

The advantage of using erythrocytes (human living cells) as the models for screening test of antioxidant properties of herbal extracts, because it is simple and low cost to perform the analysis and the given data could be extrapolated to happen in human body.

4. Acknowledgment

This work was supported by Faculty of Allied Health Sciences, Sciences, Chulalongkorn University, Bangkok.

5. References

Ames, B.; Shigena, M. & Hagen, T. (1993). Oxidants, antioxidants and the degenerative diseases of aging, *Proceedings of the National Academic Sciences of the United State of America,* Vol. 90, pp. 7915-7922, ISSN 0027-8424

Beecher, G. (2003). Overview of dietary flavonoids: nomenclature, occurrence and intake, *Journal of Nutrition,* Vol.133, pp. 3248-3254, ISSN 0022-3166

Bellezza, I.; Mierla,A. & Minelli, A. (2010). Nrf2 and NF-κB and their concerted modulation in cancer pathogenesis and progression, *Cancers,*Vol. 2, pp. 483-497, ISSN 2072-6694

Chaudhary, A. & Willett, K. (2006). Inhibition of human cytochrome CYP1 enzymes by flavonoids of St. John's wort, *Toxicology,* Vol. 217, pp. 194–205, ISSN 0300-483X.

Chaudhuri, S.; Banerjee, A.; Basu, K.; Sengupta, B. & Sengupta, P. (2007). Interaction of flavonoids with red blood cell membrane lipids and proteins: Antioxidant and antihemolytic effects, *International Journal of Biological Macromolecules* Vol. 41, pp 42-48, ISSN 0141-8130

Chen, H. & Yen G. (1997). Possible mechanisms of antimutagen by various teas as judged by their effect on mutagenesis by 2-amino-3-methylmidazo[4,5-f]quinoline and benzo(a)pyrine, *Mutation Research,* Vol.393, pp. 115-122, ISSN 0921-8262

Cook, N. & Samma, S. (1996). Flavonoids-chemistry, metabolism, cardioprotective effects and dietary sources, *Journal of Nutritional Biochemistry, Vol.* 7, pp. 66-76, ISSN 0955-2863

Graham, H. (1992). Green tea composition, consumption, and polyphenol chemistry, *Preventive Medicine,* Vol. 21, pp 334–350, ISSN 0091-7435

Greten, F. & Karin, M. (2004). The IKK/NF-kappaB activation pathway-a target for prevention and treatment of cancer, *Cancer Letters,* Vol.206, pp. 193–199, ISSN 0304-3835

Guengerich, F. (2008). Cytochrome P450 and Chemical Toxicology. *Chemical Research in Toxicology,* Vol.21, pp. 70–83, ISSN 0893-228X

Halliwell, B. & Gutteridge J. (2001) *Free radicals in biology and medicine,* Oxford Science Publication. Printed by Thomson Press (India) Ltd., pp.617-854, ISBN 0 19 850044 9 (Hbk), ISBN 0 19 850044 0 (Pbk)

Hayes, J. & McMahon M. (2001). Molecular basis for the contribution of the antioxidant responsive element to cancer chemoprevention, *Cancer Letters,* Vol.174, pp. 103–113, ISSN: 0304-3835

Hertog, M.; Kromhout, D.; Aravanis, C.; Blackburn, H.; Buzina, R., et al. (1995). Flavonoid intake and long-term risk of coronary heart disease and cancer in the seven countries study, *The Archives of Internal Medicine,* Vol.155, pp. 381-386, ISSN 0003-9926

Keli, S.; Hertog, M.; Feskens, E. & Kromhout, D. (1996). Dietary flavonoids, antioxidant vitamins, and incidence of stroke: the Zutphen study, *The Archives of Internal Medicine,* Vol. 156, pp.637–42, ISSN 0003-9926

Knekt, P.; Järvinen, R.; Seppänen, R.; Helibvaara, M.; Teppo, L.; et al. (1997). Dietary flavonoids and the risk of lung cancer and other malignant neoplasms, *American Journal of Epidemiology, Vol.*146, pp. 223–30, ISSN 0002-9262

Knekt, P.; Kumpulainen, J.; Järvinen, R.; Rissanen, H.; Heliövaara, M.; et al. (2002). Flavonoid intake and risk of chronic diseases, *American Journal of Clinical Nutrition,* Vol. 76, pp. 560-568, ISSN

Kohlmeier, L.; Weterings, K.; Steck, S. & Kok F. (1997). Tea and cancer prevention: an evaluation of the epidemiological literature, *Nutrition and Cancer,* Vol.27, pp.1-13, ISSN 0163-5581

Lambert, J.; Hong, J.; Yang, G.; Liao, J. & Yang C. (2005). Inhibition of carcinogenesis by polyphenols: evidence from laboratory investigations, *American Journal of Clinical Nutrition,* Vol.81, pp. 284S–291S, ISSN 0002-9165

Langley-Evans, S. (2001). Tea as a protective agent in cardiovascular health, *Nutrition & Food Science,* Vol. 31, pp. 75-78, ISSN 0034-6659

Lee, J. & Surh Y. (2005). Nrf2 as a novel molecular target for chemopreventio, *Cancer Letters,* Vol. 224, pp. 171–184, ISSN 0304-3835

Lin, Y.; Juan, I.; Chen, Y.; Liang, Y. & Lin, J. (1996). Composition of polyphenols in fresh tea leaves and association of their oxygen-radical absorbing capacity with antiproliferative action in fibroblast cells, *Journal of Agricultural and Food Chemistry*, Vol. 44, pp1387-1394, ISSN 0021-8561

Lin, J.; Lin, C.; Liang, Y.; Lin-Shiau, S. & Juan, I. (1998). Survey of catechins, galic acid and methylxanthines in green, oolong, pu-erh and black teas. *Journal of Agricultural and Food Chemistry*, Vol.46, pp3635-3642, ISSN 0021-8561

Lin, J. & Liang, Y. (2000). Cancer chemoprevention by tea polyphenols, *Proceedings of the National Science Council, Republic of China Part B*, Vol.24, pp 1-13, ISSN 0255-6596

Ling, W.; Wang, L. & Ma, J. (2002). Supplementation of the black rice outer layer fraction to rabbits decreases atherosclerotic plaque formation and increases antioxidant status, *Journal of Nutrition, Vol.* 132, pp. 20-26, ISSN 0022-3166

Manach, C.; Williamson, G.; Morand, C.; Scalbert, A. & Rémésy, C. (2005). Bioavailability and bioefficacy of polyphenols in humans. I. Review of 97 bioavailability studies. *American Journal of Clinical Nutrition*, Vol. 81(suppl), pp 230S-242S, ISBN 0943685443

Nakamura, H. (1997). Effect of tea flavonoid supplimentation on the susceptibility of low density lipoprotein to oxidative modification, *American Journal of Clinical Nutrition*, Vol.66, pp. 261-266, ISSN 0002-9165

Nguyen, T.; Sherratt, P. & Pickett C. (2003). Regulatory mechanisms controlling gene expression mediated by the antioxidant response element, *Annual Review of Pharmacology and Toxicology*, Vol. 43, pp. 233–60, ISSN 0362-1642

Oteiza, P.; Erlejman, A.; Verstraeten, S.; Keen, C. & Fraga, C. (2005). Flavonoid-membrane interactions: A protective role of flavonoids at the membrane surface? *Clinical and Developmental Immunology*, Vol. 12, pp19-25, ISSN 1740-2522

Packer, L. & Colman, C. (1999). *The Antioxidant Miracle*. Published by John Wiley and Sons, Inc. Canada, pp.1-248, ISBN 0-471-29768-2

Peluso, M. (2006). Flavonoids attenuate cardiovascular disease, inhibit phosphodiesterase, and modulate lipid homeostasis in adipose tissue and liver, *Experimental Biology and Medicine*, Vol.231, pp. 1287-1299, ISSN 1535-3702

Ramassamy, C. (2006). Emerging role of polyphenolic compounds in the treatment of neurodegenerative diseases: A review of their intracellular targets, *European Journal of Pharmacology, Vol.* 545, pp. 51–64, ISSN 0014-2999

Re, R.; Pellegrini, N.; Proteggente, A.; Pannala, A.; Yang, M. & Rice-Evans, C. (1999). Antioxidant activity applying an improved ABTS radical cation decolerization assay, *Free Radical in Biology and Medecine*, Vol.26, pp.1231-1237, ISSN 0891-5849

Saija, A.; Scalese, M.; Lanza, M. & et al. (1995). Flavonoids as antioxidant agents: importance of their interaction with biomembranes, *Free Radical in Biology and Medecine*, Vol.19, pp481–486, ISSN 0891-5849

Sangkitikomol, W.; Krikreangsak, N. & Klaisiri, A. (2001). The preventive effect of antioxidants in vegetables and Thai herbs on Heinz bodies formation in red blood cells induced by oxidant agent:Acetylphenylhydrazine, *Journal of Allied Health Sciences*, Vol.1, pp. 46-54, ISBN 1513-4865, Bangkok, Thailand

Sangkitikomol, W.; Tencomnao, T. & Rocejanasaroj, A. (2010a). Antioxidant effects of anthocyanins-rich extract from black sticky rice on human erythrocytes and mononuclear leukocytes, *African Journal of Biotechnology*, Vol. 9, pp. 8222-8229, ISSN 1684-5315

Sangkitikomol, W.; Tencomnao, T. & Rocejanasaroj, A. (2010 b). Effects of Thai black sticky rice extract on oxidative stress and lipid metabolism gene expression in HepG2 cells, *Genetics and Molecular Research*,Vol. 9, pp. 2086-2095, ISSN1676 -56

Sen, S.; Chakraborty, R.; Sridhar, C.; Reddy, Y. & De, B. (2010). Free radicals, antioxidants, diseases and phytomedicines: current status and future prospect, *International Journal of Pharmaceutical Sciences Review and Research*, Vol. 3, pp. 91-100, ISSN 0976-044X

Singleton, V.; Orthofer, R. & Lamuela-Raventos R. (1999). Analysis of total phenols and other oxidation substrates and antioxidants by means of Folin-Ciocalteu reagent. In *Method in Enzymology*. Packer L.(ed). Academic Press, Orlando FL. USA, Vol. 299, pp. 152-178, ISBN 0-12-182236-2

Steinmetz, K. & Potter J. (1991). Vegetables, fruit, and cancer. II. Mechanisms, *Cancer Causes and Control*, Vol.2, pp. 427–42, ISSN 0957-5243

Surh, Y. (2003). Cancer chemoprevention with dietary phytochemicals. *Nature Reviews Cancer*, Vol. 3, pp. 768–80, ISSN 1474-175X

Tarahovsky, Y.; Muzafarov, E. & Kim, Y. (2008). Rafts making and rafts braking: how plant flavonoids may control membrane heterogeneity. *Molecular and Cellular Biochemistry* Vol. 314, pp. 65-71, ISSN 0300

Thomasset, S.; Berry, D.; Garcea, G.; Marczylo, T.; Steward, W. & Gescher A. (2006). Dietary polyphenolic phytochemicals—Promising cancer chemopreventive agents in humans? A review of their clinical properties, *International Journal of Cancer*, Vol. 120, pp.451–458, ISSN 0020-7136

Tsuji, PA. & Walle, T. (2007). Benzo[*a*]pyrene-induced cytochrome P450 1A and DNA binding in cultured trout hepatocytes—Inhibition by plant polyphenols, *Chemico-Biological Interactions*,Vol. 169, pp. 25–31, ISSN 0009-2797

Virgili, F. & Marino, M. (2008). Regulation of cellular signals from nutritional molecules: a specific role for phytochemicals, beyond antioxidant activity, *Free Radical in Biology and Medecine*, Vol. 45, pp.1205–1216, ISSN 0891-5849

Wang, Q.; Han, P.; Zhang, M.; Xia, M.; Zhu, H.; et al. (2007). Supplementation of black rice pigment fraction improves antioxidant and anti-inflammatory status in patients with coronary heart disease, *Asia Pacific Journal of Clinical Nutrition*, Vol.16(Suppl 1), pp. 295-301, ISSN 0964-7058

Winterbourn, C. (1985). Free-radical production and oxidative reactions of hemoglobin. *Environmental Health Perspectives*, Vol.64, pp. 321-330, ISSN 0091-6765

Zaveri, T. (2006). Green tea and its polyphenolic catechins: Medicinal uses in cancer and noncancer applications. *Life Sciences*, Vol.78, pp. 2073–2080, ISSN 0024-3205

Zhang, Y. & Gordon, B. (2004). A strategy for cancer prevention: stimulation of the Nrf2-ARE signaling pathway, *Molecular Cancer Therapeutics*, Vol. 3, pp. 885–893, ISSN 1535-7163

Ziegler, R. (1991). Vegetables, fruits and carotenoids and risk of cancer, *American Journal of Clinical Nutrition,* Vol. 53(suppl.), pp. 251s-259s, ISSN 0002-9165

15

Modification by Aqueous Extracts of *Allium kurrat L.* and *Ricinus communis L.* of Cyanide Nephrotoxicity on Balb/C Mice

Fahmy G. Elsaid[1,2]
[1]*Zoology Department, Faculty of Science, Mansoura University, Mansoura*
[2]*Biology Department, Faculty of Science, King Khalid University, Abha,*
[1]*Egypt*
[2]*Saudi Arabia*

1. Introduction

1.1 Cyanide toxicity

Cyanides are components of electroplating solutions, fertilizers, fumigant mixtures, metal polishes and rodenticides. It is an environmental factor has been associated with many intoxication episodes in humans and animals resulting from the ingestion of foods, environmental pollution, chemical war, suicide, homicide, occupational factors and use in some drugs such as nitroprusside and laetrile (Watts 1998). In plants, cyanide can be found mainly as cyanogenic glycosides, as found in *Manihot* sp. (cassava), *Linum* sp., *Lotus* sp., *Phaseolus lunatus*, *Sorghum* sp.(Conn 1978) and the content of this substance can be high as 100-800 mg/kg of the plant material (Poulton 1993). Acute poisoning ingestion by an adult of 50 ml of liquid hydrogen cyanide or 200–300 mg of one of its salts is likely to be fatal without treatment, though death is likely to be delayed for at least 1 hour. Acute cyanide toxicity is lethal, resulting in death due to respiratory failure (Greer and Jo, 1995), while chronic cyanide exposure has been implicated in the etiology of goitre (Cliff et al., 1986), tropical ataxic neuropathy (Osuntokun, 1981). Regardless the route of exposure, cyanide is rapidly absorbed into the blood stream and distributed throughout the body. Cyanide concentrates in erythrocytes through binding to methemoglobin (Towill et al., 1978) and cause hypothyroidism that leads to goiter (Kamalu and Agharanya, 1991; Elsaid and Elkomy 2006). Subacute oral administration of cyanide in rats produced changes in several biochemical indices and pathology in various organs (Tulswani et al., 2005).

1.2 Cyanide toxicity and oxidative stress

Cyanide-induced cellular oxidative stress, i.e., increase of superoxide anions, lipid peroxide, hydroxide radicals, hydrogen peroxide and others, appears to arise through multiple pathways. At the cellular level, cyanide produces chemical hypoxia by inhibiting cytochrome c oxidase in complex IV of the mitochondrial oxidative phosphorylation chain to markedly reduce ATP (Pearce et al., 2003). As oxidative damage is mediated by free radicals, it was necessary to investigate the status of endogenous antioxidant enzymes under different treatment conditions. Aerobic organisms possess antioxidant defense systems that deal with ROS (Mates et al., 1999). SOD catalyzes the conversion of the highly reactive superoxide anion to O_2 and to H_2O_2 (Mates et al., 1999). This less damaging molecule can be converted spontaneously to highly reactive hydroxyl radicals. However, CAT and GSH-Px detoxify hydrogen peroxide by converting it to water before hydroxyl radicals can be produced. The propensity of cyanide to induce lipid peroxidation and impair antioxidant defense enzymes like CAT, SOD and GSH-Px are well known. Levels of MDA, GSH and GSSH are also correlated with lipid peroxidation (Ardelt et al., 1994). However, these effects were reversible after withdrawal of the poison. The oxygen radicals formed in these pathways may contribute to the oxidative damage of DNA and tumor formation. The inhibitors of CAT include azide, cyanide, GSH and dithiothreitol (Sun and Oberley 1989). GSH-Px inhibitors include cyanide and superoxide radicals (Blum and Fridovich1985). The decrease of cellular GSH following exposure to cyanide is likely due in part to reduced cellular ATP resulting from inhibition of cytochrome c oxidase (Prabhakaran et al., 2006). Antioxidants blocked the enhanced apoptosis produced by cyanide and this was directly linked to generation of ROS (Jones et al., 2003). However, levels of MDA, GSH and GSSH are also correlated with lipid peroxidation (Ardelt et al., 1994).Taken together the mechanism of hepatocyte injury resulting from inhibition of mitochondrial respiration includes a cytotoxic pathway that arises partly from an energy deficit but also from reductive stress which releases non protein bound iron from intracellular pools and induces cytotoxic ROS formation (Niknahad et al., 1995).

1.3 Antidotes and cyanide treatments

Antidotes are based on induction of methaemoglobin, which temporarily removes cyanide ions from solution. Anti-cyanide therapy may be acute, such as in poisonings, or prophylactic, as in military applications where a longer duration of action is required (Baskin and Fricke, 1992). Cyanide detoxification occurs *in vivo* mainly by conversion to thiocyanate. This reaction is catalysed by the enzyme rhodanese. In the presence of excess cyanide, the rate-limiting step is an adequate supply of sulphane sulphur (a divalent sulphur atom bonded to another sulphur atom), which can be supplied by thiosulphate (Beasley and glass 1998). However, the principal detoxification pathway of cyanide to thiocyanate in the presence of sulfur donor like garlic extract (Elsaid and Elkomy 2006) and sodium thiosulphate is mainly catalyzed by a liver mitochondrial enzyme, rhodanese (Cyanide: thiosulphate sulphur transferase) (Tylleskar et al., 1991). Cyanide accumulates in various body cells through binding to metalloproteins or enzymes such as CAT and cytochrome C oxidase. So, it is a potent cytotoxic agent that kills the cell by inhibiting cytochrome oxidase of the mitochondrial electron transport chain (Yen et al., 1995). The Egyptian leek contains gallic acid as a phenolic component and vitamin C, so it has antioxidant properties (Latif and ABD El-AAl, 2007). The leaves of *Ricinus communis* found to contain flavonoids like kaempferol-3-0-beta-D-rutinoside and kaempferol-3-0-beta-D-xylopyranoid and tannins (Kang et al., 1985; Khogali et al., 1992), so the present study aims to investigate the antidotal effect of *Allium* and *Ricinus* extracts.

1.4 Cyanide toxicity and renal functions

The recorded renal toxicity was also detected by the elevation in serum uric acid as well as in serum and urine creatinine and urea levels (Elsaid and Elkomy 2006). However, the protective effect of garlic on kidney indices could be attributed to its antioxidant properties because it has been found that ROS may be involved in the impairment of glomerular filtration rate (Elsaid and Elkomy 2006).

1.5 Cyanide cytotoxicity and gene expression

The rat P53 gene consists of only 10 exons. In cells with DNA injury, P53 can stop the cell cycle through p21 protein and then promote DNA repair. When DNA is seriously damaged, P53 can induce the cell to undergo programmed cell death to maintain the stability of the genome and cells. The loss of P53 function activates oncogenes and inactivates cancer suppressor genes, playing an essential role in multistage carcinogenesis (Harris 1993). Bcl-2 is constitutively expressed and localized to the outer mitochondrial membrane where it attenuates cell death signals to promote cell survival (Kroemer, 1998). Bcl-2 exerts an anti-apoptotic effect by inhibiting mitochondrial outer membrane permeabilization to suppress release of cytochrome c into the cytosol (Gross et al., 1999). Bcl-2 may also inhibit necrotic-like cell death by blocking the opening of the mitochondrial permeability transition pore to maintain cellular ATP levels within survival limits (Garland and Halestrap, 1997). Forced over-expression of Bcl-2 can block cell death produced by a variety of stimuli, including cyanide (Tsujimoto et al., 1997). Overexpression of Bcl-2 produced mitochondrial dysfunction (reduced membrane potential), caspase-independent apoptosis, and sensitization of the cells to cyanide-induced toxicity (Zhang et al., 2007). In various liver and kidney injury models, IL-4 has been shown to be both protective and deleterious. IL-4 accelerates severe hepatitis in mice deficient in suppressor of cytokine signaling proteins through activation of natural killer T cells (Naka et al., 2001), and it is believed that IL-4 plays a key role in Con A-induced hepatitis via augmentation of Vα-14 natural killer T cell-mediated cytotoxicity (Kaneko et al., 2000). The toxic effects of cyanide in kidney of rabbits have been reported after repeated exposures (Okolie and Osagie, 1999).

Although, the mechanism of toxicity of cyanide in kidney has not been delineated so far, so the study aimed to investigate the genotoxic effect of cyanide to connect the toxicity of cyanide in kidney with the physiological functions of the kidney and gene expression of some genes. Also, this work tries to discuss the antidotal effect of aqueous extracts of both *Allium kurrat* and *Ricinus communis L.*, against cyanide poisoning on kidney of Balb/c mice.

2. Material and methods

Natural Products Extracts: *Allium kurrat* (Egyptian leek), (*family Alliaceae*) and *Ricinus communis* Linn, (Castor) (*family Euphorbiaceae*) were used in this study. The leaves of leek and Castor were separately immersed in distilled water for 24 hours, filtered, stored at -20°C and freshly prepared every three days.

Animal grouping: Twenty mice weighed (45g) were used in the present experiment. The local committee approved the design of the experiments and the protocol conforms to the guidelines of the National Institutes of Health (NIH). All measures were taken to minimize the number of mice used and their suffering. Mice were divided into four groups. Animals were caged in groups of 5, allowed standard mice chow diet and water *ad libitum* as follows: Control group; mice received no treatment. Cyanide group; mice were administered with 4.5mg/kg b.w./day of potassium cyanide (0.5 of LD_{50}) in drinking water for 30days, freshly prepared every four days for the experimental periods, with some modifications (Sousa et al., 2002). Cyanide plus *Allium kurrat* group; mice were co-administered with 4.5mg/kg b.w./day KCN in drinking water for 30 days and intraperitoneal administered with *Allium kurrat* extract at 200mg/kg b.w./day. Cyanide plus *Ricinus communis* group; mice co-administered with 9mg/kg b.w./day cyanide and intraperitoneal administered with *Ricinus* extract at 200mg/kg b.w./day for the experimental period. Animals were maintained at 22–25°C and 40–60% relative humidity with 12-h light–dark cycles. At the end of the experimental period mice were sacrificed and sera and kidney tissues were collected from each group. The kidney tissues were homogenized in a phosphate buffer solution pH 7.4, centrifuged at 4°C and the supernatant was stored at -80°C.

3. Enzymatic assay

Superoxide dismutase (SOD): This assay relied on the ability of the enzyme to inhibit the phenazine methosulphate-mediated reduction of nitroblue tetrazolium dye (Nishikimi et al., 1972).

Catalase (CAT): It reacted with a known quantity of H_2O_2. The reaction is stopped after exactly one minute with CAT inhibitor.

Catalase

$$2\,H_2O_2 \longrightarrow 2\,H_2O + O_2$$

In the presence of peroxidase (HRP), the remaining H_2O_2 reacted with 3,5-Dichloro-2-hydroxybenzene sulfonic acid (DHBS) and 4-aminophenazone (AAP) to form a chromophore with a color intensity inversely proportional to the amount of CAT in the original sample, each 1unit =1µmol of H_2O_2 degraded for a minute, using a molar extinction coefficient of 43.6M^{-1} cm^{-1} (Aebi 1984).

$$2\,H_2O_2 + DHBS + AAP \xrightarrow{HRP} \text{Quinoneimine Dye} + 4H_2O$$

Glutathione peroxidase (GSH-Px): The activity was measured by the method described by (Ellman, 1959). Briefly, reaction mixture contained 0.2ml of 0.4M phosphate buffer pH 7.0, 0.1ml of 10mM sodium azide, 0.2ml of kidney homogenate (homogenate on 0.4M phosphate buffer, pH 7.0), and 0.2ml glutathione, 0.1 of 0.2mM H_2O_2. The content was incubated at 37°C for 10min. The reaction was arrested by 0.4ml of 10% TCA and centrifuged. Supernatant was assayed for glutathione content by using Ellman's reagent. The molar extinction coefficient for NADPH is 6220 mM^{-1} cm^{-1} at 340nm.

Glutathione reductase (GSH-Red): GSH-Red catalyzed the reduction of oxidized glutathione (GSSG) to reduced glutathione (GSH). GSH-Red is essential for the glutathione redox cycle that maintains adequate levels of reduced cellular GSH. According to the method of Goldberg and Spooner (1983), GSH-Red catalyzed the reduction of the oxidized glutathione (GSSG) in the presence of NADPH that oxidized into $NADPH^+$. The decrease in absorbance was measured at 340nm.

$$NADPH + H^+ + GSSG \xrightarrow{\text{GSH-Red}} NADP^+ + 2GSH$$

GSH (GSH) content: GSH served as an antioxidant, reacting with free radicals and organic peroxides, in amino acid transport, and as a substrate for the GSH-Px and glutathione-s-transferase in the detoxification of organic peroxide and metabolism of xenobiotics, respectively. Homogenize the tissue in 5–10ml cold buffer (50mM potassium phosphate, pH 7.5, 1mM EDTA) per gram tissue. Centrifuge at 100,000 x g for 15min at 4°C. Remove the supernatant for assay and store on ice and then freeze the sample at - 80°C. The method based on the reduction of 5,5` dithiobis (2-nitrobenzoic acid) (DTNB) with GSH to produce a yellow compound. The reduced chromogen directly proportional to GSH concentration and its absorbance can be measured at 405nm (Beutler et al., 1963).

Total antioxidant capacity: The determination of the total antioxidant capacity was performed by the reaction of antioxidants in the sample with a defined amount of exogenously provide hydrogen peroxide (H_2O_2).The antioxidants in the sample eliminated a certain amount of the provided hydrogen peroxide. The residual H_2O_2 was determined colorimetrically by an enzymatic reaction which involves the conversion of 3, 5, dichloro–2–hydroxyl benzensulphonate to a colored product (Koracevic et al., 2001).

Lipid peroxidation: The thiobarbituric acid reactive substances (TBARS) as malondialdehyde were estimated by the method of Ohkawa et al., (1979). Briefly, to 0.2ml of kidney homogenate, 0.2ml of 40% sodium dodecyle sulphate, 1.5ml of 20% acetic acid (prepared in 0.27M of HCl) and 1.5ml of 0.5% thiobarbituric acid were added together. The mixture was heated for 60min at 95°C in a temperature controlled water bath to give a pink color. The mixture was then centrifuged at 3500 r.p.m. for 10min. Finally, the absorbance of the supernatant layer was read spectrophotometrically at 532nm, the molar extinction coefficient factor equal $1.56 \times 10^5 M^{-1} cm^{-1}$.

Total lipid: According to the method of Zollner and Kirsch (1962), lipids reacted with sulfuric, phosphoric acids and vaniline to form pink colored complex, the absorbance was measured at 545nm.

Total protein: In the presence of an alkaline cupric sulfate, the protein produced a violet color, the intensity of which is proportional to their concentration (Gornall and Bardawill, 1949). The absorbance was read at 550nm.

Uric acid assay: According to the method of Fossati et al., (1980) by using of 3, 5-dichloro2-hydroxybenzene sulfonic acid/4-aminophenazone chromogenic system in direct enzymatic assay of uric acid in serum.

Creatinine assay: Serum creatinine level was estimated according to the method of Henry (1974).

Extraction of total RNA from kidney tissue homogenates: Total RNA was isolated from kidney homogenate using RNeasy Mini Kit according to manufacturer's instructions (QIAGEN, Germany). About 100 µl of each homogenate was subjected to RNA extraction and the resultant RNA was dissolved in DEPC-treated water, quantified spectrophotometrically and analyzed on 1.2% agarose gel. RNAs inhibitors were added to the samples during the RNA extraction process (Shati et al., 2011).

4. Real time PCR and gene expression

a- For P53 and Bcl2 genes: The extracted RNA from kidney tissues of different groups was subjected to examine the expression level of two genes using specific primers in the presence of Glyceraldehyde-3-phosphate dehydrogenase as a housekeeping gene. The Real time reaction consists of 12.5µl of 2X Quantitech SYBR® Green RT Mix (Fermentaz, USA), 2µl of the extracted RNA (50ng/ µl), 1µl of 25 pM/µl forward (F) primer, 1µl of 25 pM/µl reverse (R) primer (Table 1), 9.5 µl of RNAase free water for a total of 25 µl. Samples were spun before loading in the rotor's wells. The real time PCR program was performed as follows: initial denaturation at 95°C for 10 min.; 40 cycles of 95°C for 15 sec, annealing at 64°C for 30 sec and extension at 63°C for 30 sec. Data acquisition performed during the extension step. This reaction was performed using Rotor-Gene 6000 system (QIAGEN, USA).

Primer name	Primer sequence from 5`-`3	Annealing temp.
P53	F- AGGGATACTATTCAGCCCGAGGTG R-ACTGCCACTCCTTGCCCCATTC	64°C
Bcl-2	F-ATGTGTGTGGAGAGCGTCAACC R-TGAGCAGAGTCTTCAGAGACAGCC	63°C
IL-4	F-CTATTAATGGGTCTCACCTCCCAACT R-CATAATCGTCTTTAGCCTTTCCAAG	60°C
IL-12	F-CAGCCTTGCAGAAAAGAGAGC R-CCAGTAAGGCCAGGCAACAT	65°C
GPDH (House keeping gene)	F-ATTGACCACTACCTGGGCAA R-GAGATACACTTCAACACTTTGACCT	60 °C

GPDH: glyceraldehydes 6-phosphate dehydrogenase.

Table 1. Oligonucleotide primer sequences used in this study:

b- For interleukins (IL-4 and IL-12) genes: Real time PCR was performed using specific primers for two of cytokines genes, IL-4 and IL-12. The reaction was performed on the total RNA extracted from the kidney of all experimental groups. The PCR reaction constituents and conditions were similar to the above genes except that the annealing temperature was 60°C for IL-4 and 65°C for IL-12.

5. Molecular data analysis

Comparative quantitation analysis was performed using Rotor-Gene-6000 Series Software based on the following equation:

$$\text{Ratio target gene expression} = \frac{\textit{Fold change in target gene expresion} \quad (\frac{sample}{control})}{\text{Fold change in reference gene expresion} \quad (\frac{\text{house keeping gene}}{\text{control}})}$$

Real-time PCR data of all samples were analyzed with appropriate bioinformatics and statistical program for the estimation of the relative expression of genes using real-time PCR and the result normalized to its gene (Reference gene). The data were statistically evaluated, interpreted and analyzed using Rotor-Gene-6000 version 1.7.

6. Statistical analysis of biochemical data

The biochemical data recorded were expressed as mean±SD and statistical and correlation analyses were undertaken using the One-way ANOVA followed by a post-hoc LSD (Least Significant Difference) test. A P value < 0.05 was statistically significant. A Statistical analysis was performed with the Statistical Package for the Social Sciences for Windows (SPSS, version 10.0, Chicago, IL, USA).

6.1 Results

Table (2): ANOVA test showed high significant (P<0.001) changes between the different groups. The results showed very high significant decrease (P<0.001) in kidney total antioxidant capacity, decrease in the activities of CAT, SOD, GSH-Px, GSH-Red and a decrease in GSH content in cyanide group when compared with the control one. Also there was a very high significant increase (P<0.001) in TBARS in cyanide group. Cyanide & *Allium* and cyanide & *Ricinus* groups were improving the nephrotoxicity of cyanide when compared with cyanide group.

Groups Parameters	Control	Cyanide	Cyanide& *Allium*	Cyanide& *Ricinus*
T. Antioxidant capacity	0.58±.07	0.37±0.04***	0.47±0.09*b	0.48±0.03*b
CAT	7.7±0.58	5.4±0.83***	5.8±0.58***	6.7±0.48*b
SOD	55.3±3.9	41.7±5.7***	44.6±2.6***	50.9±3.8b
GSH-Px	7.0±0.42	4.8±0.58***	5.0±1.1***	5.77±0.85*
GSH-Red	59.1±3.6	42.7±3.2***	47.9±6.4***	52.2±3.4*b
GSH	9.3±1.8	5.4±0.76***	6.0±0.38***	7.4±0.51**b
TBARS	108.0±9.3	254.6±18.5***	162.1±33.5***c	146.4±18.4***c

In comparison control with other groups, p<0.05 *, p<0.01**, p<0.001***

In comparison cyanide group with cyanide& *Allium* and Cyanide& *Ricinus* groups, p<0.05 a, p<0.01 b, p<0.001 c

Table 2. Kidney antioxidants and oxidative stress in different Balb/C groups:

In table (3): ANOVA test showed high significant (P<0.001) changes between the different groups. serum creatinine and uric acid levels showed very high significant increase (P<0.001) in cyanide group in comparison with control group. From the results in cyanide& *Allium* and cyanide& *Ricinus* groups, the aqueous extracts of *Allium* and *Ricinus* showed good effect on renal function when compared with cyanide group. The ameliorative effect appeared in decreasing of serum creatinine and uric acid levels in cyanide& *Allium* and cyanide& *Ricinus* groups. Moreover the level of total lipids and total protein content were decreased in cyanide group when compared with the control one. The treated groups with *Allium* and *Ricinus* extracts showed a good effect in managing this decrease.

Groups Parameters	Control	Cyanide	Cyanide& *Allium*	Cyanide& *Ricinus*
Serum Creatinine	0.68±0.07	1.1±0.11***	0.84±0.05**c	0.72±0.09c
Serum Uric Acid	2.0±0.19	3.3±0.52***	2.8±0.25**a	2.3±0.0.23c
Serum T. Lipids	0.72±0.05	0.42±0.05***	0.47±0.05***	0.58±0.05***c
Kidney T. Lipids	44.1±5.4	30.0±3.01***	30.8±2.81***	39.1±4.87b
Kidney T. Protein	10.5±1.4	5.6±1.1***	6.9±0.90***	7.59±1.27***a

In comparison control with other groups, $p<0.05$ *, $p<0.01$**, $p<0.001$***

In comparison cyanide group with cyanide& *Allium* and Cyanide& *Ricinus* groups, $p<0.05$ [a], $p<0.01$ [b], $p<0.001$ [c]

Table 3. Renal biomarkers in different Balb/C groups:

6.2 Molecular data

As in figure (1), P53 expression was high in cyanide group when compared with control group as shown in figure (1). But the P53 gene expression was so low in cyanide& *Allium* and cyanide& *Ricinus* groups when compared with cyanide group. Moreover, the Bcl-2 gene was highly expressed in cyanide administered rats in comparison with control group. The treated rats with cyanide& *Allium* and cyanide& *Ricinus* extracts showed amelioration expression of this gene in kidney tissue as in fig (1).

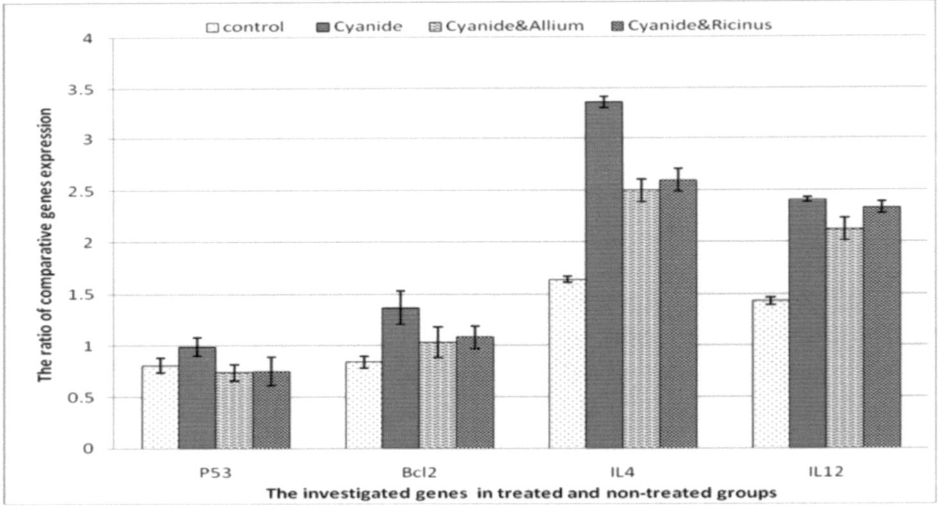

Fig. 1. The ratio of comparative gene expression in kidney of different treated groups compared with the control ones of balb/C mice, P53; Bcl-2 (B cell lymphoma 2); IL4 and IL12 (interleukin).

Also in figure (1), the expression of IL-4 and IL-12 genes was high in cyanide group in comparison with control group. But in the contrary a low gene expression was observed in cyanide& *Allium* and cyanide& *Ricinus* groups when compared with control group.

7. Discussion

7.1 Cyanide intoxication and antioxidants/oxidants balance

It was observed in KCN-treated cells that ROS generated from sources other than mitochondria represented a small portion of the total cellular ROS (Hariharakrishnan et al., 2009). Also it was indicating that mitochondria may be the primary source of ROS following cyanide (Chen et al., 2003). It is well characterized that cyanide inhibits cytochrome oxidase, which in turn stimulates ROS formation at complexes I and III. Free radicals are chemical entities that exist separately with one or more unpaired electrons. Free radicals induce damage or death of that cell of which the affected molecule is a part. The propagation of free radical formation could continue for thousands of reactions and hence the damage caused will be extensive (Campbell and Abdulla, 1995). Lipids, proteins and DNA are all susceptible to attack by free radicals. Cyanide toxicity also caused by increased generation of superoxide anion and lipid peroxidation (Daya et al., 2002) with inhibition of antioxidants enzymes (Ardelt et al., 1994). Hariharakrishnan et al., (2009) showed marked increase in total ROS and reactive nitrogen species levels and elevated lipid peroxidation after treatment with 1.25mM KCN. Reduced GSH levels render the cell more susceptible to endogenous oxidants (Slater et al., 1995).In the present study, cyanide intoxication showed increased MDA level (a marker of lipid peroxidation) with decreased CAT and SOD activities in kidney, indicating renal toxicity. In addition, the cyanide-induced oxidative stress leads to a decrease in GSH content, GSH-Red and GST activities. It exerts its effect by enhancing the non-enzymatic antioxidant as GSH and the detoxifying enzyme as GST (Saravanan et al., 2004). So it restores glutathione level and increases the activities of glutathione reductase and glutathione-S-transferase (Saravanan et al., 2004). Oxidative stress in the cells or tissues refers to enhanced generation of reactive oxygen species and/or depletion in antioxidant defense system. ROS generated in the tissues are efficiently scavenged by enzymatic antioxidant system such as GSH-Px and glutathione reductase as well as non enzymatic antioxidants such as glutathione, vitamin A, C, and E (Schlorff *et al.*, 1999). GST is a detoxification enzyme which catalyzes the conjugation of many electrophilic agents with GSH (Hayes et al., 1995) so it may be bind to CN and this explain their decrease in kidney tissues. Cyanide is known to induce oxidative stress by depleting the cellular thiol (Gunasekar et al., 1996). Similarly, in the present study, GSH, an intracellular antioxidant, was depleted after exposure to 4.5mg KCN making the cell environment less reducing and vulnerable to oxidative stress. So the changes in kidney may be induced by cyanide due to the imbalance in the antioxidant enzymes and oxidative stress as showed by (Elsaid and Elkomy 2006). In the present study, cyanide induced nephrotoxicity was reflected by the observed increases in serum uric acid and creatinine levels as well as total protein and lipid profile content (Elsaid and Elkomy 2006) as in table (3). Organosulphurs enhance the synthesis of the cellular GSH content in red blood cells (Wu et al., 2001), which is catalyzed by antioxidant enzymes (Godwin et al., 1992) also it may be alleviating the renal functions. The protective efficacy of *Allium* (Latif and ABD El-AAl, 2007) and *Ricinus* (Khogali et al., 1992) against cyanide-induced cytotoxicity can be attributed to the antioxidant property of both (Khogali et al., 1992). Also it can be interpreted by the sulphur donor properities of *Allium* (Latif and ABD El-AAl, 2007).

Moreover, cyanide is regarded as a selective neurotoxin (Pettersen and Cohen, 1993). It rapidly induces a non-competitive inhibition of the mitochondrial cytochrome c oxidase activity, which results in compromised aerobic oxidative metabolism and phosphorylation, cellular hypoxia, and lactic acidosis (Baud, 2007). Other complications of cyanide toxicity include loss of ionic homeostasis, excitotoxicity along with free radical damage (Johnson et al., 1987a). Cyanide-impaired mitochondrial energy metabolism is accompanied by severe lactic acidosis, elevated intracellular Ca^{2+} levels, decreased cellular ATP content and lipid peroxidation, leading to activation of proteases, lipases and xanthine oxidases, culminating in cell death (Bondy and Komulainen, 1988). This is may be a mechanism explains the decrease of total lipid and total protein in the kidney in cyanide group.

7.2 Cyanide intoxication and gene expression

Oxidative species such as hydrogen peroxide, superoxide, or free radical intermediates play a crucial role upstream of apoptosis activation. Oxidative stress is a key factor in cyanide-mediated apoptosis, partly by activation of redox-sensitive transcription factor NF-$_{\kappa}$B (Shou et al., 2002). Physiological effects of cyanide may be mediated through changes in the expression of gene products, either at the mRNA or protein level. Cyanide leads to stimulation of P53 expression in Cyanide group compared with the control ones. *Ricinus* and *kurrat* handle the expression of the P53 in cyanide administered mice and diminished the stimulation effect of cyanide of the P53 level in kidney cells. Activation of P53 in response to an obstruction of mitochondrial electron transport chain may additionally contribute to tissue damage (Khutornenko et al., 2010). It is worth noting that in cellular models of cyanide group, P53 was found to be conformational altered, making these cells less vulnerable to environmental stressors or genotoxic insults. But the expression of P53 was higher in cyanide& *Allium* and cyanide& *Allium* groups when compared with control one. Thus, P53 seems to play a pivotal role in cyanide detoxification, implying that modulation of cell death pathways might be of therapeutic benefit in cyanide detoxification. Apoptosis can be triggered by a variety of stimuli, including receptor ligation, growth factor withdrawal, exposure to chemotoxins, or even physical damage. Depending on the specific situation and the particular cell type, a variety of different internal signaling pathways can be initiated. However, these converge at certain control points into the execution phase of the program, where inhibitory or stimulatory cofactors like the family of Bcl-2 proteins come into action. Some members such as Bcl-2 or Bcl-xL block apoptosis, whereas others such as bax or bak promote apoptosis. Bcl-2 is apoptosis regulating gene which are functionally contradictory. The primary action of cyanide is inhibition of cytochrome c oxidase in complex IV of the oxidative phosphorylation chain, thereby blocking intracellular oxygen utilization (histotoxic hypoxia) and reducing cellular ATP generation (Leavesley et al., 2008). As a potent regulator of the Bcl-2, ROS regulate the expression of Bcl-2 via both transcription and protein degradation (Hildeman et al., 2003; Li et al., 2004). ROS-activated transcription factors such as NF-$_{\kappa}$ B often negatively regulate transcription of the Bcl-2 gene (Pugazhenthi et al., 2003). Bcl-2 expression was induced in the cyanide group compared with the control one. But when cyanide group treated with aqueous *Allium* and *Ricinus* extracts low expression of Bcl-2 was observed as a result, *Allium* and *Ricinus* extracts managed the expression of the Bcl-2 by which the cell may not directed to apoptosis. Taken together, over-expression of Bcl-2 protected against the cyanide toxicity, thus providing strong evidence that Bcl-2 down-regulation contributes to the cell death (Zhang et al., 2009). Here the data suggests that *Allium* and *Ricinus* induced P53 and suppressed Bcl-2 in the cyanide& *Allium* and cyanide& *Ricinus* groups which may help in attenuating of cyanide toxicity in the kidney cells.

Currently, it is generally believed that the role of IL-12 is a heterodimeric cytokine produced by activated blood monocytes, macrophages and glial cells. It enhances differentiation and proliferation of T cells and increases production of pro-inflammatory cytokines (Jifen et al., 2003). Cyanide makes induction for the IL-4 and IL-12 in cyanide group. But a low amount of IL-4 and IL-12 in cyanide administered rats treated with *Allium* and *Ricinus* extracts was observed as gene expression compared with control one. This confirms that both *Alluim* and *Ricinus* extract help in increasing the cyanide administered rat's immune system through modulating the interleukins expression as anti-inflammatory. To move in the end we suggest that both *Allium* and *Ricinus* play a vital role as antidotes and they are not protective but administrative.

8. Conclusion

The toxicity of cyanide associated with a decrease in the antioxidants enzymes as catalase, superoxide dismutase, glutathione and their relative enzymes and increase in the lipid peroxidation as thiobarbituric acid reactive substance in the kidney of Balb/c mice. P53 gene showed high expression in responding to cyanide toxicity as well as Bcl-2 and interleukins 4& 12 genes that showed over-expression in the kidney of mice. The watery extract of *Allium* and *Ricinus* was capable to manage the biochemical and gene expression accomplished with cyanide intoxication. So, the *Allium* and *Ricinus* have antiodotal effect at least in Balb/C mice at least in kidney tissue.

9. References

Aebi, H. (1974). Catalase. In. Bergmeyer, H.U. (Ed.), Methods of Enzymatic Analysis. Academic Press, New York, pp. 673-677.

Ardelt, B.K., Borowitz, J.L. and Isom, G.E. (1989). Brain lipid peroxidation antioxidant protectant mechanisms following acute cyanide intoxication. Toxicology 56, 147–154.

Ardelt, B.K., Borowitz, J.L., Maduh, E.U., Swain, S.L. and Isom, G.E. (1994): Cyanide-induced lipid peroxidation in different organs: subcellular distribution and hydroperoxide generation in neuronal cells, Toxicology 89, 127–137.

Baskin, S.I., and Fricke, R.F. (1992). The pharmacology of *p*-aminopropiophenone in the detoxification of cyanide. Cardiovasc. Drug Rev. 10, 358.

Baud FJ. (2007). Cyanide: critical issues in diagnosis and treatment. Hum Exp Toxicol, 26:191–201.

Beasley, D. M. G. and Glass, W. I. (1998). Cyanide Poisoning: Pathophysiology and Treatment Recommendations. *Occup Med (Lond)*. 48: 427–431.

Beutler, E., Duron, O. and Kelly, B.M. (1963).Improved method for the determination of blood glutathione. *J Lab Clin Med* 61:882–888,

Blum, J., and Fridovich, I. (1985). Inactivation of glutathione peroxidase by superoxide radical. *Arch. Biochem. Biophys.* 240:500-508.

Bondy SC, Komulainen H. (1988). Intracellular calcium as an index of neurotoxic damage. Toxicology, 35:35–41.

Campbell, I.C. and Abdulla, E.M. (1995). Strategic approaches to in vitro neurotoxicology. In: Chang, L.W., Slikker Jr., W.J. (Eds.), Approaches and Methods. Neurotoxicology Academic Press, London, pp. 495– 505.

Chen, Q., Vazquez, E. J., Moghaddas, S., Hoppel, C. L. and Lesnefsky, E. J. (2003). Production of reactive oxygen species by mitochondria: central role of complex III. J. Biol. Chem. 278:36027–36031.

Cliff J., Lundquist P., Rosling H., Sorbo B. and Wide L (1986) Thyroid function in a cassava-eating population affected by epidemic spastic paraparesis. Acta Endocrino-logica 113, 523±528.

Cliff, J., Lundquist, P. Rosling, H. Sorbo, B. and Wide, L. (1986). Thyroid function in cassava-eating population affected by epidemic spatic paraparesis. Acta Endocrinol., 113: 523-528.

Cliff, J., P. Lundquist, et al. (1986).Thyroid function in a cassava-eating population affected by epidemic spastic paraparesis. European Journal of Endocrinology 113(4): 523.

Conn, E.E. (1978). Cyanogenesis, the production of cyanide, by plants. In. Keeler, R.F., Van Kampen, K.R., James, L.F. (Eds.). Effects of Poisons Plants on Livestock. Academic Press, San Diego, pp: 301-310.

Daya, S., Walker R.B. and kumar-Dukie, S. (2002). Cyanide-induced free radical production in rat brain homogenate is reduced by asprin. Metabolic Brain disease 15: 203-210.

Ellman, G. L. (1959).Tissue sulfhydryl groups. Archives of biochemistry and biophysics 82(1): 70-77.

Elsaid, F. G. and M. M. Elkomy (2006). Aqueous Garlic Extract and Sodium Thiosulphate as Antidotes for Cyanide Intoxication in Albino Rats. Research Journal of Medicine and Medical Sciences 1(2): 50-56.

Fossati, P. Pregncipe, L. and Berti, G. (1980.) Use of 3,5-dichloro2-hydroxybenzenesulfonic acid/4 - amino-phenazone chromogenic system in direct enzymatic assay of uric acid in serum and urine. Clin. Chem., 26: 227-231.

Garland, J.M. and Halestrap, A. (1997). Energy metabolism during apoptosis. Bcl-2 promotes survival in hematopoietic cells induced to apoptose by growth factor withdrawal by stabilizing a form of metabolic arrest. J. Biol. Chem. 272, 4680–4688.

Godwin, A.K., Meister, A. O'Dwyer, P.J. Huang, C.S., Hamilton, T.C., and Anderson, M.E. (1992). High resistance to cisplatin in human ovarian cancer cell lines is associated with marked increase of glutathione synthesis, Proc. Natl. Acad. Sci. U.S.A., 89: 3070-3074.

Godwin, A.K., Meister, A., O'Dwyer, P.J., Huang, C.S., Hamilton, T.C., and Anderson, M.E. (1992). High resistance to cisplatin in human ovarian cancer cell lines is associated with marked increase of glutathione synthesis, Proc. Natl. Acad. Sci. U.S.A., 89: 3070-3074.

Goldberg, D.M. and Spooner, R.J. (1983). In Methods of Enzymatic Analysis (Bergmeyen, H.V. Ed.) 3rd ed. Vol. 3, pp 258-265, Verlog Chemie, Deerfield beach, FI .

Gornall, A. G. and Bardawill, C. J. (1949). "Determination of serum proteins by means of the biuret reaction." J. biol. Chem 177, (2): 751-766.

Greer, J. J. and Jo, E. (1995). Effects of cyanide on neural mechanism controlling breathing in neonatal rat in vivo. Neurotoxicology 16, 211-215.

Greer, J. J. and Carter, J. E. (1995). "Effects of cyanide on the neural mechanisms controlling breathing in the neonatal rat in vitro." Neurotoxicology 16(2): 211.

Gross, A., McDonnell, J.M. and Korsmeyer, S.J., (1999). BCL-2 family members and the mitochondria in apoptosis. Genes Dev. 13, 1899–1911.

Gunasekar, P.G., Sun, P., kanthasamy, A.G., Borowitz, J.L. and Isom, G.E. (1996). Cyanideinduced neurotoxicity involves nitric oxide and reactive oxygen species generation after N-methyl-d-aspartate receptor activation. J. Pharmacol. Exp. Ther. 277, 150–155.

Hariharakrishnan, J., Satpute, R., Prasad, G.B.K.S. and Bhattacharya, R. (2009). "Oxidative stress mediated cytotoxicity of cyanide in LLC-MK2 cells and its attenuation by alpha-ketoglutarate and N-acetyl cysteine." Toxicology letters 185(2): 132-141.

Harris, C.C. (1993). P53: at the crossroads of molecular carcinogenesis and risk assessment, Science 262, 1980–1981.

Hayes, J.D. and Pulford, D.J. (1995). The glutathione S-transferase supergene family. Regulation of GST and the contribution of the isoenzymes to cancer chemoprotection and drug resistance. Crit. Rev. Biochem. Mol. Biol., 30: 445-600.

Henry, R.J. (1974). Method of creatinine. Clinical Chemistry, principles and techniques 2nd Edn., Harper and row, pp: 525.

Hildeman, D.A., Mitchell, T., Aronow, B., Wojciechowski, S., Kappler, J. and Marrack, P. (2003). Control of Bcl-2 expression by reactive oxygen species. *Proc Natl Acad Sci U S A* 100:15035–15040.

Jones, D. C., Prabhakaran, K., Li, L., Gunasekar, P.G., Shou, Y., Borowitz, J.L. and Isom G.E. (2003). Cyanide Enhancement of Dopamine-Induced Apoptosis in Mesencephalic Cells Involves Mitochondrial Dysfunction and Oxidative Stress. NeuroToxicology 24, 333–342.

Johnson JD, William GC, Kevin D, Burris. Isom GE. (1987a). Peroxidation of brain lipids following cyanide intoxication in mice. Toxicology, 46:21–8.

Kamalu, B.P. and Agharanya, J.C. (1991). The effect of a nutritionally-balanced cassava (Manihot esculenta Crantz) diet on endocrine function using the dog as a model. 2-Thyroid. Br. J. Nutr., 65: 373-379.

Kaneko, Y., Harada, M., Kawano, T., Yamashita, M., Shibata, Y., Gejyo, F. , Nakayama, T. and Taniguchi. M. (2000). Augmentation of Vα14 NKT cell-mediated cytotoxicity by interleukin 4 in an autocrine mechanism resulting in the development of concanavalin A-induced hepatitis. *J. Exp. Med. 191:105.*

Kang, S.S., Cordell, A., Soejarto, D.D., Fong, H.H.S. (1985). Alkaloids and flavanoids from *Ricinus commmunis* .J Nat Prod 48: 155-156.

Khogali, A., Barakat, S., Abou-Zeid, H. (1992). Isolation and identification of the phenolics from *Ricinus communis* L . *Delta J Sci* 16: 198-211.

Khutornenko, A.A., Roudko, V.V., Chernyak, B.V., Vartapetian, A.B., Chumakov, P.M., Evstafieva , A.G. (2010). Pyrimidine biosynthesis links mitochondrial respiration to the P53 pathway. Proc Natl Acad Sci U S A. 107 (29):12828-33.

Koracevic, D., G. Koracevic, et al. (2001). "Method for the measurement of antioxidant activity in human fluids." Journal of clinical pathology 54(5): 356.

Kroemer, G. (1998). The mitochondrion as an integrator/coordinator of cell death pathways. Cell Death Differ. 5, 547.

Latif, S. S. and ABD El-AAL, H. A. (2007). Minerals profile-shlef extension and nutritive value of fresh green leafy vegetables consumed in Egypt. Afri Crop Sci conf Proceedings, 8, 1817-1826.

Leavesley, H.B., Li, L., Prabhakaran, K., Borowitz, J.L., and Isom, G.E. (2008). Interaction of cyanide and nitric oxide with cytochrome c oxidase: implications for acute cyanide toxicity. *Toxicol Sci* 101:101–111.

Li, D., Ueta, E., Kimura, T., Yamamoto, T. and Osaki, T. (2004). Reactive oxygen species (ROS) control the expression of Bcl-2 family proteins by regulating their phosphorylation and ubiquitination. *Cancer Sci* 95:644–650.

Lowry, O.H., Rosebrough, N.J., Farr, A.L. and Randall, R.J. (1951). Protein measurement with the folin phenol reagent. J. Biol. Chem. 193, 265–275.

Mates, J.M., Perez-Gomez, C. and Nunez de Castro, I. (1999). Antioxidant enzymes and human diseases. Clin. Biochem. 32, 595–603.

Naka, T., H. Tsutsui, M. Fujimoto, Y. Kawazoe, H. Kohzaki, Y. Morita, R. Nakagawa, M. Narazaki, K. Adachi, T. Yoshimoto, et al., (2001). SOCS-1/SSI-1-deficient NKT cells participate in severe hepatitis through dysregulated crosstalk inhibition of IFN and IL-4 signaling in vivo. *Immunity 14:535.*

Niknahad, H., Khan, Peter, S. and O'Brien, J. (1995). Hepatocyte injury resulting from the inhibition of mitochondrial respiration at low oxygen concentrations involves reductive stress and oxygen activation. Chemico-Biological Interactions 98, 27-44.

Nishikimi, M., Appaji Rao, N. et al. (1972). "The occurrence of superoxide anion in the reaction of reduced phenazine methosulfate and molecular oxygen." Biochemical and Biophysical Research Communications 46(2): 849-854.

Ohkawa, H., N. Ohishi, et al. (1979). "Assay for lipid peroxides in animal tissues by thiobarbituric acid reaction." Analytical biochemistry 95(2): 351-358.

Okolie, N.P. and Osagie, A.U. (1999). Liver and kidney lesions and associated enzyme changes induced in rabbits by chronic cyanide exposure. Food Chem. Toxicol. 37, 745–750.

Osuntokun B. O. (1981). Cassava diet, chronic cyanide intoxication and neuropathy in Nigerian Africans. World Review of Nutrition and Dietetics 36, 141-173.

Pettersen JC, Cohen SD. (1993). The effects of cyanide on brain mitochondrial cytochrome oxidase and respiratory activities. J Appl Toxicol, 13: 9–14.

Pearce, L. L., Bominaar, E. L., Hill, B. C. and Peterson, J. (2003). Reversal of cyanide inhibition of cytochrome c oxidase by the auxiliary substrate nitric oxide: an endogenous antidote to cyanide poisoning. J. Biol. Chem. 278: 52135–52145.

Poulton, J.E. (1983). Cyanogenic compounds in plants and their toxic effects. In. Keeler, R.F., Tu, A.T. (Eds.), Handbook of Natural Toxins. Marcel Dekker, New York, pp; 117–157.

Prabhakaran, K., Li, L., Borowitz, J.L. and Isom, G.E. (2006). Inducible nitric oxide synthase upregulation and mitochondrial glutathione depletion mediate cyanide-induced necrosis in mesencephalic cells. J. Neurosci. Res. 84, 1003–1011.

Pugazhenthi, S., A. Nesterova, et al. (2003). Oxidative stress mediated down regulation of bcl 2 promoter in hippocampal neurons, Wiley Online Library. 84: 982-996.

Saravanan, G. and Prakash, J. (2004). Effect of garlic (*Allium sativum*) on lipid peroxidation in experimental myocardial infarction in rats. J.Ethinopharmacol., 94: 155-158.

Saunders, J. A. and Conn E. E. (1978). "Presence of the cyanogenic glucoside dhurrin in isolated vacuoles from Sorghum." Plant physiology 61(2): 154.

Shati, A., Elsaid, F. G. and Hafez , E. E. (2011). Biochemical and molecular aspects of aluminium chloride-induced neurotoxicity in mice and the protective role of Crocus sativus L. extraction and honey syrup. Neuroscience.175, 66-74.

Schlorff, E. C., Husain, K. and Somani, S. M. (1999): Dose- and time-dependent effects of ethanol on plasma antioxidant system in rat. Alcohol, 17: 2, 97–105.

Shou, Y., N. Li, et al. (2002). NF B mediated up regulation of Bcl XS and Bax contributes to cytochrome c release in cyanide induced apoptosis, Wiley Online Library. 81: 842-852.

Slater,A.F., Stefan, C., Nobel, I., van den Dobblesteen, D.J. and Orrenius, S. (1995). Signalling mechanisms and oxidative stress in apoptois. Toxicol. Lett. 82–83, 149–153.

Sousa, A. B., Soto-Blanco, B. et al. (2002). "Does prolonged oral exposure to cyanide promote hepatotoxicity and nephrotoxicity? Toxicology 174 (2): 87-95.

Sousa, A.B., Soto-Blanco, B. Guerra, J.L. Kimura E.T. and G´orniak, S.L. (2002). Does prolonged oral exposure to cyanide promote hepatotoxicity and nephrotoxicity? Toxicology, 174: 87-95.

Sun, Y. and Oberley, L. W. The inhibition of catalase by glutathione.*Free Radical Biol. Med.* 7:595-602; 1989.

Towill, L.E., Drury, J.S., Whitfield, B.L., Lewis, E.B., Galyan E.L. and Hammons, A.S. (1978). Reviews of the Environmental Effects of Pollutants. V. Cyanide. U.S. Environ. Protection Agen., Rep. 600/1-78-027. 191.

Towill, L.E., Dury, J.S.,Whitfield, B.L., Lewis, E.B., Galyan, E.L., and Hammons, A.S. (1978). Reviews of the Environmental Effect of Pollutant Cyanide. Environmental Protection Agency Report No. 600/1-78-027, Ohio.

Tsujimoto, Y., Shimizu, S., Eguchi, Y., Kamiike, W. and Matsuda, H. (1997). Bcl-2 and Bcl-xL block apoptosis as well as necrosis: possible involvement of common mediators in apoptotic and necrotic signal transduction pathways. Leukemia 11 Suppl 3, 380–382.

Tulswani, P.K., Debnath, M., Pant, S.C., Kumar, O., Prakash, A.O., Vijayaraghavan R. and Bhattacharya, R. (2005). Effectsof sub-acute oral cyanide administration in rats. Protective efficacy of alphaketoglutarate and sodium thiosulphate. Chemico-Biological Interactions 156: 1-12.

Tylleskar, T., Banea, M. Bikangi, N., Fresco, L., Persson, L.A. and Rosling, H. (1991). Epidemiologicalevidence from Zaire for a dietary etiology of konzo, an upper motor neuron disease. Bull. WHO, 69, 581-589.

Watts, J. (1998). Japanese fear over festival cyanide poisoning. Lancet, 352-379.

Yen, D., J. Tsai, W. Kao, S. Hu, C. Lee and Deng, J. (1995). The clinical experience of acute CN- poisoning. Amer. J. Emergency Med., 13, 524-528.

Zhang, L., Li, L., Liu, H., Prabhakaran, K., Zhang , X., Borowitz J.L. and Isom G.E. (2007). HIF-1alpha activation by a redox-sensitive pathway mediates cyanide-induced BNIP3 upregulation and mitochondrial-dependent cell death. *Free Radic Biol Med* 43,117–127.

Zhang, X., L. Li, et al. (2009). Cyanide-induced death of dopaminergic cells is mediated by uncoupling protein-2 up-regulation and reduced Bcl-2 expression, Elsevier. 238: 11-19.

Zollner, N. and K. Kirsch (1962). "Microdetermination of lipids by the sulphophospho vanillin reaction." Z. Ges. Exp. Med 135, 545-561.

16

Dietary Antioxidants: From Micronutrients and Phytochemicals to Enzymes – Preventive Effects on Early Atherosclerosis and Obesity

Sylvie Gaillet[1], Dominique Lacan[2] and Jean-Max Rouanet[1]
[1]*UMR 204 NUTRIPASS,*
Université Montpellier 2, Montpellier,
[2]*BIONOV S.A.R.L., Montpellier,*
France

1. Introduction

Development of atherosclerosis is thought to be closely dependent upon increased oxidative stress, that is, an imbalance between reactive oxygen species (ROS).

Elsewhere, high-fat diets and sedentary lifestyles are important risk factors for obesity, which is a key feature of metabolic syndrome and which greatly predisposes individuals to liver diseases, cardiovascular disease (CVD), type 2 diabetes, dyslipidemia, hypertension, and numerous cancers and is associated with markedly diminished life expectancy.

Today, due to oxidative stress, the postulated involvement of lipid peroxidation in atherogenesis and obesity invoked intensive interest in the use of antioxidant nutritional supplements. Epidemiological evidence suggests that intake of some vitamins, minerals, and other food constituents may help to protect against heart disease, obesity, cancer and the aging process and that antioxidants may have a protective effect, in either preventing these diseases or lessening the severity of the diseases upon their onset.

Many compounds, from natural sources, newly synthesised derivatives of these compounds, and common drugs, were tested for their potency to inhibit the peroxidation of plasma lipids induced by different oxidants in vitro to reduce the plaque formation in animal models in vivo and to reduce the morbidity and mortality in human subjects, in clinical trials. Indeed, mortality from cardiovascular disease is the leading cause of death in the industrialised world. Diet is believed to play a major role in the development of this disease, and much research is being focused on identifying ways to prevent it through changes in dietary habits. A potentially important clinical corollary of the atherosclerosis oxidation theory is that inhibition of LDL oxidation may also inhibit atherosclerosis independent of lowering plasma cholesterol concentrations.

Moreover, a close correlation has been found between increased oxidative stress in accumulated fat and the pathogenic mechanism of obesity and obesity-associated metabolic syndrome. It has been reported that obesity is a strong independent predictor of systemic oxidative stress, suggesting that obesity is associated with a state of excess oxidative stress. This may be the source of several metabolic dysfunctions such as inflammation, hypertension, and impaired glucose intake in muscle and fat, which are highly related to obesity. Oxidative stress could be a potential link between fat accumulation and obesity-related morbidity such as diabetes and cardiovascular diseases. Indeed, this pathology can contribute to an increased susceptibility to ROS.

Dietary factors are thought to play a key role in the regulation of the oxidant status. Thus, it is essential to develop and utilise effective and natural antioxidants so that they can protect the human body from free radicals and retard the progress of many chronic diseases. The importance of dietary antioxidants in human health is clear and some epidemiological studies showed the potential health benefits of the consumption of antioxidant rich fruits and vegetables. Generally, dietary antioxidants are minerals, vitamins, fibers, and phytochemicals supplied by fruits and vegetables. Recently, much attention has been focused on new approaches of antioxidant therapy by providing antioxidant enzymes.

Thus, we will review the utilisation of animal models of oxidative stress and the preventive effects of some phytochemicals, micronutrients and enzymes against pathologies such as early atherosclerosis and obesity.

2. Animal models of oxidative stress and pathologies

2.1 Atherosclerosis

With the exception of a research by Adams *et al.* (2006) with a transgenic mice model, few studies have investigated the effect of plant material on atherosclerosis and oxidative stress in rodents (Nicolle *et al.*, 2003; Nicolle *et al.*, 2004); moreover, these studies were only focused on the effect of vegetables. Golden Syrian hamsters fed a fat rich diet develop dyslipidemia and atherosclerotic plaques, similar in many respects to human atheroma

Diet ingredient	Experimental diet
Casein	200
DL-methionine	3
Corne starch	393
Sucrose	154
Cellulose	50
Mineral mix [1]	35
Vitamin mix [2]	10
Lard	150
Cholesterol	5

[1] Mineral mixture contained (mg/kg diet): $CaHPO_4$, 17,200; KCl, 4000; NaCl, 4000; MgO, 420; $MgSO_4$, 2,000; Fe_2O_3, 120; $FeSO_4.7H_2O$, 200; trace elements, 400 ($MnSO_4.H_2O$, 98; $CuSO_4$. $5H_2O$, 20; $ZnSO_4.7H_2O$, 80; $CoSO_4.7H_2O$, 0.16; KI, 0.32; sufficient starch to bring to 40 g (per kg diet). [2] Vitamin mixture containing (mg/kg diet): retinol, 12; cholecalciferol, 0.125; thiamin, 40; riboflavin, 30; pantothenic acid, 140; pyridoxine, 20; inositol, 300; cyanocobalamin, 0.1; menadione, 80; nicotinic acid, 200; choline, 2720; folic acid, 10; p-aminobenzoic acid, 100; biotin, 0.6; sufficient starch to bring to 20 g (per kg diet).

Table 1. Composition of the atherogenic diet (g/kg).

(Vinson *et al.*, 2011; Auger *et al.*, 2002; Auger *et al.*, 2005 a & b). Hamsters were selected for these studies because of their responsiveness to plasma cholesterol lowering and anti-atherogenic interventions (Kowala *et al.*, 1991). Moreover, hamster has a plasma lipoprotein distribution similar to that of humans and LDL as the major plasma cholesterol carrier. To induce an oxidative stress, their high cholesterol and high fat diet was rendered deficient in vitamin C and E and in selenium (Table 1).

Thus, the arterial wall response to such a stress (fatty streak formation and aortic atherosclerosis emergence) was triggered and then the modulation of this effect by plant extracts or fruits and vegetables. In addition, for the first time in this model, we measured the modulation of oxidative stress parameters including cardiac production of superoxide anions and NAD(P)H oxidase expression.

2.2 Obesity

The high risk of oxidative stress in the obese is due to both overproduction of reactive oxygen species and low antioxidant defenses. In humans, obesity leads to increased lipid peroxidation and low activity of antioxidant enzymes such as superoxide dismutase (Cu-Zn SOD) and glutathione peroxidase (GSHPx) in plasma and erythrocytes (Olusi, 2002). These oxidative stress markers are inversely correlated with insulin sensitivity. But insulin resistance (IR) appears as a key factor for oxidative risk. Hyperinsulinemia, correlated with the fat overload, contributes to a metabolic disorder of sulfur amino acids, leading to elevated homocysteine levels (Gallistl *et al.*, 2000; Sanchez-Margalet *et al.*, 2002) which increases the risk of cardiovascular disease. Moreover, in children (Morinobu *et al.*, 2002), adolescents (Neuhouser *et al.*, 2001) and adults obese (Wallstrom *et al.*, 2001), it has been reported low plasma levels of α-tocopherol, vitamin C and carotenoids that are major micronutrients for the secondary antioxidant defense.

Obesity follows from numerous behavioral interactions, physiological and biochemical. It is associated with a reduced life expectancy and is an independent risk factor for CVD, the main cause of death worldwide. In addition, obese patients, particularly those with abdominal obesity due to accumulation of visceral fat, have a higher risk of impaired glucose tolerance and impaired fasting glucose, which often evolve into insulin resistance (IR) and type 2 diabetes mellitus (T2DM). Obesity and IR are frequently associated with hypertension, hyperhomocysteinemia and dyslipidemia (pro-atherogenic factors), nonalcoholic fatty liver disease (NAFLD) and nonalcoholic steatohepatitis (NASH), chronic inflammation and a prothrombotic state. These metabolic disorders and obesity, are grouped under the term metabolic syndrome. They are influenced by genetic and environmental factors (nutritional and/or hormonal) that can be easily studied in animal and cellular models.

The increasing morbidity from cardiovascular diseases and type 2 diabetes in the obese population suggests that obesity is associated with increased oxidative stress. Obesity can contribute to increased susceptibility to ROS. In the general population, several risk groups to high oxidative stress are identified (the elderly, smokers, those with an unbalanced diet), particularly obese patients in conjunction with reduced antioxidant defenses and/or strong radical production.

Some dietary patterns and specific dietary components have been associated with reduced prevalence of CVD in humans. For example, the Mediterranean diet, characterized by a high fiber, low glycemic index carbohydrates, unsaturated fat, vitamins and antioxidant polyphenols, has been associated with a decreased incidence of CVD, obesity and T2DM. Patients with metabolic syndrome having a Mediterranean diet, exhibited a significant reduction in body weight and IR and it seems reasonable to attribute beneficial effects to antioxidants.

Thus, we have developed an animal model of diet-induced obesity (hamster) with metabolic syndrome (and therefore with oxidative stress) receiving a high-fat diet (Table 2), and then we measured metabolic characters (adipokines plasma levels) and we investigated the action mechanisms of dietary antioxidants at target organs such as liver (hepatic steatosis characterization and activity of NAD(P)H oxidase) or the heart (expression and activity of NAD(P)H oxidase, mitochondrial dysfunction). The activity of antioxidant enzymes (SOD, glutathione peroxidase, catalase) was also assessed and blood glucose and insulin, insulin resistance (HOMA-IR) and paraoxonase activity have been measured to assess the effects and action mechanisms on the metabolic syndrome itself.

Diet ingredient	Experimental diet
% energy as protein	18.89
% energy as carbohydrate	36.75
% energy as fat	44.36
Fibre (g/100 g)	5.90
Lard (g/100 g)	21.00
Soyabean oil (g/100 g)	3.00
Cholesterol (g/100 g)	0.10
% saturated fat	9.56
% monounsaturated fat	10.61
% polyunsaturated fat	3.86
Energy density (kj/g)	20.85

Table 2. Composition of the obesogenic diet.

3. Preventive effects of dietary antioxidants against atherosclerosis

3.1 Fruit and vegetable extract

We studied the effects of a fruit and vegetable extract (Oxxynea®) on such an animal model of atherosclerosis (Sutra *et al.*, 2007). According to the manufacturer (Fytexia, Béziers, France), the powdered Oxxynea® extract was obtained from 22 varieties of antioxidant-rich fruits and vegetables including apple, asparagus, bilberry, apricot, black currant, broccoli, carrot, cherry, cucumber, garlic, grapefruit, green cabbage, olive, onion, orange, papaya, pineapple, red and white grapes, strawberry, tea, tomato, and wheat germ. Oxxynea® contains high level of catechins, that is, sum of procyanidin dimers B1, B2, B3, and B4 (1.14 g/100 g) and monomeric catechins (catechin, 0.55 g/100 g; epicatechin, 3.08 g/100 g; epichatechin-3-O-gallate, 4.10 g/100 g; epigallocatechin, 4.17 g/100 g; epigallocatechin-3-O-gallate, 21.33 g/100 g). Other phenolic compounds such as gallic acid and anthocyanins were detected in lower amounts (0.15 and 0.6 g/100 g, respectively). The extract also contained low levels of lycopene (28 mg/100 g) and vitamin C (4.92 mg/100 g).

Thirty six hamsters were divided into two groups of 18 and fed an atherogenic diet for 12 weeks. They received by gavage either water or Oxxynea® in water at a human dose equivalent of 10 fruits and vegetables per day. Oxxynea lowered plasma cholesterol and non-HDL cholesterol, but not HDL-cholesterol, and increased plasma antioxidant capacity (Table 3).

	Atherogenic diet	Atherogenic diet plus Oxxynea®
TC[2], *mmol/L*	9.54±0.20 [a]	8.42±0.16[b]
HDLC[3], *mmol/L*	6.01±0.27 [a]	5.90±0.16[a]
Non-HDLC, *mmol/L*	3.12±0.15 [a]	2.68±0.17[b]
Atherogenic index[4]	1.56±0.04 [a]	1.43±0.03[b]
PAC, *mmol/L*	1.29±0.06[a]	1.42±0.10[b]

[1] Values are means ± SEM, n = 18. Data were analyzed by one-way ANOVA followed by the least significant difference test. For each dietary treatment, means in a column with different letters differ, P < 0.05. [2] TC : total cholesterol. [3] HDLC : high-density lipoprotein cholesterol. [4] Total cholesterol/HDL-cholesterol.

Table 3. Effects of ingestion of a fruit and vegetable extract (Oxxynea®) on plasma lipid concentrations and on plasma antioxidant capacity (PAC) in hamsters fed an atherogenic diet[1].

As previously reported, a high fat diet led to an early increase in total and non-HDL cholesterol after 15 days of diet leading to lipid deposition on aortic arch at 84 days. Interestingly, the atherogenic diet-induced hypercholesterolemia is in parallel accompanied by a tendency of superoxide anion overproduction, which reaches the significance at 45 days (Figure 1 and Figure 2).

Oxxynea® reduced cardiac production of superoxide anion by 45%, and $p22^{phox}$ subunit of NAD(P)H oxidase expression by 59% (Figure 2).

Oxxynea® also strongly reduced the area of aortic fatty streak deposition by 77% (Figure 3). In agreement with oxidative hypothesis of atherosclerosis, it could be postulated that NAD(P)H oxidase expression and activity conspire with high non-HDL cholesterol level to induce foam cells and fatty streak.

The resulting improvement of atherogenic index (Table 3) obtained with fruit and vegetable extract extends the previous observation on the beneficial effects in lipid parameters obtained with grape polyphenols (Vinson *et al.*, 2001).

Our findings suggest for the first time that the fruit and vegetable antioxidant extract could prevent both NAD(P)H oxidase expression and $O_2^{\circ-}$ overproduction in the heart from hypercholesterolemic hamster. Here again, NAD(P)H oxidase inhibition could be involved in prevention of LDL oxidation and further atherosclerosis steps. Beyond the vicious circles linked to LDL oxidation, the inhibition of reactive oxigen species (ROS) production by NAD(P)H oxidase system could also prevent other early events in cardiovascular diseases such as endothelial dysfunction or arterial remodelling. A recent study has shown that endothelium-dependent vasorelaxation is impaired in the high lipidfed golden syrian hamster (Georgescu *et al.*, 2006). Our current results showing that the fruit and vegetable extract inhibits the overproduction of $O_2^{\circ-}$ by NAD(P)H system strongly suggest that Oxxynea may prevent the endothelial dysfunction. Indeed, an overproduction of superoxide

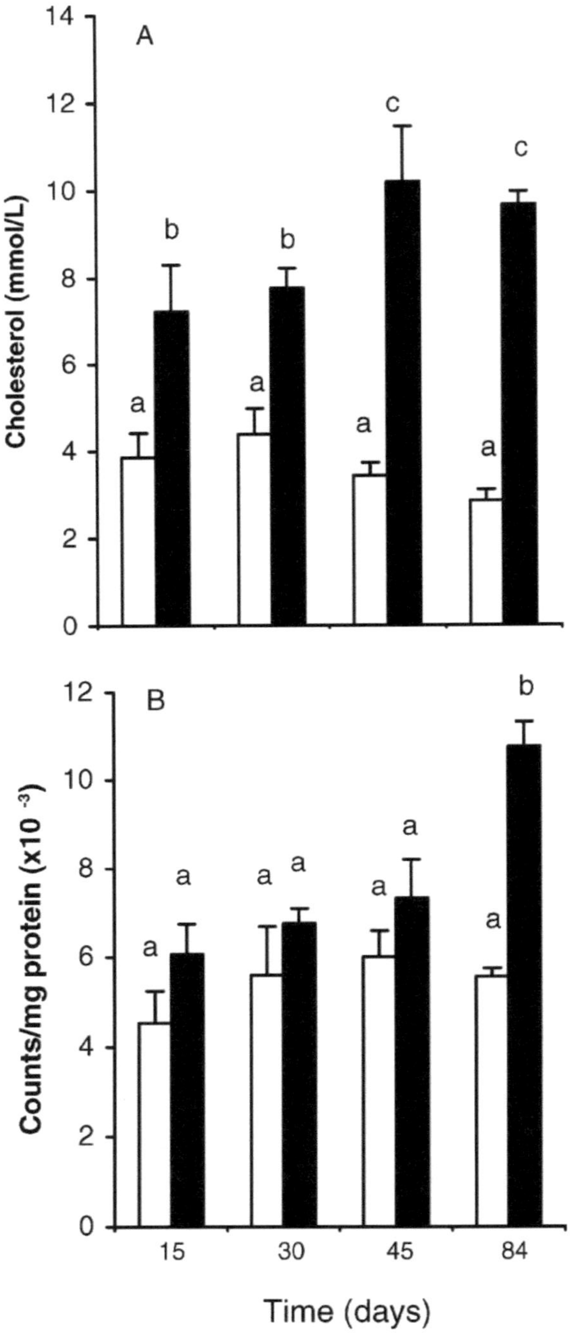

Fig. 1. Time course experiment of plasma cholesterol concentration (A) and superoxide anion production (B) in hamsters fed a standard diet (white bars) and in hamsters fed an atherogenic diet (black bars). Values are expressed as mean ± SEM of triplicate wells (n = 6). For each dietary treatment, bars with different index letters differ ($P < 0.05$).

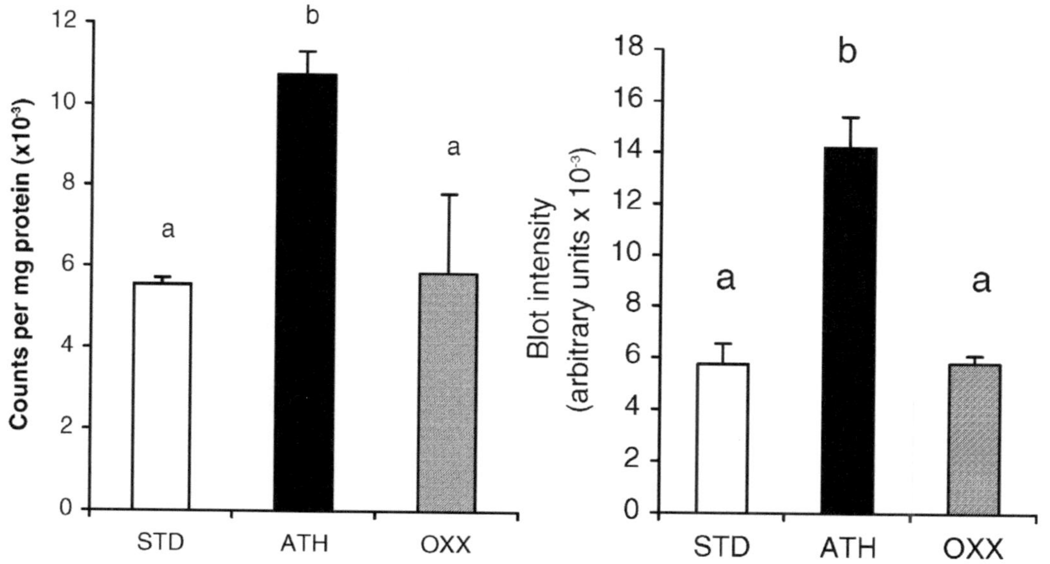

Fig. 2. Cardiac superoxide anion production (letf graph) and expression of the cardiac p22phox subunit of NAD(P)H oxidase (right graph) in hamsters fed a standard or an atherogenic diet with (OXX) or without (ATH) Oxxynea during 84 days. Values are expressed as mean ± SEM of triplicate wells (n = 6). For each dietary treatment, bars with different index letters differ ($P < 0.05$).

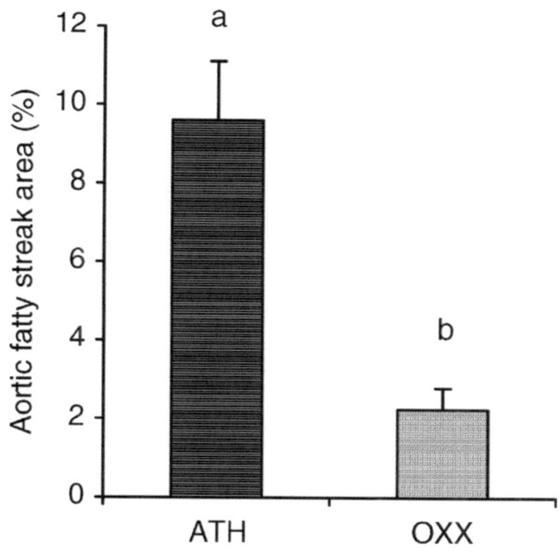

Fig. 3. Effects of ingestion of water (ATH) or a fruit and vegetable extract Oxxynea® (OXX), on aortic fatty streak area in hamsters fed an atherogenic diet for 84 days. Values are expressed as mean ± SEM (n = 12). Bars with different index letters differ ($P < 0.05$).

anion that could react with NO° to produce peroxinitrite has been involved in the hypercholesterolemiainduced impairment of vasorelaxant system (Wagner *et al.*, 2000). On the other hand, we have shown that an overproduction of ROS is strongly associated with cardiac remodelling, suggesting a pathogenic role of oxidative stress in its constitution (Al-Awwadi *et al.*, 2005). Pharmacoogical or nutritional intervention could prevent both NAD(P)H oxidase expression and activity and cardiac hypertrophy (Al-Awwadi *et al.*, 2005). Our results showing that hypercholesterolemic diet activates and that vegetable and fruit extracts inhibit NAD(P)H expression and activity in the heart reinforce the hypothesis of a nutritional modulation of ROS enzymatic producing systems.

All of these results suggest that this extract acted by mechanisms operating both inside and outside a hypolipemic effect, especially an antioxidant effect. Although the constituent(s) responsible for these effetcs remain(s) unclear, candidates such as vitamin C, vitamin E, carotenoids, selenium, and polyphenols could act synergistically or additively to prevent atherosclerosis in the hamster model.

These findings support the view that chronic consumption of antioxidants supplied by fruits and vegetables has potential beneficial effects with respect to the development of atherosclerosis. The underlying mechanism is related mainly to inhibiting pro-oxidant factors and improving the serum lipid profile.

3.2 Raspberry juices

In another recent set of studies, the effects of raspberries on early atherosclerosis in Syrian hamsters were investigated using three juices prepared from var. Cardinal, Glen Ample and Tulameen berries (Suh *et al.*, 2011). The hamsters received an atherogenic diet for 12 weeks and at the same time a juice at a daily dose corresponding to the consumption of 275 ml by a 70 kg human. A control group (CTR) received the same diet with water instead juice. The principal polyphenolic compounds in the juices were anthocyanins and ellagitannins, which were present at concentrations of 218-305 $mg.mL^{-1}$ and 45-72 $mg.mL^{-1}$, respectively. The three juices had similar but not identical effects.

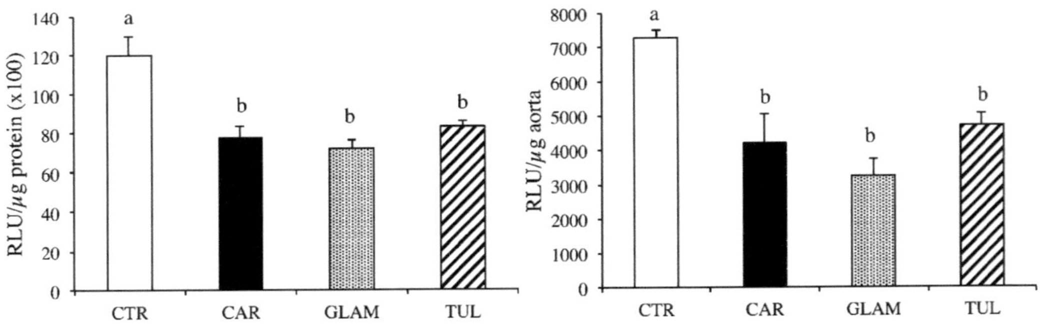

Fig. 4. Cardiac (left) and thoracic aorta (right) superoxide anion production (as Relative Luminescence Units) in hamsters fed a high-fat diet (CTR), or a CTR diet plus either Cardinal (CAR), Glen Ample (GLAM) or Tulameen raspberry juice (TUL) for 12 weeks. Values are expressed as mean ± SEM (n = 6).

They all inhibited cardiac and aortic production of superoxide anion (Figure 4) and increased hepatic glutathione peroxidase activity although only Tulameen juice brought about a significant increase in superoxide dismutase activity (Table 4). Glen Ample was the only juice to significantly increase plasma paraoxonase activity (Figure 5).

Diet	GSHPx *U/mg protein*	SOD *U/mg protein*	PON *U/mL*
Control	945±45	158±13	60±16
Cardinal	1412±76*	161± 9	70±17
Glen Ample	1352±72*	144±10	120±29*
Tulameen	1278±68*	203±12*	61±13

[1]Data expressed as mean values ± SEM (n = 12). For each dietary treatment, mean values in a column designated * are significantly different from the controls, P < 0.05. GSHPx: glutathione peroxidase. SOD: superoxide dismutase. PON: paraoxonase

Table 4. Hepatic antioxidant enzymes activities and plasma paraoxonase activity in hamsters fed an atherogenic diet plus a daily gavage of either water (control) or Cardinal, Glen Ample or Tulameen raspberry juice for 12 weeks[1].

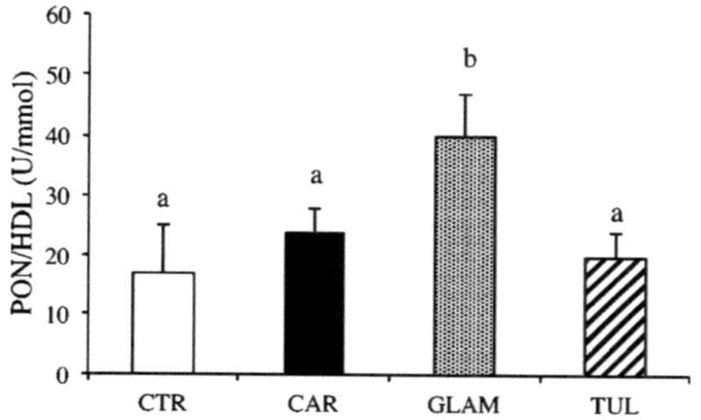

Fig. 5. Ratio paraoxonase activity (PON)/HDL concentration in hamsters fed a high-fat diet (CTR), or a CTR diet plus either Cardinal (CAR), Glen Ample (GLAM) or Tulameen raspberry juice (TUL) for 12 weeks. Values are mean ± SEM (n = 6). For each dietary treatment, bars with different index letters are statistically significantly different (P < 0.05).

All the juices lowered plasma triglyceride level while consumption of Tulameen and Cardinal, but not Glen Ample, significantly lowered plasma total cholesterol and LDL-cholesterol (not shown here). Cardinal was the sole juice to significantly increase HDL-cholesterol and likewise it also significantly reduced body weight. These findings suggest that moderate consumption of raspberry juices can help to prevent the development of early atherosclerosis, with the underlying mechanisms related to improved antioxidant status and serum lipid profiles.

3.3 Apples, purple grape, apple juice and purple grape juice

We also looked at how being fed various kinds of fruit affected the hamsters' risk of atherosclerosis (Décordé *et al.*, 2008). 40 male Syrian golden hamsters were randomly divided into five groups. The animals were fed a diet to promote the development of artery hardening, and supplemented with mashed apple or purple grape, or the same volume of apple juice or purple grape juice, or water (control group) for 12 weeks. The amount of fruit the hamsters consumed was equivalent to three apples or three bunches of grapes daily for a human. Hamsters given juice drank the equivalent of four glasses daily for a person weighing 70 kilograms.

The apples and grapes had about the same phenol content, while the purple grape juice had 2.5 times more phenols than apple juice. At the end of the study, total cholesterol levels were significantly reduced in the animals fed the fruit-supplemented diets, by 11% in apple group and 24% in the apple juice group, and 30% in the purple grape and 34 per cent in the purple grape juice group. This was attributed to the reductions in levels of non-HDL cholesterol. Compared with animals given water, those given fruit or fruit juice had less oxidative stress. The juices also outperformed the fruit for protecting against atherosclerosis, measured by the aortic fatty streak lesion area (AFSA). This value was reduced by 93% and 78% for the purple grape juice and the fruit, respectively, and by 60% and 48% for apple juice and apple, respectively (Figure 6).

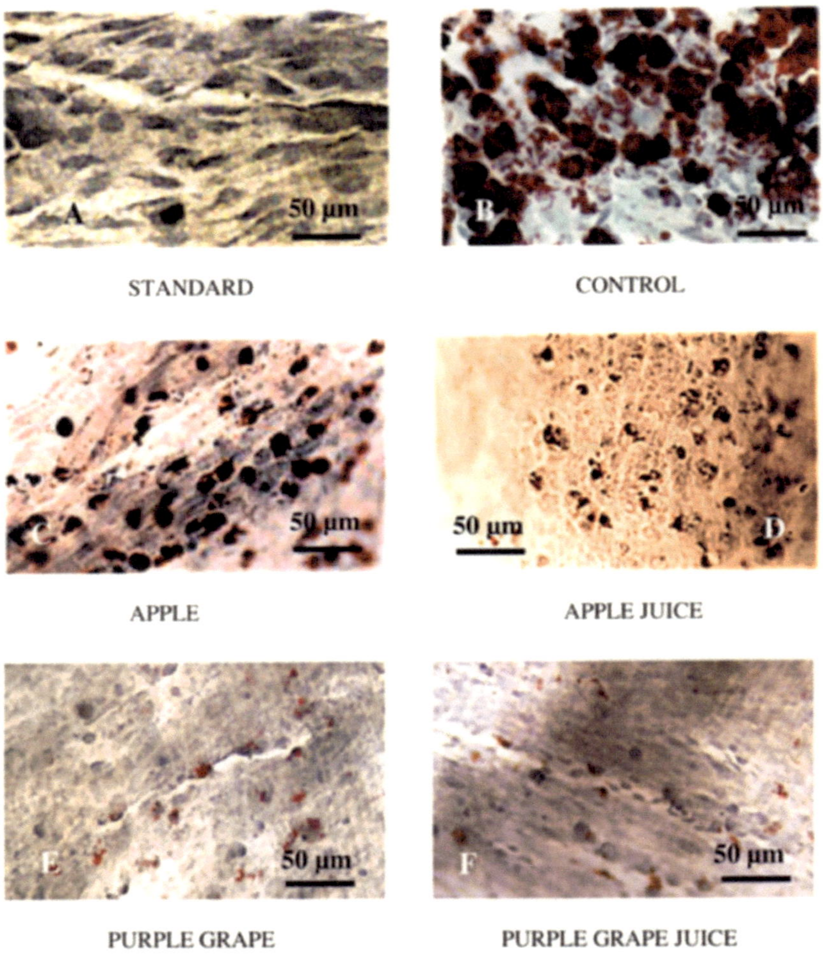

Fig. 6. Photomicrographs of typical atherosclerotic lesions of the hamsters' aortic arch (photographed at x 40) after a 12-wk experimentation period. Aortic fat deposits are stained in Oil Red O. (A) From non-atherogenic diet-fed hamsters (standard group). (B) From atherogenic diet-fed hamsters (control group). (C, D, E, F) From atherogenic diet-fed hamsters receiving either apples (C), apple juice (D), purple grape (E), or purple grape juice by gavage.

The results show for the first time that long-term consumption of antioxidants supplied by apple and purple grape, especially phenolic compounds, prevents the development of atherosclerosis in hamsters, and that processing can have a major impact on the potential health benefits of a product. Flavonoids, especially anthocyanins and catechins in purple grape and purple grape juice, generally have more hydroxyl groups than phenolic acids found in apple and apple juice. This could explain why purple grape juice and purple grape displayed a better efficacy than apple and apple juice against early atherosclerosis. Nevertheless, these beneficial effects cannot only be attributed to their phenolic contents, but to the result of the action of different antioxidant compounds present in the fruits (vitamin C, carotenoids, polyphenols) and to possible synergistic and antagonist effects still unknown. It provides encouragement that fruit and fruit juices may have a significant clinical and public health relevance.

3.4 Red berry juices and teas

Also, consuming antioxidant-rich raspberry juice or tea may prevent artery hardening, and lead to improvements in heart health (Rouanet *et al.*, 2010). Measures of atherosclerosis were reduced in hamsters with high cholesterol levels following consumption of raspberry, strawberry and bilberry juices and green and black tea, with the benefits were significantly greater for raspberry and green tea. Consuming the equivalent of 275 mL by a 70 kg human led to reductions in fat deposits in the aorta of up to 96 per cent after 12 weeks (Figure 7).

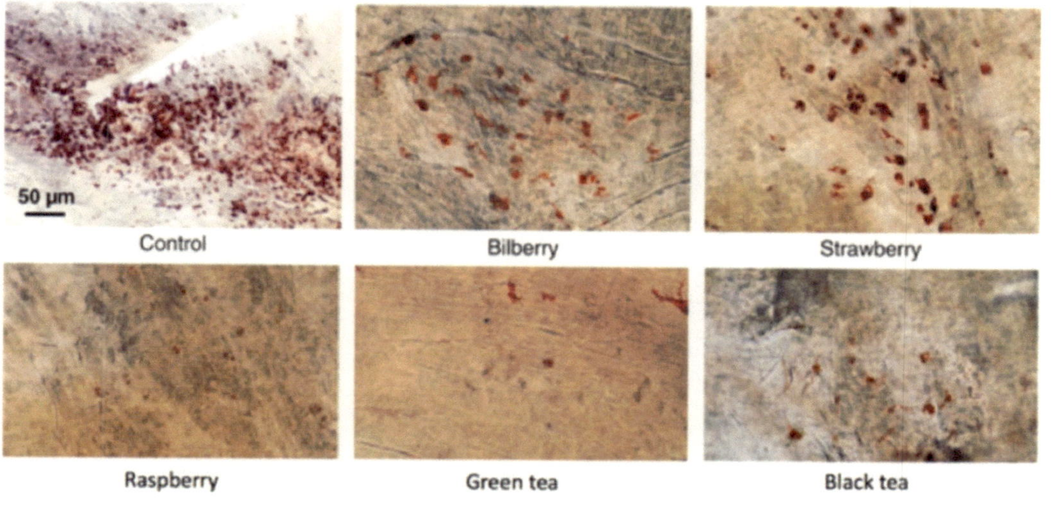

Fig. 7. Photomicrographs of hamster aortic arches after 12 weeks on an atherogenic diet (control) and 12 weeks on an atherogenic diet supplemented with either strawberry juice, bilberry juice, raspberry juice, green tea or black tea. The micrographs are examples of the aortic arch surface covered by lipid inclusion in the intima with lipids coloured red using Oil Red O stain. All micrographs have the same scale.

These findings suggest that moderate consumption of berry juices and teas can help prevent the development of early atherosclerosis. Atherosclerosis, or hardening of the arteries, is a major risk factor for cardiovascular disease, which causes almost 50 per cent of deaths in Europe, and is reported to cost the EU economy about €169bn ($202bn) per year. While all of the beverages exerted beneficial effects, the composition and concentration of individual phenolic compounds varied substantially between the five beverages. This indicates that anti-atherosclerotic effects can be induced by a diversity of phenolic compounds rather than a few specific components.

3.5 Phenolic-rich extracts from grape seed

Elsewhere, we evaluated the antiatherosclerotic effect of commercially available phenolic-rich extracts from grape seeds (ExGrape seeds, EGS; grape seed extract, GSE) and marc (ExGrape total, EGT) in cholesterol-fed hamsters and to investigate possible operating mechanisms (Auger *et al.*, 2004). These extracts fed at a moderate dose mimicking two glasses of red wine per meal reduced plasma cholesterol (-11% on average) but did not affect plasma antioxidant capacity of hamsters. The extracts prevented the development of aortic atherosclerosis by 68% (EGS), 63% (EGT), and 34% (GSE) (Figure 8).

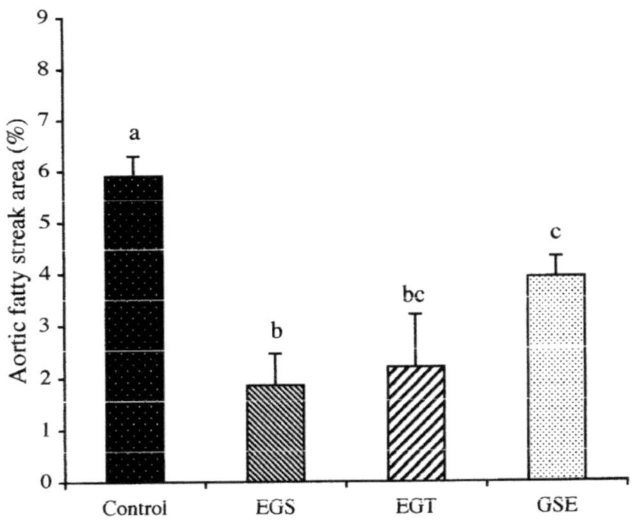

Fig. 8. Effects of force-feeding water (control), a procyanidin-rich white grape seed extract (EGS), a marc extract from red and white grapes, rich in anthocyanins (EGT), or a grape seed extract (GSE) on aortic fatty streak area (AFSA) in hamsters fed an atherogenic diet for 12 weeks. AFSA is expressed as a percentage of the total aortic area surveyed. Each bar represents mean ± SEM from eight hamsters. Bars with different letters differ, $P < 0.05$.

Elsewhere, in an *ex vivo* experiment using rat aortic rings, EGS (7 μg/mL) induced 77% endothelium-dependent relaxation, whereas EGT and GSE (30 μg/mL) induced 84 and 72%, respectively (Figure 9).

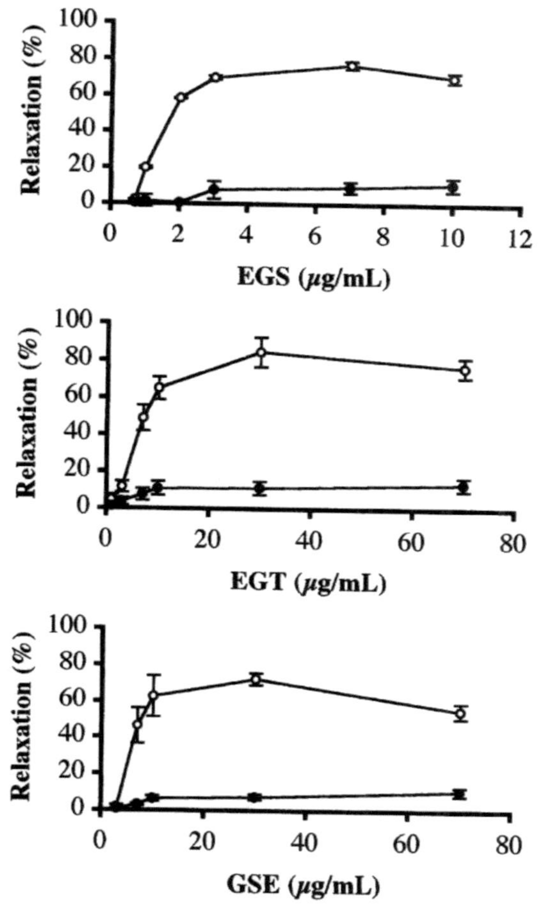

Fig. 9. Concentration-response curves for the relaxation of isolated aortic rings of rat by EGS, EGT, and GSE in the presence (O) or in the absence (•) of endothelium. Each point represents mean ± SEM from eight rats.

These results suggests that phenolic extracts from grape seeds and marc are beneficial to inhibit atherosclerosis by indirect mechanism(s).

3.6 Wines

We have thus studied the effects of a white wine enriched with polyphenols (PEWW) from Chardonnay grapes and of a sparkling red wine (SRW) from Pinot Noir and Chardonnay grapes for the first time on early atherosclerosis in hamsters (Auger *et al.*, 2005 b). Animals were fed an atherogenic diet for 12 weeks. They received by gavage PEWW, SRW, ethanol 12% (ETH), or water as control (mimicking a moderate consumption of ~2 red wine glasses per meal for a 70 kg human). Plasma cholesterol concentrations were lower in groups that consumed PEWW and SRW accompanied by an increase in the ratio apo A-1/apo B (Table 5).

	Control	Ethanol 12%	PEWW	SRW
TC[2], *mmol/L*	10.39 ± 0.65 [a]	9.65 ± 0.54 [ab]	8.96 ± 0.20[b]	8.90 ± 0.28 [b]
HDL-C[3], *mmol/L*	3.74 ± 0.37	3.71 ± 0.15	3.61 ± 0.14	3.50 ± 0.17
TG[4], *mmol/L*	1.37 ± 0.17	1.15 ± 0.10	1.19 ± 0.12	1.26 ± 0.07
ApoA-1, *g/L*	1.70 ± 0.21[a]	2.20 ± 0.12 [b]	1.98 ± 0.09[ab]	1.94 ± 0.15 [ab]
ApoB, *g/L*	0.36 ± 0.03	0.29 ± 0.02	0.32 ± 0.01	0.30 ± 0.02
ApoA-1/ApoB	4.64 ± 0.58 [a]	7.38 ± 0.66 [b]	6.07 ± 0.24 [b]	6.44 ± 0.77 [b]

[1] Values are means ± SEM, $n = 8$. Data were analyzed by one-way ANOVA followed by the least significant difference test. For each dietary treatment, means in a column with different letters differ, $P<0.05$. [2] Total cholesterol. [3] High-density lipoprotein cholesterol. [4] Triglycerides.

Table 5. Effects of force-feeding water (Control), 12% ethanol, polyphenols-enriched Chardonnay white wine (PEWW), or sparkling Pinot noir red wine (SRW) on plasma lipid and apolipoprotein concentrations of hamsters fed an atherogenic diet for 12 weeks[1].

	Control	Ethanol 12%	PEWW	SRW
		units.mg^{-1} of protein		
SOD	7.28 ± 1.15[a]	6.74 ± 0.97 [a]	10.07 ± 1.51 [b]	10.78 ± 1.82 [b]
GSHPx (x 10^{-2})	11.47 ± 1.93	9.77 ± 1.77	10.93 ± 2.37	11.32 ± 2.95
CAT	108.27 ± 4.94 [a]	100.32 ± 6.62 [a]	125.20 ± 8.49 [b]	125.04 ± 6.59 [b]
PAC, *mmol/L*	0.97 ± 0.03 [a]	1.08 ± 0.05[b]	1.12 ± 0.02 [b]	0.93 ± 0.10 [a]

[1]Values are means ± SEM, $n = 8$. Data were analyzed by one-way ANOVA followed by the least significant difference test. For each dietary treatment, means in a column with different letters differ, $P < 0.05$.

Table 6. Effects of force-feeding water (Control), 12% ethanol (ETH), polyphenols-enriched Chardonnay white wine (PEWW), or sparkling Pinot noir red wine (SRW) on liver superoxide dismutase (SOD), glutathione peroxidase (GSHPx), and catalase (CAT), and plasma antioxidant capacity (PAC) in hamsters fed an atherogenic diet for 12 weeks[1].

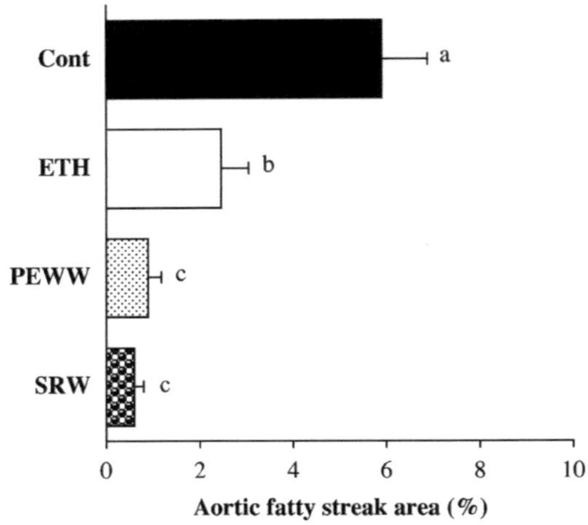

Fig. 10. Effects of daily force-feeding water (control), 12% ethanol (ETH), phenolics-enriched white wine (PEWW), or sparkling red wine (SRW) on aortic fatty streak area (AFSA) in hamsters fed an atherogenic diet for 12 weeks. Values are means ± SEM, $n = 8$. Bars with different index letters differ, $P < 0.05$.

Liver-specific activities of superoxide dismutase and catalase were significantly increased by PEWW (38 and 16%, respectively) and by SRW (48 and 15%, respectively). PEWW and ETH significantly increased plasma antioxidant capacity (Table 6). Aortic fatty streak area (AFSA) was significantly strongly reduced in the groups receiving PEWW (85%) and SRW (89%) in comparison with the control. AFSA was reduced by ethanol to a lesser extent (58%) (Figure 10).

These data suggest that tannins from the phenolics-enriched white wine induce a protective effect against early atherosclerosis comparable to that produced by sparkling red wine containing tanins and anthocyanins and dissociated from the antioxidant action of these compounds.

3.7 Pure phenolics

To get more information about the effects and mechanisms of action of polyphenols, pure phenolic compounds catechin (Cat), quercetin (Qer), and resveratrol (Res) present in red wine on early atherosclerosis were studied in hamsters (Auger *et al.*, 2005 a). Hamsters (n = 32) were divided into 4 groups of 8 and fed an atherogenic diet for 12 weeks. They received by force-feeding 7.14 mL/(kg BW/d^{-1}) Cat, Qer, or Res in water [2.86 mg/(kg BW/d^{-1}) for Cat and 0.143 mg/(kg BW/d^{-1}) for Qer and Res], mimicking a moderate consumption of alcohol-free red wine (equivalent to that supplied by the consumption of about two glasses of red wine per meal for a 70 kg human), or water as control. Plasma cholesterol concentration was lower in groups that consumed phenolics than in controls.

Exptl group	Controls	Catechin	Quercetin	Resveratrol
TC[2], mmol/L	10.0±0.25^{a}	9.10±0.31^{b}	9.00±0.23^{b}	9.25±0.12^{b}
HDLC[3], mmol/L	3.61±0.36	3.62±0.15	3.68±0.17	3.83±0.24
TG[4], mmol/L	1.37±0.19^{a}	1.20±0.05^{a}	1.11±0.05^{b}	1.45±0.08^{a}
Apo-A1, g/L	1.70±0.21^{a}	2.14±0.13^{b}	2.07±0.13^{ab}	2.02±0.09^{a}
Apo-B, g/L	0.36±0.03	0.33±0.02	0.32±0.02	0.35±0.03
Apo-A1/Apo-B	4.72±0.59^{a}	6.48±0.35^{b}	6.45±0.54^{bc}	5.80±0.42^{ac}

[1] Values are means ± SEM, n = 8. Data were analyzed by one-way ANOVA followed by the least significant difference test. For each dietary treatment, meansin a column with different letters differ, $P < 0.05$. [2] Total cholesterol. [3] High-densitylipoprotein cholesterol. [4] Triglycerides.

Table 7. Effects of daily force feeding of water (Control), Catechin, Quercetin, or Resveratrol in water on plasma lipid and apolipoprotein concentrations in hamsters fed an atherogenic diet for 12 weeks[1].

The increase in plasma apolipoprotein (Apo) A1 concentration was mainly due to Cat (26%) and Qer (22%) and to a lesser extent, but nonsignificantly, Res (19%). Apo-B was not affected (Table 7).

Plasma antioxidant capacity was not improved, and there was no sparing effect on plasma vitamins A and E. Plasma iron and copper concentrations were not modified nor were liver superoxide dismutase and catalase activities. A sparing effect of Qer on liver glutathione peroxidase activity appeared, whereas Cat and Res exhibited a smaller effect. Aortic fatty streak area was significantly reduced in the groups receiving Cat (84%) or Qer (80%) or Res (76%) in comparison with the controls (Figure 11).

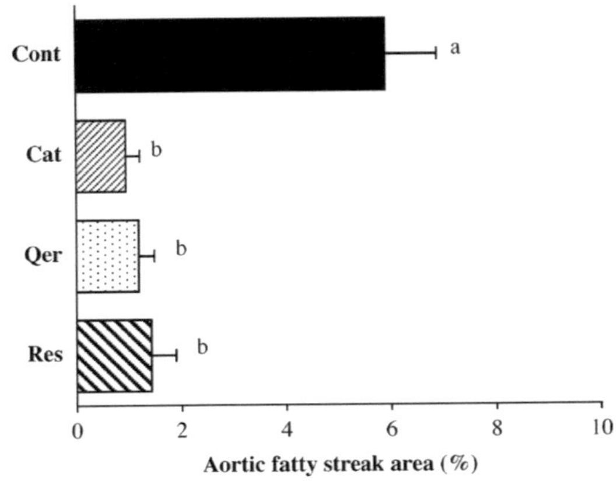

Fig. 11. Effects of daily force feeding of water (control, Cont), catechin (Cat), quercetin (Qer), or resveratrol (Res) on aortic fatty streak area (AFSA) in hamsters fed an atherogenic diet for 12 weeks. AFSA is expressed as a percentage of the total aortic area surveyed. Bars with different letters differ, $P < 0.05$.

These findings demonstrate that catechin, quercetin, and resveratrol at nutritional doses prevent the development of atherosclerosis through several indirect mechanisms.

3.8 Selenium-rich microalgae

Selenium (Se) has received considerable attention as an essential micronutrient for animals and humans. It functions in the active site of a large number of selenium-dependent enzymes such as glutathione peroxidase and is associated with anticancer and other physiological functions (Ip *et al.*, 2000; Miller *et al.*, 2001). The lack of Se in food may cause different diseases such as cardiovascular disease (Shamberger *et al.*, 1978), cancer (Schrauzer *et al.*, 1977; Salonen *et al.*, 1984), rheumatoid arthritis (Tarp *et al.*, 1985), cataract (Fecondo & Augusteyn, 1983) and anemia (Chen *et al.*, 1999). The major forms of selenium occurring in foodstuffs are the organic, protein-associated forms, selenomethionine (SeMet, plant and animal sources) and selenocysteine (SeCys, animal sources). Selenate is also present in some foodstuffs (Olson & Palmer, 1978), and in selenium-deficient areas inorganic selenium salts (selenite, selenate) are added to the food (Van Vleet, 1980). Elsewhere, it is generally believed that organic Se compounds are better and safer than inorganic Se as dietary supplements ; therefore, in response to the need for Se to support human health, selenium-enriched foods have been developed and commercialized (Stoewsand *et al.*, 1989). In the search for an economical source of organic nutritional forms of Se, we attempted to increase the normally low Se content of spirulina (*Spirulina platensis*) by growing it in Se-enriched medium (Cases *et al.*, 2001; Cases *et al.*, 2002) and suggested it represents a promising source for commercial large-scale production of organic Se, as compared with other plants (Cases *et al.*, 2002). Spirulina is a blue-green microalga belonging to the cyanobacteria family, commercially available for human consumption and used as a functional food for humans because of its high concentration of protein and other nutritional elements. Phycocyanin (PC) and allophycocyanin are blue water-soluble photosynthetic pigments derived from cyanobacteria (blue-green algae) such as spirulina. They have been used as nutrients for both human and animal consumption, as natural dyes for food and cosmetics and as pharmaceuticals (Chen *et al.*, 1996). In a recent work, we suggested that in Se-rich Spirulina, organic Se was mainly located in phycocyanin (Cases et al., 2002) ; more recently, this was corroborated by a work of Huang *et al.* (2001).

Previous reports have shown that a protean extract of *Spirulina platensis* is a potent free-radical scavenger (hydroxyl and peroxyl radicals) and inhibits microsomal lipid peroxidation ; it has also been observed that an increase in phycocyanin content was related to an increase in the antioxidant activity in different fractions, and therefore phycobiliprotein phycocyanin is the component mainly responsible for the antioxidant activity (Pinero Estrada *et al.*, 2001). Phycocyanin has also been described as a strong antioxidant (Bhat & Madyastha, 2000; Bhat & Madyastha, 2001) and anti-inflammatory (Reddy *et al.*, 2000) natural compound. It is similar in chemical structure to bilirubin, and acts as a powerful scavenger of reactive oxygen species (ROS) in various *in vitro* and *in vivo* experimental models (Stocker *et al.*, 1990).

It is now well recognized that an increased formation of oxygen radicals and other oxygen derivatives frequently accompanies tissue damage. Today, there is an explosive interest in the use of antioxidant nutritional supplements. Epidemiological evidence suggests that intake of some vitamins, minerals, and other food constituents may help to protect against heart disease, cancer and the ageing process, and that antioxidants may have a protective effect, either in preventing these diseases or lessening the severity of the diseases upon their onset. Many of their activities are mediated by ROS, which are generated during the oxidative burst (Aruoma, 1994; Imlay & Linn, 1998). Hence, it is expected here that chronic diseases, particularly atherosclerosis, may be prevented by spirulina and phycocyanin and their Se-enriched forms.

Only two studies have suggested that *Spirulina platensis* concentrate imparts a hypocholesterolemic effect in rats (Iwata *et al.*, 1990) and humans (Nakaya *et al.*, 1998), and others more recently (Nagaok *et al.*, 2005) provided the first direct evidence that phycocyanin is a novel hypocholesterolemic protein derived from *Spirulina platensis* that can powerfully influence rat serum cholesterol concentrations. Atherosclerosis may be characterized according to three theories (oxidative, inflammatory and hypercholesterolemic) ; since selenium takes part to the antioxidant defense and PC has been shown to possess antioxidant, anti-inflammatory and hypocholesterolemic properties, we hypothesized that PC and SePC might prevent this pathology.

Thus, the effects of spirulina and its chromophore phycocyanin, both without bound Se or selenium-enriched, were studied on plasma cholesterol, early atherosclerosis, cardiac production of superoxide anions, and NAD(P)H oxidase expression in hamsters fed a high-cholesterol / low antioxidant diet (Riss *et al.*, 2007). Forty hamsters were divided into 5 groups of 8 and fed an atherogenic diet for 12 weeks. They received by gavage either 7.14 mL/(kg BW/d^{-1}) phycocyanin (PC), Se-rich phycocyanin (SePC), spirulina (SP) or Se-rich spirulina (SeSP) in water, or water as control. SeSP and SePC supplied 0.4 µg of Se per 100 g body weight. Plasma cholesterol and non- HDL cholesterol concentrations were lower in group consuming SePC. HDL-cholesterol was never affected. SePC significantly increased plasma antioxidant capacity by 42% compared with controls (Table 8).

Group	Controls	PC	SePC	SP	SeSP
TC[2], *mmol/L*	7.54±0.31[a]	7.02±0.16[b]	6.79±0.28[b]	7.22±0.28[ab]	7.09±0.18[ab]
HDLC[3], *mmol/L*	4.51±0.44[a]	4.25±0.33[a]	4.58±0.43[a]	4.59±0.28[a]	4.39±0.30[a]
Non-HDLC, *mmol/L*	3.04±0.42[a]	2.78±0.29[ab]	2.21±.39[b]	2.53±0.21[ab]	2.70±0.33[ab]
Atherogenic index[4]	1.68±0.29[a]	1.66±0.15[ab]	1.48±0.11[b]	1.58±0.06[ab]	1.61±0.11[ab]
PAC, *mmol/L*	0.88±0.06[a]	1.18±0.09[bc]	1.25±0.09[b]	0.99±0.08[c]	1.05±0.07[c]
GSHPx[5], *U/mg protein*	41.07±1.23[a]	4.67±0.74[b]	5.72±1.27[b]	5.00±0.62[b]	5.59±1.41[b]
SOD[6], *U/mg protein*	6.72±0.52[a]	3.17±0.38[b]	3.00±0.32[b]	2.29±0.26[b]	3.36±0.90[b]

[1] Values are means ± SEM, $n = 8$. Data were analyzed by one-way ANOVA followed by the Least Significant Difference test. For each dietary treatment, means in a column with different superscripts differ, $P < 0.05$. [2] TC : total cholesterol. [3] HDLC : high density lipoprotein cholesterol. [4] Total cholesterol/HDL-cholesterol. [5] Glutathione peroxidase. [6] Superoxide dismutase.

Table 8. Effects of ingestion phycocyanin (PC), Se-rich phycocyanin (SePC), spirulina (SP) and Se-rich spirulina (SeSP) on plasma lipid concentrations and antioxidant capacity (PAC), and on liver antioxidant enzymes activity (GSHPx and SOD) in hamsters fed an atherogenic diet[1].

A sparing effect in liver glutathione peroxidase (87% on average) and superoxide dismutase (56% on average) activity was observed for all the groups compared to controls. Aortic fatty streak area was significantly reduced in the experimental groups (Figure 12), especially by PC (82%) and SePC (85%).

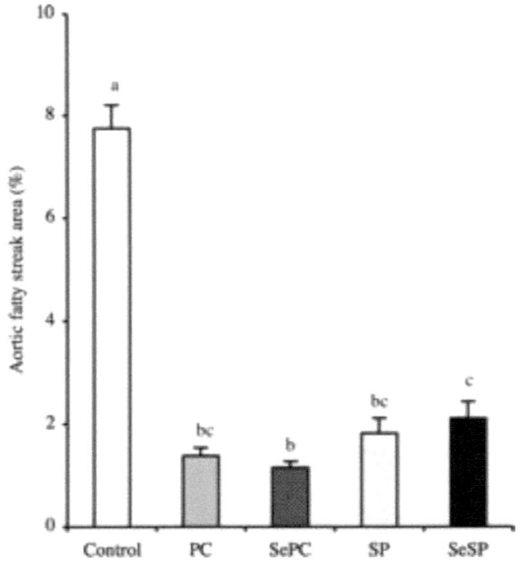

Fig. 12. Effects of daily force-feeding water (control) or phycocyanin (PC), Se-rich phycocyanin (SePC), spirulina (SP) or Se-rich spirulina (SeSP), on aortic fatty streak area in hamsters fed an atherogenic diet for 12 weeks. Values are means ± SEM, ($n = 8$). Bars with different index letters differ, P < 0.05.

Cardiac production of superoxide anion significantly decreased by ~46-76% in the four experimental groups and especially in SePC group (76%). The expression of p22phox subunit of NAD(P)H oxidase decreased by 34% after consumption of SePC (Figure 13).

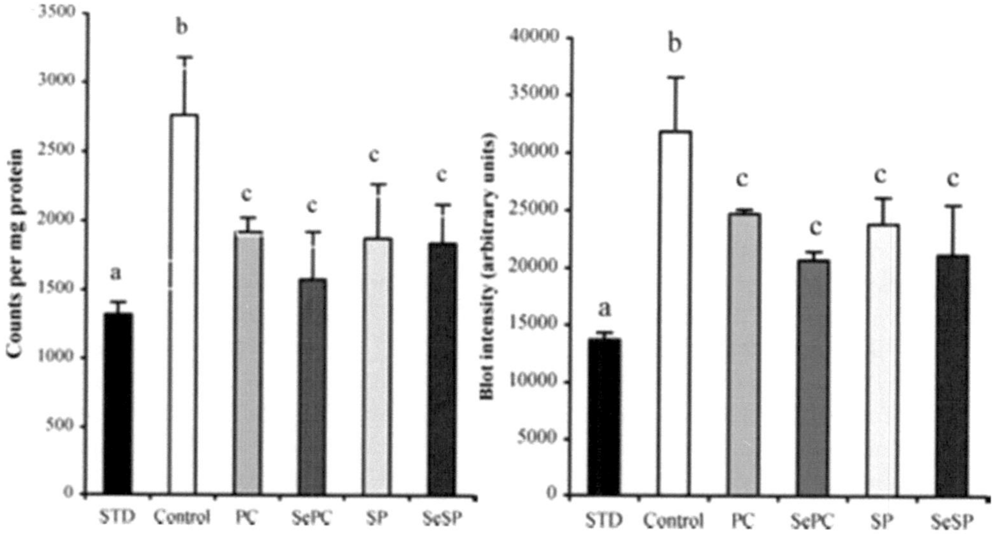

Fig. 13. Cardiac left ventricle superoxide anion production (left) measured by chemiluminescence probe (lucigenin 250 μM) and expression of cardiac p22phox subunit of NAD(P)H oxidase (right) in hamsters fed a standard diet (STD), in controls fed an atherogenic diet for 12 weeks, and in hamsters receiving an atherogenic diet plus phycocyanin (PC), Se-rich phycocyanin (SePC), spirulina (SP), or Se-rich spirulina (SeSP). Values are expressed as mean ± SEM of triplicate wells ($n = 8$). For each dietary treatment, bars with different index letters differ, $P < 0.05$.

The results indicate that chronic consumption of Se-rich spirulina phycocyanin powerfully prevents the development of atherosclerosis. The underlying mechanism is related mainly to inhibiting pro-oxidant factors and at a lesser extent improving the serum lipid profile.

3.9 Algal and fungal polysaccharides

The postulated involvement of lipid peroxidation in atherogenesis invoked intensive research on antioxidants. Many compounds, from natural sources, newly synthesised derivatives of these compounds, and common drugs, were tested for their potency to inhibit the peroxidation of plasma lipids induced by different oxidants in vitro (Rice-Evans *et al.*, 1996) to reduce the plaque formation in animal models in vivo (Barwicz *et al.*, 2000) and to reduce the morbidity and mortality in human subjects, in clinical trials (Aviram *et al.*, 2000). Indeed, mortality from cardiovascular disease is the leading cause of death in the industrialised world. Diet is believed to play a major role in the development of this disease, and much research is being focused on identifying ways to prevent it through changes in dietary habits. Oxidation of low-density lipoproteins (LDL) is traditionally accepted as initiating processes leading to the development of atherosclerosis. The earliest events in the development of the pathology are endothelial dysfunction and oxidative stress in the vascular cell wall, activation of inflammatory cells, and migration of vascular smooth muscle cells to the intima with the modification of the extracellular matrix, leading to the artery remodelling. Development of atherosclerosis is thought to be closely dependent upon increased oxidative stress, that is, an imbalance between ROS generation (chiefly superoxide anions, hydrogen peroxide, hydroxyl radicals) and natural cell antioxidant capacity in favour of the former (Frei, 1994). ROS can also regulate many signalling pathways, such as infiltration of monocytes in intima and vascular smooth muscle cell proliferation. A potentially important clinical corollary of the atherosclerosis oxidation theory is that inhibition of LDL oxidation may also inhibit atherosclerosis independent of lowering plasma cholesterol concentrations. Thus, it is essential to develop and utilise effective and natural antioxidants so that they can protect the human body from free radicals and retard the progress of many chronic diseases. Published data indicate that plant polysaccharides in general have antioxidant activities and can be explored as novel potential antioxidants (Ng *et al.*, 2004; Jiang *et al.*, 2005; Wang *et al.*, 2007). Different polysaccharide fractions extracted and purified from litchi fruit pericarp tissues exhibited strong antioxidant activities (Yang *et al.*, 2006). Tea polysaccharides were also shown to exert significant inhibitory effects on hydroxyl and superoxide radicals and lipid peroxidation. They could also improve the activity of superoxide dismutase. These results suggested that tea polysaccharides were potent antioxidants in mice (Chen *et al.*, 2007). In addition, polysaccharides extracted from mushrooms, such as *Auricularia auricular*, have also shown antioxidant properties as shown by their free radical scavenging ability (Fan *et al.*, 2007). The structure and mechanisms of the pharmaceutical effects of bioactive polysaccharides on diseases have been extensively studied, and more natural polysaccharides with different curative effects have been tested and even applied in therapies. Macroalgae have also received much attention as potential natural antioxidants (Duan *et al.*, 2006). Thus, the antioxidant activity of natural polysaccharides from the green alga *Ulva pertusa* was determined *in vitro*, including scavenging activity against superoxide and hydroxyl radicals, reducing power, and chelating ability. They showed strong scavenging activity against hydroxyl radical and chelating ability (Qi *et al.*, 2006).

Since Golden Syrian hamsters fed a fat-rich diet develop dyslipidemia and atherosclerotic plaques, similar in many respects to human atheroma (Nicolle *et al.*, 2004; Auger *et al.*, 2005 a), they were selected for a study in which we investigated the effects of Ulva, before and after processing (Godard *et al.*, 2009). Three groups of 12 hamsters were fed a high cholesterol diet for 12 wk (Control) or a high cholesterol diet where cellulose has been replaced for an equivalent fibre weight from Ulva or processed Ulva. Plasma cholesterol, non-HDL cholesterol and specially triglycerides were reduced by Ulva. Liver glutathione peroxidase activity was increased and thiobarbituric acid reactive substances were efficiently reduced by dietary treatments compared with controls, whereas plasma antioxidant capacity was increased (Table 9), cardiac $O_2°$- production decreased by 41% (Ulva) and by 31% (processed Ulva) (Figure 14) and aortic fatty streak area was decreased by 70% as compared to controls, according to the same pattern observed for TBARS (Figure 15).

	Control	Ulva	Processed Ulva
Plasma			
TC[1], *mmol/L*	7.77 ± 0.31[a]	6.65±0.39b	7.30 ± 0.19[a b]
HDLC[2], *mmol/L*	5.18 ± 0.07[a]	5.04 ± 0.12[a]	4.65 ± 0.14[a]
Non HDLC, *mmol/L*	2.59 ± 0.21[a]	1.63 ± 017[b]	2.65 ± 0.17[a]
TG[3], *mmol/L*	1.76 ± 012[a]	0.89 ± 0.10[b]	1.17 ± 0.07[c]
PAC, *mmol/L*	0.81 ± 0.05[a]	1.35 ± 0.05[b]	1.28 ± 0.04[b]
Liver			
SOD[4], *U/mg protein*	22.68 ± 1.41[a]	21.32 ± 0.91[a]	22.40±1.04[a]
GSHPx[5], *U/mg protein*	0.43 ± 0.03[a]	0.69 ± 0.13[b]	0.72±0.05[b]
TBARS, *U/mg protein*	27.05 ± 3.27[a]	.34 ± 0.29[b]	3.87±0.61[c]

Values are means ± SEM, $n = 12$. Means in a row with superscripts without a common letter differ, $P < 0.05$. [1]Total cholesterol. [2]High density lipoprotein cholesterol. [3]Triglycerides. [4]Superoxide dismutase. [5]GSHPx: glutathione peroxidase.

Table 9. Effects of ingestion of Ulva and processed Ulva on plasma lipid concentrations and plasma antioxidant capacity (PAC) and on hepatic antioxidant enzyme activities and thiobarbituric acid reactive substances (TBARS) in hamsters fed an atherogenic dietfor 12 weeks.

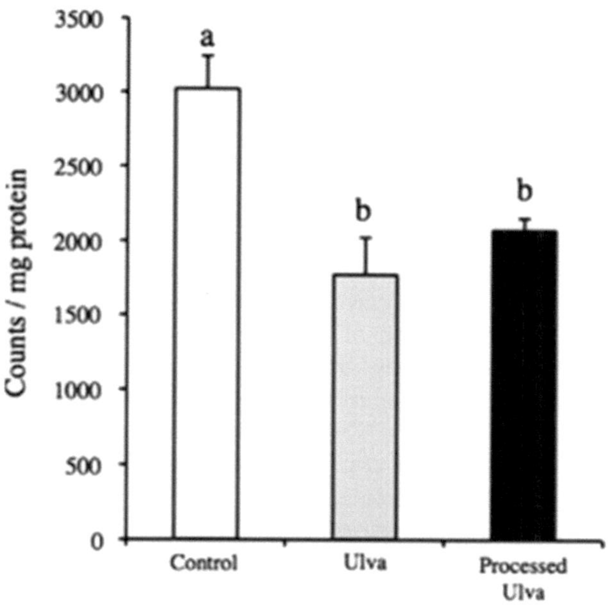

Fig. 14. Cardiac superoxide anion production in hamsters fed high cholesterol diet (control) or a high cholesterol diet where cellulose has been replaced for an equivalent fibre weight from Ulva or processed Ulva for 12 weeks. Values are expressed as mean ± SEM of triplicate wells ($n = 6$). For each dietary treatment, bars with different index letters differ ($P < 0.05$).

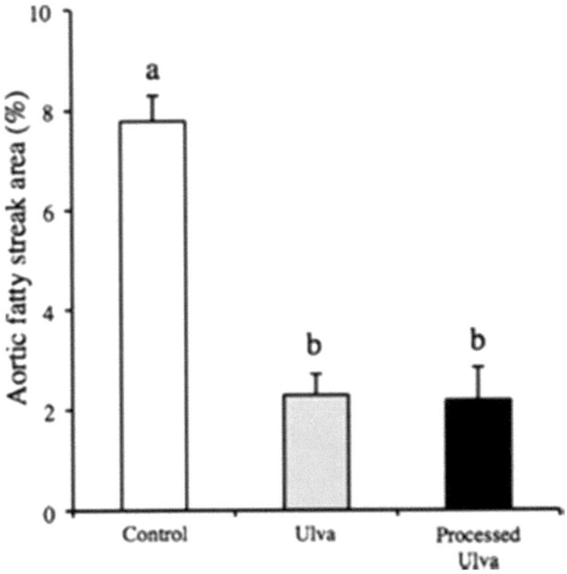

Fig. 15. Effects of feeding a high cholesterol diet (control) or a high cholesterol diet where cellulose has been replaced for an equivalent fibre weight from Ulva or processed Ulva on aortic fatty streak area in hamsters for 12 weeks. Values are expressed as mean ± SEM (n = 6). For each dietary treatment, bars with different index letters differ ($P < 0.05$).

The results show for the first time that chronic consumption of polysaccharides supplied by Ulva prevent the fall of antioxidant defences and the development of atherosclerosis in hamsters. The underlying mechanism is related mainly to increased antioxidant status although improvement of the serum lipid profile was not ruled out.

We have also investigated the effects of chitin-glucan (CG) on early atherosclerosis, cardiac production of superoxide anion, and hepatic antioxidant enzymes in the hamster model of atherosclerosis (Berecochea-Lopez *et al.*, 2009). Three groups of 12 hamsters were fed an atherogenic diet for 12 weeks. They received by gavage either water (control group) or CG in water at a dose of 21.4 mg/kg BW.d^{-1} of chitin-glucan (CG ld) or 42.8 mg/kg BW.d^{-1} (GG hd). CG did not affect plasma cholesterol but lowered triglycerides. It also strongly reduced the area of aortic fatty streak deposition by 87-97%, cardiac production of superoxide anion by 25% (Figure 16), and liver MDA by 77-85%, and enhanced liver superoxide dismutase activity by 7-45% and glutathionne peroxidase activity by 38-120% (Table 10). These findings support the view that chronic consumption of chitin-glucan has potential beneficial effects with respect to the development of atherosclerosis. The underlying mechanism is related mainly to improving the antioxidant status.

3.10 Melon juice extract rich in superoxide dismutase activity

Dietary factors are thought to play a key role in the regulation of the oxidant status. An imbalance between nutrients, and in particular those involved in antioxidant status, could explain the onset of an enhanced production of free radicals. A diet low in antioxidants contributes to the occurrence of an oxidative stress. The importance of dietary antioxidants

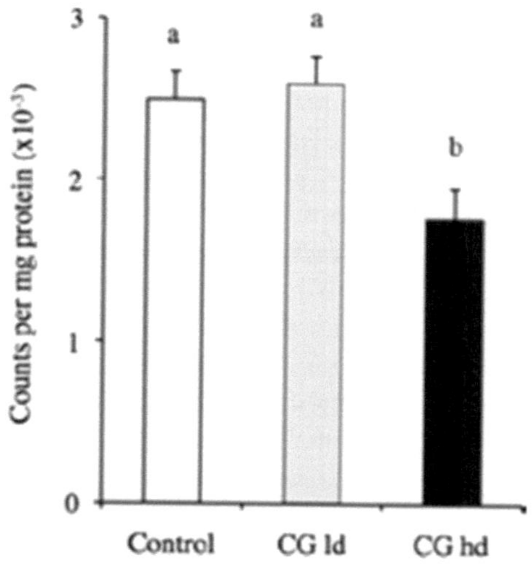

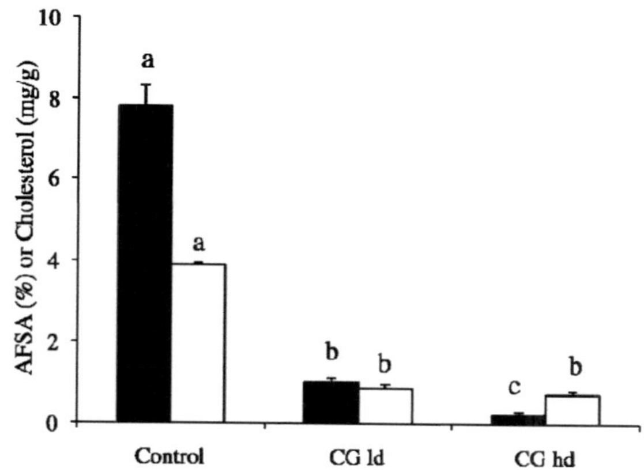

Fig. 16. Cardiac superoxide anion production (left) and aortic fatty streak area (AFSA, black bars) and aortic cholesterol level (white bars) (right) in hamsters fed an hyperlipidic diet without (Control) or with chitin-glucan at low dose (CG ld, 21.4 mg/kg $BW.d^{-1}$) and high dose (CG hd, 42.8 mg/kg $BW.d^{-1}$). Values are expressed as mean ± SEM of triplicate wells (*n* = 6). For each dietary treatment, bars with different index letters differ (P < 0.05).

	Controls	CG ld	CG hd
		units.mg protein⁻¹	
SOD[2]	42 ± 2^{a}	45 ± 2^{a}	61 ± 4^{b}
GSHPx[3] (x 10^{-2})	1.55 ± 0.32^{a}	2.14 ± 0.11^{b}	3.42 ± 0.42^{c}
		ng.mg protein⁻¹	
TBARS	27.05 ± 3.27^{a}	6.11 ± 0.39^{b}	4.07 ± 0.33^{c}

1 Values are the means ± SEM; n = 12. For each dietary treatment, means in a column with different superscripts differ, P < 0.05. 2 SOD: superoxide dismutase. 3 GSHPx: glutathione peroxidase.

Table 10. Hepatic antioxidant enzyme activities and thiobarbituric acid reactive substances (TBARS) in hamsters fed a hyperlipidic diet plus a daily gavage of water (Controls) or Chitin-Glucan at a low dose (CG ld) or Chitin-Glucan at a high dose (CG hd)[1].

in human health is clear and some epidemiological studies showed the potential health benefits of the consumption of antioxidant rich fruits and vegetables (Dauchet *et al.*, 2006). From a theoretical point of view, antioxidant enzymes could exhibit a major advantage over dietary antioxidants known as micronutrients and microconstituents: they catalyse the detoxification of their substrates (Nelson *et al.*, 2006) whereas the latter are consumed by reacting with ROS. New approaches to antioxidant therapy appeared recently. Nelson *et al.* (2006) reported that ingestion of an extract of five medicinal plants Protandim by healthy humans increased the circulating superoxide dismutase (SOD) and catalase activities and reduced the levels of thiobarbituric acid reactive-substances (TBARS), a marker of lipid peroxidation that correlates with oxidative stress and atherosclerosis (Rosenblat *et al.*, 2006). Glisodin®, a melon juice concentrate coated with gliadin containing high levels of SOD and other antioxidant enzymes, has been developed and its antioxidant properties have been demonstrated (Vouldoukis *et al.*, 2004 a). Given to mice, it exhibited properties similar to Protandim (Vouldoukis *et al.*, 2004 b) leading up to a fourfold increase of circulating and hepatic tissue antioxidant enzyme activities. In healthy volunteers, oral Glisodin® was associated with less lipid peroxidation (Muth *et al.*, 2004). Very recently, these authors demonstrated that a pre-treatment with Glisodin® may be a therapeutic option to reduce oxidative cell injury affiliated with aortic cross-clamping in pigs (Kick *et al.*, 2007). Thus, we investigated whether feeding a melon juice concentrate rich in superoxide dismutase and coated with palm oil (Extramel®) to hamsters on an atherogenic diet (HF) would prevent the development of the pathology (Décordé *et al.*, 2010). Hamsters received by gavage either water or Extramel® at 0.7, 2.8 or 5.6 mg/d. After 12 wk of oral administration, Extramel® lowered plasma cholesterol and non-HDL cholesterol and induced blood and liver SOD activities. It also strongly reduced the area of aortic fatty streak (AFSA) by 49-85% (Table 11).

	STD	HF	E1	E2	E3
Plasma					
TC[1], *mmol/L*	2.06±0.09	9.22±2.46[a]	7.25±2.09[ab]	4.62±1.43[b]	4.98±1.33[b]
HDLC[2], *mmol/L*	1.42±0.64	4.97±1.33[a]	4.33±1.28[a]	2.56±0.85[a]	2.69±1.28[a]
Non HDLC, *mmol/L*	0.64±0.03	4.25±1.13[a]	2.92±0.8[a]	1.66±0.58[b]	2.29±0.05[b]
Liver					
SOD, *IU/mg prot*	21.9±1.4	18.8±1.5[a]	20.7±0.7[a]	20.2±0.7[a]	45.1±4.6[b]
GSHPx, *IU/mg prot*	2190±120	2097±102[ab]	2517±125[a]	1640±156[b]	2401±284[a]
Aorta					
AFSA, %	0	5.8±0.5[a]	2.6±0.5[b]	1.9±0.4[b]	1.1±0.3[c]

Table 11. Effect of feeding a standard diet (STD) or high fat diet (HF), or an HF diet plus Extramel® at 0.7 (E1), 2.8 (E2) or 5.6 (E3) mg/d for 12 weeks on blood lipids, on liver antioxidant enzymes superoxide dismutase (SOD) and glutathion peroxidase (GSHPx) activities, and on aortic fatty streak area (AFSA).

Values are means ± SEM (n=12). For each dietary treatment, means in a column with different superscripts differ, $P < 0.05$. [1] TC = Total cholesterol, [2] HDL-C= High density lipoprotein cholesterol.

Cardiac (45%) and liver (67%) production of superoxide anion and liver $p22^{phox}$ subunit of NAD(P)H oxidase expression (66%) were also reduced (Figures 17 and 18).

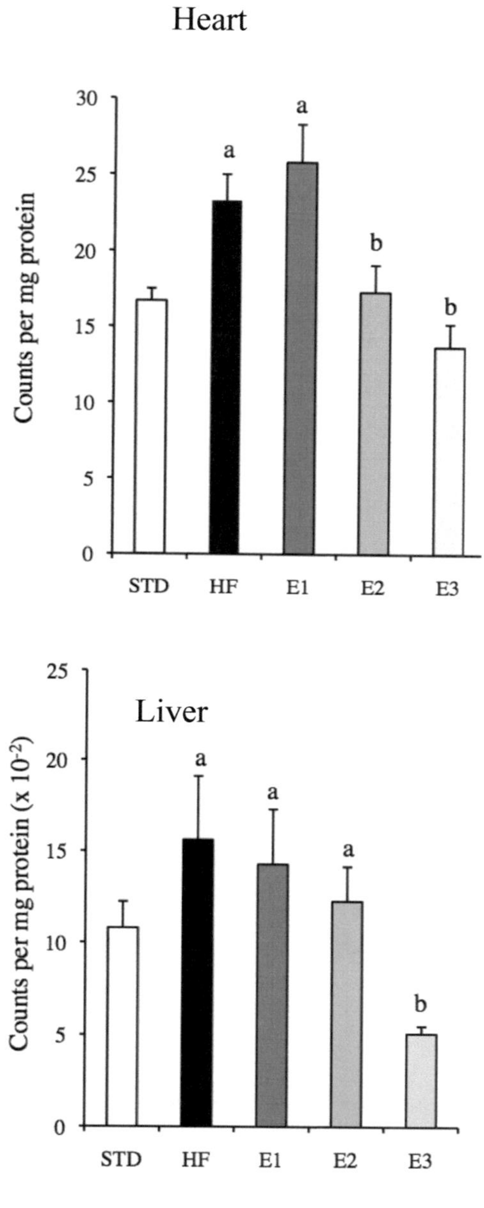

Fig. 17. Cardiac and liver superoxide anion production in hamsters fed a standard diet (STD) or a high fat diet (HF), or a HF diet plus Extramel® at 0.7 (E1), 2.8 (E2) or 5.6 (E3) mg/d for 12 weeks. Values are expressed as mean ± SEM (n = 12). For each dietary treatment, bars with different index letters differ ($P < 0.05$).

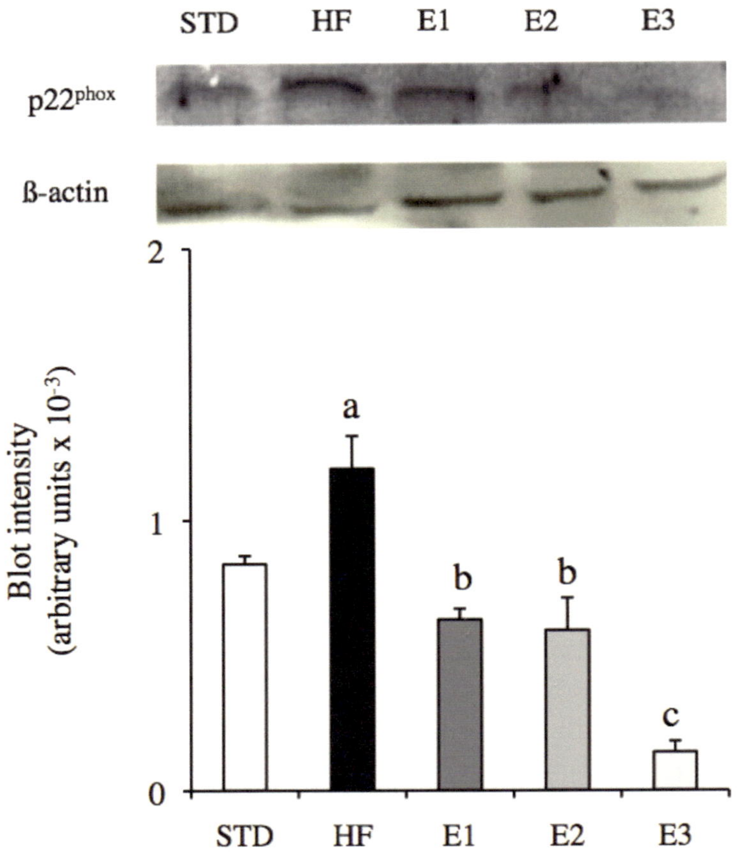

Fig. 18. Expression of hepatic p22^{phox} subunit of NAD(P)H oxidase in hamsters fed with a standard diet (STD) or a high-fat diet (HF), or an HF diet plus Extramel® at 0.7 (E1), 2.8 (E2) or 5.6 (E3) mg/d for 12 weeks. The densitometric measurement shows arbitrary area units. Values are expressed as mean ± SEM (n=6). For each dietary treatment, bars with different index letters differ ($P < 0.05$).

Finally, Extramel® attenuated the development of hepatic steatosis (Figure 19).

These findings support the view that chronic consumption of melon juice extract rich in SOD has potential beneficial effects with respect to the development of atherosclerosis and liver steatosis, emphasizing its use as potential dietary therapy.

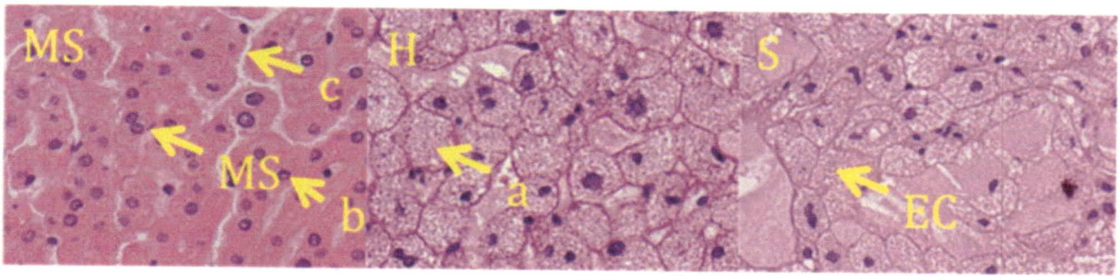

Fig. 19. Histologic evaluation of hepatic steatosis in standard (a), atherogenic (b) or Extramel®-treated (5.6 mg/d) (c) hamsters for 12 weeks. Representative liver sections (original magnification x 200) are illustrated. (a) Liver section from a hamster fed the standard diet. (b) Atherogenic hamster fed no Extramel® illustrate the development of hepatic steatosis. (c) Slight reduction in hepatic steatosis in 5.6 mg/d Extramel fed hamsters. Sinusoidal capillary (S) ; Endothelial cell (EC) ; Hepatocyte (H) ; Microvascular steatosis (MS).

4. Preventive effects of dietary antioxidants against obesity

4.1 Chardonnay grape seed procyanidin extract

According to the International Obesity Task Force, an estimated 300 millions adults worldwide are obese. Obesity, particularly that caused by visceral fat accumulation, is an important risk factor for sedentary lifestyle-related diseases, such as type 2 diabetes mellitus, coronary heart diseases and cerebrovascular events, hyperlipidemia, hypertension, and cancer (Braddon *et al.*, 1986; Esposito *et al.*, 2003; Mokdad *et al.*, 2003). Oxidative stress could be a potential link between fat accumulation and obesity-related morbidity such as diabetes and cardiovascular diseases. Indeed, this pathology can contribute to an increased susceptibility to ROS. In the general population, several groups at risk of high oxidative stress are identified and more particularly the obese subjects, in relation to reduced antioxidant defenses and/or a strong free radicals production.

High fat-feeding has commonly been used to induce visceral obesity in rodent animal models (Hansen *et al.*, 1997) because of the similar pathogenesis of obesity to that found in humans (Katagiri *et al.*, 2007). Preventive or therapeutic strategies to control most of human obesity should target these abnormalities. Anti-obesity foods and food ingredients may avert the condition, possibly leading to the prevention of lifestyle-related diseases, if they are effective in reducing the visceral fat mass (Saito *et al.*, 2005). It has been reported that the body weights of rats and their plasma triglycerides, cholesterol and LDL-cholesterol have been significantly reduced by feeding Oolong, black, and green tea leaves. Epigallocatechin gallate (EGCG) purified from green tea when given to mice in diet decreased diet-induced obesity by decreasing energy absorption and increasing fat oxidation (Klaus *et al.*, 2005). Supplementation with tea catechins resulted in a significant reduction of high-fat diet-induced body weight gain, adipose tissue mass, liver fat content, and the development of hyperinsulinemia and hyperleptinemia in C57BL/6J mice (Murase *et al.*, 2002). Recently the continuous ingestion of a green tea extract high in catechins by humans led to a reduction in body fat, systolic blood pressure, and LDL cholesterol, suggesting that the ingestion of such an extract contributes to a decrease in obesity and cardiovascular disease risks (Nagao *et al.*, 2007). More recently, Wolfram *et al.* (2005) showed that supplementation with Teavigo®, a green tea extract containing about 94% EGCG, abolishes diet-induced obesity and claimed that this was a valuable natural treatment option for obesity. Nakagawa *et al.* (2004)

demonstrated the abdominal fat lowering and hypoglycemic effects of licorice hydrophobic flavonoids in obese diabetic mice and elucidated their mechanisms of action by showing that they induce differentiation of human adipocytes. Sugiyama *et al.* (2007) suggested that oligomeric procyanidin was the main contributor to the effect of apple polyphenol extract on inhibiting triglyceride absorption in mice and humans. Apple procyanidins may relieve obesity via a lipase inhibiting activity and may be effective for obesity-related diseases. To our knowledge, the antiobesity effects of grape seed catechins have not been examined yet. Only Vadillo *et al.* (2006) showed that moderate red wine intake can prevent the increase of body weight by modulating energy intake in a rat diet-induced model of obesity. The present study was designed to examine the preventive effect of a grape seed tannin extract (GSE) on the development of obesity induced by feeding a HF diet in hamsters. Feeding hamsters a cholesterol-supplemented diet produces dyslipidemia (Nistor *et al.*, 1987). This model was also chosen because hamsters have a plasma lipoprotein distribution similar to that of humans. It has been also used by Simon *et al.* (2006) and developed obesity traits, and Leung *et al.* (2004) showed that the fat-fed Syrian Golden hamster is a good model of nutritionally induced insulin resistance. Elsewhere, GSE is a rich source of (+)-catechin, (+)-epicatechin and procyanidins (Prieur *et al.*, 1994). Keaney *et al.* (2003) reported that obesity is a strong independent predictor of systemic oxidative stress and suggested that obesity is associated with a state of excess oxidative stress. As one of the possible mechanisms of the obesity-related oxidative stress, we also examined the modulation by HF diet (with or without GSE) of cardiac production of superoxide anions and NAD(P)H oxidase expression.

We have analyzed the effects of a polyphenolic grape seed extract (GSE) on obesity and oxidative stress in hamsters receiving a high-fat diet (HF) (Décordé *et al.*, 2009 a). Three groups of hamsters received a standard diet, or a HF diet plus a daily gavage with water (Control, HF) or a solution of GSE (HF+GSE). After 12 weeks, HF diet increased abdominal fat as compared with standards. GSE avoided this feature. HF diet led to higher plasma glucose, triglycerides, insulin and greater insulin resistance (HOMA-IR) values. GSE prevented in part these effects, reducing insulinemia and leptinemia by 16.5 % and 45 % respectively, whereas adiponectin level increased by 61 % compared with obese controls. GSE lowered glycemia and HOMA-IR (Table 12) and strongly prevented cardiac production of superoxide by 74% and NAD(P)H oxidase expression by 30% (Table 13).

	STD	HF	HF+GSE
Body weight gain, *g*	14±3[c]	38±2[a]	25±2[b]
Abdominal white fat, *% BW*	1.19±0.11[c]	2.20±0.12[a]	1.55±0.03[b]
Triglycerides, *mmol/L*	0.97±0.09[b]	3.14±0.33[a]	2.76±0.19[a]
Glucose, *mmol/L*	4.76±0.41[c]	6.85±0.44[a]	5.57±0.37[b]
Insulin, *ng/mL*	0.27±0.02[b]	0.66±0.07[a]	0.55±0.07[a]
HOMA-IR	37.79±3.02[c]	132.95±11.24[a]	90.06±8.66[b]
Leptin, *ng/mL*	1.52±0.09[b]	2.63±0.22[a]	1.45±0.23[b]
Adiponectin, *ng/mL*	3.11±0.12[a]	1.43±0.15[c]	2.26±0.20[b]

Table 12. Effects of a grape seed extract (GSE) consumption on body weight, food and energy intakes, and plasma glucose, triglycerides, insulin, leptin and adiponectin of hamsters fed a standard diet (STD), a high fat diet (HF) or a high fat diet plus a grape seed extract (HF+GSE) for 12 weeks.

Plasma glucose, triglycerides, insulin, leptin and adiponectin were measured at the fasted state at the end of the 12-week treatment period. HOMA-IR : homeostatic model assessment for insulin resistance. Values are means ± SEM, n = 14. Means in a row with superscripts without a common letter differ, $P < 0$.

	STD	HF	HF+GS
$O_2^{\circ -}$, *counts/mg protein*	520 ± 30^{a}	896 ± 140^{b}	229 ± 45^{c}
NADP(H) oxidase expression[2]	5.5 ± 0.7^{a}	20.4 ± 1.8^{b}	14.3 ± 1.3^{c}

[1] Values are means ± SEM (n = *14*). Means in a row with superscripts without a common letter differ, $P < 0.05$.

[2] Blot intensity expressed as arbitrary units x 10^{-3}

Table 13. Cardiac superoxide anion production ($O_2^{\circ -}$) and expression of NADP(H) oxidase $p22^{phox}$ subunit in hamsters fed a standard diet (STD) or a high fat diet without (HF) or with a grape seed extract (HF+GSE) for 12 weeks[1].

This is the first time that chronic consumption of grape phenolics is shown to reduce obesity development and related metabolic pathways including adipokine secretion and oxidative stress.

4.2 Melon juice extract rich in superoxide dismutase activity

High-fat diets and sedentary lifestyles are important risk factors for obesity which is a key feature of metabolic syndrome and which greatly predisposes individuals to liver diseases, cardiovascular disease, type 2 diabetes, dyslipidemia, hypertension and numerous cancers and is associated with markedly diminished life expectancy (Haslam & James, 2005).

We previously investigated a hamster model of diet-induced obesity and insulin resistance that exhibits oxidative stress and some correlates to human obesity (Luft & Landau, 1995). A close correlation has been found between increased oxidative stress in accumulated fat and the pathogenic mechanism of obesity and obesity-associated metabolic syndrome, and Keaney *et al.* (2003) reported that obesity is a strong independent predictor of systemic oxidative stress, suggesting that obesity is associated with a state of excess oxidative stress. This may be the source of several metabolic dysfunctions like inflammation, hypertension, and impaired glucose intake in muscle and fat, which are highly related to obesity. Oxidative stress could be a potential link between fat accumulation and obesity-related morbidity such as diabetes and cardiovascular diseases. Indeed, this pathology can contribute to an increased susceptibility to ROS. In the general population, several groups at risk of high oxidative stress are identified and more particularly the obese subjects, in relation to reduced antioxidant defences and/or a strong free radicals production.

Of the many potential cellular sources of chronic ROS production, mitochondria and nonphagocytic NAD(P)H oxidase are the major sources under physiological conditions (Wallace, 1999; Sorescu & Griendling, 2002). Increased mitochondrial ROS generation and dysfunction are associated with cardiovascular and many other diseases (Yamauchi *et al.*, 2001). Elsewhere, adipocytes synthesize and secrete adipocytokines. Dysregulated production of these adipocytokines participates in the pathogenesis of obesity-associated metabolic syndrome. Among them, adiponectin exerts insulin-sensitizing (Okamoto *et al.*,

2002) and anti-atherogenic effects (Ozata *et al.*, 2002), and hence a decrease in plasma adiponectin is causative for insulin resistance and atherosclerosis in obesity. However, obesity has been shown to be one of the conditions that decrease antioxidant capacity (Carmiel-Haggai *et al.*, 2005), and it seems to accomplish this by lowering the levels of antioxidant enzymes (catalase, glutathione peroxidase, and glutathione reductase) (Furukawa *et al.*, 2004). In animal (Carmiel-Haggai *et al.*, 2005) and human studies (Furukawa *et al.*, 2004), obesity is associated with a decrease in tissue or plasma antioxidant capacity. In obese mice, treatment with NAD(P)H oxidase inhibitor reduced ROS production in adipose tissue, attenuated the dysregulation of adipocytokines and improved (Ozata *et al.*, 2002) diabetes, hyperlipidemia, and hepatic steatosis. Since obesity may induce systemic oxidative stress, increased oxidative stress should be an important target for the development of new therapies. Generally, dietary antioxidants are vitamins, fibers and phytochemicals supplied by fruits and vegetables. Recently, much attention has been focused on new approaches of antioxidant therapy by providing antioxidant enzymes. Thus, a melon juice concentrate containing high levels of superoxide dismutase (Lacan & Baccou, 1998) and other antioxidant enzymes has been developed and its antioxidant and anti-inflammatory properties have been demonstrated (Vouldoukis *et al.*, 2004 a). This melon concentrate combined with wheat gliadin polymer, called Glisodin®, was given for 4 weeks to mice, reducing the levels of thiobarbituric acid reactive-substances, a marker of lipid peroxidation that correlates with oxidative stress and leading up to a fourfold increase of circulating and hepatic tissue antioxidant enzyme activity (SOD and catalase) (Vouldoukis *et al.*, 2004 b).

Studies showed that dietary antioxidants could be a therapy against obesity that is associated with a state of oxidative stress. In this context, we undertook a study designed to quantitatively examine the properties of Extramel® microgranules, a melon juice extract coated with palm oil and rich in antioxidants and particularly SOD, on the development of obesity in our golden Syrian hamster model of diet-induced obesity (Décordé *et al.*, 2009 b). Five groups received a standard diet, or a high fat diet (HF) plus a daily gavage with water (Control) or extract at 0.7, 2.8 or 5.6 mg/day. After 84 days, the higher dose lowered triglyceridemia (68%), production of liver superoxide anion (12%), leptinemia (99%) and increased adiponectinemia (29%), leading to a concomitant reduction in insulinemia (39%), insulin resistance (41%) and abdominal lipids (25%) and a remarkable decrease of liver lipids (73%) (Tables 14 and 15).

	SD	HF	E1	E2	E3
Body weight gain, *g*	30.1±1.0	44.8±2.0	32.1±1.8[a]	31.6±2.3[a]	31.7±1.6[a]
Abdominal fat, *% BW*	1.10±0.07	1.34±0.04	1.22±0.01[a]	1.14±0.05[a]	1.00±0.03[a]
Plasma					
Triglycerides, *mmol/L*	0.87±0.08	6.29±0.53	6.71±0.18	1.90±0.03[a]	2.05±0.07[a]
Glucose, *mmol/L*	5.43±0.42	6.34±0.53	5.75±0.47	6.55±0.51	6.20±0.35
Insulin, *ng/mL*	0.33±0.08	0.43±0.00	0.34±0.07[a]	0.33±0.04[a]	0.26±0.02[a]
HOMA-IR	52.7±2.0	80.2±1.5	57.5±1.0[a]	63.5±1.1[a]	47.4±0.4[a]
Leptin, *ng/mL*	0.22±0.02	0.40±0.05	0.35±0.04	0.29±0.04[a]	0.28±0.04[a]
Adiponectin, *ng/mL*	7.30±1.70	4.98±0.25	5.65±0.40	6.10±0.68	6.40±0.36[a]

Table 14. Body weight, food and energy intake, plasma glucose, triglycerides, insulin, leptin and adiponectin in hamsters fed a standard diet (SD), a high fat diet (HF) or a HF diet plus Extramel® at 0.7 mg/d (E1), 2.8 mg/d (E2) or 5.6 mg/d (E3) on blood lipids for 12 weeks. Plasma glucose, triglycerides, insulin, leptin and adiponectin were measured at the fasted state at the end of the 12-week treatment period. HOMA-IR: homeostatic model assessment for insulin resistance. Values are means ± SEM (n=12). [a] means $p < 0.05$ vs HF.

	SD	HF	E1	E2	E3
Lipids, *mg/g*	6.60±2.5	21.31±1.60	12.34±1.90[a]	6.11±1.12[a]	5.73±0.61[a]
GSH-Px[1], *U/mg prot*	2191±380	4397±590	2980±268[a]	2912±314[a]	3588±590
Cu-Zn-SOD, *U/mg prot*	21.9±1.4	31.6±2.3	31.0±1.7	32.6±2.0	34.6±2.0
Mn-SOD, *U/mg prot*	3.26±0.06	2.87±0.24	3.35±0.17	3.64±0.17[a]	3.64±0.12[a]
TBARS[2], *mmol/mg prot*	6.19±1.5	11.30±2.00	8.96±1.95	7.53±3.03	7.39±1.40[a]
$O_2^{\circ -}$ production[3]	1080±143	1810±435	1750±404	1579±63[a]	1606±260

[1] Glutathion peroxidase. [2] Thiobarbituric acid reactive substances. Values are means ± SEM (n=12). [a] means $p < 0.05$ vs HF. [3] Expressed as counts per mg protein.

Table 15. Liver lipids, oxidative status of hamsters fed a standard diet (SD) and NAD(P)H-dependent anion superoxide production in hamsters fed a high fat diet (HF) or a HF diet plus Extramel® at 0.7 mg/d (E1), 2.8 mg/d (E2) or 5.6 mg/d (E3) for 12 weeks.

The extract fully prevented the steatohepatitis induced by HF diet (Figure 20).

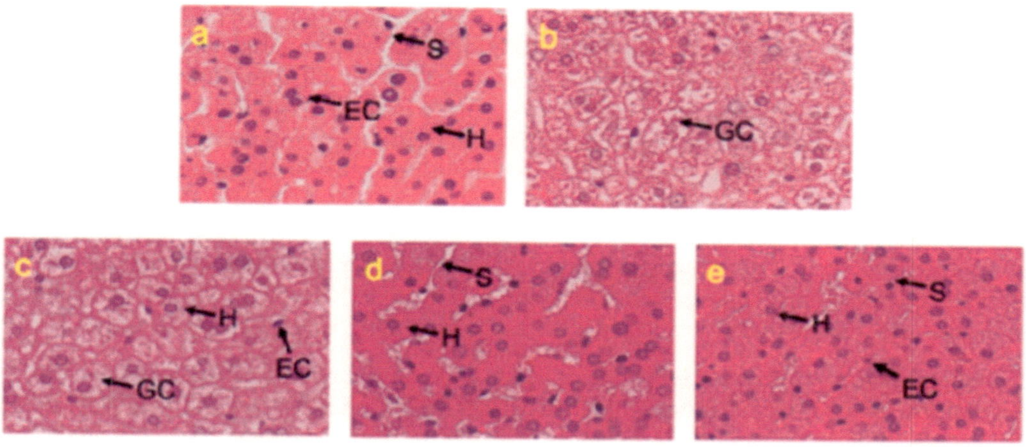

Fig. 20. Histologic evaluation of hepatic modifications in obese (b) or Extramel-treated hamsters (c, d, e) for 12 weeks. Representative liver sections (original magnification x 200) are illustrated. (a) Liver section from a hamster fed the standard diet. (b) Obese hamster fed no Extramel illustrates the development of hepatic ballooning degeneration. (c) No degeneration reduction after treatment with Extramel at 0.7 mg/d. (d, e) Gradual improvement with Extramel at 2.8 and 5.6 mg/d. Sinusoidal capillary (S); Endothelial cell (EC); Hepatocyte (H); Grainy fatty cytoplasm (GC).

These findings suggested that chronic consumption of melon extract may represent a new alternative to reduce the obesity induced by oxidative stress.

5. Discussion

Improvement of plasma lipid profile, increase in plasma antioxidant capacity (PAC), and decrease in superoxide anion production and reduction of NAD(P)H oxidase expression (p22^{phox} subunit) by Oxxynea® were associated with a total prevention of aortic fatty streak lesion area. The relative contribution of each parameter such as lipid profile, plasma antioxidant defenses and overproduction of ROS is difficult to establish. However, it is tempting to speculate on a specific role of tissular oxidative stress. Indeed, in a previous paper, it has been shown that the wine polyphenols- induced aortic fatty streak lesion area prevention was associated with lipid and plasma antioxidant capacity improvement without any effect on plasma oxidative stress markers such as MDA (malondialdehyde), AOPP (Advanced oxidation protein products), and AGEs (An advanced glycation end-product) (Auger et *et al.*, 2005 a). Taken together, these results suggest a specific role of polyphenol in vascular tissue mediated by NAD(P)H oxidase and that this extract acted by mechanisms operating both inside and outside a hypolipemic effect, especially an antioxidant effect. Although the constituent(s) responsible for these effetcs remain(s) unclear, candidates such as vitamin C, vitamin E, carotenoids, selenium, and polyphenols could act synergistically or additively to prevent atherosclerosis in the hamster model. The emphasized the complexity of numerous possible operating mechanisms of action depending on the presence and combination of different families of phenolic compounds, in both nature and quantity.

When fed daily to Syrian golden hamsters on an atherogenic diet for 12 weeks at a dose corresponding to the consumption of 275 mL by a 70 kg human, juices prepared from three varieties of raspberry, Cardinal, Glen Ample and Tulameen, lowered biomarkers of early atherosclerosis. Although the effects were not identical for all three juices, our findings suggest that moderate consumption of raspberry juices may help to prevent the development of early atherosclerosis, with the underlying mechanisms related to improved antioxidant status and serum lipid profiles. The main polyphenolic compounds in the juices were anthocyanins and ellagitannins, neither of which is likely to be absorbed into the bloodstream in sufficient quantity to induce the observed protective effects. The principal components entering the circulatory system from anthocyanins will be colonic microbiota-derived phenolic acid breakdown products while ellagitannins will be converted to urolithins in the large intestine and are which subsequently absorbed as O-glucuronides.

Regarding fruits and fruit juices (Décordé *et al.*, 2008; Rouanet *et al.*, 2010; Suh *et al.*, 2011), their antioxidant activity is known to reduce diet-induced oxidative stress. Here, we showed that apple and grape and their juices prevent early atherosclerosis, i.e. 10% foam cell coverage of aorta, in hypercholesterolemic hamsters, with AFSA development being prevented throughout an improvement of the plasma lipid profileand PAC, sparing of SOD and GSHPx activities, and a decrease in the liver TBARS level. A decreased activity of antioxidant enzymes may be a consequence of the sparing effect of dietary antioxidants, reducing the requirement for enzymatic antioxidant function when elevated concentrations of exogenous antioxidants are present in the circulatory system (Breinholt *et al.*, 1999). Elsewhere, spontaneous hepatic lipid peroxidation has been shown to decrease with an increasing level of dietary antioxidants (Leibovitz *et al.*, 1990), and supplementation with apples effectively decreases oxidative stress by decreasing malondialdehyde formation in the body (Pajk *et al.*, 2006). In this study (Décordé *et al.*, 2008), hepatic TBARS concentration decreased in each group. This improvement in the antioxidant status indicated that even a low amount of absorbed phenolics from fruits and juices might decrease lipid oxidation. Thus, our findings suggested that both apples and purple grape, as well as their juices acted by various mechanisms including not only a hypolipemic effect but also, and especially, an antioxidant effect.

Overall, the results clearly showed for the first time that apple and purple grape prevent diet-induced atherosclerosis in hamsters, and that the fruit processing can have a major impact on the potential health benefits of fruit in pathological conditions. These findings, therefore, provide encouragement that fruit and fruit juices may have a significant clinical and public health relevance.

Our study on berry juices and teas (Rouanet *et al.*, 2010) demonstrated that when fed to hamsters under atherogenic diet, they are able to facilitate a very strong inhibition of aortic fatty streaks deposition. These effects are physiologically relevant as they were induced by a daily supplement equivalent to 275 ml of beverage consumed on a daily base by a 70 kg human. There are reports that consumption of fruit juice and green tea both increase the activity of hepatic antioxidant enzymes (Lin *et al.*, 1998; Pajk *et al.*, 2006). This contrasts with the findings of the present study where berry juice and tea consumption lowered hepatic GSHPx and SOD activities, and agree with previous results in hamsters receiving either pure catechin, quercetin or resveratrol (Auger *et al.*, 2005 a) or phenolics from purple grape, apple, purple grape juice and apple juice (Décordé *et al.*, 2008). One explanation for this down regulation is that it is a consequence of dietary antioxidants being able to scavenge oxygen radicals and thus reduce the need for enzymatic endogenous antioxidants. Thus, polyphenol-rich berry juices and green and black tea intake may be of significant relevance to clinical and public health.

In our study on algal polysaccharides, the superoxide anion production induced by the atherogenic diet was prevented by feeding Ulva and processed Ulva which prevented aortic fat deposits. However, the exact explanation of the mechanism underlying the free radical scavenging activity exerted by polysaccharides is still not fully understood. The decrease of aortic fatty steak area by alga suggests that ROS-scavenging effect is a likely mechanism since the above studies have established that polysaccharides not only scavenge ROS but also act as potent antioxidants and inhibit the lipid peroxidation mediated by ROS. It is unlikely that the antioxidant effects of Ulva polysaccharides are attributable to their systemic activity because they are not absorbed. What type of antioxidant is responsible for the observed effects and by what mechanism does it act remains a major problem unresolved. At this stage, we can only propose a hypothesis. The most plausible is that these polysaccharides act only within the lumen by protecting the body against the absorption of radical species and/or by protecting algal antioxidants and nutrients against a possible oxidation before their absorption. Also, the antioxidant effect of some metabolites, produced from polysaccharide fermentation by gut flora, cannot be ruled out. Polysaccharides from other origins such as tea, wolfberry or ginseng have been shown to exert antioxidant activity. The antioxidant mechanism may be due to the supply of hydrogen by the polysaccharide, which combines with radicals and forms a stable radical to terminate the radical chain reaction. The other possibility is that the polysaccharides can combine with the radical ions, which are necessary for radical chain reaction; then the reaction is terminated. However, the exact explanation of the mechanism underlying the free radical scavenging activity exerted by polysaccharides is still not fully understood. The decrease of aortic fatty streak area by chitin-glucan (Berecochea-Lopez *et al.*, 1999) suggests that ROS-scavenging effect is a likely mechanism since the above studies have established that polysaccharides not only scavenge ROS but also act as potent antioxidants and inhibit the lipid peroxidation mediated by ROS. Moreover, improvement of hepatic SOD and GSHPx activities, lowered hepatic TBARS levels, and decreased cardiac superoxide anion production were observed, and aortic cholesterol deposit and fatty streak area development were prevented. These results suggest that chitin-glucan acted through an improvement in the antioxidant status indicating that even low amounts of ingested polysaccharides might decrease lipid oxidation. Here also, it is unlikely that the antioxidant effects of chitin-glucan polysaccharides are attributable to their systemic activity because they are not absorbed.

Additionally, a hypolipemic effect of chitin-glucan should not be rejected. To our knowledge, there are no studies reported on the preventive effets of chitin-glucan on atherosclerosis. A comparison could be made with other close nonstarch polysaccharides such as chitosan and dietary fiber components such as β-glucan. Kamil *et al.* (2002) demonstrated that among the different viscosity chitosans, 14 cP chitosan was more effective than the higher viscosity chitosans (57 cP and 360 cP) in preventing lipid oxidation in the herring flesh model system. Soluble and viscous dietary fibers, such as □-glucans are associated with two major health promoting effects, i.e., the attenuation of postprandial plasma glucose and insulin levels, and the control of cholesterol, that are important in atherosclerosis development. Increased viscosity in the intestine delays absorption of glucose and suppresses the absorption of cholesterol and the reabsorption of bile acids. Since β-glucan can appear in the digestive tract after feeding hamsters with chitin-glucan, this could be another explanation for the decreased atherosclerotic plaques in our model.

The antioxidant activity of chitosans of different MWs (30, 90, and 120 kDa) has been investigated in salmon (Kim & Thomas, 2007). In general, all chitosans exhibited antioxidative activities in salmon. The 90 kDa chitosan showed an increased free radical-scavenging activity. Another study (Youn *et al.*, 2001) showed that antioxidative effects of chitosan were increased with larger MW and revealed that the scavenging activities also depended on the MWs of chitosans. However, while the radical scavenging activity of low MW chitosans appears to play a role in their antioxidant activities, this is not the case for the high MW chitosan. In the latter case, its antioxidant activity must be explained by other mechanisms such as metal chelation or lipid binding. This could happen with chitin-glucan and fractions issuing from them. These promising results obtained by dietary supplementation with chitin-glucan from *Aspergillus niger* in a diet-induced atherosclerosis animal model give rise to its use as a potential dietary therapy. Further investigation is warranted to define the mechanisms by which chitin-glucan provides protection.

However, our recent and innovative studies (Décordé *et al.*, 2009 b; Décordé *et al.*, 2010) on dietary supplementation of a melon juice concentrate rich in SOD (Extramel®) provides evidence for the first time that such antioxidant protects against diet-induced oxidative stress, atherosclerosis and obesity in hamsters, and no toxicity or evidence of other unwanted pharmacological effects of Extramel® was noted at either levels of supplementation. Our findings suggest for the first time that Extramel® prevents both NAD(P)H oxidase expression and $O_2^{\circ-}$ overproduction in the liver from hypercholesterolemic hamster. This could be involved in the prevention of LDL oxidation and further atherosclerosis steps. High Fat-feeding activated and Extramel® reduced NAD(P)H activity (heart and liver) and expression (liver) reinforce the hypothesis of a nutritional modulation of ROS enzymatic producing systems. An additional major novel finding is that Extramel® prevented obesity in HF-fed hamsters by decreasing body weight, abdominal fat, triglyceridemia, insulinemia, insulin resistance, liver lipids, and nonalcoholic steatohepatitis (NASH) and preventing adipokine imbalance, oxidative stress is implicated in NASH. All these features due to Extramel® were associated with a huge prevention of aortic fatty streak lipid deposition and NASH. The relative contribution of each parameter is difficult to establish although it is tempting to speculate on a specific role of tissular oxidative stress. Moreover, it may be that these effects are not restricted to SOD activity that could act synergistically or additively with other antioxidant components to prevent atherosclerosis in the hamster model. This indicates that at the low doses used Extramel® is a safe nutraceutical supplement.

Our findings agree with Vouldoukis *et al.* (2004 a) who did show that feeding mice for 4 weeks with a melon extract exhibiting SOD, catalase and residual GSHPx activities led to an increase in circulating and liver SOD activity. Since Extramel® strongly prevented $O_2^{\circ -}$ production and inducted SOD activity, we can assume that it favors the antioxidative balance by decreasing the oxidative stress, thus exhibiting beneficial effects on preventing early atherosclerosis and obesity. Such an increase of liver and blood SOD activity shown by us as well as by Vouldoukis *et al.* (2004 b) is not observed with other kinds of dietary antioxidants and seems to result from this antioxidant enzyme supplementation: when encapsulated by gliadin, the same melon extract rich in SOD triggers an increase in liver and circulating SOD activity after oral administration to rats not subjected to an oxidative stress. When the melon extract rich in SOD is not protected by gliadin, no more increase in circulating and hepatic SOD activity appeared. Elsewhere, an oral supplementation of pure SOD encapsulated in liposomes promoted the circulating SOD activity; however, if not encapsulated in liposomes, and therefore destroyed at least in part during the digestive process, no longer increased circulating SOD activity can be observed (Regnault *et al.*, 1996). Such an increase in circulating SOD activity also appeared in cats infected with FIV given the extract coated with gliadin (Webb *et al.*, 2008). In the same way, SOD mimics, synthesized molecules reproducing the effect of SOD and resisting to digestive conditions, were used successfully in the treatment of steatohepatitis in murine and they also induced an increase of SOD activity in liver (Laurent *et al.*, 2004). An effect of the coating may be ruled out since an increased endogenous SOD activity appeared regardless of the type of coating used (liposomes, gliadin, palm oil). All these findings showed that increased endogenous SOD activities resulted from melon SOD and not from an active issuing peptide since nothing occurred without coating.

Finally, such an induction of the cell antioxidant status is not observed with other kinds of dietary antioxidants (Auger *et al.*, 2005 a; Riss *et al.*, 2007), suggesting a specific effect of melon SOD. It should be pointed out that circulating erythrocytes, which do not contain nuclei, are not able to induce new synthesis of SOD once they enter the bloodstream. Since their circulating life span is about 120 days, thus, during the 12-week course of the experiment about 70% of the erythrocytes would not have been still replaced by maturing reticulocytes from the bone marrow. This 11% increase therefore represents the SOD inducted by supplementation. As expected in liver, a more important induction appeared, as all cells in this tissue have nuclei and ongoing synthesis of new protein. However, as pointed out by McCord and Edeas (2005) overexpression of SOD could be lethal and dose-response curves have been established after isolated hearts were subjected to ischemia-reperfusion injury and protected by exogenous SOD (Omar *et al.*, 1990). SOD was very protective up to a point, beyond which protection was lost and injury was even exacerbated, with increased lipid peroxidation. This was corroborated by Levy *et al.* (2002) who found that high overexpression of SOD increased ischemia-reperfusion injury to the brain. Even at the higher dose of supplementation used by us, Extramel® did not lead to such upper limit dose dependent ending events, strengthening its safety as nutraceutical supplement. Even though involved mechanisms remain unclear, it is known that SOD can be absorbed despite its high MW (~35 kDa). This was shown in the liver of rats after intravenous administration of SOD conjugated with colloidal gold (Dini *et al.*, 1996) and more recently in HT-29 cells using liposomal SOD and the SOD mimic tempamine (Jubeh *et al.*, 2005).

To summarize, Extramel® prevented obesity in High Fat-fed hamsters by decreasing body abdominal and liver fat and by preventing adipokine imbalance. Furthermore, oxidative stress was reduced (decreased levels of lipids and protein oxidation products), and the main sources of ROS production (NADPH and mitochondria) were modified as NADPH-dependent $O_2^{\circ -}$ production and mitochondrial maximal activity of cytochrome oxidase decreased with Extramel®. These promising results give rise to the use of Extramel® as potential dietary therapy and to further clinical studies. Investigation is warranted to define the mechanisms by which Extramel® protects.

It is generally assumed that obesity is a prooxidant state related in part to insulin resistance. In addition, autoxidation of glucose and nonenzymatic glycation of proteins may generate superoxide (Baynes, 1991), a radical species implicated in vascular cell-mediated LDL oxidation. Hyperglycemia also induces the enzymatic production of superoxide through activation of NAD(P)H oxidase in vascular cells (Inoguchi *et al.*, 2000). Elsewhere, insulin itself promotes hydrogen peroxide generation in fat cells (Krieger-Brauer *et al.*, 1992) prompting speculation that oxidative stress is a principal mechanism of insulin resistance with chronic hyperinsulinemia. This hypothesis is further supported by Fortuno *et al.* (2006) suggesting that hyperinsulinemia may contribute to oxidative stress in metabolic syndrome patients through activation of NAD(P)H oxidase. On the other hand, it has also been suggested that lipid disorders could enhance ROS formation. The observed overproduction of superoxide anion in neutrophils from hyperlipidemic guinea pig could be linked to an induction of NAD(P)H oxidase subunit in particular gp91phox (Maeda *et al.*, 2005). It has been reported in humans that systemic oxidative stress is strongly associated with visceral fat accumulation and metabolic syndrome (Fujita *et al.*, 2006). Our observation of a HFD-induced obesity associated with insulin resistance and lipid disorders enhancing NAD(P)H oxidase expression (+270%, Table 13) in cardiac tissue extends these observations, suggesting that insulin resistance and dyslipidemia conspire in oxidative stress. NAD(P)H oxidase overexpression and activity could be also related to adipokine imbalance. In adiponectine -/- mice an overexpression of NAD(P)H oxidase subnunits has been observed in heart (Tao *et al.*, 2007) and kidneys (Ohashi *et al.*, 2007) strongly showing that adiponectine could downregulate superoxide anion production. Elsewhere, it has been recently demonstrated that leptin increases NAD(P)H oxidase protein expression and activity in isolated murine cardiomyocytes and this effect is attenuated by endothelin (ET-1) receptor antagonists (Dong *et al.*, 2006). The observation that grape seed tannin extract (GSE) both improves adiponectin and leptin levels (Table 12) and decreases NAD(P)H oxidase expression and activity (Table 13) strongly suggests a potential link between these events. Whether endothelin plays a pathogenic role in obesity is not clear, but in experimental obesity, there is an increase in gene and protein expression of endothelin in the cardiovascular system, including vasculature and kidney (Barton *et al.*, 2000). Because leptin stimulates endothelin-1 production in cardiomyocytes (Xu *et al.*, 2004) as well as in endothelial cells (Quehenberger *et al.*, 2002), these results suggest that leptin may stimulate ROS formation through endothelin and NAD(P)H oxidase dependent pathway and that phenolics present in GSE are efficient in reducing these effects. As demonstrated by Corder *et al.* (2006), phenolic compounds, principally procyanidins, are able to reduce the production of endothelin-1 by endothelial cells. Inhibition of endothelin-1 overexpression is, therefore, a further potential mechanism for the observed protective effects of GSE consumption. From this study, we can assume that NAD(P)H oxidase is upregulated by HFD due to adipokine imbalance and viceral fat accumulation (Table 12), and possibly ET-1 overexpression. GSE could prevent all these events. Interestingly, HFD increased liver antioxidant enzymes mainly SOD and GSHPx activity that could in part counteract the increase in $O_2^{\circ-}$ production; this could be due to an oxidative system-induced regulation (Maggy-Capeyron *et al.*, 2002). In this context, GSE consumption lowered hepatic SOD and GSHPx activities (not shown here). One explanation for this downregulation is that it is a consequence of the sparing effect of dietary antioxidants being able to scavenge oxygen radicals, and thus reduce the requirement for enzymatic endogenous antioxidant (Breinholt *et al.*, 1999; Auger *et al.*, 2005 a; Décordé *et al.*, 2008) Since GSE totally prevent $O_2^{\circ-}$ production (Table 13) but only partly decreased SOD and GSHPx activities, we can assume that GSE favours the antioxidative balance. Thus, grape seed tannins extract has beneficial effects on preventing diet-induced obesity by improving oxidative stress markers. All of these results suggest that grape seed tannins acted by mechanisms operating at least in part-inside an antioxidant effect and the possibility that adiponectin might modulate oxidative stress, leading to antiobesity effects. Thus, we provided insights into one mechanism, increased oxidative stress, that probably contributes to the pathological after-effects of obesity and that may have important public health implications, being a target for interventions to decrease the pathology. Therefore, the potential for antioxidant therapy to decrease obesity risks needs to be explored.

6. Conclusion

However, it should be kept in mind that the association between inflammation and oxidative stress (OS) is underlaid by molecular mechanisms leading to amplification loops between pro-inflammatory mediators and oxidant production. Thus the presence of oxidative stress will stimulate some transcription factors such as AP-1, NF-κB, PPARs or the hypoxemia induced factor, either directly by reactive forms of oxygen or indirectly through peroxidation products. Once activated, these factors control the expression of proinflammatory molecules such as iNOS, adhesion molecules and cytokines, which will in turn stimulate the NAD(P)H oxidase, originating other reactive forms. Although the initial production of $O_2^{\circ-}$ (the "spark" oxidative) may come from other sources such as the system xanthine/xanthine oxidase or mitochondrial dysfunction, the system NAD(P)H oxidase/NOX appears to play a key role in the amplification OS/inflammation. Our general hypothesis postulates that nutrition can modulate either positively or negatively amplification loops OS/inflammation. The association between OS and inflammation is observed during chronic diseases and may be considered as an aggravating factor. Thus, such an association may be involved in the conversion between insulin resistance and diabetes or cardiovascular complications of obesity.

Then, dietary antioxidants can play a key role in the regulation of the oxidant status and it seems essential to develop and utilise natural antioxidants so that they can retard the progress of many chronic diseases. The consumption of antioxidant rich fruits and vegetables that generally supplies minerals, vitamins, fibers, phytochemicals such as phenolics but also provides antioxidant enzymes is therefore advised.

7. References

Adams, M. R., Golden, D. L., Chen, H., Register, T. C., Gugger, E. T. A diet rich in green and yellow vegetables inhibits atherosclerosis in mice. *J. Nutr.* 2006, 136, 1886-9.

Al-Awwadi, N. A., Araiz, C., Bornet, A., Delbosc, S., Cristol, J.P., Linck, N., Azay, J., Teissedre, P.L., Cros, G. Extracts enriched in different polyphenolic families normalize increased cardiac NADPH oxidase expression while having differential effects on insulin resistance, hypertension, and cardiac hypertrophy in high-fructose-fed rats. *J. Agric. Food Chem.* 2005, 53, 151-7.

Aruoma, O.I. Nutrition and health aspects of free radicals and antioxidants. Food Chem. Toxicol. 1994, 32, 671-83. Auger, C., Caporiccio, B., Landrault, N., Teissèdre, P. L., Laurent, C., Cros, G., Besançon, P., Rouanet, J. M. Red wine phenolic compounds reduce plasma lipids and apolipoprotein B, and prevent early aortic atherosclerosis in hypercholesterolemic Golden Syrian hamsters (Mesocricetus auratus). *J. Nutr.* 2002, 132, 1207-13.

Auger, C., Gérain, P., Laurent-Bichon, F., Portet, K., Bornet, A., Caporiccio, B., Cros, G., Teissèdre, P.L., Rouanet, J.M. Phenolics from commercialized grape extracts prevent early atherosclerotic lesions in hamsters by mechanisms other than antioxidant effect. *J. Agric. Food Chem.* 2004, 52, 5297-302.

Auger, C., Teissèdre, P.L., Gérain, P., Lequeux, N., Bornet, A., Serisier, S., Besançon, P., Caporiccio, B., Cristol, J.P., Rouanet, J.M. Dietary wine phenolics catechin, quercetin and resveratrol efficiently protect hypercholesterolemic hamsters against aortic fatty streak accumulation. *J. Agric. Food Chem.* 2005 a, 53, 2015-21.

Auger, C., Rouanet, J.M., Vanderlinde, R., Bornet, A., Décordé, K., Lequeux, N., Cristol, J.P., Teissèdre, P.L. Polyphenols-enriched chardonnay white wine and sparkling pinot noir red wine identically prevent early atherosclerosis in hamsters. *J. Agric. Food Chem.* 2005 b, 53, 9823-9.

Aviram, M., Dornfeld, L., Rosenblat, M., Volkova, N., Kaplan, M., Coleman, R., et al.. Pomegranate juice consumption reduces oxidative stress, atherogenic modifications to LDL, and platelet aggregation: Studies in humans and in atherosclerotic apolipoprotein E-deficient mice. *Am. J. Clin. Nutr.* 2000, 71, 1062-76.

Barton, M., Carmona, R., Morawietz, H., d'Uscio, L.V., et al. Obesity is associated with tissue-specific activation of renal angiotensin-converting enzyme in vivo: Evidence for a regulatory role of endothelin. *Hypertension* 2000, 35, 329-36.

Barwicz, J., Gruda, I., Tancrede, P. A kinetic study of the oxidation effects of amphotericin B on human low-density lipoproteins. *FEBS Letters* 2000, 465, 83-6.

Baynes, J. W. Role of oxidative stress in the development of complications in diabetes. *Diabetes* 1991, 40, 405-12.

Berecochea-Lopez, A., Décordé, K., Ventura, E., Godard, M., Bornet, A., Teissèdre, P.L., Cristol, J.P., Rouanet, J.M. Fungal chitin-glucan from Aspergillus niger efficiently reduces aortic fatty streak accumulation in the high-fat fed hamster, an animal model of nutritionally induced atherosclerosis, *J. Agric. Food Chem.* 2009, 57, 1093-8.

Bhat, V.B., Madyastha, K.M. C-phycocyanin: a potent peroxyl radical scavenger in vivo and in vitro. *Biochem. Biophys. Res. Commun.* 2000, 275, 20-5.

Bhat, V.B., Madyastha, KM. Scavenging of peroxynitrite by phycocyanin and phycocyanobilin from Spirulina platensis: protection against oxidative damage to DNA. *Biochem. Biophys. Res. Commun.* 2001, 286, 262-6.

Braddon, F.E., Rodgers, B., Wadsworth, M.E. Onset of obesity in a 36 year birth cohort study. *Br. Med. J.* 1986, 293, 299-303.

Breinholt, V., Lauridsen, S.T., Dragsted, L.O. Differential effects of dietary flavonoids on drug metabolizing and antioxidant enzymes in female rat. *Xenobiotica* 1999, 29, 1227-40.

Carmiel-Haggai, M., Cederbaum, A.I., Nieto, N. A high-fat diet leads to the progression of non-alcoholic fatty liver disease in obese rats. *FASEB J.* 2005, 19, 136-8.

Cases, J., Vacchina, V., Napolitano, A., Caporiccio, B., Besançon, P., Lobinski, R., Rouanet, J.M. Selenium from selenium-rich Spirulina is less bioavailable than selenium from sodium selenite and selenomethionine in selenium-deficient rats. *J. Nutr.* 2001, 131, 2343-50.

Cases, J., Wysocka, I.A., Caporiccio, B., Jouy, N., Szpunar, J., Rouanet, J.M. Assessment of selenium bioavailability from high-selenium Spirulina subfractions in selenium-deficient rats. *J. Agric. Food Chem.* 2002, 50, 3867-73.

Chen, C., Zhang, P., Hou, X., Chai, Z. Subcellular distribution of selenium and Se-containing proteins in human liver. *Biochim. Biophys. Acta* 1999, 1427, 205-15.

Chen, F., Zhang, Y.M., Guo, S.Y. Growth and phycocyanin formation of Spirulina platensis in photoheterotrophic culture. *Biotechnol. Lett.* 1996, 18, 603-8.

Chen, H., Zhang, M., Xie, B. Components and antioxidant activity of polysaccharide conjugate from green tea. *Food Chem.* 2005, 90, 17-21.

Corder, R., Mullen, W., Khan, N. Q., Marks, S. C., et al. Red wine procyanidins and vascular healt. *Nature* 2006, 444, 566-70.

Dauchet, L., Amouyel, P., Hercberg, S., Dallongeville, J. Fruit and vegetable consumption and risk of coronary heart disease: a meta-analysis of cohort studies. *J. Nutr.* 2006, 136, 2588-93.

Décordé, K., Teissèdre, P.L., Auger, C., Cristol, J.P., Rouanet, J.M. Phenolics from purple grape, apple, purple grapejuice and apple juice prevent early atherosclerosis induced by an atherogenic diet in hamsters. *Mol. Nutr. Food Res.* 2008, 52, 400-7.

Décordé, K., Teissèdre, P.L., Sutra, T., Ventura, E., Cristol, J.P., Rouanet, J.M. Chardonnay grape seed procyanidin extract supplementation prevents high-fat diet-induced obesity in hamsters by improving adipokine imbalance and oxidative stress markers. *Mol. Nutr. Food Res.* 2009 a, 53, 659-66.

Décordé, K., Agne, A., Lacan, D., Ramos, J., Fouret, G., Ventura, E., Feillet-Coudray, C., Cristol, J.P., Rouanet, J.M. Preventive effect of a melon extract rich in superoxide scavenging activity on abdominal and liver fat and adipokine imbalance in high-fat-fed hamsters. *J. Agric. Food Chem.* 2009 b, 57, 6461-7.

Décordé, K., Ventura, E., Lacan, D., Ramos, J., Cristol, J.P., Rouanet, J.M. A SOD rich melon extract Extramel® prevents aortic lipids and liver steatosis in diet-induced model of atherosclerosis. *Nutr. Metabol. Cardiovasc. Dis.* 2010, 20, 301-7.

Dini, L., Falasca, L., Rossi, L., Rotilio, G. In vivo uptake of Cu,Zn superoxide dismutase. Morphological evidence for preferential endocytosis and accumulation by sinusoidal liver cells. *Cell Mol. Biol.* 1996, 42, 269-77.

Dong, F., Zhang, X., Renx, J. Leptin regulates cardiomyocyte contractile function through endothelin-1 receptor-NADPHoxidase pathway. *Hypertension* 2006, 47, 222-9.

Duan, X. J., Zhang, W. W., Li, X. M., Wang, B. G. Evaluation of antioxidant property of extract and fractions obtained from a red alga, Polysiphonia urceolata. *Food Chem.* 2006, 95, 37-43.

Esposito, K., Pontillo, A., Di Palo, C., Giugliano, G. et al. Effect of weight loss and lifestyle changes on vascular inflammatory markers in obese women: a randomized trial. *JAMA* 2003, 282, 1799-1804.

Fan, L., Zhang, S., Yu, L., Ma, L. Evaluation of antioxidant property and quality of breads containing Auricularia auricula polysaccharide flour. *Food Chem.* 2007, 101, 1158-63.

Fecondo, J.V., Augusteyn, R.C. Superoxide dismutase, catalase and glutathione peroxidase in the human cataractous lens. *Exp. Eye Res.* 1983, 36, 15-23.

Fortuno, A., San Jose, G., Moreno, M. U., Beloqui, O., et al. Phagocytic NADPH oxidase overactivity underlies oxidative stress in metabolic syndrome. *Diabetes* 2006, 55, 209-15.

Frei, B. Reactive oxygen species and antioxidant vitamins: Mechanisms of action. *Am. J. Med.* 1994, 97, 5S-13S. 22S-28S.

Fujita, K., Nishizawa, H., Funahashi, T., Shimomura, I., Shimabukuro, M. Systemic oxidative stress is associated with visceral fat accumulation and the metabolic syndrome. *Circ. J.* 2006, 70, 1437-42.

Furukawa, S., Fujita, T., Shimabukuro, M., Iwaki, M., et al. Increased oxidative stress in obesity and its impact on metabolic syndrome. *J. Clin. Invest.* 2004, 114, 1752-61.

Gallistl, S., Sudi, K., Mangge, H., Erwa, W., Borkenstein, M. Insulin is an independant correlate of plasma homocysteinelevels in obese children and adolescents. *Diabetes Care* 2000, 23, 1348-52.

Georgescu, A., Alexandru, M., Constantinescu, E., Popo, B. Effect of gap junction uncoupler heptanol on resistance arteries reactivity in experimental model of diabetes, hyperlipemia and hyperlipemia-diabetes. *Vasc. Pharmacol.* 2006, 44, 513-8.

Godard, M., Décordé, K., Ventura, E., Soteras, G., Baccou, J.C., Cristol, J.C., Rouanet, J.M. Polysaccharides from the green alga Ulva rigida improve the antioxidant status and prevent fatty streak lesions in the high cholesterol fed hamster, an animal model of nutritionally-induced atherosclerosis. *Food Chem.* 2009, 115, 176-80.

Hansen, P.A., Han, D.H., Nolte, L.A., Chen, M., Holloszy, J.O. DHEA protects against visceral obesity and muscle insulin resistance in rats fed a high-fat diet. *Am. J. Physiol.* 1997, 273, 1704-8.

Haslam, D.W., James, W.P.T. Obesity. *Lancet* 2005, 366, 1197-1209.

Huang, Z., Guo, B.J., Wong, R.N.S., Jiang, Y. Characterization and antioxidant activity of selenium-containing phycocyanin isolated from Spirulina platensis. *Food Chem.* 2007, 100, 1137-43.

Imlay, J.A., Linn, S. DNA damage and oxygen radical toxicity. *Science* 1998, 240, 1302-9.

Inoguchi, T., Li, P., Umeda, F., Yu, H. Y., et al. High glucose level and free fatty acid stimulate reactive oxygen species production through protein kinase C-dependent activation of NAD(P)H oxidase in cultured vascular cells. *Diabetes* 2000, 49, 1939-45.

Ip, C., Lisk, D.J., Ganther, H.E. Chemoprevention with triphenylselenonium chloride in selenium-deficient rats. *Anticancer Res.* 2000, 20, 4179-82.

Iwata, K., Inayama, T., Kato, T. Effects of *Spirulina platensis* on plasma lipoprotein lipase activity in fructose-induced hyperlipidemic rats. *J. Nutr. Sci. Vitaminol.* 1990, 36, 165-71.

Jiang, Y. H., Jiang, X. L., Wang, P., Hu, X. K. In vitro antioxidant activities of water-soluble polysaccharides extracted from Isaria farinosa B05. *J. Food Biochem.* 2005, 29, 323-35.

Jubeh, T.T., Antier, S., Haupt, S., Barenholz, Y., Rubinstein, A. Local prevention of oxidative stress in the intestinal epithelium of the rat by adhesive liposomes of superoxide dismutase and tempamine. *Mol. Pharm.* 2005, 2, 2-11.

Kamil, Y.V.A.J., Jeon, Y.J., Shahidi, F. Antioxidative activity of chitosans of different viscosity in cooked comminuted flesh of herring (Clupea harengus). *Food Chem.* 2002, 79, 69-77.

Katagiri, K., Arakawa, S., Kurahashi, R., Hatano, Y. Impaired contact hypersensitivity in diet-induced obese mice. *J. Dermatol. Sci.* 2007, 46, 117-26.

Keaney, J.F., Larson, M.G., Vasan, R.S., Wilson, P.W.F. et al. Obesity and systemic oxidative stress : clinical correlates of oxidative stress in the Framingham Study. *Arterioscler. Thromb. Vasc. Biol.* 2003, 23, 434-9.

Kick, J., Hauser, B., Bracht, H., Albicini, M., Öter, S., Simon, F., et al. Effects of a cantaloupe melon extract/wheat gliadin biopolymer during aortic cross-clamping. *Int. Care Med.* 2007, 33, 694-702.

Kim, K.W., Thomas, R.L. Antioxidative activity of chitosans with varying molecular weights. *Food Chem.* 2007, 101, 308-3.

Klaus, S., Pultz, S., Thone-Reineke, C., Wolfram, S. Epigallocatechin gallate attenuates diet-induced obesity in mice by decreasing energy absorption and increasing fat oxidation. *Int. J. Obesity* 2005, 29, 615-23.

Kowala, M. C., Nunnari, J. J., Durham, S. K., Nicolosi, R. J. Doxazosin and cholestyramine similarly decrease fatty streak formation in the aortic arch of hyperlipidemic hamsters. *Atherosclerosis* 1991, 91, 35-49.

Krieger-Brauer, H. I., Kather, H. Human fat cells possess a plasma membrane-bound H_2O_2-generating system that is activated by insulin via a mechanism bypassing the receptor kinase, *J. Clin. Invest.* 1992, 89, 1006-13.

Lacan, D., Baccou, J. C. High levels of antioxidant enzymes correlate with delayed senescence in nonnetted muskmelon fruits. *Planta* 1998, 204, 377-82.

Laurent, L., Nicco, C., Tran Van Nhieu, J., Borderie, D., Chéreau, C., Conti, F., et al. Pivotal role of superoxide anion and benefical effect of antioxidant molecules in murine steatohepatitis. *Hepatology* 2004, 39, 1277-85.

Leibovitz, B., Hu, M.L., Tappel, A.L. Dietary supplements of vitamin E, □-carotene, coenzyme Q10 and selenium protect tissues against lipid peroxidation in rat tissue slices. *J. Nutr.* 1990, 120, 97-104.

Leung, N., Naples, N., Uffelman, K., Szeto, L. et al. Rosglitazone improves intestinal lipoprotein overproduction in tha fat-fed Syrian Golden hamster, an animal model of nutritionally-induced insulin resistance. *Atherosclerosis* 2004, 174, 235-41.

Levy, R., Glozman, S., Milman, D., Seruty, C., Hagay, Z., Yavin, E., et al. Ischemic reperfusion brain injury in fetal transgenic mice with elevated levels of copperezinc superoxide dismutase. *J. Perinat. Med.* 2002, 30, 158-65.

Lin, Y.L., Cheng, C.Y., Lin, Y.P., Lau, Y.W., Juan, I.M., Lin, J.K. Hypolipidemic effect of green tea leaves through induction of antioxidant andPhase II enzymes including superoxide dismutase, catalase, and glutathione S transferase in rats. *J. Agric. Food Chem.* 1998, 46, 1893-9.

Luft, R., Landau, B. R. Mitochondrial medicine. *J. Intern. Med.* 1995, 238, 405-21.

Maeda, K., Yasunari, K., Sato, E. F., Inoue, M. Enhanced oxidative stress in neutrophils from hyperlipidemic guinea pig. *Atherosclerosis* 2005, 181, 87-99.

Maggy-Capeyron, M.F., Cases, J., Badia, E., Cristol, J.P., et al. A diet high in cholesterol and deficient in vitamin E induces lipid peroxidation but does not enhance antioxidant enzyme expression in rat liver. *J. Nutr. Biochem.* 2002, 13, 293-301.

McCord, J.M., Edeas, M.A. SOD, oxidative stress and human pathologies: a brief history and a future vision. *Biomed. Pharmacother.* 2005, 59, 139-42.

Miller, S., Walker, S.W., Arthur, J.R., Nicol, F., Pickard, K., Lewin, M.H., Howie, A.F., Beckett, G.J. Selenite protects human endothelial cells from oxidative damage and induces thioredoxin reductase. *Clin Sci.* 2001, 100, 543-50.

Mokdad, A.H., Ford, E.S., Bowman, B.A., Dietz, W.H. et al. Prevalence of obesity, diabetes, and obesity-related health risk factors, 2001. *JAMA* 2003, 289, 76-9.

Morinobu,T., Murata, T., Takaya, R., Tamai, H. Nutritional status of β-carotene, α-tocopherol and retinol in obese children. *Int. J. Vitam. Nutr. Res.* 2002, 72, 119-23.

Murase, T., Nagasawa, A., Suzuki, J., Hase, T., Tokimitsu, I. Beneficial effects of tea catechins on diet-induced obesity: stimulation of lipid catabolism in the liver. *Int. J. Obesity* 2002, 26, 1459-64.

Muth, C.M., Glenz, Y., Klaus, M., Radermacher, P., Speit, G., Leverve, X., et al. Influence of an orally effective SOD onhyperbaric oxygen-related cell damage. *Free Radic. Res.* 2004, 38, 927-32.

Nagao, T., Hase, T., Tokimitsu, I.A green tea extract high in catechins reduces body fat and cardiovascular risks in humans. *Obesity* 2007, 15, 1473-83.

Nagaoka, S., Shimizu, K., Kaneko, H., Shibayama, F., Morikawa, K., Kanamaru, Y., Otsuka, A., Hirahashi, T., Kato, T. A novel protein C-Phycocyanin plays a crucial role in the hypocholesterolemic action of *Spirulina platensis* concentrate in rats. *J. Nutr.* 2005, 135, 2425-30.

Nakagawa, K., Kishida, H., Arai, N., Nishiyama, T., Mae, T. Licorice flavonoids suppress abdominal fat accumulation and increase in blood glucose level in obese diabetic KK-A^y mice. *Biol. Pharm. Bull.* 2004, 27, 1775-8.

Nakamura, T., Tokunaga, K., Shimomura, I., Nishida. M. et al. Contribution of visceral fat accumulation to the development of coronary artery disease in non-obese men, *Atherosclerosis* 1994, 107, 239-46.

Nakaya, N., Homma, Y., Goto, Y. Cholesterol lowering effect of spirulina. *Nutr. Rep. Int.* 1988, 37, 1329-37.

Nelson, S.K., Bose, S.K., Grunwald, G.K., Myhill, P., McCord, J.M. The induction of human superoxide dismutase and catalase in vivo: a fundamentally new approach to antioxidant therapy. *Free Radic. Biol. Med.* 2006, 40, 341-7.

Neuhouser, M.L., Rock, C.L., Eldridge, A.L., Kristal, A.R., Patterson, R.E., et al. Serum concentrations of retinol, α -tocopherol and the carotenoids are influenced by diet, race and obesity in a sample of healthy adolscents. *J. Nutr.* 2001, 131, 2184-91.

Ng, T. B., He, J. S., Niu, S. M., Zhao, L., Pi, Z. F., Shao, W., et al.. A gallic acid derivative and polysaccharides with antioxidative activity from rose (Rosa rugosa) flowers. *J. Pharm.Pharmacol.* 2004, 56, 537-45.

Nicolle, C., Cardinaut, M., Aprikian, O., Busserolle, J., Grolier, P., Rock, E., Demigné, C., Mazur, A., Sclabert, A., Amouroux, P., Rémézy, C. Effect of a carrot intake on cholesterol metabolism and on anti-oxidant status in cholesterol-fed rat. *Eur. J. Nutr.* 2003, 42, 254-61.

Nicolle, C., Cardinaut, M., Gueux, E., Jaffrelo, L., Rock, E., Mazur, A., Amouroux, P., Rémézy, C. Health effect of vegetablebased diet : lettuce consumption improves cholesterol metabolism and antioxidant status in the rat. *Clin. Nutr.* 2004, 23, 605-14.

Nistor, A., Bulla, A., Filip, D.A., Radu, A. The hyperlipidemic hamster as a model of experimental atherosclerosis. *Atherosclerosis* 1987, 68, 159-73.

Ohashi, K., Iwatani, H., Kihara, S., Nakagawa, Y., et al. Exacerbation of albuminuria and renal fibrosis in subtotal renal ablation model of adiponectin-knockout mice. *Arterioscler. Thromb. Vasc. Biol.* 2007, 27, 1910-7.

Okamoto, Y., Kihara, S., Ouchi, N., Nishida, M., et al. Adiponectin reduces atherosclerosis in apolipoprotein E-deficient mice. *Circulation* 2002, 106, 2767-70.

Olson, O.E. , Palmer, S. Selenium in food consumed by South Dakotans. *Proc. S. D. Acad Sci.* 1978, 57, 113-21.

Olusi, S.O. Obesity is an independant risk factor for plasma lipid peroxidation and depletion of erythrocyte cytoprotectic enzymes in humans. *Int. J. Obes. Relat. Metab. Disord.* 2002, 26, 1159-64.

Omar, B.A., Gad, N.M., Jordan, M.C., Striplin, S.P., Russell, W.J., Downey, J.M., et al. Cardioprotection by Cu, Zn-superoxide dismutase is lost at high doses in the reoxygenated heart. *Free Radic. Biol. Med.* 1990,9, 465-71.

Ozata, M., Mergen, M., Oktenli, C., Aydin, A., et al. Increased oxidative stress and hypozincemia in male obesity. *Clin. Biochem.* 2002, 35, 627-31.

Pajk, T., Rezar, V., Levart, A., Salobir, J. Efficiency of apples, strawberries, and tomatoes for reduction of oxidative stress in pigs as a model for humans. *Nutrition* 2006, 22, 376–384.

Pinero Estrada, J.E., Bermejo Bescos, P., Villar del Fresno, A.M. Antioxidant activity of different fractions of Spirulina platensis protean extract. *Il Farmaco* 2001, 56, 497-500.

Prieur, C., Rigaud, J., Cheynier, V., Moutounet, M. Oligomeric and polymeric procyanidins from grape seeds. *Phytochemistry* 1994, 36, 781-4.

Qi, H., Zhang, Q., Zhao, T., Hu, R., Zhangc, K., Li, Z. In vitro antioxidant activity of acetylated and benzoylated derivatives of polysaccharide extracted from Ulva pertusa (Chlorophyta). *Bioorg. Med. Chem. Lett.* 2006, 16, 2441-5.

Quehenberger, P., Exner, M., Sunder-Plassmann, R., Ruzicka, K., et al. Leptin induces endothelin-1 in endothelial cells in vitro. *Circ. Res.* 2002, 90, 711-8.

Reddy, C.M., Bhat, V.B., Kiranmai, G., Reddy, M.N., Reddanna, P., Madyastha, K.M. Selective inhibition of cyclooxygenase-2 by C-phycocyanin, a biliprotein from Spirulina platensis. *Biochem. Biophys. Res. Commun.* 2000, 277, 599-603.

Regnault, C., Soursac, M., Roch-Arveiller, M., Postaire, E., Hazebroucq, G. Pharmacokinetics of superoxide dismutase in rats after oral administration. *Biopharm. Drug Dispos.* 1996, 17, 165-74.

Rice-Evans, C., Leake, D., Bruckdorfer, K. R., & Diplock, A. T. Practical approaches to low density lipoprotein oxidation: Whys, wherefores and pitfalls. *Free Radic. Res.* 1996, 25, 285-311.

Riss, J., Décordé. K., Sutra, T., Delage, M., Baccou, J.C., Jouy, N., Brune, J.P., Oréal, H., Cristol, J.P., Rouanet, J.M. The phycobiliprotein C-Phycocyanin from *Spirulina platensis* is powerfully responsible for reducing oxidative stress and NADPH oxidase expression induced by an atherogenic diet in hamsters. *J. Agric. Food Chem.* 2007, 55, 7962-7.

Rosenblat, M., Hayek, T., Aviram, M. Anti-oxidative effects of pomegranate juice (PJ) consumption by diabetic patients on serum and onmacrophages. *Atherosclerosis* 2006, 187, 363-71.

Rouanet, J.M., Décordé, K., Del Rio, D., Auger, C., Borges, G., Cristol, J.P., Lean, M.E.J., Crozier, A. Berry juices, teas, antioxidants and the prevention of atherosclerosis in hamsters. *Food Chem.* 2010, 118, 266-71.

Saito, M., Ueno, M., Ogino, S., Kubo, K. et al. High dose of Garcinia cambogia is effective in suppressing fat accumulation in developing male Zucker obese rats, but highly toxic to the testis. *Food. Chem. Toxicol.* 2005, 43, 411-9.

Salonen, J.T., Alfthan, G., Huttunen, J.K., Puska, P. Association between selenium status and the risk of cancer. *Am. J. Epidemiol.* 1984, 120, 342-9.

Sanchez-Margalet, V., Valle, M., Ruz, FJ., Gascon, F., Mateo, J., et al. Elevated plasma total homocysteine levels in hyperinsulinemic obese subjects. *J. Nutr. Biochem.* 2002, 13, 75-9.

Schrauzer, G.N., White, D.A., Schneider, C.J. Cancer mortality correlation studies. III. Statistical associations with dietary selenium intakes. *Bioinorg. Chem.* 1977, 7, 23-34.

Shamberger, R.J., Willis, C.E., McCormack, L.J. Selenium and heart disease. III. Blood selenium and heart mortality in 19 states. In: Trace Substances in Environmental Health (Hemphill, D. D., ed.), 1978 ; vol. 12, pp. 59-63. University of Missouri Press, Columbia, MO.

Simon, E., Macarulla, M.T., Churruca, I., Fernandez-Quintela, A. et al. Trans-10, cis-12 conjugated linoleic acid prevents adiposity but not insulin resistance induced by an atherogenic diet in hamsters. *J. Nutr. Biochem.* 2006, 17, 126-31.

Sorescu, D., Griendling, K.K. Reactive oxygen species, mitochondria, and NAD(P)H oxidases in the development and progression of heart failure. *Congest. Heart Fail.* 2002, 8, 132-40.

Stocker, R., McDonagh, A.F., Glazer, A.N., Ames, B.N. Antioxidant activities of bile pigments: biliverdin and bilirubin. *Methods Enzymol.* 1990, 186, 301-9.

Stoewsand, G.S., Anderson, J.L., Munson, L., Lisk, D.J. Effect of diatery Brussels sprouts with increased selenium content on mammary carcinogenesis in the rat. *Cancer Lett.* 1989, 45, 43-8.

Sugiyama, H., Akazome, Y., Shoji, T., Yaamaguchi, A. et al. Oligomeric procyanidins in apple polyphenol are main active components for inhibition of pancreatic lipase and triglyceride absorption. *J. Agric. Food Chem.* 2007, 55, 4604-9.

Suh, J.H., Romain, C., Gonzalez-Barrio, R., Cristol, J.P., Teissèdre, P.L., Crozier, A., Rouanet, J.M. Raspberry juice consumption, oxidative stress and reduction of atherosclerosis risk factors in hypercholesterolemic golden Syrian hamsters. *Food Funct.* 2011, 2, 400-5.

Sutra, T., Décordé, K., Riss, J., Dallas, C., Cristol, J.P., Rouanet, J.M. A commercial extract of fruits and vegetables, oxxynea, acts as a powerful antiatherosclerotic supplement in an animal model by reducing cholesterolemia, oxidative stress, and NADPH oxidase expression. *J. Agric. Food Chem.* 2007, 55, 4258-63.

Tao, L., Gao, E., Jiao, X., Yuan, Y., et al. Adiponectin cardioprotection after myocardial ischemia/reperfusion involves the reduction of oxidative/nitrative stress. *Circulation* 2007, 115, 1408-16.

Tarp, U., Overvad, K., Hansen, J.C, Thorling, E.B. Low selenium level in severe rheumatoid arthritis. *Scand. J. Rheumatol.* 1985, 14, 97-101.

Vadillo, M., Ardevol, A., Fernàndez-Larrea, J., Pujadas, G. et al. Moderate red-wine consumption partially prevents body weight gain in rats fed a hyperlipidic diet. *J. Nutr. Biochem.* 2006, 17, 139-42.

Van Vleet, J.F. Current knowledge of selenium-vitamin E deficiency in domestic animals. *J. Am. Vet . Med. Assoc.* 1980, 176, 321-32.

Vinson, J. A., Teufel, K., Wu, N. Red wine, dealcoholized red wine, and espacially grape juice, inhibit atherosclerosis in a hamster model. *Atherosclerosis* 2001, 156, 67-72.

Vouldoukis, I., Conti, M., Krauss, P., Kamate, C., Blazquez, S., Tefit, M., et al. Supplementation with gliadin-combined plant superoxide dismutase extract promotes antioxidant defences and protects against oxidative stress. *Phytother. Res.* 2004, 18, 957-62.

Vouldoukis, I., Lacan, D., Kamate, C., Coste, P., Calenda, A., Mazier, D., et al. Antioxidant and anti-inflammatory properties of a Cucumis melo LC extract rich in superoxide dismutase activity. *J. Ethnopharmacol.* 2004, 94, 67-75.

Wagner, A. H., Kohler, T., Ruckschloss, U., Just, I., Hecker, M. Improvement of nitric oxide-dependent vasodilatation by HMGCoA reductase inhibitor through attenuation of endothelial superoxide anion formation. *Arterioscler. Thromb. Vasc. Biol.* 2000, 20, 61-9.

Wallace, D.C. Mitochondrial diseases in man and mouse. *Science* 1999, 283, 1482-8.

Wallstrom, P., Wirfalt, E., Lahmann, P.H., Gullberg, B., Janzon, L., et al. Serum concentrations of beta-carotene and alpha-tocopherol are associated with diet, smoking, and general and central adiposity. *Am. J. Clin. Nutr.* 2001, 73, 777-85.

Wang, Z., Luo, D. Antioxidant activities of different fractions of polysaccharide from Gynostemma pentaphyllum Makino. *Carbohyd. Polym.* 2007, 68, 54-8.

Webb, C.B., Lehman, T.L., McCord, K.W. Effects of an oral superoxide dismutase enzyme supplementation on indices of oxidative stress, proviral load, and CD4:CD8 ratios in asymptomatic FIV-infected cats. *J. Feline Med. Surg.* 2008, 10, 423-30.

Wolfram, S., Raederstorff, D., Wang, Y., Teixera, S.R. et al. TEAVIGO (Epigallocatechin Gallate) supplementation prevents obesity in rodents by reducing adipose tissue mass. *Ann. Nutr. Metab.* 2005, 49, 54-63.

Xu, F.P., Chen, M.S., Wang, Y.Z., Yi, Q., et al. Leptin induces hypertrophy via endothelin-1-reactive oxygen species pathway in cultured neonatal rat cardiomyocytes. *Circulation* 2004, 110, 1269-75.

Yamauchi, T., Kamon, J., Waki, H., Terauchi, Y., et al. The fat derived hormone adiponectin reverses insulin resistance associated with both lipoatrophy and obesity. *Nat. Med.* 2001, 7, 941-46.

Yang, B., Wang, J., Zhao, M., Liu, Y., Wang, W., Jiang, Y. Identification of polysaccharides from pericarp tissues of litchi (Litchi chinensis Sonn.) fruit in relation to their antioxidant activities. *Carbohyd. Res.* 2006, 341, 634-8.

Youn, S.K., Kim, Y.J., Ahn, D.H. Antioxidative effects of chitosan in meat sausage. *J. Korean Soc. Food Sci. Nutr.* 2001, 30, 477-81.

Young, J.F., Nielsen, S.E., Haraldsdottir, J., Daneshvar, B., Lauridsen, S. T., Knuthsen, P., et al. Effect of fruit juice intake on urinary quercetin excretion and biomarkers of antioxidative status. *Am. J. Clin. Nutr.* 1999, 69, 87-94.

17

Effects of NK-4, a Cyanine Dye with Antioxidant Activities: Attenuation of Neuronal Deficits in Animal Models of Oxidative Stress-Mediated Brain Ischemia and Neurodegenerative Diseases

Hitomi Ohta, Kenji Akita and Tsunetaka Ohta
R&D Center, HAYASHIBARA Co., Ltd.,
Japan

1. Introduction

Cyanine photosensitizing dyes have been studied for over 150 years and continue to be of interest in their biology and medicine. They have been shown to possess various biological activities, including antimicrobial, antioxidant, macrophage activating, and oxidative phosphorylation uncoupling activities (Hayami, 1984; Ushio et al., 2009; Ishihara & Fujisawa, 2007; Kunikata et al., 2002; Takeguchi et al., 1985). Some dyes have been used as immunomodulators to treat allergy and rheumatoid arthritis, cancer, and to promote wound healing (Trauner et al., 1998; Motoyoshi et al., 1991). Since cumulative evidence suggests an involvement of oxidative stress and neuroinflammation as the common feature in the pathogenesis of neurodegenerative disorders (Amor et al., 2010; Wolozin & Behl, 2000a, 2000b), it would be reasonable to expect that cyanine dyes with anti-oxidative and anti-inflammatory properties could protect the central nervous system from neuroinflammatory-related brain insults.

Recently, we screened more than 250 cyanine dyes for their neurotrophin-like activity and found that NK-4 and some other related compounds are potent neurotrophic agents for the promotion of growth and differentiation of neuronal rat adrenal pheochromocytoma cell line PC12. NK-4 is a divalent cationic pentamethine trinuclear cyanine dye that contains three quinolinium rings, N-alkyl side chains, and two iodine anions (Fig. 1A). Addition of NK-4 into the culture at nanomolar concentrations significantly augmented cell growth of PC12 cells in 3-day cultures (Fig. 1B). NK-4 also promoted nerve growth factor (NGF)-primed neurite-outgrowth at micromolar and submicromolar concentrations (Fig. 1C). Since the intervention using neurotrophic or neuroprotective small molecules is thought to have potential for treating neurodegenerative disorders, we investigated the neuroprotective effects of NK-4 against neurotoxic insults *in vitro* and further evaluated its pharmacological effects using animal models of neurodegeneration, including ischemic stroke, cerebellar ataxia, and Alzheimer's disease (AD).

In this chapter, we focus on the neuroprotective effects of NK-4 against oxidative damage *in vitro* and *in vivo*. We introduce the multipotent properties of NK-4, which may act in concert to attenuate the common pathological pathways of neurodegeneration, and discuss the potential use of NK-4 for neurodegenerative disease therapy.

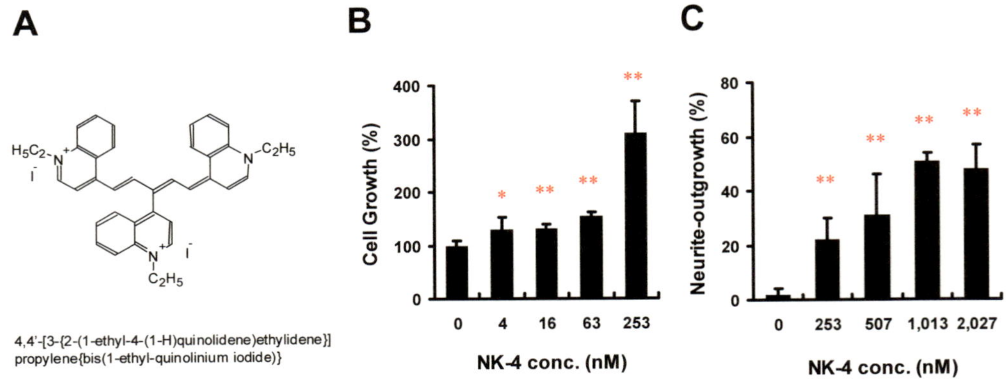

(A) Chemical structure of NK-4. It was synthesized in HAYASHIBARA Co., Ltd. (B) The proliferative effect of NK-4 on PC12 cells. Cell growth was assessed by AlamarBlue® assay in 3-day cultures. (C) The promotive effect of NK-4 on NGF (5 ng/ml)-primed neurite-outgrowth in PC12 cells. Neurite-outgrowth was evaluated by counting neurite-outgrowth positive cells in 3-day cultures. The cells were defined as positive when the length of the longest neurite was >2-fold longer than that of the cell body. Data are means ± SD (n=3). *P<0.05 and ** P<0.01 vs. control (no NK-4).

Fig. 1. Chemical structure and neurotrophin-like effects of NK-4.

2. *In Vitro* Properties of NK-4

2.1 Free radical-scavenging activity

As it has been known that some cyanine dyes show significant antioxidative property due to the extended π-electron conjugated system contained within the structure (Ishihara & Fujisawa, 2007), the free radical-scavenging capacity of NK-4 for superoxides, hydroxyl, and peroxy radicals was evaluated (Koya-Miyata et al., 2010). These radicals are produced *in vivo* in many experimental models of ischemia and reperfusion, and they are therefore generally regarded as the primary free radicals involved in oxidative stress-mediated injury. Half-maximal inhibitory concentrations (IC_{50}) of NK-4 against these radicals are shown in Table 1.

Ascorbate, a well-characterized free radical-scavenging agent, and edaravone were used as positive controls. Edaravone, also known as MCI-186, is a hydroxy radical scavenger, and has been used for acute phase stroke therapy in Japan (Abe & Kogure, 1988). NK-4 displayed a direct and powerful hydroxyl radical-scavenging activity, significantly greater than those of ascorbate and edaravone. The scavenging activity for peroxy radicals was also potent in NK-4 and to a lesser extent in ascorbate. NK-4 and ascorbate acted on superoxides to a comparable extent; however, edaravone did not act on superoxides as previously reported (Tanaka, 2002). The highly reactive hydroxyl radical oxidizes cellular lipids, proteins, and DNA, leading to cell death, and is therefore highly detrimental (Gilgun-Sherki et al., 2002). Accordingly, the antioxidants that effectively scavenge hydroxyl radicals and other free radicals should be able to eliminate oxidative injury (Barinaga, 1996), and NK-4 could be one of such molecules.

Compound	IC_{50} (μM)		
	Hydroxyl Radical	Peroxy Radical	Superoxide
NK-4	7.6 ± 0.6	5.2 ± 2.5	107 ± 2.2
Ascorbate	47.6 ± 0.2	30.5 ± 1.7	130 ± 24
Edaravone	779 ± 18	181 ± 19	> 3,800

Free radical-scavenging efficacy was evaluated in a cell-free *in vitro* system using ESR. Superoxides, hydroxyl radicals, and peroxy radicals were produced and stabilized in DMPO-hypoxanthine/XOD, DMPO-DTPA/H_2O_2/$FeSO_4$ and DMPO-AAPH systems, respectively. The signal intensity of each radical was recorded in the presence or absence of test samples. The scavenging activity of the samples was expressed as IC_{50} against each radical. Means ± SD (n=3). Reproduced in part with permission from Biological & Pharmaceutical Bulletin Vol.33 No.11. Copyright [2010] Pharmaceutical Society of Japan.

Table 1. Free radical-scavenging effects of NK-4 and other antioxidants.

2.2 Neuroprotective effects against various cytotoxic stresses

Neurotrophins, such as NGF or brain-derived neurotrophic factor (BDNF) regulate the growth, survival, and differentiation of central neurons (Blum & Konnerth, 2005). For example, NGF protects PC12 cells against 6-hydroxydopamine (6-OHDA)- and hydrogen peroxide (H_2O_2)-induced oxidative stress (Salinas et al., 2003; Wang et al., 2001). Since NK-4 displayed a remarkable neurotrophin-like activity, namely the promotion of cell growth and NGF-primed neurite-outgrowth in PC12 cells (Fig. 1B, 1C), we next examined whether NK-4 protects neuronal cells from oxidative or starvation stress (Ohta et al., 2011). An acute oxidative stress challenge was induced by 2-hr treatment with H_2O_2 or 24-hr treatment with 6-OHDA. These treatments cause a significant decrease in metabolic capacity, which reflects a decline in cell viability, and this is accompanied by an enhancement of apoptotic markers (Franco et al., 2010). Particularly, 6-OHDA has been widely used in experimental models of Parkinsion's disease, and its neurotoxicity involves oxidative damage to catecholaminergic neurons via the generation of hydroxyl radicals, monoamine oxidase-mediated formation of H_2O_2, and mitochondrial generation of superoxide (Blum et al., 2001).

Treatment of PC12 cells with H_2O_2 decreased cell viability to approximately 50% of control. In contrast, NK-4 at concentrations of 6 μM or above significantly protected the cells from the acute oxidative damage (Fig. 2A). NK-4 also attenuated 6-OHDA-induced oxidative stress in PC12 cells at nanomolar concentrations in a dose-dependent manner (Fig. 2B). Next we tested the effect of NK-4 on trophic factor depletion-induced cell death. Extended serum depletion in PC12 cells triggers ATP shortage, which results in mitochondrial reactive oxygen species (ROS) generation and activation of apoptotic pathways (Troy et al., 2001). NK-4 at submicromolar concentrations significantly increased cellular viability of PC12 cells over 3 days of serum starvation (Fig. 2C). These results suggest that the antioxidative effects of NK-4, together with its neurotrophic properties, engender strong survival signals in PC12 cells.

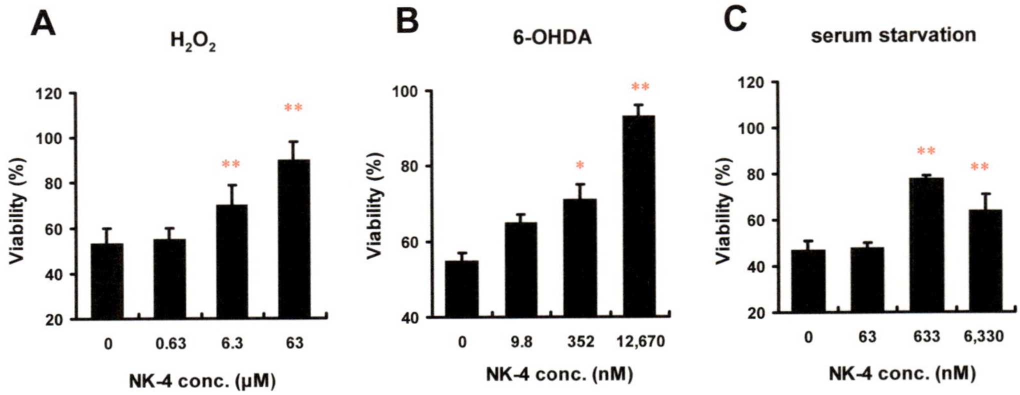

PC12 cells were seeded at a density of 2 x 10^4 cells/well and cultured in serum supplemented D-MEM for 24-hr at 37°C in collagen-coated microplates. (A) H_2O_2 (200 μM) or (B) 6-OHDA (100 μM) was applied in the presence or absence of indicated concentrations of NK-4 and incubated for 2-hr or 24-hr, respectively. (C) After the pre-culture, serum-containing D-MEM was replaced with serum-free medium and the cells were cultured for additional 3 days. Cell viability was assessed by the AlamarBlue® method. Results are shown as means ± SD (n=3). *P<0.05 and **P<0.01 vs. no NK-4.

Fig. 2. Protective effect of NK-4 from cytotoxic stresses in PC12 cells.

2.3 Neuroprotective effect against β-amyloid (Aβ) toxicity

Next, we determined whether NK-4 was effective against β-amyloid (Aβ)-induced neurotoxicity *in vitro* using PC12 cells (Ohta et al., 2010a). PC12 cells are reported to be highly sensitive to Aβ peptide or the aggregated $A\beta_{25\text{-}35}$ fragment (Shearman et al., 1994). The viability of PC12 cells treated with 50 μM of aged $A\beta_{25\text{-}35}$ for 72-hr was about 25% compared with controls without $A\beta_{25\text{-}35}$ (Fig. 3A). NK-4 dose-dependently attenuated the cytotoxic effect of $A\beta_{25\text{-}35}$ and the effects were significant at doses over 60 nM. Aβ peptides induce morphological changes associated with apoptotic cell death, such as somal shrinkage, plasma membrane blebbing, chromatin condensation, and nuclear fragmentation (Ivins et al., 1999). Nuclear staining by Hoechst 333442 dye demonstrated the typical nuclear fragmentation in $A\beta_{25\text{-}35}$-treated PC12 cells and this was clearly inhibited by the addition of NK-4 (Fig. 3B).

Since PC12 cells undergoing Aβ-mediated apoptosis produce large amounts of ROS due to deficits in mitochondrial function (Kadowaki et al., 2005), we next examined ROS generation in $A\beta_{25\text{-}35}$-treated PC12 cells (Fig. 3C). Compared with the control, the amounts of intracellular ROS significantly increased after addition of 50 μM $A\beta_{25\text{-}35}$ in 48-hr cultures, and 300 nM of NK-4 almost totally suppressed the ROS induction. Under normal culture conditions, NK-4 also decreased basal ROS levels in normal PC12 cells; however, the effect was not statistically significant. These data suggest that NK-4 attenuates Aβ-mediated cytotoxicity by reducing the unfavorable ROS accumulation in PC12 cells, probably through the restoration of mitochondrial function.

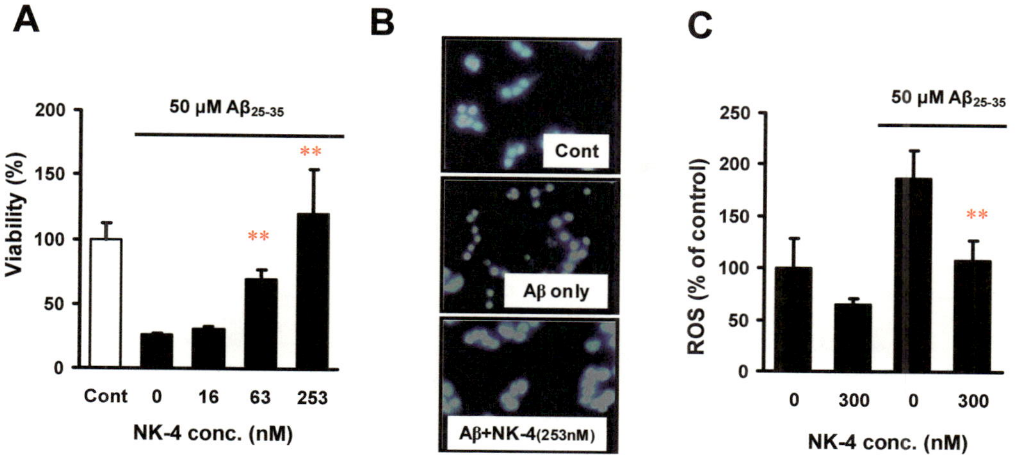

(A) $A\beta_{25\text{-}35}$ was dissolved in saline and incubated at 37 °C for 4 days, which is termed an "aging" process. PC12 cells were treated with 50 μM aged $A\beta_{25\text{-}35}$ for 72-hr in the absence or presence of the indicated concentrations of NK-4. Cell viability was assessed by AlamarBlue® assay. Results are shown as means ± SD (n=3). **P<0.01 vs. no NK-4. (B) PC12 cells were treated as (A) and stained with Hoechst 33342 dye. (C) PC12 cells were treated as (A), but for 48 hr and the amount of intracellularly produced ROS was assessed by H_2DCF-DA dye fluorescence (ex.492nm/em.527nm). Results are shown as means ± SD (n=3). **P<0.01 vs. no NK-4.

Fig. 3. Cytoprotective effects of NK-4 on $A\beta_{25\text{-}35}$-induced cytotoxicity in PC12 cells.

2.4 Intracellular signaling

Neurotrophins act by binding to two kinds of plasma membrane receptors, the Trk receptor tyrosine kinases (Trks) and the p75 pan-neurotrophin receptor ($p75^{NTR}$). There are several subtypes of Trk receptor kinases characterized by their specific affinities for different neurotrophins. NGF binds preferentially to TrkA, whereas BDNF and neutrophin-4/5 show a high affinity for TrkB (Berg et al., 1992). To address whether the effects of NK-4 are mediated by Trk activity, K252a, a non-specific inhibitor of Trks, was applied in growth assays of PC12 cells. Pretreatment with K252a dose-dependently inhibited NGF-induced cell growth, but did not inhibit growth induced by NK-4 (Fig. 4A). This suggests that NK-4 acts independently of TrkA activation in PC12 cells. A similar inhibitory profile of K252a was also found in the neurite-outgrowth of PC12 cells (data not shown).

We next examined whether NK-4 activates phosphatidylinositol 3-kinase (PI3K) and its downstream signaling effector Akt. This cascade is implicated in survival signaling mediated by NGF in serum-deprived PC12 cells, and in neuritogenesis in PC12 cells (Martin et al., 2004; Kim et al., 2004). LY294002, a specific PI3K inhibitor (Vlahos et al., 1994), blocked the cell growth-promoting activity of both NK-4 and NGF (Fig. 4B). Also, as shown in figure 4C, NK-4 strongly induced phosphorylation of Akt at Ser 473 in a time-dependent manner. The dose of NK-4 required for the induction of Akt phosphorylation was consistent with that required for promotion of PC12 cell growth. These results suggest that sequential activation of PI3K and its downstream signaling effector Akt are important for NK-4-induced neurotrophic effects in PC12 cells.

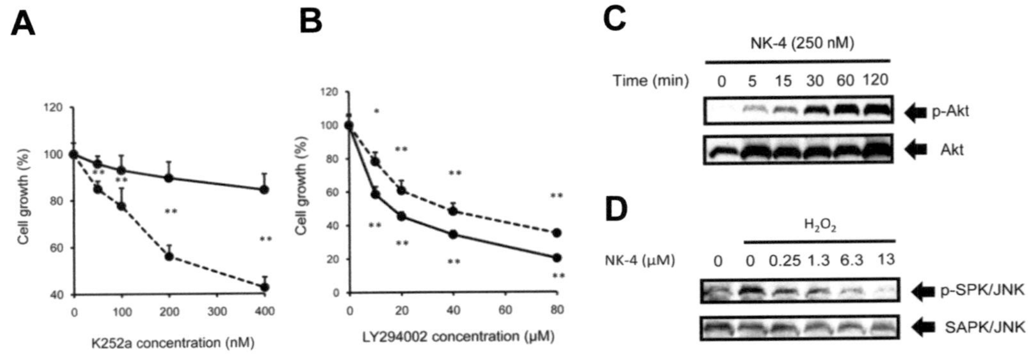

(A, B) Effects of K252a or LY294002 on PC12 cell growth induced by NK-4 (bold line) or NGF (broken line). PC12 cells were preincubated in serum-free D-MEM with the indicated concentrations of K252a (A) or LY294002 (B) for 15 min. NK-4 (250 nM) or NGF (50 ng/ml) was added and the cells were further incubated for 72-hr. Cell viabilities were assessed by AlamarBlue® assay. Values are means ± SD (n=3). $^{*}P<0.05$, $^{**}P<0.01$ vs. control. (C) PC12 cells were treated with 250 nM NK-4 for the indicated times. Whole cell lysates were analyzed by Western blotting using an anti-phospho-Akt (Ser 473) mAb (upper) or an anti-Akt mAb (lower). (D) PC12 cells were treated with 400 μM H_2O_2 for 2-hr in the presence of the indicated concentrations of NK-4. Phospho-SAPK/JNK (upper) and SAPK/JNK (lower) were analyzed by Western blotting using whole cell lysates.

Fig. 4. NK-4 alters phosphorylation of Akt and SAPK/JNK in a Trk-independent manner.

On the other hand, we also found that the SAPK/JNK stress-induced signaling pathway was altered by NK-4. SAPK/JNK is a major cellular stress-responsive protein activated by oxidative stress. Activated SAPK/JNK translocates from the cytosol to the nucleus and regulates the activity of transcription factors such as c-Jun, ATF-2, p53, SMAD4, and Elk-1 (Kyriakis & Avruch, 2001). PC12 cells treated with cytotoxic levels of H_2O_2 displayed augmented phosphorylation of SAPK/JNK, and this was attenuated by NK-4 in a concentration-dependent manner (Fig. 4D). This observation suggests that NK-4 might attenuate H_2O_2-induced cellular stress upstream of SAPK/JNK.

2.5 Anti-aggregative effect on Aβ and tau

To examine the mechanisms of NK-4-mediated protection from amyloid toxicity in neuronal cells, we evaluated the effect of NK-4 on fibrillization of Aβ (Fig. 5), since oligomerized or aggregated Aβ seems more toxic to neuronal cells (Lesne et al., 2006, Shearman et al., 1994). ThioflavinT (ThT) fluorescence is enhanced upon binding to Aβ fibrils in proportion to the amount of fibrils in solution (Lashuel et al., 2002). Therefore, the effect of NK-4 on Aβ fibril formation was evaluated using ThT (Fig. 5A). In the absence of NK-4, both $A\beta_{1-40}$ and $A\beta_{1-42}$ solutions displayed greatly enhanced emission at 482 nm, which is characteristic for ThT bound to amyloid fibrils. Significant decreases in ThT fluorescence were detected in the presence of NK-4 at concentrations of 10 μM or higher against 100 μM of Aβ. An equimolar concentration of NK-4 almost totally inhibited the increase in ThT fluorescence. Under electron microscopic observation, $A\beta_{1-42}$ formed long and dense fibrils after 3 days of incubation (Fig. 5B, upper). In contrast, $A\beta_{1-42}$ incubated with 10 μM of NK-4 produced fewer and shorter filaments in soluble assemblies (Fig. 5B, lower).

In addition to Aβ deposition, intraneuronal tau aggregates in the brain is a hallmark of AD. Since the mutated tau transgenic mouse P301L exhibited a clear correlation between tau aggregates and neurodegeneration (Lewis et al., 2000), small molecule inhibition of tau aggregation is considered a potential therapeutic strategy, alongside Aβ anti-aggregative strategy (Pickhardt et al., 2007). Self-aggregation of tau protein occurs in the presence of polyanions such as heparin or poly(Glu), and can be quantitatively assessed by thioflavin S (ThS) fluorescence (Friedhoff et al., 1998). Therefore, we evaluated the inhibitory effect of NK-4 on tau aggregation using recombinant human tau-352 protein (0N3R variant, 15 μM) and heparin (2.5 μM) (Fig. 5C). As a result, NK-4 reduced ThS fluorescence in a dose-dependent manner and 10 μM of NK-4 substantially inhibited ThS fluorescence of tau aggregation. This result, together with the inhibition of Aβ aggregation, strongly suggests that NK-4 is an effective inhibitor of protein aggregation *in vitro*. Probably, NK-4 associates at the interface of the β-sheet domains of proteins by its planar structure and would thus act as a "β-sheet breaker" (Suh & Checler, 2002).

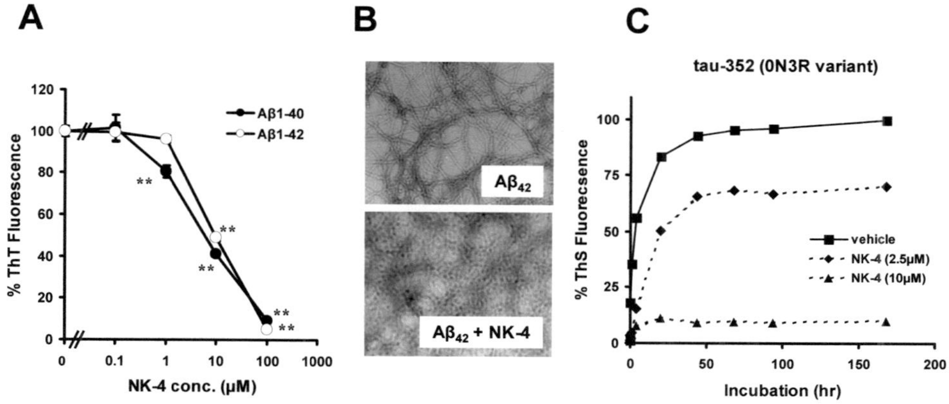

(A) Effects of NK-4 on Aβ fibril formation. Solutions of $A\beta_{1\text{-}40}$ and $A\beta_{1\text{-}42}$ (100 μM each) were incubated at 37°C for 3 days in the absence or presence of NK-4 at the indicated concentrations. Thioflavin T fluorescence was measured. Values are means ± SD (n=3). **P<0.01 vs. control. (B) Electron microscopy of $A\beta_{1\text{-}42}$ fibrils. The solution of 100 μM $A\beta_{1\text{-}42}$ was incubated for 3 days alone (upper) or with 10 μM of NK-4 (lower). Incubated Aβ was adsorbed on copper grid and negatively stained by uranyl acetate, then observed with EM at 80 keV. (C) Effects of NK-4 on tau-352 aggregation. Solutions of tau-352 (15 μM) were incubated with 2.5 μM of heparin at 37°C in the absence or presence of NK-4 (2.5 or 10 μM) and thioflavin S fluorescence was measured.

Fig. 5. NK-4 inhibits Aβ fibril formation as well as tau aggregation.

2.6 Cholinesterase (ChE) inhibitory activity

As mentioned above, NK-4 showed remarkable neurotrophic and neuroprotective activities *in vitro*, and it might have potential to modulate Aβ and tau pathologies, which is implicated in AD. In relation to a possible application of NK-4 for AD, we examined whether NK-4 has an inhibitory effect on acetylcholinesterase (AChE) and butyrylcholinesterase (BChE) since the progressive deterioration of cholinergic innervation in the cerebral cortex leads to cognitive deficits in AD, and AChE inhibitors are still the primary choice for treatment of AD (Ellis, 2005).

A comparison of IC_{50} values between NK-4 and other ChE inhibitors suggests that NK-4 is a potent and selective inhibitor of AChE (Table 2). The inhibitory effect of NK-4 against AChE was weaker than that of donepezil, the most popular drug worldwide for the treatment of dementia, but almost comparable to tacrine, and stronger than galantamine (Ohta et al., 2010a). These results suggest that NK-4 might have potential for the amelioration of disturbed cholinergic transmission in AD patients.

Compound	IC_{50} (nM)		
	AChE	BChE	Specificity (B/A)
NK-4	88 ± 2.5	> 6,333	> 70
Donepezil	6.7 ± 0.35	7,400 ± 130	1100
Tacrine	77 ± 1.4	69 ± 1.4	0.90
Galantamine	1,200 ± 33	18,000 ± 333	15

Inhibitory activities toward AChE and BChE by NK-4 and other ChE inhibitors are shown. IC_{50} values of NK-4 for AChE and BChE were determined based on the Ellman method using purified enzymes of human origin. Other data are cited from a review article (Sugimoto, 2004). Values are means ± SD.

Table 2. ChE inhibitory activities and specificity of NK-4 and other ChE inhibitors.

3. Anti-neurodegenerative effects of NK-4

NK-4 exhibits a neurotrophin-like activity and a potent free radical-scavenging capacity *in vitro*. Moreover, it confers significant stress tolerance to neuronal cells via activation of survival signaling pathways. Therefore, we examined whether NK-4 produces effects in animal models of neurodegenerative disorders, including models of ischemic stroke, cerebellar ataxia, and AD.

3.1 Stroke model (MCAO Rats)

Stroke is the second most common cause of death and a major cause of long-lasting disability worldwide. In relation to the massive socio-economic impact of ischemic stroke, neuroprotective agents with different modes of action and/or extended application time windows are therefore urgently needed. We evaluated the effect of NK-4 on ischemic stroke using a focal and transient ischemia model in rats. The middle cerebral artery occlusion (MCAO) method in rats is an animal model of focal brain ischemia (Bederson et al., 1986) frequently used to evaluate drug efficacy.

In the brains of MCAO rats, infarct volume increased mainly in the ipsilateral cerebral cortex and stratum (sites affected by MCAO; Fig. 6B). Administration of NK-4 substantially decreased infarct size (Fig. 6C), especially in the cortex penumbra region, a zone of incomplete cerebral ischemia. Edaravone also reduced infarct volume by approximately 25% compared to the vehicle control, but the difference was not statistically significant (Fig. 6C). Regarding the edema ratio, both NK-4 and edaravone significantly attenuated brain swelling compared to the vehicle control (Fig. 6D). Although edaravone (1,000 μg/kg) and a lower dose of NK-4 (25 μg/kg) failed to significantly reduce infarct size, they still significantly decreased brain edema.

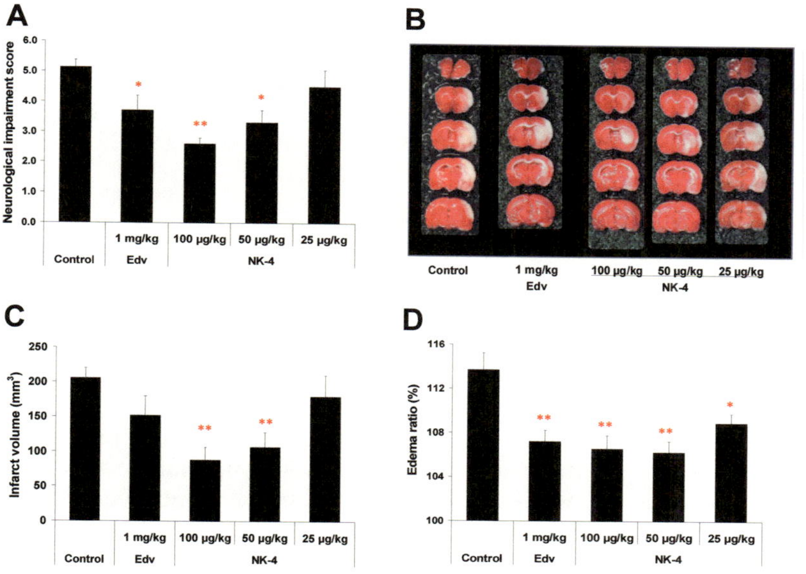

The rat right middle cerebral artery was occluded by a silicon embolizer for 2-hr and then reperfused for 24-hr. Drugs were administered intravenously twice at the indicated doses 1-hr after the occlusion and at the start of reperfusion. (A) After the 24-hr reperfusion, neurological impairment scores of rats were rated with a maximum impairment score of 6.0. (B) Representative serial brain sections (2 mm thickness) stained with 2,3,5-triphenyltetrazolium chloride (TTC) after 2-hr MCAO and 24-hr reperfusion. (C) Quantification of hemispheric lesion volumes. Infarct volumes were calculated from serial brain sections by Scion Image software. (D) Cerebral edema ratios were calculated based on the values from volumetric analysis. NK-4 groups, n=5 each; edaravone group, n=8; control group, n=8. Data are means ± S.E.M. *P<0.05, **P<0.01 vs. vehicle control group. Reproduced in part with permission from Biological & Pharmaceutical Bulletin Vol.33 No.11. Copyright [2010] Pharmaceutical Society of Japan.

Fig. 6. Suppressive effect of NK-4 on MCAO-induced infarct formation in rats.

We demonstrated that the efficacy of NK-4 was superior to that of an existing drug (edaravone) in scavenging free radicals *in vitro* (Table 1) and in protecting MCAO model rats from brain injury *in vivo* (Fig. 6). Thus the antioxidative function of NK-4 may foster an overall protective environment against neuronal damage after focal brain ischemia. Since NK-4 elicits neurotrophic activity *in vitro* in PC12 cells aside from its antioxidant function, it may also enhance endogenous neurotrophic signals *in vivo* and thereby play an important role in the protection of neuronal cells from ischemic damage.

3.2 Ataxia model (PCD hamsters)

In the ischemic stroke model, intravenously administered NK-4 showed a significant neuroprotective effect (Fig. 6). This model reflects an acute oxidative insult at MCA-affected areas in the brain. We next tested the effects of NK-4 on progressive chronic neurodegeneration using a genetic ataxia model in the Syrian hamster characterized by Purkinje cell degeneration (hmPCD model). This animal model was established in HAYASHIBARA Co., Ltd. and is thought to be homologous to the well-characterized *pcd* mutant mice. Both animal models display a suppressed expression of the brain *nna1* gene, which encodes a putative zinc carboxypeptidase originally identified by its induction in spinal motor neurons during axonal regeneration (Akita & Arai, 2009; Mullen et al., 1997; Fernandez-Gonzalez et al., 2002; Harris et al., 2000). Homozygous mutants of hmPCD show a moderate ataxia beginning at 7 weeks of age and exhibit an adult-onset degenerative loss of cerebellar Purkinje cells (PCs) followed by a slow, mild reduction in granule cell (GC) density (Akita et al., 2007).

To examine the effect of NK-4 on cerebellar ataxia in the hmPCD model, animals were administered 20 or 100 μg/kg/day of NK-4 (intraperitoneally) for 8 weeks, starting at 3 weeks of age. Motor coordination in ataxic and non-ataxic animals was evaluated weekly with a rota-rod test. As shown in figure 7A, a low dose of NK-4 (20 μg/kg) elicited a moderate, but significant, effect in attenuating the deterioration of motor function. A high dose of NK-4 (100 μg/kg) produced a considerable improvement in their rota-rod performance that lasted for the entire test period. NK-4 could not halt or reverse the disease symptoms in hmPCD, however; it profoundly delayed the progress of disease. At the end of the study (10 weeks of age), the motor ability of NK-4-treated animals was evaluated by counting the frequency of falling. The hmPCD animals began to fall frequently in their cage

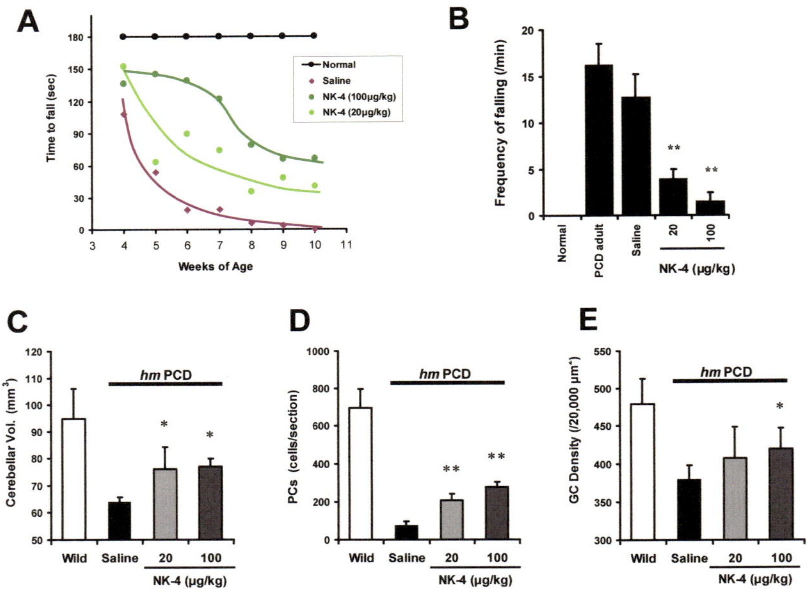

(A) Effect of NK-4 on motor performance in the rota-rod test. Animals were tested weekly for the ability to remain on the rotating rod at a constant speed (6 rpm), and the time spent on the rod was recorded. (B) Effect of NK-4 on frequency of falling in hmPCDs (10 weeks of age). Spontaneous falling of each animal was counted for 60 s. (C) Effect of NK-4 on cerebellar size in hmPCDs. The volume of the hamster cerebella at 10 weeks of age was calculated as an approximate oval sphere. (D) Effect of NK-4 on the number of PCs in hmPCDs. PCs were counted in the mid-sagittal section of the cerebellum from hmPCDs. (E) GC density in the cerebellum from hmPCDs and wild type controls. H&E stained sections of cerebellum cortex from hmPCDs and wild type controls were counted for GCs in an area of 20,000 μm². Graphs show the mean ± S.E.M. of 6 hamsters at 10 weeks of age. *P<0.05, **P<0.01 vs. hmPCD control. Adapted from Ohta H *et al*, *PLoS ONE* 6(2):e17137 (Ohta et al., 2011).

Fig. 7. Effects of NK-4 on motor coordination and cerebellar degeneration in hmPCDs.

from around 7 weeks of age and the frequency of falling increased with age. Animals treated with the low and high doses of NK-4 showed a significant reduction in falling frequency compared to saline-treated controls (Fig. 7B). These observations demonstrate that NK-4 is effective in treating motor discoordination associated with cerebellar ataxia in the hmPCD model.

The mutants also showed severe cerebellar atrophy and volumetric reduction at 10 weeks of age (Fig. 7C). Animals treated with low or high doses of NK-4 had a significantly larger cerebellum volume compared with saline-treated controls. H&E staining and calbindin immunohistochemistry of cerebellar cortical sections revealed a large reduction in the number of cerebellar PCs in the brain of hmPCDs at 10 weeks of age (Fig. 7D). The PC dendrites in high dose NK-4-treated hmPCDs were significantly longer and thicker than those in surviving PCs in saline-treated hmPCDs (data not shown). In the brain of hmPCDs, the cerebellar GC density was moderately reduced compared to wild type controls. NK-4 dose-dependently attenuated cellular atrophy and prevented the reduction in GCs (Fig. 7E).

In this model, daily intraperitoneal injection of NK-4 at a dose of 20 or 100 μg/kg for 8 consecutive weeks was effective in attenuating motor discoordination and degenerative loss of both PCs and GCs with no detectable adverse events. This suggests that NK-4 can attenuate neurodegeneration in the central nervous system via a peripheral route of administration. PCs are susceptible to ischemic damage because of their reduced capacity to isolate glutamate and reduced ability to generate energy during anoxia (Welsh et al., 2002). GCs are also vulnerable to a variety of toxins that decrease glutathione levels and this makes the cells more vulnerable to cellular damage from ROS (Fonnum & Lock, 2004). Direct scavenging of free radicals by NK-4 may protect these cells. In addition, NK-4 appears to activate survival-signaling pathways in degenerating cerebellar neurons.

3.3 Alzheimer's Disease (AD) mouse models

AD, the most common form of dementia, is characterized clinically by ongoing declines in cognitive and functional ability and emergence of behavioral and psychological symptoms. More than 35 million people living with dementia worldwide in 2010, increasing to 65.7 million by 2030 and 115.4 million by 2050 (Wimo & Prince, 2010). However, no effective disease-modifying therapy is available. A report from Alzheimer's association predicts the total costs of care for AD patients will increase five-fold per year, and the treatments which delay the onset of Alzheimer's disease or slow the progression of this condition dramatically reduce the costs of Medicare or Medicaid (Alzheimer's association, 2010). Presently, the only approved therapies for AD are ChE inhibitors and an N-methyl-D-aspartate (NMDA) receptor antagonist, and the beneficial effects of these symptomatic treatments appear limited and are not long lasting (Lanctôt et al., 2009). Thus, more evidence-based effective therapies, whether they are symptomatic treatment or disease-modifying strategies, are urgently needed.

NK-4 shows potent neuroprotective and neurotrophin-like activities *in vitro* (Fig. 1~3) and also *in vivo* (Fig. 6, 7). Furthermore, it remarkably and selectively inhibits AChE *in vitro* (Table 2). These multiple properties of NK-4 raised the possibility that it could halt or slow the progression of AD. To date, several mouse models for AD based on the amyloid hypothesis or the tau hypothesis have been developed (Bloom et al., 2005). Therefore, we tested the potential effects of NK-4 on AD using *in vivo* studies incorporating the amyloid precursor protein (APP) transgenic mice Tg2576 (Hsiao et al., 1996) and the Aβ-induced amnesia model (Maurice et al., 1996).

3.3.1 Effects of NK-4 in APP-Tg (Tg2576) mice

The Tg2576 mouse is the most thoroughly characterized AD mouse model and is considered to reflect human amyloid pathology most closely among mouse models. This mouse develops a considerable amount of Aβ deposits in the brain with age and this is accompanied by gradual declines in cognitive function (Hsiao, 2001). We examined whether chronic administration of NK-4 produced an effect on the cognitive deficits in Tg2576 mice (Ohta et al., 2010b). Tg2576 mice were administered NK-4 as doses of 100 or 500 μg/kg/day once a day, 5 times a week for 9 months, starting at 3 months of age. Donepezil, an existing drug for dementia based on its AChE inhibitory properties (Sugimoto, 2004) was used as a control drug and was administered at a dose of 200 μg/kg/day.

Wild type and Tg2576 mice were tested for spatial learning ability in a Morris water maze test (Morris, 1984) at ages 6 and 12 months. There were significant differences in escape latency between the wild type controls and saline-treated Tg2576 mice at 6 and 12 months of age (Fig. 8A). For the saline-treated Tg2576 group, the latency to reach the hidden platform did not shorten during the whole test period at 12 months of age, suggesting that the spatial learning ability of saline-treated Tg2576 mice became impaired with age. We found significant differences in spatial learning ability between the NK-4-treated Tg2576 mice (at both low and high doses) and saline-treated Tg2576 mice. Saline-treated Tg2576 mice consistently exhibited longer escape latencies compared to the NK-4-treated Tg2576 group at 6 and 12 months of age. Donepezil-treated mice showed a partial improvement in spatial memory deficits at 6 months of age, although the drug became less effective over time. At 12 months of age, the escape latencies of donepezil-treated and saline-treated Tg2576 mice showed no significant difference. These results suggest a long-lasting ameliorative effect of NK-4 on Aβ-mediated spatial memory impairment.

We also evaluated the object recognition memory of Tg2576 mice in a novel object recognition test (Nagai et al., 2003). Long-term recognition memory was tested at ages of 6, 9, and 12 months (Fig. 8B). Saline-treated Tg2576 mice displayed significantly decreased object recognition memory compared to wild type mice at the same ages. Tg2576 mice treated with low and high doses of NK-4 or donepezil spent a longer time exploring the novel object than did the saline-treated controls at 6 months of age. However, at 9 and 12 month of age, the exploration preference values of low dose NK-4 and donepezil-treated groups decreased, and the differences between these two groups and saline-treated group became insignificant. In contrast, the high dose of NK-4 group spent a significantly longer time exploring the novel object versus the saline-treated mice throughout the test period, and the exploratory preference was comparable to that of wild type controls at 12 months of age. This suggests that NK-4 administration for a longer period and at high dosing might be more effective for the improvement of object recognition memory.

At the end of experiment (12 months of age; after 9-month treatment), the effects of NK-4 on plasma and brain Aβ levels in Tg2576 mice were evaluated (Fig. 9). Dose-dependent increases in $A\beta_{1-40}$ and $A\beta_{1-42}$ were observed in plasma from the NK-4-treated mice (Fig. 9A). Although donepezil treatment also increased both plasma $A\beta_{1-40}$ and $A\beta_{1-42}$ levels, the effect was lower compared to NK-4 treatments. Brain levels of Aβ were estimated separately in detergent-insoluble and -soluble fractions. Both detergent-insoluble and soluble $A\beta_{1-40}$ levels were significantly decreased by NK-4 treatment (Fig. 9B, upper graphs). Donepezil did not affect Aβ concentrations in brain. Because the levels of $A\beta_{1-42}$ were very low in all groups compared to those of $A\beta_{1-40}$ (Fig. 9B, lower graphs), the levels of $A\beta_{1-40}$ seem to reflect the total amount of Aβ in the brain.

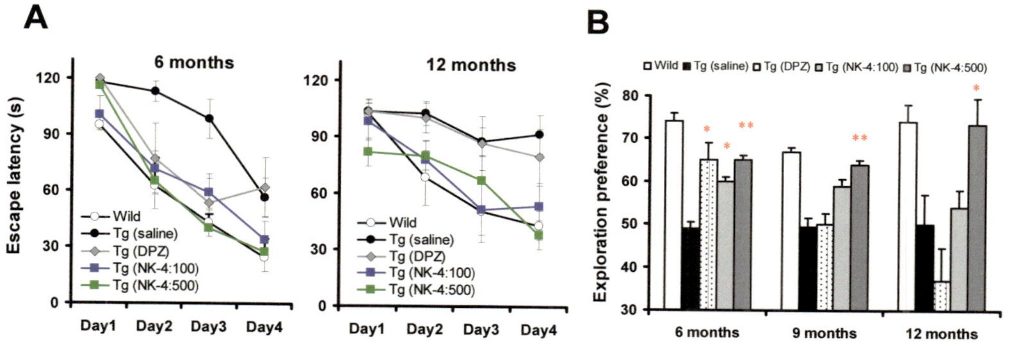

NK-4 was injected intraperitoneally to Tg2576 mice at a dose of 100 or 500 μg/kg once a day, five times a week for 9 months, beginning at 3 months of age (n=10 each). Donepezil (DPZ) was administered at a dose of 200 μg/kg (n=5). Control mice received 200 μl of saline (wild type, n=10; Tg2576, n=10). (A) Morris water maze test at 6 and 12 months of age. The escape latency represents the average time to find a hidden platform placed in a fixed location in a circular pool (ø130 cm). Two trials per day were conducted for 4 consecutive days with an upper cutoff time of 120 sec. (B) Novel object recognition test at 6, 9 and 12 months of age. Two objects X and Y were placed in the test box and the exploration behavior of mice was recorded. A retention session was performed 24 hr after the training session. One of the familiar objects Y was replaced by a novel object Z. Mice were allowed to explore freely for 10 min, with the time spent in exploring each of the two objects recorded. The exploratory preference was expressed as a ratio of the time spent exploring the novel object (T_Z) over that spent on the two objects (T_X+T_Z). Data are expressed as means ± S.E.M. *$P<0.05$, **$P<0.01$ vs. saline-treated Tg2576 group. Wild: non-transgenic mice. Tg(saline), saline-treated Tg2576 mice; Tg(DPZ), donepezil-treated Tg2576 mice; Tg(NK-4: 100), low-dose NK-4-treated Tg2576 mice; Tg2576 (NK-4: 500), high-dose NK-4-treated Tg2576 mice.

Fig. 8. Effect of NK-4 on recognition memory in Tg2576 mice.

We next surveyed brain Aβ deposition in Tg2576 mice by Aβ immunohistochemistry (Fig. 9C). Aβ-immunoreactive small diffuse plaques were abundantly present in the cortex of saline-treated Tg2576 mice (upper panel), and they were visibly reduced by high dose NK-4 treatment both in size and in quantity (bottom panel). The donepezil-treated group showed a slight decrease in Aβ-immunoreactivity (middle panel).

Nine months of NK-4 administration to Tg2576 mice significantly attenuated the cognitive decline as assessed by behavioral tests (Fig. 8), and decreased the levels of Aβ in the brain (Fig. 9B, C) while augmenting levels in the plasma (Fig. 9A). These results imply that NK-4-induced cognitive improvement was attributable to decreased Aβ brain levels. Although there is still no consensus as to the mechanism by which some drugs, including anti-Aβ antibodies, alter amyloid deposition in the brain (Levites et al., 2006), Aβ may be cleared across the blood-brain barrier to the blood in NK-4-treated Tg2576 mice.

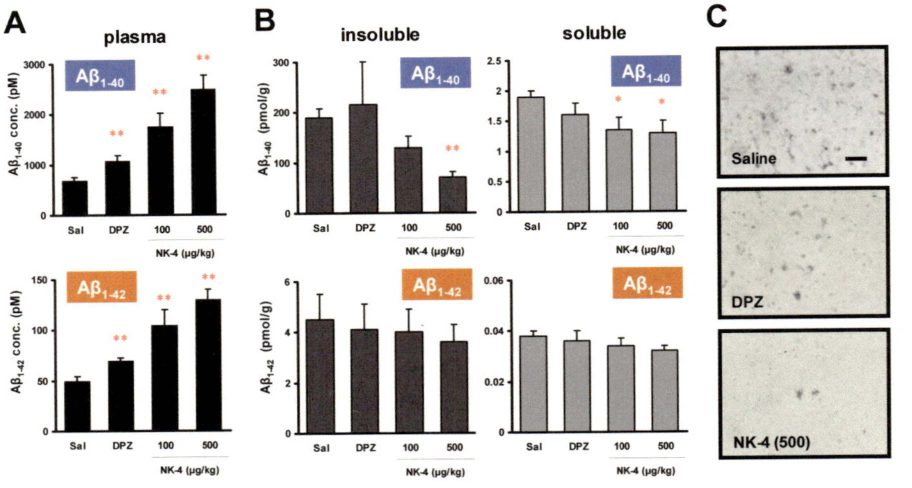

(A) Plasma Aβ concentrations of Tg2576 mice at 12 months of age. Aβ levels were separately measured by ELISA for $A\beta_{1\text{-}40}$ and $A\beta_{1\text{-}42}$. (B) Brain Aβ levels of detergent-insoluble or soluble fractions at 12 months of age. Hemi-brains from Tg2576 mice were homogenized and separated into detergent-insoluble and soluble fractions. $A\beta_{1\text{-}40}$ and $A\beta_{1\text{-}42}$ in individual samples were measured and are expressed as picomoles/g wet brain weight. Values are the means ± S.E.M. Saline (Sal), n=10; Donepezil (DPZ), n=5; NK-4 (100), n=9; and NK-4 (500), n=8. *P<0.05, **P<0.01 vs. saline-treated group. (C) Aβ-immunohistochemistry of the cerebral cortex from Tg2576 mice at 12 months of age. Hemi-brains were fixed with 10% buffered formalin and embedded in paraffin. Coronal sections of cerebral cortex with 5-μm thickness were immunostained with rabbit anti-Aβ-peptide polyclonal antibody. Bar represents 50-μm.

Fig. 9. Effect of NK-4 on the plasma and brain concentrations of Aβ in Tg2576 mice.

Oxidative mechanisms are thought to be involved in cell loss and other neuropathologies associated with AD (Zhu et al., 2001; Cutler et al., 2004). During AD pathogenesis, ROS impair mitochondrial redox activity and further increase ROS generation (Shearman et al., 1994; Hensley et al., 1994). Aβ induces the production of ROS and leads to apoptotic neuronal cell death that can be inhibited by antioxidants (Behl et al., 1994; Mattson & Goodman, 1995). Pathological and biochemical studies suggest that ROS induced by fibrillar Aβ produce neurotoxic effects (Gevais et al., 1999). Since NK-4 is a potent scavenger of ROS (Table 1) and is an inhibitor of Aβ aggregation (Fig. 2), these features might work in concert to attenuate Aβ toxicity in AD.

3.3.2 Effects of NK-4 in Aβ-icv mice

To further determine whether the effect of NK-4 was directly attributable to mitigation of Aβ pathology, an Aβ-induced amnesia model was employed (Ohta et al., 2010a). In this model, ICR mice received intracerebroventricular (icv) administration of aggregated $A\beta_{25\text{-}35}$ peptide to induce memory deficits due to cholinergic dysfunction (Maurice et al., 1996). Long-term recognition memory was evaluated by the step-through type passive avoidance test 12 days after $A\beta_{25\text{-}35}$ injection. In this behavioral assay, daily NK-4 treatment significantly and dose-dependently improved memory deficits induced by $A\beta_{25\text{-}35}$ (Fig. 10A). There was a significant difference in step-through latencies between sham operated mice and $A\beta_{25\text{-}35}$-injected ICR mice. Both low (50 μg/kg) and high (500 μg/kg) doses of NK-4 successfully prolonged the step-through latencies with statistically significant differences ($p<0.05$ for the low dose and $p<0.01$ for the high dose, respectively; Fig. 10A). Improved long-term memory retention following NK-4 administration was also confirmed using this animal model in a novel object recognition test (data not shown).

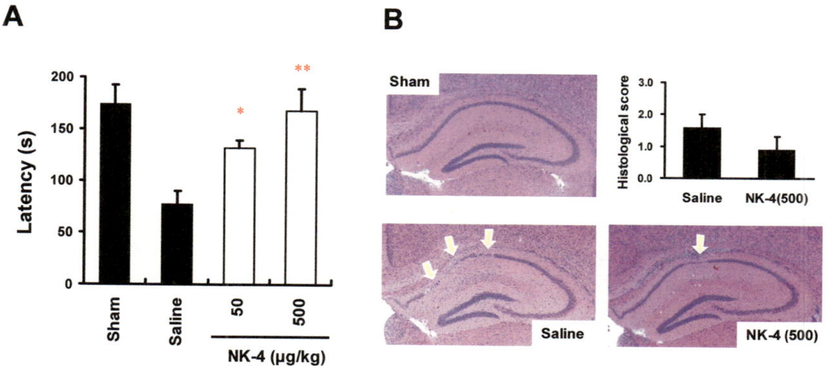

(A) Passive avoidance test of $A\beta_{25\text{-}35}$–icv mice. Male ICR mice were injected aged $A\beta_{25\text{-}35}$ (9 nmol/mouse) into the left lateral ventricle using the following coordinates from Bregma: 0.5 mm posterior, 1.0 mm lateral, and 2.0 mm ventral. A low (50 μg/kg) or high (500 μg/kg) dose of NK-4 was administered intraperitoneally to mice for twelve consecutive days starting from the next day of Aβ injection. Then, mice were tested for step-through passive avoidance during days 9-12. The step-through latency in the retention session (24-hr after the training session) was recorded. Values are means ± S.E.M. (n=10). Sham, sham-operated group; saline, saline-treated $A\beta_{25\text{-}35}$–icv group; NK-4, NK-4-treated $A\beta_{25\text{-}35}$–icv groups. *$P<0.05$, **$P<0.01$ vs. saline-treated group. (B) Histological damage scores in the hippocampal CA1 region. Hippocampal neuronal cell loss was assessed in coronal H&E sections. Scores are represented as means ± S.E.M. (n=10). Arrows indicate the sites of degeneration.

Fig. 10. Effects of NK-4 on $A\beta_{25\text{-}35}$-induced cognitive impairments in ICR mice.

These beneficial effects of NK-4 on recognition memory are in agreement with histological findings shown in figure 10B. A single icv injection of $A\beta_{25\text{-}35}$ caused significant hippocampal neuronal loss mainly in the CA1 region. The difference in histological scores was not statistically significant between the saline-treated (1.6 ± 0.41) and the high dose NK-4–treated mice (0.9 ± 0.41, $p = 0.067$). However, NK-4 ameliorated the $A\beta_{25\text{-}35}$-induced injury of pyramidal neurons in the hippocampus (Fig. 10B). Reportedly, associative learning in the passive avoidance test strongly depends on hippocampal function (Phillips & LeDoux, 1992). In this context, neuronal injury in the hippocampal CA1 region following $A\beta_{25\text{-}35}$-icv injection provides a reasonable explanation for the impaired long-term memory. These results, in combination with the results from Tg2576 mice, and AChE inhibitory property of NK-4 suggest that NK-4 treatment effectively improves cognitive deficits in AD model mice by both attenuating Aβ pathology and enhancing cholinergic transmission in the brain.

4. Conclusion

In this chapter, we showed that NK-4, a type of cyanine dye, exerts a wide spectrum of biochemical and biological activities implicated in neuroprotection. NK-4 significantly ameliorated neurological and cognitive deficits, as well as neurodegeneration, in four distinct animal models (MCAO rats, ataxic hamsters, APP-transgenic mice, and Aβ-icv mice). Molecular mechanisms by which NK-4 acts against neurodegeneration remain unclear, although it was found that NK-4 activates the PI3K-Akt pathway independently of Trk receptors. Thus, activated Akt may be a key mediator of the beneficial effects on neural cell survival. In addition, since activated Akt is required for PI3K-mediated synaptic plasticity and memory consolidation via activation of the downstream regulator CREB (Brightwell et al., 2007), induction of Akt phosphorylation might play a critical role in NK-4-mediated memory improvement in animal models of dementia. In addition to the modification of intracellular signaling, a direct antioxidative property of NK-4 may also be involved in survival and functional maintenance of neurons.

Regarding the safety profile, there were no specific adverse events in mice that received intraperitoneal injections of NK-4 at doses of up to 500 μg/kg/day, 5 days a week, for 9 months (3.3.1. Tg2576 study). Furthermore, NK-4 was well tolerated in rats up to 4,000 mg/kg in an acute toxicity study and 100 mg/kg in a subacute toxicity study (both via the oral rout of administration) based on mortality, clinical observations, body weight, hematology, blood chemistry, organ weights, and histological examination of a complete tissue list (Ohmori et al., 1983). Furthermore, NK-4 has been used as an active ingredient of over the counter (OTC) medicine in Japan for treating allergy, and for promoting wound healing since the 1950s (Suzue, 1969). Additionally, NK-4 was not mutagenic in the standard Ames test (our unpublished data). These observations suggest that NK-4 is a safe compound that will not cause serious adverse reactions.

Application of small neurotrophic molecules that modulate neuronal survival and synaptic function is a promising and valid therapeutic approach for neurodegenerative disorders. Lines of evidence described here strongly suggest the potential utility of NK-4 as a treatment for neurodegenerative disease. The evaluations of pharmacokinetics, bioavailability, as well as the efficacy of orally administered NK-4 are ongoing. In addition, assessment of NK-4 in human neurodegenerative therapy will require further studies.

5. Acknowledgment

We would like to thank Dr. S. Arai, Ms. C. Ushio, and Dr. M. Takeuchi (HAYASHIBARA Co., Ltd.) for their helpful support.

6. References

Abe, K. & Kogure, K. (1988) Strong attenuation of ischemic and postischemic brain edema in rats by a novel free radical scavenger. *Stroke,* 19, pp.480-485, ISSN 0039-2499

Akita, K., Arai, S., Ohta, T., Hanaya, T. & Fukuda, S. (2007) Suppressed Nna1 gene expression in the brain of ataxic Syrian hamsters. *J Neurogenet,* 21, pp.19–29, ISSN 0167-7063

Akita, K. & Arai, S. (2009) The ataxic Syrian hamster: an animal model homologous to the pcd mutant mouse? *Cerebellum,* 8, pp.202–210, ISSN 1473-4222

Alzheimer's association (2010) Changing the trajectory of Alzheimer's disease: A national imperative. 16.06.2010, Available from: http://www.alz.org/documents_custom/trajectory.pdf

Amor, S., Puentes, F., Baker, D. & van der Valk, P. (2010) Inflammation in neurodegenerative diseases. *Immunology* 129, pp.154-169, ISSN 1365-2567

Barinaga, M. (1996) Finding new drugs to treat stroke. *Science,* 272, pp.664-666, ISSN 0036-8075

Bederson, J. B., Pitts, L. H, Tsuji, M., Nishimura, M. C., Davis, R. L. & Bartkowski, H. (1986) Rat middle cerebral artery occlusion: Evaluation of the model and development of a neurologic examination. *Stroke,* 17, pp.472-476, ISSN 0039-2499

Behl, C., Davis, J., Lesley, R. & Schubert, D. (1994) Hydrogen peroxide mediates amyloid beta protein toxicity. *Cell,* 77, pp.817-827, ISSN 0092-8674

Berg, M., Sternberg, D., Parada, L. & Chao, M. (1992) K-252a inhibits nerve growth factor-induced trk proto-oncogene tyrosine phosphorylation and kinase activity. *J Biol Chem,* 267, pp.13–16, ISSN 0021-9258

Bloom, F. E., Reilly, J. F., Redwine, J. M., Wu, C. C., Young, W. G. & Morrison J. H. (2005) Mouse models of human neurodegenerative disorders: Requirements of medication development. *Arch Neurol,* 62, pp.185-187, ISSN 0003-9942

Blum, D., Torch, S., Lambeng, N., Nissou, M., Benabid, A. L., et al. (2001) Molecular pathways involved in the neurotoxicity of 6-OHDA, dopamine and MTPT: contribution to the apoptotic theory in Parkinson's disease. *Prog Neurobiol,* 65, pp.135-172, ISSN 0301-0082

Blum, R. & Konnerth, A. (2005) Neutrophin-mediated rapid signaling in the central nervous system: Mechanisms and functions. *Physiology,* 20, pp.70-78, ISSN 1548-9213

Brightwell, J., Smith, C., Neve, R. & Colombo, P. (2007) Long-term memory for place learning is facilitated by expression of cAMP response element-binding protein in the dorsal hippocampus. *Learn Mem,* 14, pp.195-199, ISSN 1072-0502

Cutler, R., Kelly, J., Storie, K., Pedersen, W., Tammara, A., et al. (2004) Involvement of oxidative stress-induced abnormalities in ceramide and cholesterol metabolism in brain aging and Alzheimer's disease. *Proc Natl Acad Sci USA,* 101, pp.2070-2075, ISSN 0027-8424

Ellis, J. M. (2005) Cholinesterase inhibitors in the treatment of dementia. *J Am Osteopath Assoc,* 105, pp.145-158, ISSN 0098-6151

Fernandez-Gonzalez, A., La Spada, A., Treadaway, J., Higdon, J., Harris, B., et al. (2002) Purkinje cell degeneration (pcd) phenotypes caused by mutations in the axotomy-induced gene, Nna1. *Science,* 295, pp.1904–1906, ISSN 0036-8075

Fonnum, F. & Lock, E. (2004) The contributions of excitotoxicity, glutathione depletion and DNA repair in chemically induced injury to neurons: exemplified with toxic effects on cerebellar granule cells. *J Neurochem,* 88, pp.513–531, ISSN 0022-3042

Franco, J. L., Posser, T., Gordon, S. L., Bobrovskaya, L., Schneider, J. J. et al., (2010) Expression of tyrosine hydroxylase increases the resistance of human neuroblastoma cells to oxidative insults. *Toxicol Sci,* 113, pp.150-157, ISSN 1096-6080

Friedhoff, P., Schneider, A., Mendelkow, E. M. & Mendelkow E. (1998) Rapid assembly of Alzheimer-like paired helical filaments from microtubule-associated protein tau monitored by fluorescence in solution. *Biochemistry,* 37, pp.10223-10230, ISSN 0006-2960

Gevais, F., Xu, D., Robertson, G., Vaillancourt, J., Zhu, Y., et al. (1999) Involvement of caspases in proteolytic cleavage of Alzheimer's amyloid-β precursor protein and amyloidogenic Aβ-peptide formation. *Cell,* 97, pp.395-406, ISSN 0092-8674

Gilgun-Sherki, Y., Rosenbaum, Z., Melamed, E. & Offen, D. (2002) Antioxidant therapy in acute central nervous system injury: current state *Pharmacol Rev,* 54, pp.271-284, ISSN 0031-6997

Harris, A., Morgan, J. I., Pecot, M., Soumare, A., Osborne, A. & Soares, H. D. (2000) Regenerating motor neurons express Nna1, a novel ATP/GTP-binding protein related to zinc carboxypeptidase. *Mol Cell Neurosci* 16, pp.578-596, ISSN 1044-7431

Hayami, M. (1984) Kanko-so and its antimicrovial action. *J Jpn Cosmetic Sci Soc*, 8, pp.43-59, ISSN 1880-2532

Hensley, K., Carmey, J., Mattson, M., Aksenova, M., Harris, M., et al. (1994) A model for β-amyloid aggregation and neurotoxicity based on free radical generation by the peptide: relevance to Alzheimer disease. *Proc Natl Acad Sci USA*, 91, pp.3270-3274, ISSN 0027-8424

Hsiao, K., Chapman, P., Nilsen, S., Eckman, C., Harigaya, Y., et al. (1996) Correlative memory deficits, A beta elevation, and amyloid plaques in transgenic mice. *Science*, 274, pp.99-102, ISSN 0036-8075

Hsiao, K. (2001) Learning and memory in transgenic mice modeling Alzheimer's disease. *Learn Mem*, 8, pp.301-308, ISSN 1072-0502

Ishihara, M. & Fujisawa, S. (2007) Photooxygenation, photodegradation and antioxidative activity of platonin, a cyanine photosensitizing dye. *In Vivo*, 21, pp.163-174, ISSN 0258-851X

Ivins, K. J., Ivins, J. K., Sharp, J. P. & Cotman, C. W. (1999) Multiple pathways of apoptosis in PC12 cells: CrmA inhibits apoptosis induced by β-amyloid. *J Biol Chem*, 274, pp.2107-2112, ISSN 0021-9258

Kadowaki, H., Nishitoh, H., Urano, F., Sadamitsu, C., Matsuzawa, A., et al. (2005) Amyloid β induces neuronal cell death through ROS-mediated ASK1 activation. *Cell Death Deffer*, 12, pp.19-24, ISSN 1350-9047

Kim, Y., Seger, R., Suresh, B. C., Hwang, S. & Yoo, Y. (2004) A positive role of the PI3-K/Akt signaling pathway in PC12 cell differentiation. *Mol Cell*, 18, pp.353–359, ISSN 0270-7306

Koya-Miyata, S., Ohta, H., Akita, K., Arai, S., Ohta, T., et al. (2010) Cyanine photosensitizing dyes attenuate cerebral ischemia and reperfusion injury in rats. *Biol Pharm Bull*, 33, pp.1872-1877, ISSN 0918-6158

Kunikata, T., Ishihara, T., Ushio, S., Iwaki, K., Ikeda, M., et al. (2002) Lumin, a cyanine dye, enhances interleukin 12-dependent interferon gamma production by lipopolysaccharide-stimulated mouse splenocytes. *Biol Pharm Bull*, 25, pp.1018-1021, ISSN 0918-6158

Kyriakis, J. M. & Avruch, J. (2001) Mammalian mitogen-activated protein kinase signal transduction pathways activated by stress and inflammation. *Physiol Rev*, 81, pp.807-869, ISSN 0031-9333

Lanctôt, K. L., Rajaram, R. D. & Herrmann, N. (2009) Therapy for Alzheimer's disease: how effective are current treatments? *Ther Adv Neurol Dis*, 2, pp.163-180, ISSN 1756-2856

Lashuel, H. A., Hartley, D. M., Balakhaneh, D., Aggarwal, A., Teichberg, S., et al. (2002) New class of inhibitors of amyloid-β fibril formation. Implications for the mechanism of pathogenesis in Alzheimer's disease. *J Biol Chem*, 277, pp.42881-42890, ISSN 0021-9258

Lesne, S., Koh, M., Kotilinek, L., Kayed, R., Glabe, C., et al. (2006) A specific amyloid-beta protein assembly in the brain impairs memory. *Nature*, 440, pp.352-357, ISSN 0028-0836

Levites, Y., Smithson, L. A., Price, R. W., Dakin, R. S., Yuan, B., et al. (2006) Insights into the mechanisms of action of anti-Aβ antibodies in Alzheimer's disease mouse models. *FASEB J*, 20, pp.2576-2578, ISSN 0892-6638

Lewis, J., McGowan, E., Rockwood, J., Melrose, H., Nacharaju, P., et al. (2000) Neurofibrillary tangles, amyotrophy and progressive motor disturbance in mice expressing mutant (P301L) tau protein. *Nat Genet,* 25, pp.402-405, ISSN 1061-4036

Martin, D., Rojo, A. I., Salinas, M., Diaz, R., Gallardo, G., et al. (2004) Regulation of heme oxygenase-1 expression through the phosphatidylinositol 3-kinase/Akt pathway and the Nrf2 transcription factor in response to the antioxidant phytochemical carnosol. *J Biol Chem,* 279, pp.8919–8929, ISSN 0021-9258

Mattson, M. & Goodman, Y. (1995) Different amyloidogenic peptides share a similar mechanism of neurotoxicity involving reactive oxygen species and calcium. *Brain Res,* 676, pp.219-224, ISSN 0006-8993

Maurice, T., Lockhart, B. P. & Privat, A. (1996) Amnesia induced in mice by centrally administered β-amyloid peptides involves cholinergic dysfunction. *Brain Res,* 706, pp.181-193, ISSN 0006-8993

Morris, R. (1984) Developments of a water-maze procedure for studying spatial learning in the rat. *J Neurosci Methods,* 11, pp.47-60, ISSN 0165-0270

Motoyoshi, F., Kondo, N., Ono, H. & Orii, T. (1991) The effect of photosensitive dye platonin on juvenile rheumatoid arthritis. *Biotherapy,* 3, pp.241-244, ISSN 1573-8280

Mullen, R. J., Eicher, E. M. & Sidman, R. L. (1976) Purkinje cell degeneration, a new neurological mutation in the mouse. *Proc Natl Acad Sci USA,* 73, pp.208–212, ISSN 0027-8424

Nagai, T., Yamada, K., Kim, H., Kim, Y., Noda, Y., et al. (2003) Cognition impairment in the genetic model of aging klotho gene mutant mice: a role of oxidative stress. *FASEB J,* 17, pp.50-52, ISSN 0892-6638

Ohmori, M., Kobayashi, S., Hirata, K., Miki, H. & Ohmori, E. (1984) Toxicity test of lumin: On the acute and subacute toxicity to rats by oral administration. *Kankohshikiso,* 89, pp.44-53, ISSN 0461-5956

Ohta, H., Akita, K., Arai, S., Ohta, T. & Fukuda, S. (2010a) Prevention of β-amyloid induced memory impairment by cyanine photosensitizing dyes: a potential novel therapeutic agent for Alzheimer's disease. *Proceeding of 130th annual congress of The Pharmaceutical Society of Japan,* Suppl 3, pp.175 (in Japanese), Okayama, Japan, Mar 28-30, 2010

Ohta, H., Akita, K., Arai, S., Ohta, T., Kawada, T. & Fukuda, S. (2010b) NK-4, a photosensitizing cyanine dye, prevented beta-amyloid-induced cognitive impairment in a transgenic mouse model of Alzheimer's disease. *Alzheimer Dement,* 6, Suppl 1, pp.S-568, ISSN 1552-5260

Ohta, H., Arai, S., Akita, K., Ohta, T. & Fukuda, S. (2011) Neurotrophic Effects of a Cyanine Dye via the PI3K-Akt Pathway: Attenuation of Motor Discoordination and Neurodegeneration in an Ataxic Animal Model. *PLoS ONE,* 6, e17137, ISSN 1932-6203

Petullo, D., Masonic, K., Lincoln, C., Wibberley, L., Teliska, M. & Yao, D. (1999) Model development and behavioral assessment of focal cerebral ischemia in rats. *Life Sci,* 64, pp.1099-1108, ISSN 0024-3205

Phillips, R. & LeDoux, J. E. (1992) Differential contribution of amygdala and hippocampus to cued and contextual fear conditioning. *Behav Neurosci,* 106, pp.274-285, ISSN 0735-7044

Pickhardt, M., Larbig, G., Khlistunova, I., Coksezen, A., Meyer, B., et al. (2007) Phenylthiazolyl-hydrazide and its derivatives are potent inhibitors of tau aggregation and toxicity in vitro and in cells. *Biochemistry,* 46, pp.10016-10023, ISSN 0006-2960

Salinas, M., Diaz, R., Abraham, N. G., Ruiz de Galarreta, C. & Caudrado, A. (2003) Nerve growth factor protects against 6-hydroxydopamine-induced oxidative stress by increasing expression of heme oxygenase-1 in a phosphatidylinositol 3-kinase-dependent manner. *J Biol Chem,* 278, pp.13898-13904, ISSN 0021-9258

Shearman, M., Ragan, C. & Iversen, L. (1994) Inhibition of PC12 cell redox activity is a specific, early indicator of the mechanism of β-amyloid-mediated cell death. *Proc Natl Acad Sci USA,* 91, pp.1470-1474, ISSN 0027-8424

Sugimoto, H. (2004) Scope and limitation of acetylcholinesterase inhobotors. *Folia Pharmacol Jpn,* 124, pp.163-170, ISSN 0015-5691

Suh, Y. H. & Checler, F. (2002) Amyloid precursorprotein, presenilins, and α-synuclein: Molecular pathogenesis and pharmacological applications in Alzheimer's disease. *Pharmacol Rev,* 54, pp.469-525, ISSN 0031-6997

Suzue, K. (1969) Medical research for photosensitizing dyes. *Kankohshikiso,* 71, pp.22-42, ISSN 0461-5956

Takeguchi, N., Saitoh, T., Morii, M., Yoshikawa, K., Terada, H. (1985) Formation of a leakage-type ion pathway in lipid bilayer membranes by divalent cationic cyanine dyes in cooperation with inorganic phosphate. *J Biol Chem,* 260, pp.9158-9161, ISSN 0021-9258

Tanaka, M. (2002) Pharmacological and clinical profile of the free radical scavenger edaravone as a neuroprotective agent. *Folia Pharmacol Jpn,* 119, pp.301-308, ISSN 0015-5691

Trauner, K., Gandour-Edwards, R., Bamberg, M., Shortkroff, S., Sledge, C., et al. (1998) Photodynamic synovectomy using benzoporphyrin derivative in an antigen-induced arthritis model for rheumatoid arthritis. *Photochem Photobiol,* 67, pp.133-139, ISSN 0031-8655

Troy, C. M., Rabacchi, S., Hohl, J. B., Angelastro, J. M., Greene, L. A. & Shelanski, M. L. (2001) Death in the balance: alterative participation of the caspase-2 and –9 pathways in neuronal death induced by nerve growth factor deprivation. *J Neurosci,* 21, pp.5007-5016, ISSN 0270-6474

Ushio, C., Ariyasu, H., Ariyasu, T., Arai, S., Ohta, T., et al. (2009) Suppressive effects of a cyanine dye against herpes simplex virus (HSV)-1 infection. *Biomed Res,* 30, pp.365-368, ISSN 0388-6107

Wang, W., Dow, K. E., Riopelle, R.J. & Ross, G. M. (2001) The common neurotrophin receptor p75NTR enhances the ability of PC12 cells to resist oxidative stress by trkA-dependent mechanism. *Neurotox Res,* 3, pp.485-499, ISSN 1029-8428

Welsh, J., Yuen, G., Placantonakis, D., Vu, T., Haiss, F., et al. (2002) Why do Purkinje cells die so easily after global brain ischemia? Aldolase C, EAAT4, and the cerebellar contribution to posthypoxic myoclonus. *Adv Neurol,* 89, pp.331–359, ISSN 0091-3952

Wimo, A. & Prince, M. (21 Sep. 2010) The global economic impact of dementia. In: *World Alzheimer Report 2010,* 12.12.2010, Available from:
http://www.alz.co.uk/research/files/World AlzheimerReport2010.pdf

Wolozin, B. & Behl, C. (2000a) Mechanisms of neurodegenerative disorders. Part 1: Protein Aggregates. *Arch Neurol,* 57, pp.793-796, ISSN 0003-9942

Wolozin, B. & Behl, C. (2000b) Mechanisms of neurodegenerative disorders. Part 2: Control of cell death. *Arch Neurol,* 57, pp.801-804, ISSN 0003-9942

Zhu, X., Raina, A., Rottkamp, C., Aliev, G., Perry, G., et al. (2001) Activation and redistribution of c-jun N-terminal kinase/stress activated protein kinase in degenerating neurons in Alzheimer's disease. *J Neurochem,* 76, pp.435-441, ISSN 0022-3042